Shape Optimization and Free Boundaries

NATO ASI Series

Advanced Science Institutes Series

A Series presenting the results of activities sponsored by the NATO Science Committee, which aims at the dissemination of advanced scientific and technological knowledge, with a view to strengthening links between scientific communities.

The Series is published by an international board of publishers in conjunction with the NATO Scientific Affairs Division

A Life Sciences	Plenum Publishing Corporation
B Physics	London and New York
C Mathematical	Kluwer Academic Publishers
and Physical Sciences	Dordrecht, Boston and London
D Behavioural and Social Sciences	
E Applied Sciences	
F Computer and Systems Sciences	Springer-Verlag
G Ecological Sciences	Berlin, Heidelberg, New York, London,
H Cell Biology	Paris and Tokyo
I Global Environmental Change	

NATO-PCO-DATA BASE

The electronic index to the NATO ASI Series provides full bibliographical references (with keywords and/or abstracts) to more than 30000 contributions from international scientists published in all sections of the NATO ASI Series.
Access to the NATO-PCO-DATA BASE is possible in two ways:

– via online FILE 128 (NATO-PCO-DATA BASE) hosted by ESRIN,
Via Galileo Galilei, I-00044 Frascati, Italy.

– via CD-ROM "NATO-PCO-DATA BASE" with user-friendly retrieval software in English, French and German (© WTV GmbH and DATAWARE Technologies Inc. 1989).

The CD-ROM can be ordered through any member of the Board of Publishers or through NATO-PCO, Overijse, Belgium.

Shape Optimization and Free Boundaries

edited by

Michel C. Delfour
Centre de recherches mathématiques,
Université de Montréal, Montréal, Québec, Canada

and

Gert Sabidussi
Technical Editor

Springer-Science+Business Media, B.V.

Proceedings of the NATO Advanced Study Institute and
Séminaire de mathématiques supérieures on
Shape Optimization and Free Boundaries
Montréal, Canada
June 25 – July 13, 1990

ISBN 978-94-010-5201-6   ISBN 978-94-011-2710-3 (eBook)
DOI 10.1007/978-94-011-2710-3

Table of Contents

Preface

The Université de Montréal was particularily fortunate to host two major and complementary events in the Summer of 1990: the 29th session of the Séminaire de mathématiques supérieures (SMS) on "Shape Optimization and Free Boundaries" and the International Free Boundary Conference organized by J. Chadam and H. Rasmussen. The SMS is a NATO Advanced Study Institute (ASI) with a long and well-established worldwide reputation. This book contains the notes of the lectures given by the main speakers.

The three main themes of the ASI were Shape Optimization, Free Boundary Problems and Microstructures. Their common feature is that the variable of interest is a geometric entity such as the shape of a domain, the nature and position of a boundary or the microscopic structure of a domain. Parametrization by a vector of real parameters or functions is often possible and useful. However only a deep understanding of the fundamental interactions between the problem and its geometry can yield significant results, good agreement with the physics of the underlying problems, and innovative and efficient numerical algorithms.

This is a difficult family of problems both from the modelization and mathematical viewpoints. The range of applications is impressively broad and includes extremely important problems. Shape optimization includes traditional Engineering applications to domain and structural designs, but also original ones to image segmentation, geometric control theory (localization of sensors and actuators), stabilization of membranes and plates by boundary variations, etc. Some of the techniques of microstructures originated in the design of plates and composite materials. They can be considered as limit cases of shape optimization problems when the distributions of two materials in a fixed domain does not yield distinct subdomains seperated by well-defined boundaries. We have also included in the topic of microstructures the theory of Liquid Crystals and expect that the special tools used in their analysis can eventually impact on the other topics.

The objective and originality of this ASI was to bring together Shape Optimization and Free and Moving Boundary Problems since systems of equations describing many free boundary problems can be obtained from shape variational principles of appropriate energy functionals. Roughly speaking, to solve a free boundary problem can be equivalent to finding the appropriate boundary, that is, the appropriate shape of the domain for which all the boundary conditions and constraints are satisfied.

In the choice of the twelve series of lectures we attempted to provide a broad coverage of the three main themes and keep a balance between the mathematics, the numerical methods and the applications. Basic material on shape derivatives and their computation via theorems on the differentialiblity of extrema with respect to a parameter was presented by M. Delfour and classical applications to Aerodynamics by O. Pironneau. J. Sokolowski covered the challenging problems governed by variational inequalities on polyhedric convex sets.

J. Chadam introduced Free Boundary Problems in Geochemistry and special techniques to analyze moving reaction fronts. This material will certainly be a source of motivation and inspiration for mathematicians in the coming years. We are especially grateful to him for his participation in the SMS shortly after the heavy organizational work he had been doing at the Free Boundary Conference. A. Fasano chose two industrial applications involving an electrochemical machining process and a Bingham flow. I. Stakgold gave a lively treatment of diffusion with nonlinear absorption and J.L. Vazquez an original introduction to the mathematical theory of porous medium equations which find applications in mathematical biology, water filtration, lubrication, boundary layer theory, etc.

Interactions between techniques in Shape Optimization and those in Free Boundary Problems were provided in the lectures of J.P. Zolésio and M. Fortin. J.P. Zolésio, who laid down the mathematical foundations of Shape Optimization in 1979, introduced pioneering and fundamental results and methods for the existence and characterization of free boundaries for several important problems. M. Fortin dealt with free surface problems in fluid mechanics. He introduced a general framework which emphasizes the similarities and differences between fluids and elastic bodies. He also discussed the use of Newton and quasi Newton methods such as GMRES in the computation of the displacement of free boundaries. R. Temam presented his recent powerful methods for the approximation and localization of attractors (possibly fractal) which are connected with the asymptotic state of the underlying free or moving boundary problems.

The appearance of microstructures in the design of composite materials or in the modelization of Liquid Crystals has been the motivation for the development and the mastering of special mathematical techniques. R.V. Kohn lectured on minimum compliance problems in Structural Optimization via relaxed formulations. The lectures of H. Brézis on the mathematical theory of Liquid Crystals provided another important perspective.

It is fair to say that we have met our original objective of bringing together specialists and ideas in neighbouring fields. This ASI has helped to create interesting interactions between its various themes and a strong potential for cooperation and scientific developments in this general area of research.

We wish to express our sincere thanks to all lecturers and participants for having helped to make this ASI a success. Special thanks go to Aubert Daigneault, the director of the ASI, and Ghislaine David, its efficient and charming secretary, for the high quality and the smoothness of the organization of the SMS. We would also like to thank Gert Sabidussi and acknowledge his patient and careful work of editing and assembling the manuscripts of this book.

The ASI was funded in large part by NATO, with additional support from the Natural Sciences and Engineering Research Council of Canada, the Ministère de l'Éducation du Québec, and the Université de Montréal. We would like to thank all three organizations for their support. For their efforts on behalf of this ASI we are especially grateful to the Scientific and Environmental Affairs Division of NATO, particularly to Dr. Luis V. da Cunha, the Director of the ASI programme.

Michel Delfour
Scientific Director

Participants

Frédéric ABERGEL
Laboratoire d'Analyse Numérique
Bâtiment 425
Université de Paris-Sud
91405 Orsay Cédex
France

Paul ARMINJON
Département de mathématiques
et de statistique
Université de Montréal
C.P. 6128, Succ. A
Montréal, Qué., H3C 3J7
Canada

Jacques BÉLAIR
Département de mathématiques
et de statistique
Université de Montréal
C.P. 6128, Succ. A
Montréal, Qué., H3C 3J7
Canada

Alain BÉLIVEAU
Département de mathématiques
et de statistique
Université de Montréal
C.P. 6128, Succ. A
Montréal, Qué., H3C 3J7
Canada

Hichem BEN-EL-MECHAIEKH
Department of Mathematics
Brock University
St. Catharines, Ont., L2S 3A1
Canada

Hakima BOUHAR
178, rue Giraudeau
37000 TOURS
France

Mustapha BOUHAR
Dép. de math. et Inform. Scient.
Fac. des Sciences et Techniques
Université de Tours
Parc de Grandmont
37200 Tours
France

Haïm BRÉZIS
Laboratoire d'Analyse Numérique
Tour 55-65 – 5e étage
Univ. Pierre et Marie Curie
4, pl. Jussieu
75252 Paris Cédex 05
France

Alain CHALIFOUR
Département de chimie-biologie
Univ. du Québec à Trois-Rivières
C.P. 500
Trois-Rivières, Qué., G9A 5H7
Canada

Yin CHANG
Chang's Engineering Inc.
P.O. Box 391
Don Mills, Ont., M3C 2S7
Canada

Karen CLARK
Courant Inst. of Mathematical Sciences
New York University
251 Mercer Street
New York, NY 10012
USA

Elena COMPARINI
Istituto Matematico "U. Dini"
Univ. degli Studi di Firenze
Viale Morgagni 67/A
50134 Firenze
Italy

Olivier COULAUD
Département de mathématiques
Université de Nancy I
B.P. 239
54506 Vandoeuvre les Nancy Cédex
France

Arnaud DEBUSSCHE
Laboratoire d'Analyse Numérique
Bâtiment 425
Université de Paris-Sud
91405 Orsay Cédex
France

Françoise DEMENGEL
Laboratoire d'Analyse Numérique
Bâtiment 425
Université de Paris-Sud
91405 Orsay Cédex
France

Jean DETEIX
Département de mathématiques
et de statistique
Université de Montréal
C.P. 6128, Succ. A
Montréal, Qué., H3C 3J7
Canada

François DUBEAU
Département de mathématiques
Coll. Militaire Royal de Saint-Jean
Richelain, Qué., J0J 1R0
Canada

Saïd EL HAJJI
Département de mathématiques
et de statistique
Université Laval
Québec, Qué., G1K 7P4
Canada

Mohamed FARHLOUL
Département de mathématiques
et de statistique
Université Laval
Québec, Qué., G1K 7P4
Canada

Nick FIROOZYE
Courant Inst. of Mathematical Sciences
New York University
251 Mercer Street
New York, NY 10012
USA

André FORTIN
Département de mathématiques
appliquées
École Polytechnique
Case Postale 6079, Succ. A
Montréal, Qué., H3C 3A7
Canada

Abdou GARBA
Département de Mathématiques
Université de Nice
Parc Valrose
06034 Nice Cédex
France

Touria GHEMIRES
Département de mathématiques
Faculté des Sciences
Université Mohammed V
B.P. 1014
Rabat
Morocco

Roberto GIANNI
Istituto Matematica "U. Dini"
Univ. degli Studi di Firenze
Viale Morgagni 67/A
50134 Firenze
Italy

Raul GONZALEZ DE PAZ
Departamento de Matemáticas
Universidad del Valle de Guatemala
Apdo Postal No 82
01901 Guatemala
Guatemala

Mohammed GUEDDA
U.E.R. de Mathématiques et
d'Informatique
Université de Picardie
33 rue St. Leu
80039 Amiens Cédex
France

Robert GUÉNETTE
Département de mathématiques
et de statistique
Université Laval
Québec, Qué., G1K 7P4
Canada

Faruk GÜNGÖR
Department of Mathematics
Faculty of Science
Istanbul Technical University
80626 Maslak-Istanbul
Turkey

Ernesto GUTIERREZ-MIRAVETE
Hartford Graduate Center
275 Windsor Street
Hartford, CT 06120-2991
USA

Rejeb HADIJI
Laboratoire d'Analyse Numérique
Tour 55-65 – 5e étage
Univ. Pierre et Marie Curie
4, pl. Jussieu
75252 Paris Cédex 05
France

Jean-Pierre HAEBERLY
Department of Mathematics
Fordham University
Bronx, NY 10458-5165
USA

Shen HAI
Département de chimie
Université de Sherbrooke
Sherbrooke, Qué., J1K 2R1
Canada

Antoine HENROT
Département de mathématiques
Université de Nancy I
B.P. 239
54506 Vandoeuvre les Nancy Cédex
France

Shaoyun HUANG
Department of Mathematics
Peking University
Beijing 100871
People's Republic of China

Mekki IDRISSI
Département de mathématiques
et de statistique
Université de Montréal
C.P. 6128, Succ. A
Montréal, Québec, H3C 3J7
Canada

Yorgo ISTEFANOPULOS
Dept. of Electrical &
Electronic Engineering
Bogaziçi University
80815 Bebek-Istanbul
Turkey

Jamila KARRAKCHOU
Faculté des Sciences
Université Mohammed V
École Mohammadia d'Ingénieurs
Ave Ibn Sina – B.P. 765
Rabat
Morocco

Walter T. KYNER
Dept. of Mathematics & Statistics
University of New Mexico
Albuquerque, NM 87131
USA

Robert LIPTON
Department of Mathematics
University of California at Berkeley
Berkeley, CA 94720
USA

Wenbin LIU
School of Mathematics
University of Leeds
Leeds LS2 9JT
United Kingdom

George MEJAK
Inst. of Math., Physics & Mechanics
University of Ljubljana
Jadranska c. 19
61000 Ljubljana
Slovenia

Zoubida MGHAZLI
Faculté des Sciences
Université Ibn Tofail
Kénitra
Morocco

Andrzej MYSLINSKI
Systems Research Institute
Polish Academy of Sciences
ul. Newelska 6
01-447 Warszawa
Poland

Klaus NOLTE
Department of Mathematics
University of Ottawa
Ottawa, Ont., K1N 6N5
Canada

Gonca ONARGAN
Div. of Applied Mathematics
Fac. of Engineering & Architecture
Dokuz Eylül University
B-5000 Bornova-Izmir
Turkey

Abdellatif OUANSAFI
Faculté des Sciences
Université Ibn Tofail
Kénitra
Morocco

Guy PAYRE
Département de génie mécanique
Université de Sherbrooke
Sherbrooke, Qué., J1K 2R1
Canada

Nadia RAÏSSI
Faculté des Sciences
Université Ibn Tofail
Kénitra
Morocco

Jean Michel RAKOTOSON
Département de Mathématiques
Université de Poitiers
40 ave du Recteur Pineau
86022 Poitiers
France

Jean ROCHE
Département de Mathématiques
Université de Nancy I
B.P. 239
54506 Vandoeuvre les Nancy Cédex
France

Georg SCHMIDT
Dept. of Mathematics & Statistics
Burnside Hall
McGill University
805 Sherbrooke St. West
Montréal, Qué., H3C 2K6
Canada

Sankatha P. SINGH
Dept. of Mathematics & Statistics
Memorial University
St. John's, Newfoundland, A1C 5S7
Canada

Mohammed SLAOUI
Lab. de physique mathématique
Univ. des Sci. et Tech. du Languedoc
Pl. Eugène-Bataillon
34060 Montpellier Cédex
France

Georgios STRAVROULAKIS
Inst. of Steel Structures
School of Technology
Aristotle University
54006 Thessaloniki
Greece

Srdjan STOJANOVIC
Dept. of Mathematical Sciences
Old Chemistry Bldg. (ML 25)
University of Cincinnati
Cincinnati, OH 45221-0025
USA

Domingo A. TARZIA
PROMAR (CONICET-UNR)
Inst. de Matemática "Bepo Levi"
Universidad Nacional de Rosario
Avda. Pellegrini 250
2000 Rosario
Argentina

Timo TIIHONEN
Department of Mathematics
University of Jyväskylä
Seminaarinkatu 15
40100 Jyväskylä
Finland

Cristina TURNER
Department of Mathematics (2-130)
Massachussetts Inst. of Technology
Cambridge, MA 02139
USA

Rémi VAILLANCOURT
Département de mathématiques
Université d'Ottawa
Ottawa, Ont., K1N 6N5
Canada

Nguyen VAN HIEN
Département de Mathématique
Facultés Universitaires de Namur
Rempart de la Vierge 8
5000 Namur
Belgium

Italo VECCHI
Seminar für Angewandte Mathematik
ETH-Zentrum
8092 Zürich
Switzerland

Andreas VOSSINIS
I.N.R.I.A.
Domaine de Voluceau-Rocquencourt
B.P. 105
78153 Le Chesnais Cédex
France

Günther WIRSCHING
Mathematisch-Geographische Fakultät
Katholische Universität Eichstätt
Ostenstr. 28
8078 Eichstätt
Germany

Günter WÖRSCHING
Lehrstuhl für Angewandte Mathematik 1
Universität Augsburg
Universitätsstr. 8
8900 Augsburg
Germany

Gülgün YALCINKAYA
Dept. of Aerospace Engineering &
Mechanics
University of Minnesota
107 Akerman Hall
Minneapolis, MN 55455
USA

Fahuai YI
Department of Mathematics
Suzhou University
Suzhou (Jian Su)
People's Republic of China

Contributors

John CHADAM
Dept. of Mathematics & Statistics
McMaster University
Hamilton, Ont., L8S 4K1
Canada

Michel C. DELFOUR
Centre de recherches mathématiques
Université de Montréal
C.P. 6128, Succ. A
Montréal, Qué., H3C 3J7
Canada

Antonio FASANO
Istituto Matematico "U. Dini"
Univ. degli Studi di Firenze
Viale Morgagni 67/A
50134 Firenze
Italy

Michel FORTIN
Dép. de mathématiques & statistique
Université Laval
Québec, Qué., G1K 7P4
Canada

Robert V. KOHN
Courant Inst. of Mathematical Sciences
New York University
251 Mercer Street
New York, NY 10012
USA

Olivier PIRONNEAU
I.N.R.I.A.
Domaine de Voluceau-Rocquencourt
B.P. 105
78153 Le Chesnais Cédex
France

Keith PROMISLOW
Inst. for Applied Mathematics
Indiana Univ. at Bloomington
618 East 3rd Street
Bloomington, IN 47405
USA

Riccardo RICCI
Dip. di Matematica "F. Enriques"
Università di Milano
Via C. Saldini 50
21033 Milano
Italy

Jan SOKOLOWSKI
Systems Research Institute
Polish Academy of Sciences
ul. Newelska 6
01-447 Warszawa
Poland

Ivar STAKGOLD
Dept. of Mathematical Sciences
University of Delaware
501 Ewing Hall
Newark, DE 19716
USA

Roger TEMAM
Laboratoire d'Analyse Numérique
Bâtiment 425
Université de Paris-Sud
91405 Orsay Cédex
France

Juan Luis VAZQUEZ
Departamento de Matemáticas
Univ. Autónoma de Madrid
28049 Madrid
Spain

Juan J.L. VELÁZQUEZ
Departamento de Matemática Aplicada
Facultad de Matemáticas
Univ. Conplutense de Madrid
28040 Madrid
Spain

Jean-Paul ZOLÉSIO
Institut Non Linéaire de Nice
Université de Nice
Parc Valrose
06034 Nice Cédex
France

Free Boundary Problems in Geochemistry

John CHADAM
Department of Mathematics
McMaster University
Hamilton, Ontario, L8S 4K1
Canada

With an Appendix by
Riccardo RICCI
Dipartimento di Matematica "F. Enriques"
Università di Milano
Via C. Saldini 50
I-20133 Milano
Italy

Abstract

Two important examples arising in geochemistry are modelled as moving free boundary problems. The shape stability of these moving reaction fronts is studied using matched asymptotics, functional analysis, bifurcation and stability theory and numerical simulation.

1 Introduction

In these lectures we shall study two important examples arising in geochemistry – solidification of amphorous solids and reactive flows in porous media. In both of these examples free boundaries arise in a natural manner as the reaction interfaces. The shape instabilities in these moving free boundaries are important issues for each of the examples. In the first, they are the basis for the cellular structure of solidification interfaces (see Fig. 1.1) as well as, perhaps, for dendritic growth. In the second, these instabilities can lead to scalloped fronts (Fig. 1.2) as well as to fingering of the flow. Understanding these shape instabilities qualitatively and quantitatively can be a useful tool in many applications such as locating roll-front redox mineral deposits, avoiding fingering of flows in enhanced oil recovery and breakout from chemical and nuclear waste repositories, controlling in situ coal gasification, etc.

1

M. C. Delfour and G. Sabidussi (eds.), Shape Optimization and Free Boundaries, 1–34.
© 1992 *Kluwer Academic Publishers.*

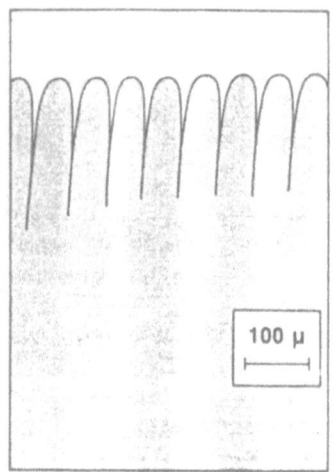

Fig. 1.1

Trace from a photograph of the cellular structure in succinonitrile-salol (after Venugopalan and Kirkaldy, McMaster University preprint).

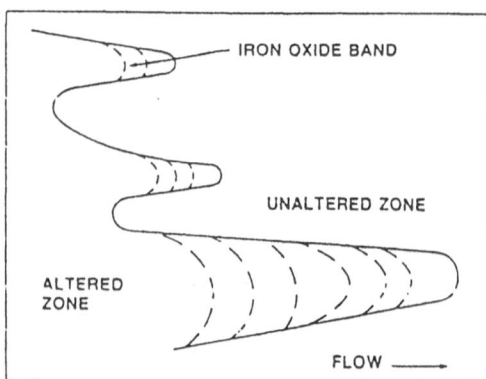

Fig. 1.2

Trace from a photograph of a loose block of the Bordon Group, Southern Indiana, containing a scalloped reaction front. The wavelengths of the scallops is 5 to 6 cm. Because of the iron oxide bands indicated by the dashed lines, the water is inferred to have flowed to the right.

For each example we shall begin by describing it as a self-organization phenomenon; i.e., we shall delineate a destabilizing feedback mechanism which competes with a restabilizing process to autonomously select the evolving shapes of the front. We model these examples mathematically as moving free boundary problems which capture the essential competing mechanisms. Using modern methods of applied mathematics (matched asymptotics, similarity solutions, functional analysis, bifurcation and stability theory) we shall study basic questions of existence and uniqueness, asymptotic behavior including finite-time blow-up, as well as the shape stability questions posed above. We conclude with some of our numerical simulations of the second problem. They show how the unstable modes, obtained analytically using bifurcation methods, compete through the nonlinear dynamics to generate an amazingly rich class of steady and dynamic structures for the evolving reaction interface.

2 Solidification instabilities

2.1 The model

In this section we study the growth of an amorphous solid from a diffusing growth material which surrounds it (for example, a precipitate particle growing from a supersaturated solute or ice freezing in undercooled water – the two problems can be related by a change of variables and our treatment alternates between the two examples). Our mathematical model is similar to the classic Mullins-Sekerka model [1,2] (cf., in addition, the review article of Langer [3] and the references therein), but does not impose the quasi-steady-state assumption whereby the diffusion equation is replaced by Laplace's equation. Specifically, if $c(\vec{x}, t)$ is the concentration of the solute and the solid-solute interface is $S(\vec{x}, t) = 0$, then the model problem (no temperature, impurity, etc. effects) is to find these two quantities subject to

$$c_t = D\Delta c \text{ in } \{S(\vec{x}, t) > 0\}, \tag{2.1a}$$

$$c = c_{eq}[1 + \gamma K(S)] \text{ on } \{S(\vec{x}, t) = 0\}, \tag{2.1b}$$

$$D\frac{\partial c}{\partial n} = (\rho - c)V_n \text{ on } \{S(\vec{x}, t) = 0\}, \tag{2.1c}$$

$$c \to c_\infty \text{ as } |\vec{x}| \to \infty, \tag{2.1d}$$

$$c(\vec{x}, 0) = c_0(\vec{x}) \text{ in } \{S(\vec{x}, 0) = S_0(\vec{x}) > 0\}, \tag{2.1e}$$

for given initial data $c_0(\vec{x})$ and $S_0(\vec{x})$. Here, ρ is the density of the solid, c_{eq} is the equilibrium concentration on a planar interface, c_∞ is the ambient concentration far from the solid, and is the normal to $S(\vec{x}, t) = 0$ with $V_n = -S_t/|\nabla S|$ the normal velocity to the interface. The function $K(S)$ is the mean curvature of the surface $S = 0$, and $\gamma \geq 0$

is a measure of the interfacial energy. That is, the concentration at $S = 0$ is maintained at equilibrium as prescribed by the Gibbs-Thomson relation (2.1b). Problem (2.1) is a generalization of the classic one-phase Stefan problem [4–7] in that it includes the surface tension $\gamma K(S)$ in (2.1b) and that $(\rho - c)$ in the Stefan condition (2.1c) is not a constant (which is a less dramatic modification since a constant would result by taking $\rho \gg c$). The most important difference, however, is that we will study solidification solutions of problem (2.1) $(c_\infty > c_{eq})$ rather than those associated with melting. This is the basis for the instabilities we intend to analyze as can be anticipated from the following heuristic arguments. The diffusive nature of the solute causes an increasing concentration profile to develop from the interface to the distant ambient value. If a protrusion should develop in the interface, its tip would thus find itself in a region of higher concentration than the rest of the interface causing its growth to accelerate. Thus diffusion can be considered as a shape destabilizing mechanism which could lead to fingering or dendritic growth. On the other hand, when surface tension is present $(\gamma \neq 0)$, equation (2.1b) indicates that a higher concentration is required to maintain the highly curved tip in equilibrium. Thus the surface tension's flattening action mollifies that of diffusion which tends to make the solid send out fingers to distant regions of higher concentration. The competition of these two mechanisms selects, in an autonomous way (self-organizes), the preferred growing shapes of the solid. In the remainder if this section we shall study the stability of these shapes. In particular we shall study the stability with respect to shape perturbations of planar solidification fronts. It turns out that finding the global attractor in the class of planar solutions (the correct planar solution off which to do a perturbation analysis) leads to many interesting problems involving global existence, asymptotic behavior and possible finite-time blow-up. This will be the subject of section 2.2. The stability analysis done in section 2.3 will be complicated by the fact that the base planar solution is not a travelling wave solution.

2.2 Planar development

With $S(\vec{x}, t) = x - R(t)$ the planar version of problem (2.1) is:

$$c_t = Dc_{xx}, \quad \text{in } x > R(t), \quad t > 0, \tag{2.2a}$$

$$c = c_{eq}, \quad \text{on } x = R(t), \quad t > 0, \tag{2.2b}$$

$$Dc_x = (\rho - c_{eq})\dot{R}(t), \quad \text{on } x = R(t), \quad t > 0, \tag{2.2c}$$

$$c \to c_\infty, \quad \text{as } x \to \infty, \tag{2.2d}$$

$$c(x,0) = c_0(x) \quad \text{in } x = R(0) = R_0 \tag{2.2e}$$

The complexity of this one dimensional problem is greatly reduced because the curvature term disappears from (2.2b) on the planar interface. Letting $u(x,y) = c(x,t) - c_{eq}$, $L = \rho - c_{eq} > 0$ and $D = 1$ (by rescaling distance with a typical length scale ℓ and time with $\frac{\ell^2}{D}$) we obtain

$$u_t = u_{xx} \quad \text{in } x > R(t), \quad t > 0, \tag{2.3a}$$

$$u = \quad 0 \quad \text{on } x = R(t), \qquad t > 0, \tag{2.3b}$$

$$u_x = \quad L\dot{R}(t) \quad \text{on } x = R(t), \qquad t > 0, \tag{2.3c}$$

$$u \to \quad u_\infty \quad \text{as } x \to \infty, \tag{2.3d}$$

$$u(x,0) = \quad u_0(x) \quad \text{in } x > R(0) \equiv R_0, \tag{2.3e}$$

where $u_\infty = c_\infty - c_{eq}$ and $u_0(x) = c_0(x) - c_{eq}$. If $u_0(x) \geq 0$ (i.e. the initial concentration $c_0(x)$ and the ambient concentration c_∞ both exceed the planar equilibrium concentration c_{eq}) the maximum principle implies that $u_x \geq 0$ on the interface. Thus $\dot{R}(t) \geq 0$ so that this corresponds to the solidification process. Similarly, $u_0(x) \leq 0$ corresponds to the melting process.

One might begin by looking for travelling wave solutions to problem (2.3a - d); i.e. solutions of the form $R(t) = at, u(x,t) = U(x - at)$ which ignore the initial data. Direct substitution into equation (2.3) gives (with $\xi = x - at$ and $\frac{d}{d\xi}$ denoted by a prime)

$$U'' + aU' = 0, \xi > 0, \tag{2.4a}$$

$$U = 0, \xi = 0, \tag{2.4b}$$

$$U' = aL, \xi = 0, \tag{2.4c}$$

$$U \to u_\infty, \xi \to \infty. \tag{2.4d}$$

Solving (2.4a,b,d) directly gives

$$U(\xi) = u_\infty(1 - e^{-a\xi}), \tag{2.5}$$

which when substituted into the Stefan condition (2.4c) gives

$$au_\infty = aL. \tag{2.6}$$

This is entirely unsatisfactory because (2.6) requires that $U_\infty = L$ (i.e. $c_\infty - c_{eq} = \rho - c_{eq}$ or $c_\infty = \rho$!!!) and then permits all velocities $a > 0$ (i.e., a specific velocity is not selected by the data). For diffusion problems there is another class of similarity solutions of the form $R(t) = \alpha\sqrt{t}$ with $u(x,t) = U(x/2\sqrt{t})$. A straightforward calculation [8,p. 287] gives

$$U(\eta) = -\alpha L e^{\alpha^2/4} \int_\eta^\infty e^{-v^2} dy + u_\infty, \tag{2.7}$$

which when substituted into (2.3b) gives

$$\alpha e^{\alpha^2/4} \int_{\alpha/2}^\infty e^{-v^2} dv = u_\infty/L. \tag{2.8}$$

The left hand side of (2.8) can be seen to be a monotone increasing function of α going from $-\infty$ for $\alpha = -\infty$, zero when $\alpha = 0$ and going to 1 for $\alpha \to +\infty$ [9]. Thus (2.8) gives a unique value of α for each u_∞/L, negative when $u_\infty < 0$, positive when $u_\infty > 0$ but only for the "Stefan number" $u_\infty/L < 1$, a cutoff which appeared previously in the travelling wave calculation.

These results for special solutions (with $R(0) = 0$ and special u_0) motivate the following results for arbitrary initial data.

Theorem 2.1 [10] *Suppose $u_0 \in C^1[R(0), \infty), u_0(R(0)) = 0$ and $u_0(x) - u_\infty \in L^1(R(0), \infty)$.*

a) *Problem (2.3) has a unique global (in t) solution if either*

    i) $0 \geq u_0(x) \geq u_\infty$ *(the strictly melting case) or,*

    ii) $\|u_0\|_\infty/L < 1/8$ *if u_0 is anywhere positive (the solidifying case).*

b) *If $\|u_0\|_\infty/L > 1$ and $0 \leq u_0(x) \leq u_\infty, u \leq 0$, then there exists a $0 < T < \infty$ such that $\lim_{t \to T-0} \dot{R}(t) = +\infty$.*

c) *If a global solidifying solution exists, then there exist $n, N > 0$ such that for $t > 0$, one has $(1 - \epsilon)\alpha\sqrt{t} - n \leq R(t) \leq (1 + \tilde{\epsilon})\alpha\sqrt{t} + N$ where α is given by (2.8) and $\epsilon, \tilde{\epsilon} \to 0$ as the Stephan number $u_\infty/L \to 0$.*

Remark Part a) ii) of the theorem indicates that solidification solutions with arbitrary, but sufficiently small data, exist globally in time. Part c) suggests that they tend asymptotically to the $t^{1/2}$ solution with the same value of u_∞. This justifies the use of the $t^{1/2}$ solution as the base planar solution in a shape stability analysis. The condition $\|u_0\|_\infty/L \leq 1/8$ is a technical dificiency from the method we use. On the other hand, part b) shows that we cannot expect global solutions for $\|u_0\|_\infty/L > 1$, the cutoff suggested by the previous special solutions. A more complete analysis by Fasano and Primicerio [11], Ricci and Xie [12], and Dewynne, Howison, Ockendon and Xie [13] closes the gap in part a) ii) and provides a stronger version of part c). This is summarized in an Appendix to these lectures by Ricci.

Sketch of the Proof We begin by reformulating the problem as an integral equation following Friedman [5]. This method works independent of whether it is a melting or solidification problem. Proceeding formally, if $E(x - z, t - s)$ is the fundamental solution of the heat equation, then one has

$$u(x, t) = \int_{R(0)}^\infty E(x - z, t)u_0(z)dz - \int_0^t E(x - R(s), t - s)\frac{\partial u}{\partial x}(R(s), s)ds. \qquad (2.9)$$

Now, differentiating with respect to x, taking the limit as $x \to R(t) + 0$ noting the jump discontinuity in the second integral, one obtains for $U(t) = u_x(R(t), t)$,

$$U(t) = 2\left[\int_{R(0)}^\infty E_x(R(t) - z, t)u_0(z)dz - \int_0^t E_x(R(t) - R(s), t - s)U(s)ds\right] \qquad (2.10a)$$

$$= 2\left[-\int_{R(0)}^{\infty} E(R(t) - z, t)u_0'(z)dz - \int_0^t E_x(R(t) - R(s), t - s)U(s)ds\right] \qquad (2.10b)$$

since u_0 is differentiable and $u_0(R(0)) = 0$. Integrating (2.3c) one obtains

$$R(t) = R(0) + L^{-1}\int_0^t U(s)ds. \qquad (2.11)$$

By a straightforward modification of the classical proof [5] one obtains the basic result of this approach – that the differential equations (2.3) and the integral equations (2.10, 2.11) are equivalent.

Scholium 2.2 [5, p. 221] $u(x,t), R(t)$ *is a classical solution of problem* (2.3) *over the interval* $0 < t < T$ *if and only if* $U(t)$ *is a continuous solution of the integral equation* (2.10) *over* $0 < t < T$, *where* $R(t)$ *is given by equation* (2.11).

The hypotheses on u_0 guarantee that the integral involving the initial data in (2.10b) is continuous in $0 < t < T$ and so the problem (2.10, 2.11) can be solved by a contraction argument [5, p. 222] in the space $C[0,T]$ with T sufficiently small and can be extended globally (i.e., $T = \infty$) if $|U(t)|$ remains finite [5, p. 223]. The following analysis to establish this provides one of the simplest and most striking differences between melting and solidification in problem (2.3).

Proof of Theorem 2.1, Part a) i) In the interval of existence, $U(t)$ is bounded from above by the maximum principle as follows: $u_0(x) \le 0$ on $R(0) < x < \infty$ and $u = 0$ on $x = R(t)$ for $0 < t < T$, so that $u(x,t) \le 0$ in $R(t) < x < \infty, 0 < t < T$, which implies $U(t) = u_x(R(t), t) \le 0$ since $U(R(t), t) = 0$. In addition, the lower bound can be obtained immediately via the integral equation in this melting case. Indeed, from (2.10b, 2.11) one has

$$U(t) = -2\int_{R(0)}^{\infty} E(R(t) - z, t)u_0'(z)dz$$

$$- 2\int_0^t -\frac{2(R(t) - R(s))}{8\sqrt{\pi}(t - s)^{3/2}} U(s)\exp\left[-\frac{(R(t) - R(s))^2}{4(t - s)}\right]ds. \qquad (2.12)$$

Because $U(t) \le 0, R(t) - R(s) = L^{-1}\int_s^t U(w)dw \le 0$ so that the last term is manifestly non-negative. Thus

$$U(t) \ge -2\int_{R(0)}^{\infty} E(R(t) - z, t)dz||u_0'||_\infty \ge -2||u_0'||_\infty.$$

Thus extension of T to $+\infty$ is a simple consequence of the maximum principle in the melting case. We leave it as an exercise to show that in the solidifying case (i.e., $u_0(x) \ge 0$ which implies $U(t) \ge 0$), the upper bound cannot be similarly obtained through the integral equation (2.10b, 2.11) by ignoring the second term in (2.10b). Instead one must estimate this last term. Here we give the original proof which is technically deficient in that it is valid only for small values of $||u_0||_\infty/L$ ($< 1/8$, in particular). It does capture two important features: that the sign of $u_0(x)$ need not be constant and that the condition on the data involves only $||u_0||_\infty$ not $||u_0'||_\infty$. The sketch of a more complete proof for $||u_0||_\infty/L \leq 1$ is given by Ricci in the Appendix.

Proof of Theorem 2.1, Part a) ii) The integral involving the initial data in (2.10a) can be estimated directly to give

$$|2\int_{R(0)}^{\infty} E_x(R(t)-z,t)u_0(z)dz| \leq \frac{2||u_0||_\infty}{\sqrt{\pi}\sqrt{t}}. \tag{2.13}$$

Thus, from (2.10a) one has

$$|U(t)| \leq a/\sqrt{t} + b\int_0^t \int_s^t \frac{|U(w)||U(s)|}{(t-s)^{3/2}}dwds, \tag{2.14}$$

where $a = 2||u_0||_\infty/\sqrt{\pi}$ and $b = L^{-1}/2\sqrt{\pi}$. It turns out that the central function to be estimated is

$$V(r) = \int_0^r \frac{|U(t)|}{(r-t)^{1/2}}dt. \tag{2.15}$$

Multiplying (2.14) by $(r-t)^{1/2}$ and integrating one has

$$\begin{aligned} V(r) &\leq a\pi + b\int_0^r \int_0^t \int_s^t \frac{|U(w)||U(s)|}{(t-s)^{3/2}(r-t)^{1/2}}dwdsdt = \\ &\leq a\pi + b\int_0^r |U(w)| \int_0^w |U(s)| \int_w^r (t-s)^{-3/2}(r-t)^{-1/2}dtdsdw. \end{aligned} \tag{2.16}$$

Letting $v = (t-s)(r-s)^{-1}$, the internal integral becomes

$$(r-s)^{-1}\int_{(w-s)/(r-s)}^1 v^{-3/2}(1-v)^{-1/2}dv = -\frac{2}{r-s}\left(\frac{1-v}{v}\right)^{1/2}\Big|_{(w-s)/r-s}^1 \tag{2.17}$$

Thus, one obtains

$$V(r) \leq a\pi + 2b\int_0^r |U(w)| \int_0^w \frac{|U(s)|}{r-s}\frac{(r-w)^{1/2}}{(w-s)^{1/2}}dsdw. \tag{2.18}$$

But, $s \leq w$ implies that $(r-s)^{-1} \leq (r-w)^{-1}$ so that one obtains

$$V(r) \leq a\pi + 2b\int_0^r \frac{|U(w)|}{(r-w)^{1/2}} \int_0^w \frac{|U(s)|}{(w-s)^{1/2}}dsdw; \tag{2.19}$$

so that if we define

$$m(r) = \max_{0 \leq t \leq r} V(t), \tag{2.20}$$

one has

$$m(r) \leq a\pi + 2bm(r)^2. \tag{2.21}$$

A standard method now can be used to obtain a uniform bound for m and hence V. Graph the quadratic in (2.21), obtaining two positive real roots if $4(a\pi)(2b) \leq 1$ or, using the definitions of a and b, $8||u_0||_\infty/L \leq 1$. Now $m(0) = 0$ and $m(r)$ is continuous so it must remain in the interval between 0 and the smallest root of (2.21) where it initially began. Thus

$$m(r) \leq \frac{1 - (1 - 4a\pi 2b)^{1/2}}{4b} = \frac{\sqrt{\pi}}{2}L\left[1 - \left(1 - \frac{8||u_0||_\infty}{L}\right)^{1/2}\right]. \tag{2.22}$$

Returning to equation (2.10b) one has

$$\begin{aligned}
|U(t)| &\leq 2||u_0'||_\infty + b\int_0^t \int_s^t \frac{|U(w)||U(s)|}{(t-s)^{3/2}}\,dw\,ds \\
&\leq 2||u_0'||_\infty + bM(t)V(t),
\end{aligned} \tag{2.23}$$

where we have defined

$$M(t) = \max_{0 \leq w \leq t} |U(w)|. \tag{2.24}$$

Thus,

$$M(t) \leq 2||u_0'||_\infty + bm(t)M(t). \tag{2.25}$$

But, $bm(t) \leq \left(\frac{L^{-1}}{2\sqrt{\pi}}\right)\left(\frac{\sqrt{\pi}}{2}L\right) = \frac{1}{4}$ from estimate (2.22), giving $\frac{3}{4}M(t) \leq 2||u_0'||_\infty$ or

$$M(t) \leq \frac{8}{3}||u_0'||_\infty, \tag{2.26}$$

thus concluding the proof if the smallness condition on the Stefan number $||u_0||_\infty/L \leq 1/8$ is satisfied.

It is clear that a more careful analysis would allow for larger Stefan numbers (see the Appendix by Ricci). On the other hand, the next part of Theorem 2.1 shows that if $||u_0||_\infty/L > 1$ it is possible to preclude global existence by having blow-up in finite time (i.e., there is a $T < \infty$ such that $\lim_{t \to T-0} = L\lim_{t \to T-0}\dot{R}(t) = +\infty$). A more complete analysis of the asymptotic behavior is given in [11–14]. The first instance of such blow-up was given by Sherman [14] in a finite geometry and was referred to in the lectures of Fasano.

Proof of Theorem 2.1, Part b) The central quantity in this analysis is

$$q(t) = \int_{R(t)}^\infty [u_\infty - u(x,t)]dt. \tag{2.27}$$

In the interval of existence guaranteed in part a), one may differentiate this quantity to obtain

$$\dot{q}(t) = L(1 - u_\infty/L)\dot{R}(t). \tag{2.28}$$

Additionally, since $0 \leq u(x,t) \leq u_\infty$ and $u_{xx}(x,t) \leq 0$ for $x > R(t)$ (the second of which, as we shall see, also follows from the maximum principle), $q(t)$ can be bounded from below by the area of the triangle in Fig. 2.1.

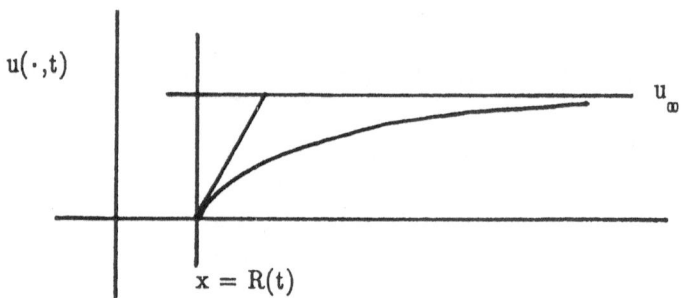

$$u(\cdot,t)$$

$$u_\infty$$

$$x = R(t)$$

Fig. 2.1   Graph of u as a function of x at an arbitrary fixed time t.

Using equation (2.3c) in calculating the slope, one obtains

$$q(t) \geq \frac{u_\infty^2}{2L\dot{R}(t)}. \tag{2.29}$$

Noting that $q(t) \geq 0$ (since $u_\infty - u(x,t) \geq 0$) and $\dot{R}(t) = L^{-1}u_x(R(t),t) \geq 0$ (since $u(R(t),t) = 0$ and $u(x,t) \geq 0$ for $x > R(t)$) the above inequalities can be combined to give

$$\frac{d}{dt}(q^2(t)) = 2q\dot{q} \leq u_\infty^2(1 - u_\infty/L), \tag{2.30}$$

which upon integration gives

$$q^2(t) \leq q^2(0) + u_\infty^2(1 - u_\infty/L)t. \tag{2.31}$$

Since $(1 - u_\infty/L) < 0$, this is clearly impossible after

$$T = q^2(0)[u_\infty^2(u_{\infty/L} - 1)]^{-1}. \tag{2.32}$$

Before this critical time one has from (2.29) and (2.31) that

$$\dot{R}(t) \geq \frac{u_\infty^2}{2L}(q^2(0) - u_\infty(u_\infty/L - 1)t)^{-1/2}. \tag{2.33}$$

All that remains is to establish $u_{xx}(x,t) \leq 0$ for $x > R(t)$. Notice that u_{xx} satisfies the heat equation and is taken to be non-positive at $t = 0$. The result follows then from the maximum principle if $u_{xx}(R(t),t) \leq 0$. To this end

$$\begin{aligned} u_{xx}(R(t),t) = u_t(R(t),t) &= \frac{d}{dt}[u(R(t),t)] - u_x(R(t),t)\dot{R}(t) \\ &= -L[\dot{R}(t)]^2 \leq 0 \end{aligned} \tag{2.34}$$

because $\frac{d}{dt}[u(R(t),t)] = 0$ from (2.3b).

The proof of part c) is given in the Appendix. Combining the results of Theorem 2.1 we see that when solutions to problem (2.3) with arbitrary $R(0)$ and u_0 exist for all time, they tend, as $t \to +\infty$, to the special $t^{1/2}$ solution with the same value of u_∞ and L when $\|u_0\|_\infty/L < 1$. This justifies the use of this special solution as the base planar solution in the shape stability analysis in the next section which was our main purpose in this subsection.

Before proceeding to that analysis we digress for a moment to try to understand the significance of the finite-time blow-up for $\|u_0\|_\infty/L > 1$. The physical meaning of this condition in the context of the motivating model is that $c_\infty > \rho$, that is, the ambient concentration exceeds the density of the solid being formed – a clearly non-realistic situation. On the other hand, if the problem is transformed to that for the temperature of supercooled water from which ice is forming, it makes perfectly good sense. Specifially, if $T = -u$ in (2.3), then

$$T_t = T_{zz}, \quad x > R(t), \ t > 0, \tag{2.35a}$$

$$T = 0, \quad x = R(t), \ t > 0, \tag{2.35b}$$

$$T_x = -L\dot{R}, \quad x = R(t), \ t > 0. \tag{2.35c}$$

$$T \to T_\infty, \quad x \to +\infty \tag{2.35d}$$

$$T(x,0) = T_0(x) \quad \text{for} \quad x > R(0) = R_0, \tag{2.35e}$$

where $T_\infty = -u_\infty < 0$. Now the Stefan number $-T_\infty/L$ can be made larger that 1 in a physical realistic manner simply by lowering the ambient temperature past the latent heat L. Mathematically, problem (2.35) allows for finite-time blow-up *if* $-T_\infty/L > 1$ but physically this can never occur. One way to resolve this dilemma is to suggest that in this hypercooled situation ($-T_\infty/L > 1$) the attachment is occurring so fast that the temperature at the interface does not reach equilibrium. Specifically, one replaces (2.35b), and more generally the analogue of (2.1b), with the non-equilibrium condition [15]

$$V_n = -q(T + \gamma K(S)). \tag{2.36}$$

Thus when $q \simeq \infty$ then, in order to obtain finite normal velocities, one must have the boundary temperature maintained at the equilibrium, Gibbs-Thomson value $-\gamma K(S)$. If $q < \infty$, then the boundary temperature from (2.36) is

$$T = -\gamma K(S) - V_n/q; \tag{2.37}$$

i.e., it is undercooled by the dynamical quantity V_n/q. In the planar case, (2.36) becomes

$$\dot{R}(t) = -qT \tag{2.35b'}$$

which with the rest of (2.35) has a $R(t) = \alpha t^{1/2}$ solution for $-T_\infty/L < 1$ for which α satisfies precisely the same condition (2.8) as in the case when $q = \infty$ [18], the only change being that the temperature profile corresponding to (2.7) depends on q. A more surprising fact is that for $-T_\infty/L > 1$, this kinetically undercooled model (2.35') has a stable travelling wave solution [15]. Specifically, following along the lines of the argument in (2.4 – 6), if $R(t) = at$ and $\xi = x - at$ is the coordinate moving along with the front, then the resulting temperature profile is

$$T(\xi) = T_\infty + (T_b - T_\infty)e^{-a\xi}, \tag{2.38}$$

where

$$T_b = T_\infty + L, \tag{2.39}$$

and, most importantly, the velocity is selected by the desiderata through

$$a = -qT_b = -q(T_\infty + L). \tag{2.40}$$

This transition from $t^{1/2}$-to t-growth as the Stefan number $-T_\infty/L$ crosses 1 had been observed experimentally and numerically by Schaefer and Glicksman [16] and has been studied mathematically more recently by Ricci and Xie [12] and Dewynne, Howison, Ockendon and Xie, [13]. The shape stability of this travelling wave for $-T_\infty/L > 1$ will be discussed at the end of the next section [15].

2.3   Shape stability of planar fronts

We now do a linearized stability analysis about the $t^{1/2}$ solutions discussed in the previous section [8, sect. 4]. This t-dependence will give rise to some computational complication but it is instructional to do this since the stability of planar travelling wave solutions has been discussed in this series and will be seen again in the next section. Denoting the special planar solution in (2.7 and 8) by $\bar{R}(t)$ and $\bar{U}(x,t)$ we introduce non-planar perturbations of the form

$$u(x,y,t) = \bar{U}(x,y) + \epsilon u_1(x,y,t) + O(\epsilon^2), \tag{2.41a}$$

$$x = R(y,t) = \bar{R}(t) + \epsilon R_1(y,t) + O(\epsilon^2). \tag{2.41b}$$

These will now be substituted into the transformed versions of equations (2.1):

$$u_t = \Delta u \qquad\qquad \text{in}\quad x > R(y,t),\ t > T \tag{2.42a}$$

$$u = \gamma c_{eq}\left[-\frac{1}{2}\frac{\partial}{\partial y}\left(\frac{R_y}{(1+R_y^2)^{1/2}}\right)\right] \quad \text{on}\quad x = R(y,t),\ t > T \tag{2.42b}$$

$$u_x - u_y R_y = LR_t, \qquad\qquad \text{on}\quad x = R(y,t),\ t > T \tag{2.42c}$$

$$u \to u_\infty \qquad\qquad \text{as}\quad x \to +\infty \tag{2.42d}$$

$$u(x,y,T) = u_0(x,y) \qquad \text{in}\quad x > R(y,t) = R_0(y) \tag{2.42e}$$

where we have explicitly written the mean curvature in (2.42b), the normal derivative and normal velocity in (2.42c), and have taken the time of the initial perturbation at

$t = T$ for generality. All terms are linear except the right side of (2.42b) and the left side of (2.42c) which to $O(\epsilon)$ can be seen to contribute $-\gamma \frac{c_{eq}}{2} R_{1yy}$ and the $O(\epsilon)$ term from $u_x(R(y,t)t)$ which is slightly complicated by the fact that the evaluation is on the unknown, perturbed boundary. Thus

$$
\begin{aligned}
u_x(R(y,t),t) &= \bar{U}_x(R(y,t),t) + \epsilon u_{1x}(R(y,t),t) + O(\epsilon^2) \\
&= \bar{U}_x(\bar{R}(t) + \epsilon R_1(y,t) + O(\epsilon^2),t) \qquad\qquad (2.43) \\
&+ \epsilon u_{1x}(\bar{R}(t) + \epsilon R_1(y,t) + O(\epsilon^2),t) + O(\epsilon^2)
\end{aligned}
$$

which at $O(\epsilon)$ contributes

$$
u_x(R(y,t),t)|_{O(\epsilon)} = \bar{U}_{xx}(\bar{R}(t),t) R_1(y,t) + u_{1x}(\bar{R}(t),t). \qquad (2.44)
$$

Similarly, the $O(\epsilon)$ contribution from the left side of (2.42b) is

$$
u(R(y,t),t)|_{O(\epsilon)} = \bar{U}_x(\bar{R}(t),t) R_1(y,t) + u_1(\bar{R}(t),t). \qquad (2.45)
$$

In summary, the linearized equations for the perturbations (dropping the subscript 1) are:

$$
\begin{aligned}
u_t &= \Delta u & \text{in} \quad & x > \bar{R}(t), \ t > T & (2.46a) \\
u &= -\bar{U}_x R - \frac{\gamma c_{eq}}{2} R_{yy} & \text{on} \quad & x = \bar{R}(t), \ t > T, & (2.46b) \\
u_x &= L R_t - \bar{U}_{xx} R & \text{on} \quad & x = \bar{R}(t), \ t > T, & (2.46c) \\
u &\to 0 & \text{as} \quad & x \to +\infty, & (2.46d) \\
u(x,y,T) &= \phi(x,y) & \text{on} \quad & x > \bar{R}(T), & (2.46e)
\end{aligned}
$$

where $\bar{U}_x = \bar{U}_x(\bar{R}(t),t) = \frac{L\alpha}{2} t^{-1/2}$ from (2.3c) and $\bar{U}_{xx} = \bar{U}_{xx}(\bar{R}(t),t) = -\frac{L\alpha^2}{4} t^{-1}$ from (2.34). Thus the coefficients in (2.46) are functions of t and the boundary is the special planar one with $\bar{R}(t) = \alpha t^{1/2}$ so that the standard Laplace transform or Green's function methods do not apply to this albeit linear problem (2.46). On the other hand, problem (2.46) can be decomposed into Fourier modes

$$
R(y,t) - r_m(t) e^{imy} \qquad (2.47)
$$

and using asymptotic analysis we obtain the following result for the amplitude $r_m(t)$ of these modes.

Theorem 2.2 [17] *If $|m| > 0, T > 0, r_m(T) \neq 0$, then $r_m(t)$ has the following asymptotic behavior:*

$$
r_m(t) \approx r_m(T) \exp\left[\left(\frac{2m^2\alpha}{3\sqrt{\pi}\sqrt{T}} - \frac{2\gamma c_{eq} L^{-1} m^4}{3\sqrt{\pi}} \right) (t-T)^{3/2} \right] \qquad (2.48)
$$

as $t \to T + 0$.

Remarks Because (2.46) was a variable coefficient, variable domain problem, the spectrum (whose change of sign gives the onset of instability) cannot be computed. Instead one must examine expressions like (2.48) to determine if the size of the perturbation is initially growing or shrinking. In our case this condition for stability is

$$\frac{2m^2\alpha}{3\sqrt{\pi}\sqrt{T}} - \frac{2\gamma c_{eq}m^4}{(\rho - c_{eq})3\sqrt{\pi}} < 0 \tag{2.49}$$

which, noting that the velocity of the planar front is $V = \dot{\bar{R}}(T) = \frac{\alpha}{2\sqrt{T}}$, can written as

$$m^2 > \frac{2V}{\gamma}\left(\frac{\rho - c_{eq}}{c_{eq}}\right). \tag{2.50}$$

As expected by heuristic reasoning, formula (2.50) shows that in the melting situation ($V < 0$) all modes are stable, while in the solidifying case, without surface tension ($\gamma = 0$), all modes are unstable and only high frequency modes are stabilized by the inclusion of surface tension.

Sketch of the Proof of Theorem 2.2. Step 1 (to obtain an integral equation formulation for $r_m(t)$). To obtain $u = 0$ in (2.46b), we change variables to

$$v = u + \bar{U}_x R + \frac{\gamma c_{eq}}{2} R_{yy}, \tag{2.51}$$

and decomposing the resulting problem into Fourier components via

$$v(x, y, t) = w_m(x, t)e^{-m^2t}e^{imy}, \tag{2.52}$$

we obtain (dropping the subscript m),

$$w_t - w_{xx} + [\dot{r}(t) + m^2r(t)]\left[-\bar{U}_x + \frac{\gamma c_{eq}}{2}m^2\right] = 0 \quad \text{in} \quad x > \bar{R}(t), t > T \tag{2.53a}$$

$$\begin{aligned}
w &= 0 &&\text{on} && x = \bar{R}(t), \ t > T &&\text{(2.53b)}\\
w_x &= L\dot{r}(t)e^{m^2t} &&\text{on} && x = \bar{R}(t), \ t > T &&\text{(2.53c)}\\
w &\to -\frac{\gamma c_{eq}}{2}m^2r(T)e^{m^2T} &&\text{as} && x \to \infty &&\text{(2.53d)}\\
w(x, T) &= \chi(x) &&\text{in} && x > \bar{R}(T). &&\text{(2.53e)}
\end{aligned}$$

If $E(x - \xi, t - s)$ denotes the fundamental solution of the heat equation, then, as before, one can write the solution of (2.53) in integral form and taking into account the discontinuity of E_x, one obtains

$$w_x(\bar{R}(t), t) = 2\int_{\bar{R}(T)}^{\infty} \chi(\xi)E_x(\bar{R}(t) - \xi, t - T)d\xi$$

$$+2\int_T^t ds \int_{\bar{R}(s)}^{\infty} d\xi \left\{\left[\bar{U}_x(\xi, s) - \frac{\gamma c_{eq}m^2}{2}\right][\dot{r}(s) + m^2r(s)]e^{m^2s}E_x(\bar{R}(t) - \xi, t - s)\right\}$$

$$-2\int_T^t w_x(\bar{R}(s), s)E_x(\bar{R}(t) - \bar{R}(s), t - s)ds. \tag{2.54}$$

Following a method of Rubinstein [9] we write (2.54) in a more convenient form as follows. Denoting the kernel $2\,E_x(\bar{R}(t) - \bar{R}(s), t - s)$ by $k(t,s)$ and the associated operator by K, $w_x(\bar{R}(t), t)$ by $g(t)$ and the first three terms on the right side of (2.54) by $h_i(t), h_p(t)$ and $h_r(t)$ respectively, then equation (2.54) can be written as

$$g(t) = h_i(t) + h_p(t) + h_r(t) - (K_g)(t) \tag{2.55}$$

or, formally solving,

$$g(t) = [(I + K)^{-1}(h_i + h_p + h_r)](t). \tag{2.56}$$

The first term of (2.56), $I(t) = (I + K)^{-1}h_i(t)$ satisfies

$$
\begin{aligned}
I(t) &= h_i(t) - (KI)(t) \\
&= h_i(t) - 2\int_T^t I(s)E_x(\bar{R}(t) - \bar{R}(s), t - s)ds.
\end{aligned}
\tag{2.57}
$$

The other terms of (2.56) can be similarly treated to obtain

$$g(t) = I(t) + \int_T^t [\dot{r}(s) + m^2 r(s)]e^{m^2 s}r(s)]e^{m^2 s}[P(t,s) + \Gamma(t,s)]ds \tag{2.58}$$

where the inhomogeneous term $I(t)$ and the kernels P and Γ arising from $\bar{U}_x$ and $-\gamma\frac{c_{eq}}{2}m^2$, in (2.54), respectively satisfy

$$
\begin{aligned}
I(t) &= 2\int_{\bar{R}(T)}^\infty \chi(\xi)E_x(\bar{R}(t) - \xi, t - T)d\xi \\
&\quad - 2\int_T^t I(s)E_x(\bar{R}(t) - \bar{R}(s), t - s)ds
\end{aligned}
\tag{2.59a}
$$

$$
\begin{aligned}
P(t,s) &= 2\int_{\bar{R}(s)}^\infty \bar{U}_x(\xi, s)E_x(\bar{R}(t) - \xi, t - s)d\xi \\
&\quad - 2\int_s^t P(\tau, s)E_x(\bar{R}(t) - \bar{R}(\tau), t - \tau)d\tau
\end{aligned}
\tag{2.59b}
$$

$$
\begin{aligned}
\Gamma(t,s) &= -\gamma c_{eq}m^2 \int_{\bar{R}(s)}^\infty E_x(\bar{R}(t) - \xi, t - s)d\xi \\
&\quad - 2\int_T^t \Gamma(\tau, s)E_x(\bar{R}(\tau), t - \tau)d\tau.
\end{aligned}
\tag{2.59c}
$$

In terms of these, and using (2.53c) for $g(t)$ we obtain finally the desired integro-differential equation for the amplitude $r(t)$:

$$
\begin{aligned}
\dot{r}(t) &= L^{-1}I(t)e^{-m^2 t} \\
&\quad + L^{-1}\int_T^t [\dot{r}(s) + m^2 r(s)]e^{-m^2(t-s)}[P(t,s) + \Gamma(t,s)]ds
\end{aligned}
\tag{2.60}
$$

Step 2 (the asymptotic estimates for P, Γ and $I(t)$). Using a combination of Laplace's method for evaluating integrals and iteration one can show [17, section 3] the following for I, P, Γ defined by (2.59).

Lemma 2.3 *As* $t \to s + 0$,

$$\text{a)} \quad P(t,s) \sim \frac{L\alpha}{2\sqrt{\pi}\sqrt{s}(t-s)^{1/2}} - \frac{7L\alpha^2}{8s} + O((t-s)^{1/2}), \qquad (2.61a)$$

$$\text{b)} \quad \Gamma(t,s) \sim \frac{\gamma c_{eq} m^2}{2\sqrt{\pi}(t-s)^{1/2}} - \frac{\alpha \gamma c_{eq} m^2}{8\sqrt{s}} + O((t-s)^{1/2}). \qquad (2.61b)$$

c) *If* $\chi \in C^1[\bar{R}(T), \infty), \chi(\bar{R}(T)) = 0$ *and* χ *is bounded as* $x \to \infty$, *then*

$$I(t) \sim O((t - T^{-1/2}) \quad as \quad t \to T + 0. \qquad (2.61c)$$

The *final step* is to combine the above estimates. Using the first term of (2.61a,b) in (2.60), one obtains

$$\dot{r}(t) \approx L^{-1} I(t) e^{-m^2 t} + L^{-1} \int_T^t F(s) \left[\frac{D_1 e^{-m^2(t-s)}}{\sqrt{s}(t-s)^{1/2}} + \frac{D_2 e^{-m^2(t-s)}}{(t-s)^{1/2}} \right] ds \qquad (2.62)$$

where $F(s) = \dot{r}(s) + m^2 r(s)$ and D_1 and D_2 are the first constants in (2.61a) and (2.61b) respectively. By the intermediate value theorem $F(s)$ can be passed outside the integral as $F(t)$ and the resulting integrals can be estimated to obtain

$$\int_T^t \frac{e^{-m^2(t-s)}}{\sqrt{s}(t-s)^{1/2}} ds \sim \frac{2}{\sqrt{T}}(t-T)^{1/2} \quad as \quad t \to T + 0, \qquad (2.63a)$$

$$\int_T^t \frac{e^{-m^2(t-s)}}{(t-s)^{1/2}} ds \sim 2(t-T)^{1/2} \quad as \quad t \to T + 0. \qquad (2.63b)$$

Thus (2.62) becomes, as $t \to T + 0$,

$$\dot{r}(t) \approx L^{-1} I(t) e^{-m^2 t} + 2L^{-1} [\dot{r}(t) + m^2 r(t)] \left(\frac{D_1}{\sqrt{T}} - D_2 \right) (t-T)^{1/2}. \qquad (2.64)$$

By elementary differential equation methods,

$$r(t) \approx \exp \left[\frac{2m^2}{L} \left(\frac{D_1}{\sqrt{T}} - D_2 \right) \int (t-T)^{1/2} dt \right] \times$$
$$\times \left[L^{-1} \int I(t) e^{-m^2 t} e^{-[\ldots]} dt + Const. \right] \qquad (2.65)$$

where $[\ldots]$ is the same exponent as in the previous line, giving

$$r(t) \approx \exp \left[\frac{4}{3} \frac{m^2}{L} \left(\frac{D_1}{\sqrt{T}} - D_2 \right) \int (t-T)^{3/2} \right] \times$$
$$\times \left[L^{-1} e^{-m^2 T} \int I(t) dt + Const. \right]. \qquad (2.66)$$

But $\int I(t) dt = O((t-T)^{1/2}$ from (2.61c) and the Const is $r(T)$ by evaluation so the result follows from (2.66).

2.4   Concluding remarks

For the wide class of data in the hypotheses of Lemma 2.3, the above is a complete linearized stability analysis of planar solidification and melting front when the Stefan number is less than 1. It is important to note that the effect of the initial data is subdominant; i.e., there are no transient effects from the data which obscure the true picture of the stability.

An analysis similar to that in section 2.3 had originally been carried out [9] for $T \to \infty$. This current work [17] is more satisfying because it eleminates questions of whether other instabilities could have developed before the asymptotic behavior obtains. Moreover, this work subsumes the previous results [8] as follows. If $T \to \infty$ and the position of the interface, $\bar{R}(T) = \alpha\sqrt{T}$, is to remain finite, then $\alpha \to 0$. In this case the stability criterion (2.50) becomes

$$m^2 > \frac{2V}{\gamma}\left(\frac{\rho - c_{eq}}{c_{eq}}\right) \tag{2.67}$$

which is precisely the result of [9]. It should also be noted that (2.67) was obtained by Ockendon [18] using the travelling wave solutions when the Stefan number is 1. This computation is quite short and instructive but cannot be considered as rigorous in view of the complicated stability dependence of the travelling wave solution (2.5,6) on initial data [12,13].

The shape stability for the travelling wave solution (2.38–40) for the kinetically undercooled problem (2.35') when the Stefan number is larger than 1 was obtained [15] with slightly unexpected results. One would expect, since the kinetic undercooling in (2.35b') is heuristically a stabilizing term, that the stability interval (2.67) would be enlarged. In fact it turns out to be precisely the same [15] except that the amplitude of the instabilities are decreased.

Finally, the global existence and finite time blow-up of spherical solutions has a development similar to that in section 2.2 [19] even though these governing equations include surface tension in the geometry ($\gamma c_{eq}\kappa(s) = \gamma c_{eq}/R(t)$). Moreover, the shape stability analysis [19, section 3] gives values for critical radii of the growing sphere, R_ℓ^*, for which the $Y_{\ell m}(\theta, \phi)$ modes lose stability, which are less than those predicted by the classic Mullins and Sekerka theory [1,2] when the growth is slow (Stefan number near zero) and substantially smaller when the growth is rapid (Stefan number near unity). That is, the more complete model which uses the diffusion equation suggests that the solidification is more unstable than previously thought.

2.5   Appendix (by R. Ricci)

The one-dimensional stability analysis of the one-phase Stefan problem is considerably simplified by transforming the problem (2.3) into a new problem where the free boundary appears a posteriori as the boundary of the support of the solution.

Let $u(x,t)$ denote a solution of problem (2.3). Using a Baiocchi-type transform introduced in [11], we define

$$c(x,t) = \int_{R(t)}^{x} dy \int_{R(t)}^{y} (L - u(\xi,t))d\xi. \qquad (A.1)$$

The function $c(x,t)$ is defined for any $x > R(t)$. It is straightforward to verify that it solves the following free boundary problem:

$$\begin{align}
c_t - c_{xx} + L &= 0, \quad R(t) < x < \infty, t > 0, & (A.2a) \\
c(R(t),t) &= 0, \quad t > 0, & (A.2b) \\
c_x(R(t),t) &= 0, \quad t > 0, & (A.2c) \\
c(x,0) &= c_0(x), R(0) = 0 \leq x < \infty, & (A.2d)
\end{align}$$

where $c_0(x)$ is the transform of $u(x,0)$ by (A.1).

Problem (A.2) is known as the oxygen consumption problem. It has the mathematically pleasant property of admitting a variational formulation as long as $c(x,t)$ is non-negative, see [7]. In fact c is the non-negative solution of the equation

$$c_t - c_{xx} + LH(c) = 0, -\infty < x < \infty, t > 0, \qquad (A.3)$$

with initial datum $c(x,0) = c_0(x), x > 0$, and $c \equiv 0, x < 0$, and where the H is the Heaviside function. Conversely, if equation (A.3) has a sufficiently regular solution, such that $c(x,t) > 0$ for $x > R(t)$, and $c(x,t) = 0$ for $x < R(t)$, then the function $u(x,t) = -c_t(x,t)$ and the free boundary $x = R(t)$ solve the Stefan problem.

The main advantage of this formulation is the possibility of comparing two solutions of equation (A.3) without referring to the free boundary. The following comparison principle holds.

Lemma 1   *Let c_1 and c_2 be solutions of (A.3). Then $c_1(x,0) \geq c_2(x,0)$ for any x implies $c_1(x,t) \geq c_2(x,t)$ for any $x, t > 0$.*

The proof of this lemma is an easy consequence of the monotonicity of the sink term $LH(c)$ in (A.3). A second property of the solutions of (A.3) is of interest for our purpose.

Lemma 2   *Let $c_1(x,t)$ and $c_2(x,t)$ be two solutions of (A.3) such that $c_1(x,0) - c_2(x,0)$ vanishes at infinity. Then $c_1(x,t) - c_2(x,t) \to 0$ uniformly in x as $t \to \infty$.*

This lemma is also easy to prove since $c_1(x,t) - c_2(x,t)$ is controlled in L^∞ by the solutions of the heat equation with initial data plus and minus $|c_1(x,0) - c_2(x,0)|$. Consider now two solutions c_1 and c_2 like in Lemma 2, both of them with support of the form $\{x : R_i(t) < x\}$. Then an estimate of the relative behavior of the free boundary R_i is given by:

Proposition A *If* $\sup_x |c_1(x,t) - c_2(x,t)| = \delta(t) \to 0$ *as* $t \to \infty$, *then* $|R_1(t) - R_2(t)| \leq C\sqrt{\delta(t)}$.

For details of the proof we refer to [12]. This estimate is essentially due to the local behavior of the solution of (A.3) near the free boundary, where the second derivative c_x jumps from 0 to L.

Now the asymptotic analysis can be done quite easily. Suppose that we know the behavior of a particular solution u_p of the problem (2.3), with free boundary R_p. Then we defined the corresponding solution c_p of (A.3) according to (A.1). This, in turn, will select an entire family of initial data c_0 such that $|c_p(x,0) - c_0(x)|$ vanishes at infinity. The initial data $u(x,0)$ whose transforms according to (A.1) are members of this family, will give rise to free boundaries $R(t)$ such that $|R(t) - R_p(t)| \to 0$ as $t \to \infty$.

In particular for similarity solutions of the type $R_\alpha = \alpha\sqrt{t}$, corresponding to initial data $u_\alpha(x,0) = u_\infty < L$ for $x > 0$, where u_∞ and α are related by (2.8), we have:

Theorem A.1 *Suppose that* $\lim_{x\to\infty} u(x,0) = u_\infty$ *and both* $u(x,0) - u_\infty$ *and* $x(u(x,0) - u_\infty)$ *belong to* L^1. *Then*

$$\lim_{t\to\infty} \left(R(t) - \alpha\sqrt{t - t_0} - x_0 \right) = 0.$$

where t_0 *and* x_0 *are solutions of a system of two transcendental equations (see [12] for details).*

Notice that an initial datum $u(x,0)$ selects a special solution among the infinitely many solutions obtained by spatial and temporal shifts of u_0 and R_0.

If we only know that $\lim_{x\to\infty} u(x,0) = u_\infty$, then we still have a weaker stability result, namely

$$\lim_{t\to\infty} \frac{R(t) - \alpha\sqrt{t}}{\sqrt{t}} = 0.$$

Similar results are obtained in the case of travelling waves $(u(x,0) \to L$ as $x \to \infty$, see equations (2.5) and (2.6)).

3 Reactive flows in porous media

3.1 The models

In this section we shall consider reactive fluids flowing through a porous medium which can dissolve some of the minerals and cause changes in the porosity. Through Darcy's law this alters the flow pattern and can give rise to a reaction-infiltration feedback mechanism which can cause instabilities in the shape of the porosity level curves. Consider an aquifer consisting of an insoluble porous matrix (e.g., quartz sandstone) with some soluble mineral (e.g., calcite) partially filling the pores. If water is forced through this porous medium, the soluble component will be dissolved out upstream and the water will become saturated sufficiently far downstream. Between these extremes there is a dissolution (reaction) zone across which the soluble mineral content – and hence the porosity, permeability and diffusion coefficient – changes from its original downstream value to the final, altered value upstream. The question of interest is whether the shape of this dissolution zone is stable. Notice that if a protrusion (in the porosity level curves) in the reaction zone exists at some time, the flow of the undersaturated (hence reactive) fluid tends to be focused to the tip of the protrusion via Darcy's law since "inside" the protrusion (on the upstream side) the permeability is greater than in the neighboring regions (see Fig. 3.1). Since

Fig. 3.1   Focusing of flow to the tip of a porosity-level curve.

additional reactive fluid arrives at the tip, it tends to advance more rapidly. This is the so-called reaction-infiltration instability mechanism. On the other hand, diffusion from the sides of the protrusion raises the concentration of the solute in the water (causing it to be less reactive) which is focusing at the tip and hence decelerates its advancement. (Thus diffusion is stabilizing in this situation as opposed to the solidification model in which it was the destabilizing force.) The competition between these two mechanisms can lead to decay of the protrusion, restabilization to a morphologically more complicated advancing dissolution zone (fingering) or the temporal development of successively more complicated patterns (tip splitting, budding, etc). This shape selection (self-organization) process most certainly is important in many geochemical situations (e.g., the diagenesis and evolution of mineral, oil and gas reservoirs, the dynamics of breakout from nuclear and chemical waste repositories, in situ coal gasification, enhanced oil recovery, the occurence of roll-front mineral deposits [20–22]).

For the remainder of this subsection we shall derive two mathematical models of this phenomenon, the first a coupled set of pde's which, by taking a certain distinguished limit (using matched asymptotics) reduces to the second, a moving free boundary problem.

The rate of increase of porosity, φ, (equivalently, the rate of dissolution of the soluble mineral) is proportional to the reaction rate:

$$\varphi_t = -k(\varphi_f - \varphi)^{2/3}(c - c_{eq}), \tag{3.1a}$$

where k is the reaction rate constant, φ_f is the final porosity after complete dissolution and c is the concentration of solute in water with its equilibrium concentration being c_{eq}. The 2/3-power indicates that we are considering surface reactions. The solute concentration per rock volume, φc, satisfies a mass conservation equation:

$$(\varphi c)_t = \nabla \cdot [\varphi D(\varphi)\nabla c + \varphi c \kappa(\varphi)\nabla p] + \rho \varphi_t, \tag{3.1b}$$

where D and κ are the porosity-dependent diffusion and permeability and ρ is the density of the mineral being dissolved. In (3.1b) p is the pressure and Darcy's law for the velocity $\vec{v}$ has been used in the form (the viscosity is taken to be constant)

$$\vec{v} = -\kappa(\varphi)\nabla p. \tag{3.2}$$

Finally the conservation of water implies

$$(\varphi \rho_w)_t + \nabla \cdot (\varphi \rho_w \vec{v}) = 0, \tag{3.3}$$

which with the constancy of the density of water, ρ_w, and Darcy's law implies

$$\varphi_t = \nabla \cdot (\varphi \kappa(\varphi)\nabla p). \tag{3.1c}$$

The three equations (3.1a,b,c) are to be solved for the three unknowns φ, c, p subject to the imposed asymptotic conditions;

$$c \to 0, \varphi \to \varphi_f \text{ and } p_x \to p_f' \text{ as } x \to -\infty \tag{3.1d}$$

and

$$c \to c_{eq}, \varphi \to \varphi_0 \text{ and } p_x \to ? \text{ as } x \to +\infty \tag{3.1e}$$

and initial data. The condition (3.1d) say that at the inlet $(x = -\infty)$, the water is free of solute, the porous medium has reached its final altered state in which all the soluble mineral has been previously dissolved out and a horizontal pressure gradient (equivalently velocity, through Darcy's law) has been imposed. Similarly, the conditions (3.1e) say that at the outlet far downstream $(x = +\infty)$, the water is saturated with solute, the porous medium is in its original state and the pressure gradient is to be determined as part of the solution. We take zero flux boundary conditions on the transverse boundaries of the aquifer.

Problem (3.1) is an example of the generalization which Professor Stakgold mentioned in his lectures – the situation in which the reaction changes the medium. Notice the power of c in the reaction is less than 1 so it would be an intersting exercise to see how much of that analysis can be carried over to this case. One might ask for a planar travelling wave solution. Because (3.1b and c) are in divergence form, an integration from $-\infty$ to $+\infty$ gives algebraic equations which are soluble for the velocity of the travelling wave and the unknown $p_x(+\infty, t)$. On the other hand it appears impossible to obtain the profiles of φ, c, p explicitly. Indeed, in the simpler case when there is no change in the porosity ($\varphi_f = \varphi_0$) the existence of these profiles could only be obtained by a non-constructive argument [23]. This situation is quite unsatisfactory since, to do a stability analysis, one requires the base planar solution explicitly. Fortunately there is a way out of the predicament since in most geological examples of interest the (transverse) size of the reaction front is several orders of magnitude larger than the thickness of the reaction zone and the details inside this zone are not of interest per se except in the manner in which the cumulative effect provides the dynamics for the evolution of the reaction front on the larger scale. Mathematically, this means scaling problem (3.1) with the objective of finding a parameter which, in typical geological situations, is small. To this end, let [24]

$$\epsilon = c_{eq}/\rho, \tag{3.4}$$

and examine the limit $\epsilon \to 0$ (typical values of ϵ range from 10^{-3} to 10^{-10}). One expects the thickness of the resulting reaction front separating the two values of φ to be very thin (indeed, it is $O(\epsilon^{1/2})$) and the front itself to move very slowly. Thus we introduce a slow dimensionless time $\tilde{t}$ by

$$\tilde{t} = \epsilon(\kappa c_{eq})t, \tag{3.5a}$$

and space variables by

$$\tilde{\vec{v}} = [\kappa c_{eq}/D(\varphi_f)]^{1/2}\vec{v}. \tag{3.5b}$$

The dependent variables are scaled by

$$\tilde{c} = c/c_{eq}, \tilde{p} = \frac{\kappa(\varphi_f)}{D(\varphi_f)}p, \tag{3.6}$$

and the diffusion and permeability by

$$\tilde{D}(\varphi) = D(\varphi)/D(\varphi_f), \tilde{\kappa}(\varphi) = \kappa(\varphi)/\kappa(\varphi_f). \tag{3.7}$$

Then, equations (3.1) can be written in the form (dropping the tildes)

$$\epsilon(\varphi c)_t = \nabla \cdot (\varphi D(\varphi)\nabla c + \varphi\kappa(\varphi)c\nabla p) + \varphi_t, \tag{3.8a}$$

$$\epsilon\varphi_t = -(\varphi_f - \varphi)^{2/3}(c - 1), \tag{3.8b}$$

$$\epsilon\varphi_t = \nabla \cdot (\varphi\kappa(\varphi)\nabla p), \tag{3.8c}$$

$$c \to 0, \varphi \to \varphi_f, p_x \to p'_f \quad \text{as} \quad x \to -\infty \tag{3.8d}$$

$$c \to 1, \varphi \to \varphi_0, p_x \to p'_0 \quad \text{as} \quad x \to +\infty \tag{3.8e}$$

where φ_f and p'_f are given and taken to be constant, the situation least likely to cause patterning, and p'_0 is to be determined.

We end this subsection by sketching how taking the limit $\epsilon \to 0$ (large solid density asymptotics) gives rise to a moving free boundary problem which captures the geological intuition discussed above [24]. Problem (3.8) in the current variables will give rise to the outer solution. At $O(\epsilon^0)$, equation (3.8b) is

$$(\varphi_f - \varphi)^{2/3}(c - 1) = 0, \tag{3.9}$$

which implies $\varphi = \varphi_f$ to the left of the front, $x < R(y,t)$ and $c = 1$ to the right of the front $x > R(y,t)$. At $O(\epsilon)$, equation (3.8b) is

$$\varphi_t = 0 \tag{3.10}$$

which is trivially satisfied in $x < R(y,t)$ where $\varphi = \varphi_f$ and implies $\varphi = \varphi_0$ in $x > R(y,t)$. At $O(\epsilon^0)$ in (3.8c) one obtains

$$\nabla \cdot (\varphi\kappa(\varphi)\nabla p) = 0 \tag{3.11}$$

and since $\varphi = const$ on either side, this results in

$$\Delta p = 0, \tag{3.12}$$

on either side of the front. Likewise, at $O(\epsilon^0)$ in (3.8a), using (3.10), one obtains

$$\nabla \cdot (\varphi D(\varphi)\nabla c + \varphi\kappa(\varphi)c\nabla p) = 0. \tag{3.13}$$

Using the constancy of φ and (3.12), this results in

$$D(\varphi_f)\Delta c + \kappa(\varphi_f)\nabla c \cdot \nabla p = 0 \tag{3.14}$$

in $x < R(y,t)$ and is identically satisfied in $x > R(y,t)$ because $c = 1$. In summary, then, the outer problem is, using $D(\varphi_f) = D_f$ and $\kappa(\varphi_f) = \kappa_f$:

$$\varphi \equiv \varphi_f, x < R(y,t), \tag{3.15a}$$

$$D_f\Delta c + \kappa_f\nabla c.\nabla p = 0, x < R(y,t), \tag{3.15b}$$

$$\Delta p = 0, x < R(y,t), \tag{3.15c}$$

and to the right of the interface,

$$\varphi \equiv \varphi_0, \quad x > R(y,t), \tag{3.15d}$$
$$c \equiv 1, \quad x > R(y,t), \tag{3.15e}$$
$$\Delta p = 0, \quad x > R(y,t). \tag{3.15f}$$

Here $x = R(y,t)$ is the unknown reaction interface whose evolution will be governed by jump conditions on c and p across the interface and these conditions will be obtained by matching the outer solution to (3.15) to the inner solutions obtained from (3.8).

Let σ denote a coordinate normal to $x = R(y,t)$ and τ a coordinate in the tangent line (at some generic point), then the gradient in the x,y coordinates is

$$\nabla = \frac{\partial}{\partial\sigma}\vec{n} + \frac{\partial}{\partial\tau}\vec{t}, \tag{3.16}$$

where $\vec{n}$ and $\vec{t}$ are respectively the unit normal and tangent vectors to $x = R(y,t)$. Furthermore, the time derivative becomes

$$\frac{\partial}{\partial t} = \frac{\partial}{\partial t} - V\frac{\partial}{\partial\sigma} \tag{3.17}$$

where V is the normal velocity of the front. As is quite standard in such layer problems, the thickness of the front turns out to be $O(\epsilon^{1/2})$ suggesting the following inner variable expansions:

$$c = c_0\left(\frac{\sigma}{\epsilon^{1/2}}, \rho, t\right) + \epsilon^{1/2}c_1\left(\frac{\sigma}{\epsilon^{1/2}}, \tau, t\right) + \dots , \tag{3.18a}$$

$$p = p_0\left(\frac{\sigma}{\epsilon^{1/2}}, \tau, t\right) + \epsilon^{1/2}p_1\left(\frac{\sigma}{\epsilon^{1/2}}, \tau, t\right) + \dots , \tag{3.18b}$$

$$\varphi = \varphi_0\left(\frac{\sigma}{\epsilon^{1/2}}, \tau, t\right) + \epsilon^{1/2}\varphi_1\left(\frac{\sigma}{\epsilon^{1/2}}, \rho, t\right) + \dots , \tag{3.18c}$$

$$V = V_0\left(\frac{\sigma}{\epsilon^{1/2}}, \tau, t\right) + \epsilon^{1/2}\varphi_1\left(\frac{\sigma}{\epsilon^{1/2}}, \tau, t\right) + \dots . \tag{3.18d}$$

Denoting the stretched variable $\frac{\sigma}{\epsilon^{1/2}}$ by ξ, inserting the above into (3.8), one obtains to leading order $O(\epsilon^{-1})$:

$$\frac{\partial}{\partial\xi}\left(\varphi_0 D(\varphi_0)\frac{\partial c_0}{\partial\xi} + \varphi_0\kappa(\varphi_0)\frac{\partial p_0}{\partial\xi}\right) = 0, \tag{3.19a}$$

$$\frac{\partial}{\partial\xi}\left(\varphi_0\kappa(\varphi_0)\frac{\partial p_0}{\partial\xi}\right) = 0. \tag{3.19b}$$

Equation (3.19b) implies that

$$\varphi_0\kappa(\varphi_0)\frac{\partial p_0}{\partial\xi} = \pi(\tau, t). \tag{3.20}$$

If the function $\pi(\tau, t)$ were not identically zero infinite pressures would be obtained at the interface $\xi \to \pm\infty$. Thus, since $\varphi_0, \kappa(\varphi_0)$ are not zero, it follows that

$$\frac{\partial p_0}{\partial\xi} = 0. \tag{3.21}$$

Thus p_0 is constant across the interface and from the matching condition

$$\lim_{x \to R(y,t)\pm 0} p^{outer}(x, y, t) = \lim_{\xi \to \pm\infty} p_0(\xi, \tau, t) \tag{3.22}$$

we obtain the continuity of the outer solution for p across the interface; i.e.

$$\lim_{z \to R(y,t)-0} p(x,y,t) = \lim_{z \to R(y,t)+0} p(x,y,t). \tag{3.23}$$

or, in terms of the jump notation,

$$[f] = \lim_{z \to R(y,t)+0} f(x,y,t) - \lim_{z \to R(y,t)-0} f(x,y,t), \tag{3.24}$$

we have

$$[p] = 0, \quad \text{on} \quad x = R(y,t). \tag{3.25a}$$

Similarly, inserting (3.19b) into (3.19a) one can then deduce in exactly the same way the continuity of the concentration across the interface; i.e.,

$$[c] = 0. \tag{3.25b}$$

At the next order, $O(\epsilon^{-1/2})$, we have

$$\frac{\partial}{\partial \xi}\left(\varphi_0 D(\varphi_0)\frac{\partial c_1}{\partial \xi} + \varphi_0 \kappa(\varphi_0)\frac{\partial p_1}{\partial \xi}\right) - V_0 \frac{\partial \varphi_0}{\partial \xi} = 0 \tag{3.26a}$$

$$\frac{\partial}{\partial \xi}\left(\varphi_0 \kappa(\varphi_0)\frac{\partial p_1}{\partial \xi}\right) = 0. \tag{3.26b}$$

By (3.26b) one sees again that $\varphi_0 \kappa(\varphi_0)\frac{\partial p_1}{\partial \xi}$ is a constant across the front and since

$$\lim_{\xi \to \pm\infty}\frac{\partial p_1}{\partial \xi} = \lim_{\xi \to R(y,t)\pm 0}\frac{\partial p^{outer}}{\partial n}, \tag{3.27}$$

one has

$$\lim_{\xi \to -\infty}\varphi_0 \kappa(\varphi_0)\frac{\partial p_1}{\varphi \xi} = \varphi_f \kappa_f \lim_{z \to R(y,t)-0}\frac{\partial p}{\partial n} = \lim_{z \to R(y,t)-0}\varphi \kappa(\varphi)\frac{\partial p}{\partial n} \tag{3.28a}$$

and similarly

$$\lim_{\xi \to +\infty}\varphi_0 \kappa(\varphi_0)\frac{\partial p_1}{\varphi \xi} = \varphi_0 \kappa_0 \lim_{z \to R(y,t)+0}\frac{\partial p}{\partial n} = \lim_{z \to R(y,t)+0}\varphi \kappa(\varphi)\frac{\partial p}{\partial n} \tag{3.28b}$$

giving, because of the constancy of $\varphi_0 \kappa(\varphi_0)\frac{\partial p_1}{\partial \xi}$

$$\left[\varphi\kappa(\varphi)\frac{\partial p}{\partial n}\right] = 0, \quad \text{on} \quad x = R(y,t). \tag{3.29}$$

Similarly, using (3.26b) in (2.36a), one obtains the following jump in the concentration gradient:

$$\left[\varphi D(\varphi)\frac{\partial c}{\partial n}\right] - V_0[\varphi] - 0, \tag{3.30}$$

which can be written, using the fact that $c = 1$ for $x > R(y, t)$,

$$\varphi_f D_f \frac{\partial c}{\partial n} = V_0(\varphi_f - \varphi_0) = \frac{R_t}{(1 + R_y^2)^{1/2}}(\varphi_f - \varphi_0). \qquad (3.31)$$

Furthermore, $\frac{\partial c}{\partial n} = (c_x - c_y R_y)(1 + R^2 y)^{-1/2}$, which gives in (3.21)

$$\varphi_f D_f(c_x - c_y R_y) = (\varphi_f - \varphi_0) R_t. \qquad (3.32)$$

In summary then, the large solid density limit ($\epsilon = c_{eq}/\epsilon \to 0$) turns problem (3.8) into the following free boundary problem (the pressure on the right is denoted by q):

$$
\begin{aligned}
D_f \Delta c + \kappa_f \nabla c . \nabla p &= \ 0, \quad x < R(y, t), & (3.33a) \\
\Delta p &= \ 0, \quad x < R(y, t), & (3.33b) \\
\Delta q &= \ 0, \quad x > R(y, t), & (3.33c)
\end{aligned}
$$

while on the moving, unknown interface $x = R(y, t)$,

$$
\begin{aligned}
c &= \ 1 & x &= R(y, t) & (3.33d) \\
p &= \ q, & x &= R(y, t), & (3.33e) \\
p_x - p_y R_y &= \ \Gamma(q_x - q_y R_y), & x &= R(y, t), & (3.33f) \\
\varphi_f D_f(c_x - c_y R_y) &= \ (\varphi_f - \varphi_0) R_t, & x &= R(y, t) & (3.33g)
\end{aligned}
$$

along with the asymptotic conditions

$$c \to 0, \quad \frac{\partial p}{\partial x} \to p_f' \quad \text{as} \quad x \to -\infty \qquad (3.33h)$$

$$c \to 1, \quad \frac{\partial p}{\partial x} \to p_0' \quad \text{as} \quad x \to +\infty \qquad (3.33i)$$

where $\Gamma = \varphi_0 \kappa_0 / \varphi_f \kappa_f$ is a measure of the medium change and p_0' is to be determined as part of the solution.

3.2   Planar solution

The planar version of problem (3.33) with interface $x = R(t)$ is:

$$D_f c_{xx} + \kappa_f c_x p_x = 0, \quad x < R(t), \qquad (3.34a)$$

(3.34a)

$$p_{xx} = 0, \quad x < R(t), \qquad (3.34b)$$

$$q_{xx} = 0, \quad x > R(t), \qquad (3.34c)$$

while on the interface

$$c = \quad 1, \quad x = R(t) \qquad (3.34\text{d})$$
$$p = \quad q, \quad x = r(t), \qquad (3.34\text{e})$$
$$p_x = \quad \Gamma q_x, \quad x = R(t), \qquad (3.34\text{f})$$
$$D_f c_x = \quad \gamma \dot{R}, \quad x = R(t), \qquad (3.34\text{g})$$

along with the asymptotic conditions

$$c \to 0, p_x \to p_f', \quad \text{as} \quad x \to -\infty, \qquad (3.34\text{h})$$

$$c \to 1, p_x \to p_0', \quad \text{as} \quad x \to +\infty, \qquad (3.34\text{i})$$

where $\gamma = \left(1 - \frac{\varphi_0}{\varphi_f}\right)$ is another measure of the porosity change which, as we shall see, will enter the calculation in an inessential way, not affecting the interval of instability which will be determined by Γ and the Peclet number $v_f L/D_f = -\kappa_f p_f' L/D_f$ (L is the transverse width of the aquifer on which we assume zero-flux boundary conditions). Letting $\xi = x - R(t)$, all x–derivatives become ξ–derivatives and the solutions of (3.34b and c) subject to the boundary and interface conditions are

$$p(\xi) = p_f' \xi + A, \xi < 0, \qquad (3.35\text{a})$$

$$q(\xi) = p_0' \xi + A, \xi > 0, \qquad (3.35\text{b})$$

where, by (3.34f),

$$p_0' = \frac{1}{\Gamma} p_f'. \qquad (3.36)$$

Substituting into (3.34a) with $-\kappa_f p_f'/D_f = \alpha > 0$ we obtain

$$c_{\xi\xi} - \alpha c_\xi = 0, \qquad (3.37)$$

whose solution subject to (3.34d and h) is

$$c(\xi) = e^{\alpha\xi}. \qquad (3.38)$$

The Stefan condition (3.34g) then gives

$$D_f \alpha = -\kappa_f p_f' = \gamma \dot{R}(t); \qquad (3.39)$$

i.e., the velocity is forced to be a constant and hence the explicit solution (3.35, 3.38) is a travelling wave solution. As a result the shape stability analysis of this solution will be more standard and easier that that of section 2.3.

3.3   Shape stability analysis

Another approach to linearizing equation (3.34) about the planar solution (3.35, 3.38, 3.39) is to first transform the unknown boundary to a fixed one at $\xi = 0$ at the expense of making the equations quite complicated. To this end let $\xi = x - at - r(y,t), \eta = y, \tau = t$ where $a = -\kappa_f p_f'/\gamma = \dot{R}(t)$ the velocity of the planar front from (3.39) and $r(y,t)$ is the perturbation off the planar front. Equations (3.34) then become, in $\xi < 0$,

$$D_f[(1 + r_\eta^2)c_{\xi\xi} - 2r_y c_{\xi\eta} + c_{\eta\eta} - r_{\eta\eta}c_\xi]$$

$$+\kappa_f[(1 + r_\eta^2)c_\xi p_\xi - r_\eta(c_\xi p_\eta + c_\eta p_\xi) + c_\eta p_\eta] = 0, \tag{3.40a}$$

$$(1 + r_\eta^2)p_{\xi\xi} - 2r_\eta p_{\xi\eta} + p_{\eta\eta} - r_{\eta\eta}p_\xi = 0, \tag{3.40b}$$

in $\xi > 0$,

$$(1 + r_\eta^2)q_{\xi\xi} - 2r_\eta q_{\xi\eta} + q_{\eta\eta} - r_{\eta\eta}q_\xi = 0, \tag{3.40c}$$

on the interface, $\xi = 0$,

$$c = 1, \tag{3.40d}$$

$$p = q, \tag{3.40e}$$

$$(1 + r_\eta^2)p_\xi - r_\eta p_\eta = \Gamma[(1 + r_\eta^2)q_\xi - r_\eta q_\eta], \tag{3.40f}$$

$$D_f[(1 + r_\eta^2)c_\xi - r_\eta c_\eta] = \gamma R_\tau, \tag{3.40g}$$

along with the asymptotic conditions,

$$c \to 0, p_\xi \to p_f' \quad \text{as} \quad \xi \to -\infty,$$

$$\tag{3.40h}$$

$$(c \to 1), p_\xi \to p_0' \quad \text{as} \quad \xi \to +\infty.$$

In order to treat the shape stability of the planar solution (3.35, 3.38, 3.39) (which will be denoted by superbars in what follows), we consider perturbations of the form

$$c(\xi,\eta,\tau) = \bar{c}(\xi) + \delta c_1(\xi)e^{\sigma(m)\tau}\cos m\eta, \tag{3.41a}$$

$$p(\xi,\eta,\tau) = \bar{p}(\xi) + \delta p_1(\xi)e^{\sigma(m)\tau}\cos m\eta, \tag{3.41b}$$

$$q(\xi,\eta,\tau) = \bar{q}(\xi) + \delta q_1(\xi)e^{\sigma(m)\tau}\cos m\eta, \tag{3.41c}$$

$$r(\eta,\tau) = 0 + \delta e^{\sigma(m)\tau}\cos m\eta. \tag{3.41d}$$

This could be thought of as linearizing equations (3.40), decomposing into the spatial eigenfunctions $\cos m\eta$ (since we are assuming zero flux conditions on the transverse boundaries $\eta = 0$ and $\eta = \pi$) and then taking the Laplace transform in τ. The dual variable $\sigma(m)$ determines the growth or decay (instability or stability, respectively) of the $\cos m\eta$ mode. From a bifurcation theory viewpoint $\sigma(m)$ is just the spectrum of the linearized problem for the $\cos m\eta$ mode. It is a straightforward exercise to find the equations for

c_1, p_1, q_1 be retaining only the $O(\delta)$ terms. Denoting ξ–derivatives by primes and substituting the actual forms for $\bar{c}, \bar{p}, \bar{q}$ from (3.35, 3.38 and 3.39), these equations are:

$$c_1'' - \alpha c_1' - m^2 c_1 + \alpha m^2 e^{\alpha\xi} + \frac{\kappa_f}{D_f}\alpha e^{\alpha\xi} p_1' = 0, \xi < 0, \tag{3.42a}$$

$$p_1'' - m^2 p_1 + m^2 p_f' = 0, \xi < 0, \tag{3.42b}$$

$$q_1'' - m^2 q_1 + m^2 p_f'/\Gamma = 0, \xi < 0, \tag{3.42c}$$

subject to the interface and asymptotics conditions

$$c_1 = 0, p_1 = q_1, p_1' = \Gamma q_1', D_f c_1' = \gamma\sigma(m), \xi = 0, \tag{3.42d}$$

$$c_1 \to 0, p_1' \to 0 \quad \text{as} \quad \xi \to -\infty, \tag{3.42e}$$

$$c_1 \to 0, q_1' \to 0 \quad \text{as} \quad \xi \to +\infty. \tag{3.42f}$$

It is straightforward to solve these equations:

$$p_1(\xi) = p_f' \left[\frac{1-\Gamma}{1+\Gamma} e^{|m|\xi} + 1 \right], \xi < 0 \tag{3.43a}$$

$$q_1(\xi) = -\frac{p_f'}{\Gamma} \left[\frac{1+\Gamma}{1-\Gamma} e^{|m|\xi} - 1 \right], \xi > 0 \tag{3.43b}$$

$$c_1(\xi) = \alpha \left[-\frac{2}{1+\Gamma} e^{\pi_m\xi} + e^{\alpha\xi} + \frac{1-\Gamma}{1+\Gamma} e^{(\alpha+|m|)\xi} \right], \xi < 0, \tag{3.43c}$$

where $\pi_m = [\alpha + (\alpha^2 + 4m^2)^{1/2}]/2$ is the positive root of the auxiliary equation of (3.42a). Substituting the expression for c_1 into the Stefan condition in (3.42d) one obtains the spectrum of the linearized problem

$$\sigma(m) = \frac{-\kappa_f p_f'}{\gamma(1+\Gamma)} [-(\alpha^2 + 4m^2)^{1/2} + \alpha + (1-\Gamma)|m|]. \tag{3.44}$$

Fig. 3.2 shows σ as a function of $|m|$, revealing clearly that the planar state is linearly unstable to long wavelength perturbations and is stable to short wavelength perturbations.

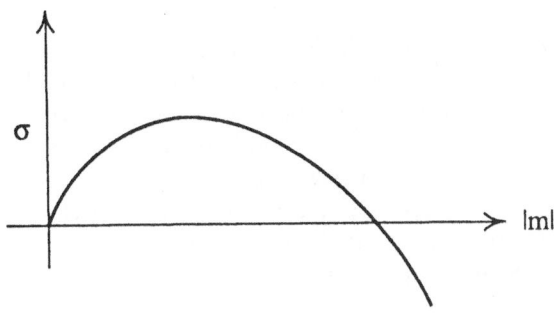

Fig. 3.2   Graph of the spectrum $\sigma(m)$.

The critical wave number ($|m_0|$ at which $\sigma(m) = 0$) is given by

$$|m_0| = \frac{2(1 - \Gamma)}{(3 - \Gamma)(1 + \Gamma)}\alpha. \tag{3.45}$$

On the other hand, the frequency can only assume integer values (since the transverse width was scaled to π). Thus, from (3.45) the critical value of α at which the $|m_0| = 1$ mode loses stability is

$$\alpha_c = \left(\frac{v_f}{D_f}\right)_c = \frac{(3 - \Gamma)(1 + \Gamma)}{2(1 - \Gamma)}. \tag{3.46}$$

It is clear from (3.46) that the critical value becomes large as $\Gamma \to 1$. This indicates that if there is no change in the porosity, the planar front is stable to all modes since the flow cannot be changed (see also [24]). On the other hand, as the inlet velocity v_f gets large or the diffusion D_f becomes small the planar front becomes more unstable.

A weakly nonlinear analysis was done [25] in the neighborhood of the bifurcation point given in (3.46). One finds at the nonlinear level the competing mechanisms restabilize the front to one with a more complicated geometry. Indeed, one finds that the amplitude $r(t)$ of the critical $\cos y$ mode satisfies a Landau equation

$$\dot{r} = \sigma_2 r - \Lambda r^3, \tag{3.47}$$

where $\sigma^2 > 0$ and the Landau constant $\Lambda > 0$ indicating this restabilization at the (next) nonlinear level. We have also examined the situation in which the dissolved solute increases the viscosity of the fluid [26]. Since in this situation fresh (less viscous) fluid is being forced into saturated (more viscous) fluid, this corresponds to coupling the Saffman-Taylor instability with the reaction-infiltration instability. The results [26] show that the reaction-infiltration instability is dominant in this situation and we completely quantify the increased instability from the viscosity dependence.

3.4   Numerical simulations

The free boundary problem (3.33) which was so useful for the analytical treatment is unsuitable for a numerical treatment since it would require complicated front-tracking methods. Instead we use the reaction-transport model (3.1) with ϵ small (but not so small that the reaction front is excessively thin giving rise to large gradients and stiff numerical problems). In the figures below, the position of the front is approximated by the porosity level curve $\varphi = (\varphi_f + \varphi_0)/2$ whose position is shown at successive time intervals. At time zero the domain is uniformly cemented except for a small high porosity "bump" located near the inlet on the left.

In Fig. 3.3a one sees that for $v_f \ll v_c = v_{c,1}$ (i.e., all modes are linearly stable) the perturbation dies out leaving a stable evolving planar front. In Fig. 3.3b where $v_f \simeq v_c$, the planar front eventually loses stability to a constant velocity $\cos y$ front. In Fig. 3.3c where $\nu_f \simeq 5\nu_c$, the many unstable modes couple through the nonlinear dynamics to produce an elongating finger.

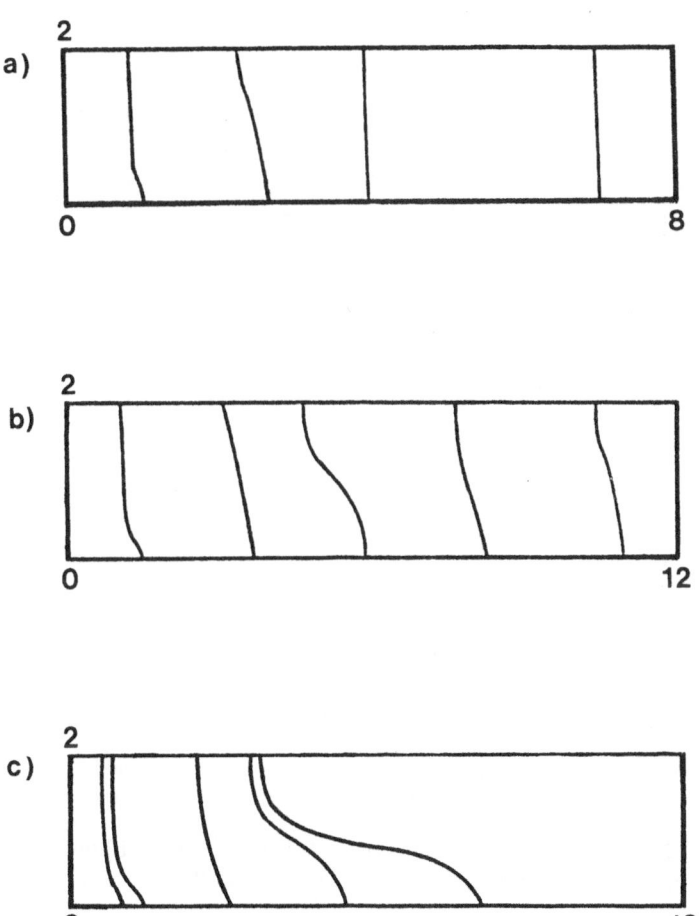

Fig. 3.3   Numerical simulations of the evolution of an initial bump in the reaction zone when (a) $\nu_f \ll \nu_c$, (b) $\nu_f \simeq \nu_c$ and (c) $\nu_f \gg \nu_c$.

Higher precision calculations were performed at Indiana University [26] allowing for much longer time evolution of the fronts and displaying new dynamic self-organization. In Fig. 3.4a one sees that for faster inlet flows or wider domains, the advancing finger can destabilize causing "tip splitting". For different values of the parameters, the central finger can dominate the side fingers giving rise to "budding" as in Fig. 3.4b. The buds arise periodically in time in a frame moving with the advancing central finger. This suggests that a critical value of the bifurcation parameter exists at which the steady advancing nonplanar state (finger) undergoes a (secondary) Hopf bifurcation to a temporally oscillatory budding state. With nonsymmetric data as in Fig. 3.4c, this process may change the course of the main branch leading to the "meandering" of the dissolution channel.

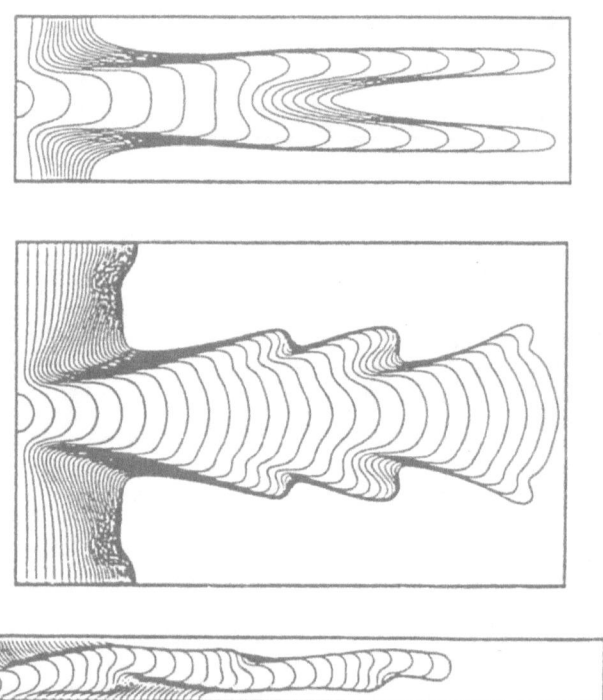

Fig. 3.4   Numerical simulations exhibiting (a) tip splitting, (b) budding and
(c) meandering (from Chen and Ortoleva, [26]).

References

[1] W.W. Mullins and R.F. Sekerka, Morphological stability of a particle growing by diffusion or heat flow, *J. Appl. Phys.* **34** (1964), 323–329.

[2] W.W. Mullins and R.F. Sekerka, Stability of a planar interface during solidification of a dilute alloy, *J. Appl. Phys.* **35** (1964), 444–451.

[3] J.S. Langer, Instabilities and pattern formation in crystal growth, *Rev. Modern Phys.* **52** (1980), 1–28.

[4] L.I. Rubinstein, *The Stefan Problem*, Transl. Math. Monographs 27, Amer. Math. Soc., Providence (1971).

[5] A. Friedman, *Partial Differential Equations of Parabolic Type*, Prentice Hall, Englewood Cliffs (1964).

[6] A. Friedman, *Variational Principles and Free Boundary Problems*, Wiley, New York (1982).

[7] C.M. Elliott and J.R. Ockendon, *Weak and Variational Methods in Moving Boundary Problems*, Pitman, London (1982).

[8] H.S. Carslaw and J.C. Jaeger, *Conduction of Heat in Solids*, Clarendon Press, Oxford (1959).

[9] J. Chadam and P. Ortoleva, The stabilizing effect of surface tension on the development of the free boundary in a planar, one-dimensional, Cauchy-Stefan problem, *IMA J. Appl. Math.* **30** (1983), 57–66.

[10] J. Chadam, J.-B. Baillon, M. Bertsch, P. Ortoleva and L.A. Peletier, Existence, uniqueness and asymptotic behaviour of solutions of the planar, supersaturated solidification, Cauchy-Stefan problem, Séminaire Lions-Brézis, Vol. VI, Collège de France, *Research Notes in Mathematics* **109**, Pitman, London (1984), 27–47.

[11] A. Fasano and M. Primicerio, A critical case for the solvability of Stefan-like problems, *Math. Methods Appl. Sci.* **5** (1983), 84–96.

[12] R. Ricci and W. Xie, On the stability of some solutions of the Stefan problem, preprint, *European J. Appl. Math.* **2** (1991), 1–15.

[13] N.N. Dewynne, S.D. Howison, J.R. Ockendon and W. Xie, Asymptotic behavior of solutions to the Stefan problem with a kinetic condition at the free boundary, *J. Austral. Math. Soc. Ser. B* **31** (1989), 81–96.

[14] B. Sherman, A general one-phase Stefan problem, *Quart. Appl. Math.* **28** (1971), 377–382.

[15] G. Caginalp and J. Chadam, Stability of interfaces with velocity correction term, *Rocky Mountain J. Math.* **21** (1991), 617–629.

[16] R.J. Schaefer and M.E. Glicksman, Fully time-dependent theory for the growth of spherical crystal nuclei, *J. Crystal Growth* **5** (1969), 44–58.

[17] Q. Zhu, A. Peirce and J. Chadam, Initiation of shape instabilities of free boundaries in planar Cauchy-Stefan problems, McMaster preprint, submitted for publication in *European J. Math.* (1991).

[18] J. Ockendon, Linear and nonlinear stability of a class of moving boundary problems, *Proc. Sem. on Free Boundary Problems*, E. Magenes, ed., Pavia 1979, Istituto Nazionale di Alta Matematica, Rome (1980).

[19] J. Chadam, S.D. Howison and P. Ortoleva, Existence and stability for spherical crystals growing in a supersaturated solution, *IMA J. Appl. Math.* **39** (1987), 1–15.

[20] J. Chadam, J. Hettmer, E. Merino, C. Moore and P. Ortoleva, Geochemical self-organization I: Feedback mechanisms and modelling approach, *Amer. J. Sci.* **287** (1987), 977–1007.

[21] J. Chadam, E. Merino, P. Ortoleva and A. Sen, Self-organization in water-rock interaction systems II: The reactive-infiltrate instability, *Amer. J. Sci.* **287** (1987), 1008–1040.

[22] G. Auchmuty, J. Chadam, E. Merino, P. Ortoleva and E. Ripley, The structure and stability of moving redox fronts, *SIAM J. Appl. Math.* **46** (1986), 588–604.

[23] J. Chadam, P. Ortoleva and A. Sen, Reactive percolation instability, *IMA J. Appl. Math.* **36** (1987), 207–220.

[24] J. Chadam, P. Ortoleva and A. Sen, Weakly nonlinear stability of reaction-percolation interfaces, *SIAM J. Appl. Math.* **48** (1988), 1362–1378.

[25] J. Chadam, P. Ortoleva and A. Peirce, Stability of reactive flows in porous media: Coupled porosity and viscosity changes, *SIAM J. Appl. Math.* **51** (1991), 684–692.

[26] W. Chen and P. Ortoleva, Reaction front fingering in carbonate-cemented sandstones, *Earth-Sci. Rev.* **29** (1990), 183-198.

Shape Derivatives and Differentiability of Min Max

Michel C. DELFOUR
Centre de recherches mathématiques
Université de Montréal
C.P. 6128, succ. A
Montréal, Qué., H3C 3J7
Canada

Abstract

One of the central issues in Shape Analysis and Optimization is the choice of a good definition of the shape derivative which will play the same role as the directional semiderivative in vector spaces. In these notes we introduce a derivative which is based on the Velocity (Speed) Method and show the various connections with other methods. The second part of the notes deals with shape problems where the shape cost functional is a function of the solution of a boundary value problem over the domain. We promote the use of Lagrangian techniques coupled with the use of simple theorems on the differentiability of a Min, a saddle point or a Min Max with respect to a parameter. Their main advantage is to avoid the study of the shape derivative of the solution of the boundary value problem.

Orientation

In §1 we briefly review the notion of Gâteaux and Hadamard semiderivatives over a (vector) Banach space, discuss their analogues for the space of subsets (domains) of $\mathbf{R}^N$ which is not a vector space, and give several classical examples of perturbations of a domain. In §2 we establish the precise equivalence between perturbations by non-autonomous vector fields and transformations. §3 is devoted to the precise definition of first and second order shape derivatives for unconstrained and constrained (subsets of a fixed hold-all) families of domains. In each case we give the Hadamard-Zolésio structure theorems. In so doing we recover an intrinsic characterization of the famous class of Finite Perimeter or Cacciopoli sets which have played a key role in the solution of the Plateaus problem. Basic elements of Shape calculus are given in §4. For more details the reader is referred to J.P. Zolésio [1,2] and the forthcoming book of Sokolowski-Zolésio [1].

The last three sections of the notes deal with shape problems involving the solution $y(\Omega)$ of an associated boundary value problem over the optimization domain Ω. §5 looks at the so-called "compliance problems" where the cost function to be minimized over Ω

35

M. C. Delfour and G. Sabidussi (eds.), Shape Optimization and Free Boundaries, 35–111.
© 1992 *Kluwer Academic Publishers.*

is itself the minimum of an energy functional over a given set of functions defined on Ω. Stationary conditions with respect to the set of functions and with respect to the family of domains Ω typically yield conditions which are characteristic of a Free Boundary problem. We show that the priviledged tool to get the shape derivative of the infimum with respect to the domain Ω is the use of theorems on the derivative of an infimum with respect to a parameter. To get around the dependence of the underlying function spaces with respect to the parameter, we use the technique of Function Space Parametrization.

In the general case when the solution (state variable) of the boundary value problem (state equation) is not related to the shape functional to be minimized, the state equation is incorporated in the cost function via a "Lagrange multiplier" and a MinMax or saddle point formulation. Such techniques have been promoted by J. Céa [3] and Fortin-Glowinski [1]). In §6 and §7 we show how the technique of §5 can be extended. Here theorems on the differentiability of a MinMax or a saddle point are used in conjunction with Function Space Parametrization (Neumann problem) and Function Space Embedding (Dirichlet problem). The fundamental difference with the "compliance problem" of §5 is the presence of an "adjoint state" which is the solution of a dual boundary value problem.

We have chosen to limit ourselves to simple examples to describe and illustrate the basic techniques. They readily extend to more complex and interesting examples. For more material and examples the reader is referred to the notes of J.P. Zolésio in this book and the forthcoming book by Sokolowski-Zolésio [1]. Unilateral problems are treated by J. Sokolowski in this book and innovative techniques can be found in Delfour-Zolésio [6,14]. Most of the material in this set of notes is based on recent joint work with J.P. Zolésio (cf. Delfour-Zolésio [1] to [14].) Extensions of the differentiability theorems for an Infimum, a saddle point or a MinMax to ϵ-solutions have been considered by Delfour-Morgan [1, 3].

1   Construction of a shape derivative

The object of this section is the choice of perturbations of a domain Ω and the construction of a *Shape derivative* with respect to such perturbations. The basic idea is to adapt the constructions used in the definitions of derivatives and semiderivatives in linear function spaces.

The strongest notion is of course the *Fréchet* derivative which consists in the characterization of the tangential linear mapping. At the other end of the spectrum, *Nonsmooth Analysis* provides several extremely weak notions of semiderivatives. We shall focus our attention on *Gâteaux* and *Hadamard* derivatives and semiderivaties. Roughly speaking the Gâteaux derivative is a sufficiently weak notion which generalizes the idea of directional derivative. As for the Hadamard derivative it is still sufficiently weak for our purposes but strong enough to preserve the very useful *chain rule* of the classical calculus.

The analogue of the Gâteaux semiderivative for shapes will be the *Method of Pertur-bation of the Identity Operator* while the analogue of the Hadamard semiderivative will be the *Velocity (Speed) Method*. The Hadamard notion is extremely important in situations where the shape is parametrized and the chain rule is necessary. In §1.3 we shall complete the picture by showing that for functionals from $\mathbf{R}^N$ to $\mathbf{R}$ the semiderivative obtained by the Velocity Method coincides with the Hadamard semiderivative. These basic notions and constructions will be formally extended to deal with domain functionals. The velocity and transformations viewpoints will be emphasized and a series of examples will be given in §1.4.

In the next two sections we shall distinguish between the case where the domains are *unconstrained* and the case where they are *constrained* to lie in a fixed *hold-all D* of $\mathbf{R}^N$. In both cases we shall show that under reasonable conditions, starting from a family $\{T_t\}$ of transformations of $\mathbf{R}^N$ (resp. $\overline{D}$) is equivalent to starting from a family of vector fields $\{V(t)\}$ on $\mathbf{R}^N$ (resp. $\overline{D}$). This is an extremely important result which throws a bridge between two points of view which have undergone parallel developments in the literature. In particular we shall show that Lipschitzian perturbations of the identity operator fall in that category.

In §1.5 we shall establish that, under very general conditions, methods using a family of transformations are equivalent to the Velocity (Speed) Method. This will be true for both unconstrained domains and constrained domains, that is, a family of domains Ω which are subsets of a fixed subset or *hold-all D* of $\mathbf{R}^N$.

In view of this equivalence the subsequent developments of these notes will be based on the Velocity (Speed) Method. We shall see that this method offers certain advantages over others. In particular it will naturally yield a *canonical shape Hessian* while the other methods will not.

In the sequel we shall adopt some notation and terminology used in Mechanics. Capital X will be a *Lagrangian* or *material* coordinate while lower case x will be the *Eulerian* or *actual* coordinate.

1.1 Derivatives in Banach spaces

Prior to studying the notions of shape derivative it is useful to briefly review the situation in Banach spaces. In that context we shall start with the weaker notion of Gâteaux semiderivative and emphasize the key role played by the Hadamard semiderivative in the semiderivative of the composition of two functions, since in most applications the extension of the *chain rule* will be an essential ingredient of a good differential calculus.

Definition 1.1 Let f be a real continuous function defined on an open subset of a real Banach space E.

(i) We say that f has a *Gâteaux semiderivative* at a point $x \in U$ in the direction $v \in E$ if the following limit exists

$$\lim_{\epsilon \searrow 0} \frac{1}{\epsilon}[f(x + \epsilon v) - f(x)]; \tag{1.1}$$

whenever it exists it will be denoted by $df(x; v)$. If $df(x; v)$ exists for all v in E we say that f is Gâteaux *semidifferentiable* at x.

(ii) We say that f has a *Hadamard semiderivative* at a point $x \in U$ in the direction $v \in E$ if the following limit exists

$$\lim_{\substack{\epsilon \searrow 0 \\ w \to v}} \frac{1}{\epsilon}[f(x + \epsilon w) - f(x)]; \tag{1.2}$$

whenever it exists it will be denoted by $d_H f(x; v)$. If $d_H f(x; v)$ exists for all v in E we say that f is Hadamard *semidifferentiable* at x.

It is clear that $df(x; v)$ exists and is equal to $d_H f(x; v)$ whenever $d_H f(x; v)$ exists but the converse is not true without additional hypotheses as can be seen from the following example.

Example 1 Consider the function $f : \mathbf{R}^2 \to \mathbf{R}$

$$f(x, y) = \begin{cases} \dfrac{x^6}{(y - x^2)^2 + x^8} & \text{if } (x, y) \neq (0, 0) \\ 0 & \text{if } (x, y) = (0, 0). \end{cases} \tag{1.3}$$

It is readily seen that f has a Gâteaux semiderivative at $(0,0)$ in all directions v in $\mathbf{R}^2$ and that

$$df((0,0); v) = 0, \quad \forall v \in \mathbf{R}^2, \tag{1.4}$$

which is trivially linear and continuous with respect to v. However if for $\epsilon > 0$ we choose the directions

$$w(\epsilon) = (1, \epsilon) \to v = (1, 0) \quad \text{as} \quad \epsilon \to 0,$$

then

$$\frac{f(\epsilon w(\epsilon)) - f(0,0)}{\epsilon} = \frac{1}{\epsilon^3} \to +\infty$$

and $d_H f((0,0); (1,0))$ does not exist.   □

Proposition 1.1 *Given a function $f : U \to \mathbf{R}$ which is Lipschitzian in a neighborhood U of x in E,*

$$\exists c(x) > 0, \ \forall y, z \in U, \ |f(y) - (z)| \leq c\|y - z\|_E, \tag{1.5}$$

then $d_H f(x; v)$ exists if $df(x; v)$ exists and they are equal.

Proof For any w in E there exists $\bar{\epsilon} > 0$ such that

$$\forall \epsilon, \ 0 < \epsilon \leq \bar{\epsilon}, \ x + \epsilon w \in U \ \text{and} \ x + \epsilon v \in U.$$

Then

$$\frac{f(x+\epsilon w) - f(x)}{\epsilon} = \frac{f(x+\epsilon w) - f(x+\epsilon v)}{\epsilon} + \frac{f(x+\epsilon v) - f(x)}{\epsilon}$$

and

$$\left| \frac{f(x+\epsilon w) - f(x)}{\epsilon} - df(x;v) \right| \leq \left| \frac{f(x+\epsilon v) - f(x)}{\epsilon} - df(x;v) \right| + c\|w - v\|.$$

So as $\epsilon \to 0$ and $w \to v$, $d_H f(x;v) = df(x;v)$. $\qquad\square$

The Hadamard semiderivative is continuous with respect to the direction but not necessarily linear as can be seen from the next example.

Example 2 The Hadamard semiderivative of the norm $f(x) = \|x\|_E$ at $x = 0$ is given by

$$d_H f(0;v) = \|v\|_E, \quad \forall v \in E. \tag{1.6}$$

$\qquad\square$

As for the Gâteaux semiderivative it is generally neither linear nor continuous even in finite dimension.

Example 3 Consider the following function $f : \mathbf{R}^2 \to \mathbf{R}$

$$f(x,y) = \begin{cases} \dfrac{x^5}{(y - x^2)^2 + x^8}, & \text{if } (x,y) \neq (0,0) \\ 0 & , \text{if } (x,y) = (0,0). \end{cases} \tag{1.7}$$

It has a Gâteaux semiderivative at $(0,0)$ in all directions $v = (v_1, v_2)$ in $\mathbf{R}^2$ and

$$df((0,0); (v_1, v_2)) = \begin{cases} v_1, & \text{if } v_2 = 0 \\ 0, & \text{if } v_2 \neq 0 \end{cases}. \tag{1.8}$$

$\qquad\square$

But the main difference between the above two semiderivatives is that the composition of two functions which are semidifferentiable in the sense of Hadamard is semidifferentiable in the sense of Hadamard.

Theorem 1.1 *Let E and F be two Banach spaces and let h be the composition of two mappings f and g*

$$h(x) = f(g(x)) \tag{1.9}$$

in a neighborhood U of a point x in E where

$$g : U \subset E \to F \quad \text{and} \quad f : g(U) \to F. \tag{1.10}$$

Assume that

(i) g has a Gâteaux (resp. Hadamard) semiderivative at x in the direction v,

$$\begin{cases} dg(x;v) = \lim_{\epsilon\searrow 0} \dfrac{g(x+\epsilon v) - g(x)}{\epsilon} & \text{in } F \\[2mm] (resp.\ d_H g(x;v)) = \lim_{\substack{\epsilon\searrow 0 \\ w\to v}} \dfrac{g(x+\epsilon w) - g(x)}{\epsilon} & \text{in } F), \end{cases} \qquad (1.11)$$

(ii) and $d_H f(g(x); dg(x;v))$ exists.

Then

$$\begin{cases} dh(x;v) = d_H f(g(x); dg(x;v)) \\ (resp.\ d_H h(x;v)) = d_H f(g(x); d_H g(x;v))). \end{cases} \qquad (1.12)$$

Proof (a) For $\epsilon > 0$ small enough let

$$m(\epsilon) = \frac{g(x+\epsilon v) - g(x)}{\epsilon} - dg(x;v) \text{ in } F.$$

By hypothesis

$$m(\epsilon) \to 0 \text{ as } \epsilon \to 0.$$

By definition of $dh(x;v)$ we want to find the limit of the differential quotient

$$d(\epsilon) = \frac{f(g(x+\epsilon v)) - f(g(x))}{\epsilon},$$

which can be rewritten as

$$d(\epsilon) = \frac{f(g(x) + \epsilon(dg(x;v) + m(\epsilon))) - f(g(x))}{\epsilon}$$

where

$$dg(x;v) + m(\epsilon) \to dg(x;v) \text{ as } \epsilon \to 0.$$

So by definition of $d_H f$,

$$\lim_{\epsilon\searrow 0} d(\epsilon) = d_H f(g(x); dg(x;v)).$$

(b) When g is Hadamard semidifferentiable we replace $m(\epsilon)$ and $d(\epsilon)$ by

$$m(\epsilon, w) = \frac{g(x+\epsilon w) - g(x)}{\epsilon} - dg(x;v)$$

and

$$d(\epsilon, w) = \frac{f(g(x) + \epsilon(dg(x;v) + m(\epsilon, w))) - f(g(x))}{\epsilon}$$

and proceed as in part a).   □

In general we cannot improve the derivative of h by improving the derivative of g when f is not Hadamard semidifferentiable.

Example 4 Consider the composition of the function $f : \mathbf{R}^2 \to \mathbf{R}$ in Example 1 and the map

$$g : \mathbf{R} \to \mathbf{R}^2, \quad g(x) = (x, x^2). \tag{1.13}$$

The map f is Gâteaux but not Hadamard semidifferentiable and the map g is infinitely differentiable. However the composition

$$h(x) = f(g(x)) = \frac{1}{x^2} \tag{1.14}$$

is not even Gâteaux semidifferentiable at 0. $\qquad\qquad\qquad\qquad\qquad\qquad\square$

It is important to reiterate that the Hadamard semidifferentiability is a key property for the "chain rule". In general the composition $h = f \circ g$ of two maps will fail to have a semiderivative unless f has a Hadamard semiderivative and this even if the map $g : E \to F$ is Fréchet differentiable at the point x:

$$\begin{cases} \exists\, Dg(x) \in \mathcal{L}(E, F) \text{ such that} \\ \displaystyle\lim_{\|v\|_E \to 0} \frac{\|g(x + v) - g(x) - Dg(x)v\|_F}{\|v\|_E} = 0. \end{cases}$$

Finally the class of functions which are Hadamard semidifferentiable is not restrictive since it contains the classical continuously differentiable functions and the convex continuous functions.

Theorem 1.2 *Let $f : U \subset E \to \mathbf{R}$ be a convex function defined in a convex neighborhood U of a point x of a Banach space E.*

(i) *There exists a neighborhood V of x such that*

$$\forall y \in V, \quad \forall v \in E, \quad df(y; v) \ \text{exists}.$$

(ii) *If, in addition, f is continuous at x, then there exists a neighborhood W of x such that*

$$\forall y \in W, \quad \forall v \in E, \quad d_H f(y; v) \ \text{exists}.$$

Proof Part (ii) is a consequence of part (i) and the fact that a convex function which is continuous at x is locally Lipschitzian in a neighborhood of x (cf. Ekeland-Temam [1, pp. 11–12]) by direct application of Proposition 1.1. The proof of part (i) is not difficult, but we give it for completeness. Let $\theta \in]0, 1]$. Notice that for fixed x and v

$$\begin{aligned} &\exists\, \alpha, \quad 0 < \alpha < 1, \quad \text{such that} \ x - \alpha v \in U, \\ &\exists\, \theta_0, \quad 0 < \theta_0 < 1, \quad \text{such that} \ x + \theta v \in U, \ \forall \theta \in]0, \theta_0]. \end{aligned}$$

For this fixed α, we show that

$$\forall \theta \in]0, \theta_0[, \ \frac{f(x) - f(x - \alpha v)}{\alpha} \leq \frac{f(x + \theta v) - f(x)}{\theta}. \tag{1.15}$$

This follows from the identity

$$x = \frac{\alpha}{\alpha + \theta}(x + \theta v) + \frac{\theta}{\alpha + \theta}(x - \alpha v)$$

and the convexity of f

$$f(x) \le \frac{\alpha}{\alpha + \theta}f(x + \theta v) + \frac{\theta}{\alpha + \theta}f(x - \alpha v).$$

This can be rewritten

$$\frac{\theta}{\theta + \alpha}[f(x) - f(x - \alpha v)] \le \frac{\alpha}{\theta + \alpha}[f(x + \theta v) - f(x)]$$

and yields (1.15). Define

$$\varphi(\theta) = \frac{f(x + \theta v) - f(x)}{\theta}, \quad 0 < \theta < \theta_0.$$

Now we show that φ is a monotone increasing function of $\theta > 0$. For all θ_1 and θ_2, $0 < \theta_1 < \theta_2 < \theta_0$

$$
\begin{aligned}
f(x + \theta_1 v) - f(x) &= f\left(\frac{\theta_1}{\theta_2}(x + \theta_2 v) + (1 - \frac{\theta_1}{\theta_2})x\right) - f(x) \\
&\le \frac{\theta_1}{\theta_2}f(x + \theta_2 v) + (1 - \frac{\theta_1}{\theta_2})f(x) - f(x) \\
\Rightarrow \varphi(\theta_1) &\le \varphi(\theta_2).
\end{aligned}
$$

As φ is a monotone increasing function which is bounded below, its limit exists as θ goes to 0. By definition it is equal to $df(x; v)$. $\qquad \square$

1.2 Choice of deformations and derivatives of a domain

We choose to work in a framework where the variable domains Ω are the subsets of a fixed underlying smooth domain D. The set of all such domains is denoted by

$$\mathcal{P}(D) = \{\Omega : \Omega \subset D\}. \tag{1.16}$$

A *domain functional* is a map

$$J : \mathcal{A} \to \mathbf{R} \tag{1.17}$$

where $\mathcal{A} \subset \mathcal{P}(D)$ is a set of *admissible domains*. Since there is no vector space structure on the set $\mathcal{P}(D)$, it is not possible to directly use a differential quotient. However we can introduce an infinitesimal deformation of the domain Ω and compute the variation of the domain functional.

For convenience let $D = \mathbf{R}^N$. To define a deformation of a smooth domain Ω, we can assume the existence of a given vector field $V : \mathbf{R}^N \to \mathbf{R}^N$ which moves each point X of $\mathbf{R}^N$ to a new point

$$x(t) = X + tV(X) \tag{1.18}$$

for a small parameter $t > 0$ which will go to 0.

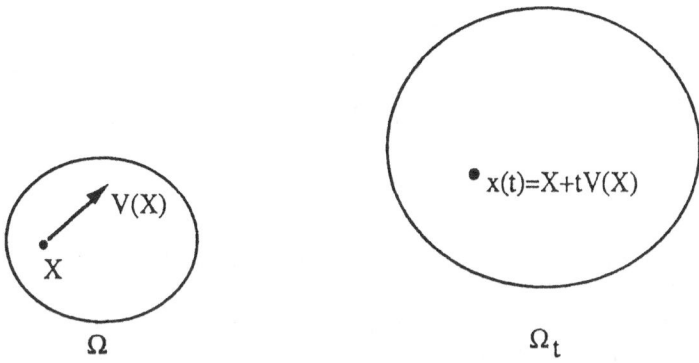

Figure 1. Perturbation of Ω by V

In this construction it is important that the vector field V be dependent on the position of the point X: a constant field would only translate the domain Ω into $\Omega + tV$ without deformation. This method is called *perturbation of the identity* since the associated transformation T_t of $\mathbf{R}^N$,

$$T_t = I + tV, \quad (T_t X = X + tV(X), \ \forall X), \tag{1.19}$$

is a small perturbation of the identity operator I by the vector field V. In this approach V plays the role of a *direction* and we can speak of the shape semiderivative of J at Ω in the direction V by looking at the limit of the differential quotient

$$\lim_{t \searrow 0} \frac{J(\Omega_t) - J(\Omega)}{t} \tag{1.20}$$

where Ω_t is the image of Ω by T_t,

$$\Omega_t = T_t(\Omega) = \{T_t X : X \in \Omega\}. \tag{1.21}$$

Nevertheless, this first approach to the definition of a shape semiderivative is not completely satisfactory since the perturbation of the identity is a *non-local* method: the point $x(t)$ depends on the velocity field V evaluated at X and not at $x(t)$. In general this will not affect first order semiderivatives of J, but it will introduce an artificial "acceleration term" in the definition of second order semiderivatives (cf. §3.3 and §3.4 for more details). Fortunately this can be easily fixed and a more natural (or physical) approach can be followed. Go back to the initial choice of deformation in (1.18) and consider $t \geq 0$ as an *artificial time*. Subsequently rewrite expression (1.21) as a difference

$$x(t) - x(0) = (t - 0)V(x(0)), \quad x(0) = X$$

and in differential form
$$dx(t) = dtV(x(0)) \cong dtV(x(t)).$$
In the limit this naturally yields a new *local deformation* $T_t : \mathbf{R}^N \to \mathbf{R}^N$

$$T_t X = x(t), \quad t \geq 0, \tag{1.22}$$

where $x(t)$ is the solution of the differential equation

$$\frac{dx}{dt}(t) = V(x(t)), \quad t > 0, \quad x(0) = X. \tag{1.23}$$

This is the basis of the *Velocity (Speed)* method. Here the velocity of the point $x(t)$ at time t is equal to the velocity field V evaluated at $x(t)$: it is a local deformation. The choice of the word "Velocity" to describe this method is accurate but may become ambiguous in problems where the variables involved are themselves "physical velocities": this situation is commonly encountered in Fluid Mechanics and Material Sciences.

In such cases it may be necessary to distinguish between the "artificial velocity" and the "physical velocity". This was at the origin of the terminology "*Speed Method*" which has often been used in the literature. This last terminology is convenient but not as accurate and descriptive as the *Velocity Method*. We shall keep both terminologies and conveniently use the one which is most suitable in the context of the problem under investigation.

The shape semiderivative of J at Ω in the direction V will be defined as

$$dJ(\Omega; V) = \lim_{t \searrow 0} \frac{J(\Omega_t) - J(\Omega)}{t} \tag{1.24}$$

(whenever the limit exists), where

$$\Omega_t = T_t(\Omega) = \{T_t X : X \in \Omega\} \tag{1.25}$$

and T_t is the transformation of $\mathbf{R}^n$ obtained by the Velocity Method as defined by (1.22)–(1.23). *In these notes we have chosen to give all the basic definitions, constructions and theorems within the context of the Velocity Method.* This choice has been motivated by the fact that all the results based on some form of perturbation of the identity can by easily recovered by direct application of the Velocity Method extended to *non-autonomous fields* $V(t, x)$, that is

$$T_t X = x(t), \quad t > 0, \quad x(0) = X, \tag{1.26}$$

where $x(t)$ is the solution of the non-autonomous differential equation

$$\frac{dx}{dt}(t) = V(t, x(t)), \quad t > 0, \quad x(0) = X, \tag{1.27}$$

and $V(t, x)$ is now a function of the time $t \geq 0$ and the space variable x in $\mathbf{R}^N$. To illustrate this last assertion we construct the non-autonomous velocity field $W(t, x)$ associated with the perturbation of the identity

$$T_t X = x(t) = X + tV(X). \tag{1.28}$$

The field W is to be chosen such that the function x in (1.15) is the solution of the differential equation

$$\frac{dx}{dt}(t) = W(t, x(t)), \quad x(0) = X. \tag{1.29}$$

It is readily seen that a good choice for W is

$$W(t, x) = V(T_t^{-1}x), \quad \forall x \in \mathbf{R}^N, \ \forall t \geq 0, \tag{1.30}$$

for $t > 0$ sufficiently small. Now under appropriate continuity and differentiability hypotheses:

$$W(0, x) = V(x), \quad \forall x \in \mathbf{R}^N, \tag{1.31}$$

$$\frac{\partial W}{\partial t}(0, x) = -[DV(x)]V(x), \quad \forall x \in \mathbf{R}^N, \tag{1.32}$$

where $DV(x)$ is the Jacobian matrix of V at the point x. In compact notation

$$W(0) = V \quad \text{and} \quad \overset{\bullet}{W}(0) = -[DV]V \tag{1.33}$$

where $W(0)$ is the velocity field at $t = 0$

$$[W(0)](x) = W(0, x), \quad \forall x \in \mathbf{R}^N \tag{1.34}$$

and $\overset{\bullet}{W}(0)$ is the acceleration field at $t = 0$

$$\left[\overset{\bullet}{W}(0)\right](x) = \frac{\partial W}{\partial t}(0, x), \quad \forall x \in \mathbf{R}^N. \tag{1.35}$$

This last exercise is extremely informative since it shows that at time 0, the points of the domain Ω are simultaneously affected by a velocity field $W(0) = V$ and an acceleration field

$$\overset{\bullet}{W}(0) = -[DV]V.$$

In most problems the two methods will produce the same first order semiderivative. However second order semiderivatives will differ by an acceleration term which will appear in the expression obtained by the method of perturbation of the identity.

1.3 Equivalence between Hadamard semiderivatives and semiderivatives in the velocity sense for functionals on $\mathbf{R}^N$

In the previous section we have introduced the method of perturbation of the identity operator as an extension of the notion of Gâteaux semiderivative for the derivative of a domain functional. To complete the picture we now show that the velocity method is an extension of Hadamard semiderivative for the derivative of a domain functional.

Theorem 1.3 *Let $f : N_X \to \mathbf{R}$ be a functional defined in a neighbourhood N_X of a point X in $\mathbf{R}$. Then f is Hadamard semidifferentiable at (X, v) if and only if there exists $\tau > 0$ such that for all velocity fields $V : [0, \tau] \to \mathbf{R}$ satisfying the hypotheses*

a) $\forall x \in N_X$, $V(\cdot, x) \in C^0([0, \tau]; \mathbf{R}^N)$,

b) $\exists c > 0$, $\forall x, y \in N_X$, $\|V(\cdot, y) - V(\cdot, x)\|_{C^0([0,\tau];\mathbf{R}^N)} \le |y - x|$,

c) $\lim_{t \searrow 0} V(t, X) = v$,

the limit of the differential quotient

$$\frac{f(x(t; X)) - f(X)}{t}$$

exists, where $x = x(\cdot; X)$ is the solution of the differential equation

$$\begin{cases} \dfrac{dx}{dt}(t) = V(t, x(t)), \quad 0 < t < \tau \\[2mm] x(0) = X. \end{cases}$$

Proof ($\Rightarrow$) Let V be a vector field satisfying conditions a), b) and c). Define

$$w(t) = \frac{1}{t} \int_0^t V(s, x(s)) ds, \quad 0 < t \le \tau.$$

It is continuous on $]0, \tau]$ and

$$w(t) - v = \frac{1}{t} \int_0^t [V(s, x(s)) - V(0, X)] ds.$$

Therefore

$$|w(t) - v| \le c \max_{[0,t]} |x(s) - X| + \max_{[0,t]} |V(s, X) - V(0, X)|$$

and

$$\lim_{t \searrow 0} w(t) = v.$$

So

$$\lim_{t \searrow 0} \frac{f(x(t; X)) - f(X)}{t} = \lim_{\substack{w(t) \to v \\ t \searrow 0}} \frac{f(X + tw(t)) - f(X)}{t}$$

$$= d_H f(X; v),$$

since f is Hadamard semidifferentiable at (X, v).

($\Leftarrow$) Conversely we first prove that to each w in $C^1([0, \tau]; \mathbf{R}^N)$ we can associate the velocity field

$$V(t, X) = \frac{d}{dt}[tw(t)] = w(t) + tw'(t), \quad t > 0,$$

which trivially satisfies assumptions a), b) and c). But

$$\begin{cases} \dfrac{dx}{dt}(t) = \dfrac{d}{dt}[tw(t)] \\[2mm] x(0) = X \end{cases}$$

and

$$x(t) = X + tw(t).$$

Therefore by hypothesis the limit

$$\lim_{t \searrow 0} \frac{f(X + tw(t)) - f(X)}{t}$$

exists. This is true for any w in $C^1([0, \tau]; \mathbf{R}^N)$. So given any sequence $\{w_n, t_n\}$ in $\mathbf{R}^N \times \mathbf{R}^+$ such that $w_n \to v$ and $t_n \to 0$, it is always possible to construct a $C^1([0, \tau]; \mathbf{R}^N)$ function w such that

$$w(t_n) = w_n, \quad \forall n \geq 1, \quad w(0) = v$$

and apply the previous results. This proves that f is Hadamard semidifferentiable at (X, v):

$$\lim_{\substack{w_n \to v \\ t_n \searrow 0}} \lim \frac{f(X + t_n w_n) - f(X)}{t_n}$$

exists. □

Remark 1.1 The equivalence Theorem 1.3 is extremely important. It indicates how to generalize the Hadamard semiderivative from linear spaces to differentiable manifolds. □

1.4 Examples of families of transformations of domains

In §1.2 we have given a formal abstract presentation of some of the key steps leading to the definition of shape derivatives. In particular we have emphasized the method of perturbation of the identity operator and the Velocity (Speed) Method. One important conclusion was that we can always associate with a perturbation of the identity a non-autonomous velocity field which generates the same family of transformations of $\mathbf{R}^N$.

In fact, as we shall see in the more technical §2, there is a general equivalence between families of transformations $\{T_t : 0 \leq t \leq \tau\}$ of $\mathbf{R}^N$ and families of non-autonomous velocity fields $\{V(t) : 0 \leq t \leq \tau\}$. Perturbations of the identity are just special cases in which this equivalence holds.

Before proceeding with more abstract statements and proofs of theorems, we give a few typical examples of shape derivatives which often occur in the literature. In each case we construct an underlying family (not necessarily unique) of tranformations $\{T_t : 0 \leq t \leq \tau\}$ corresponding to the chosen shape derivative. We shall show how those various constructions fit in the general framework of §1.2 and motivate the material in §2 and §3 on unconstrained and constrained domains.

We specifically consider special classes of domains (C^∞, C^k, Lipschitzian) and special applications to Cartesian graphs, polar coordinates and level curves which provide classical examples of parametrized and/or constrained deformations.

We first recall the definitions of C^k and C^∞ domains. Given a domain Ω in $\mathbf{R}^N$ with boundary Γ, we choose a finite subcover of open neighbourhoods $\mathcal{U}_1, \ldots, \mathcal{U}_m$ of Γ and a local representation of the boundary $\Gamma_j = \Gamma \cap \mathcal{U}_j$, $1 \leq j \leq m$, which is not necessarily a Cartesian graph. Introduce the notation $\varsigma = (\varsigma', \varsigma_N)$ for a point $\varsigma = (\varsigma_1, \ldots, \varsigma_N)$ in $\mathbf{R}^N$ where $\varsigma' = (\varsigma_1, \ldots, \varsigma_{N-1})$. Denote by B the unit ball in $\mathbf{R}^N$ and define the sets

$$B_0 = \{\varsigma \in B : \varsigma_N = 0\}$$
$$B_+ = \{\varsigma \in B : \varsigma_N > 0\}, \quad B_- = \{\varsigma \in B : \varsigma_N < 0\}.$$

Assume the existence of bijective mappings

$$c_j : \mathcal{U}_j \to B, \quad 1 \leq j \leq m,$$

with the following properties: for all j, $1 \leq j \leq m$,

$$c_j \in C^k(\mathcal{U}_j, B) \text{ and } h_j = c_j^{-1} \in C^k(B, \mathcal{U}_j)$$
$$\Omega \cap \mathcal{U}_j = h_j(B_+)$$
$$\Gamma_j = \Gamma \cap \mathcal{U}_j = h_j(B_0), \quad B_0 = c_j(\Gamma_j).$$

Let $\{e_1, \ldots, e_N\}$ be the standard unitary orthonormal basis in $B \subset \mathbf{R}^N$. Then

$$\{e_1, \ldots, e_{N-1}\} \subset B_0$$

and the tangent space $T_x\Gamma$ at x to Γ is the vector space spanned by the $N-1$ vectors

$$\{Dh_j(\varsigma)e_i : 1 \leq i \leq N-1\}, \quad \varsigma = c_j(x) \in B_0,$$

where $Dh_j(\varsigma)$ is the Jacobian matrix of h_j at the point ς. So an outward normal vector field to Γ at x is given by

$$m_j(x) = - \, {}^*Dh_j(\varsigma)^{-1}e_N$$

$({}^*Dh_j(\varsigma)$ is the transposed matrix $Dh_j(\varsigma))$, since

$$m_j(x) \cdot Dh_j(\varsigma)e_i = e_N \cdot e_i = 0, \quad 1 \leq i \leq N-1,$$

where "$\cdot$" and $|\cdot|$ denote the inner product and norm in the Euclidean finite dimensional space $\mathbf{R}^N$. The density term

$$\omega_j(x) = |m_j(x)|$$

will play an important role in the boundary integration on Γ. It also enters in the definition of the outward unitary normal field $n(x)$ at $x \in \Gamma$

$$n(x) = -\omega_j(x)^{-1}m_j(x), \quad \text{if } x \in \Gamma_j. \tag{1.36}$$

It can be verified that n is uniquely defined on Γ by checking that for $x \in \Gamma_j \cap \Gamma_{j'}$, n is uniquely defined by (1.36).

The boundary integration on Γ is obtained by using a partition of unity $\{r_j : 1 \leq j \leq m\}$ for the set of open neighborhoods $\{\mathcal{U}_j : 1 \leq j \leq m\}$ of Γ:

$$\begin{cases} r_j \in C_{\text{comp}}^{\infty}(\mathcal{U}_j), \quad 0 \leq r_j(x) \leq 1 \\[2ex] \displaystyle\sum_{j=1}^{m} r_j(x) = 1 \text{ in a neighborhood of } \Gamma. \end{cases}$$

If $g \in C^0(\Gamma)$, that is $g : \Gamma \to \mathbf{R}$ is a mapping such that

$$(gr_j) \circ h_j \in C^0(B_0), \quad 1 \leq j \leq m,$$

we define

$$\int_{\Gamma_j} gr_j \, d\Gamma = \int_{B_0} (gr_j) \circ h_j(\varsigma')\omega_j(\varsigma')d\varsigma'$$

and

$$\int_{\Gamma} g \, d\Gamma = \sum_{j=1}^{m} \int_{\Gamma_j} gr_j \, d\Gamma.$$

The case of C^{∞} domains is a special case of C^k for $k = \infty$. It will be necessary to work at that level when the solution $y(\Omega)$ of the state equation is chosen in $C^{\infty}(\overline{\Omega})$. In practice this situation is academic but mathematically convenient to short-circuit many mathematical technicalities and quickly obtain expressions for derivatives of cost functions with respect to the domain Ω.

1.4.1 C^{∞} domains

Let Ω be a smooth bounded domain in D and assume that its boundary Γ is a C^{∞} manifold. Its outward unitary normal n is defined by (1.36) for C^k domains,

$$n(x) = \omega_j(x)^{-1}m_j(x), \quad \text{if } x \in \Gamma_j, \quad 1 \leq j \leq m. \tag{1.37}$$

For C^{∞} domains n belongs to $C^{\infty}(\Gamma; \mathbf{R}^N)$, that is

$$\forall j, \ 1 \leq j \leq m, \quad (nr_j) \circ hj \in C^{\infty}(B_0; \mathbf{R}^N).$$

For any ρ in $C^{\infty}(\Gamma)$ and small $t \geq 0$, consider the following perturbation Γ_t of Γ along the normal n

$$\Gamma_1 = \{x \in \mathbf{R}^N : x = X + t\rho(X)n(X), \ \forall X \in \Gamma\}.$$

It is clear that

$$\exists \, t_1 > 0 \text{ such that } \forall t \in [0, t_1], \quad \Gamma_t \subset \overline{D},$$

if and only if either $\Gamma \cap \partial D = \emptyset$ or

$$\forall X \in \text{ neighborhood of } \Gamma \cap \partial D, \quad \rho(X) \leq 0.$$

Therefore for t_1 sufficiently small and all t, $0 \le t \le t_1$, Γ_t will be the boundary of a C^∞ domain Ω_t. To be more specific we shall now build a transformation T_t of $\overline{D}$ which maps Ω onto Ω_t and Γ onto Γ_t. The first step is the construction of a smooth extension $N \in C^\infty(\overline{D}; \mathbf{R}^N)$ of the normal field n on Γ. Let m_j be the analytic normal field constructed in (1.8) on $\Gamma_j = \Gamma \cap \mathcal{U}_j$ and define the field

$$m = \sum_{j=1}^{m} m_j \, r_j \ \text{ on } \Gamma.$$

By construction

$$m \in C^\infty(\Gamma; \mathbf{R}^N)$$

and this definition can be extended to a neighbourhood of Γ since the mapping h_j is defined in a tubular neighbourhood of c_j ([supp r_j]$\cap\Gamma$) in B. The field m is different from zero on Γ and by compactness of Γ there exists a neighborhood $\mathcal{U}$ of Γ in $\overline{D}$ such that

$$|m(x)|_{\mathbf{R}^N} > 0, \quad \forall x \in \mathcal{U}.$$

Now construct a function r_0 in $C^\infty(\mathcal{U}), 0 \le r_0(x) \le 1$, and a neighborhood $\mathcal{U}_1$ of Γ such that

$$r_0(x) = 1, \ \forall x \in \mathcal{U}_1 \ \text{ and } \ \Gamma \subset \mathcal{U}_1 \subset \overline{\mathcal{U}}_1 \subset \mathcal{U}.$$

The vector field

$$N(x) = r_0(x)|m(x)|^{-1}m(x) \tag{1.38}$$

belongs to $C^\infty(\overline{D}; \mathbf{R}^N)$ and coincides with n on $\mathcal{U}_1 \cap \Gamma$. Similarly, with the help of n, extend the function ρ in $C^\infty(\Gamma)$ to a function $\tilde{\rho}$ in $C^\infty(\overline{D})$.

Define the following transformation of $\overline{D}$

$$T_t(X) = X + t\tilde{\rho}(X)N(x), \quad 0 \le t \le t_1.$$

It can easily be verified that for t small enough and admissible ρ's, T_t is a one-to-one mapping from $\overline{D}$ (resp. ∂D) onto $\overline{D}$ (resp. ∂D). So the domain Ω_t associated with the boundary Γ_t defined in (1.17) can be chosen as

$$\Omega_t = T_t(\Omega) = (I + t\tilde{\rho}N)(\Omega).$$

Since the domain Ω_t is specified by its boundary Γ_t, it only depends on ρ and not on its extension $\tilde{\rho}$. The special transformation T_t which is introduced here is of class C^∞, that is $T_t \in C^\infty(\overline{D}, \overline{D})$, and $\frac{1}{t}[T_t - I]$ is proportional to the normal field n on Γ, but it is not proportional to the normal n_t on Γ_t for $t > 0$. In other words at $t = 0$ the deformation is along n but at $t > 0$ the deformation is generally not along n_t.

If $J(\Omega)$ is a shape functional defined on C^∞ domains in D, the semiderivative (if it exists) is defined as follows:

$$\begin{cases} \forall \rho \in C^\infty(\Gamma) \\[2mm] d_n J(\Omega; \tilde{\rho}) = \lim_{t \searrow 0} \dfrac{J((I + t\tilde{\rho}N)\Omega) - J(\Omega)}{t}. \end{cases}$$

It can be shown that this limit only depends on ρ and not on its extension $\tilde{\rho}$.

1.4.2 C^k domains

When Ω is a smooth domain in D with a boundary Γ which is a C^k-manifold, the normal field n belongs to $C^{k-1}(\Gamma; \mathbf{R}^N)$. Therefore taking deformations along the normal would yield transformations $\{T_t\}$ mapping C^k-domains Ω onto C^{k-1}-domains $\Omega_t = T_t(\Omega)$. The obvious way to deal with C^k-domains is to relax the constraint that the perturbation be carried by the normal and choose vector fields $\Theta \in C^k(\overline{D}; \mathbf{R}^N)$ such that for all $x \in \partial D$

$$
\begin{cases}
\Theta(x) \cdot \nu(x) = 0 & \text{if } \nu(x) \text{ exists} \\
\Theta(x) = 0 & \text{otherwise,}
\end{cases}
$$

where $\nu(x)$ is the outward unit normal at x to the piecewise smooth boundary ∂D of D. Consider the family of transformations

$$
T_t = I + t\Theta, \quad \Omega_t = T_t(\Omega), \quad t \geq 0.
$$

This is a generalization of (1.36) in §1.4.1 corresponding to $\Theta = \tilde{\rho}N$. For sufficiently small t's, T_t is a one-to-one mapping from $\overline{D}$ (resp. ∂D) onto itself.

A more restrictive approach to get around the lack of sufficient smoothness of the normal n to Γ would be to introduce a transverse field p on Γ with the following properties

$$
p \in C^\infty(\Gamma; \mathbf{R}^N), \quad p(x) \cdot n(x) > 0, \quad \forall x \in \Gamma.
$$

Then for p, and arbitrary $t_1 \geq 0$ and $\rho \in C^\infty(\Gamma)$ define

$$
\Gamma_t = \{x \in \mathbf{R}^N : x = X + t\rho(X)p(X), \quad X \in \Gamma\}.
$$

Choosing C^k extensions $\tilde{\rho}$ and P of ρ and p we can go back to the case where

$$
\Theta = \tilde{\rho}P \in C^k(\overline{D}; \mathbf{R}^N).
$$

Now define for any $\Theta \in C^k(\overline{D}; \mathbf{R}^N)$ the following semiderivative

$$
d_k J(\Omega; \Theta) = \lim_{t \searrow 0} \frac{J((I + t\Theta)(\Omega)) - J(\Omega)}{t}.
$$

In §1.4.1 and §1.4.2 the perturbed domain Ω_t always appears in the form $\Omega_t = T_t(\Omega)$, where T_t is a one-to-one transformation of $\overline{D}$ leaving ∂D invariant. So far T_t was of the form $I + t\Theta$, but many other examples of such transformations can be given in specific situations as will be seen in the following sections.

1.4.3 Cartesian graphs

In many problems we are dealing with domains Ω which are the hypograph of some positive function γ in Cartesian coordinates. Such domains are typically of the form

$$
\Omega = \{(x', x_N) \in \mathbf{R}^N : x' \in U \subset \mathbf{R}^{N-1} \text{ and } 0 < x_N < \gamma(x')\},
$$

where U is a connected open set in $\mathbf{R}^{N-1}$ and $\gamma \in C^0(\overline{U}; \mathbf{R}_+)$ is a positive function. Many Free Boundary and contact problems are formulated over such domains. Usually the domains Ω (and hence the functions γ) will be constrained. The function γ can be specified on $\overline{U} \backslash U$ or not. In some examples the derivative of γ could also be specified, that is $\partial \gamma / \partial \nu = g$ on ∂U, where U is smooth and ν is the outward normal field along ∂U.

When γ (resp. $\partial \gamma / \partial \nu$) is specified along ∂U, the directions of deformation are chosen in $C^0(U; \mathbf{R}_+)$ such that

$$\mu = 0 \quad \left(\text{resp.} \ \frac{\partial \mu}{\partial \nu} = 0\right) \quad \text{on } \partial U.$$

For small $t \geq 0$ and each such μ define the perturbed domain

$$\Omega_t = \{(x', x_N) \in \mathbf{R}^N : 0 < x_N < \gamma(x') + t\mu(x')\}$$

and the obvious family of transformations

$$(x', x_N) \mapsto T_t(x', x_N) = \left(x', x_N + \frac{x_N}{\gamma(x')} t\mu(x')\right) : \Omega \to \Omega_t$$

which can be extended to a neighborhood D of Ω containing the perturbed domains $\Omega_t, 0 \leq t \leq t_1$, for some small $t_1 > 0$. In general D will be such that $\overline{D} = \overline{U} \times [0, L]$ for some $L > 0$.

This construction is appropriate for Lipschitzian domains as well as C^k domains $(1 \leq k \leq \infty)$ depending on whether γ is a Lipschitzian or a C^k function. Again the transformation T_t is equal to $I + t\Theta$ where

$$\Theta(x', x_N) = x_N \frac{\mu(x')}{\gamma(x')}.$$

But of course this Θ is not the only choice for which $\Omega_t = T_t(\Omega)$. For instance let $\lambda : \mathbf{R}_+ \to \mathbf{R}_+$ by any smooth increasing function such that $\lambda(0) = 0$ and $\lambda(1) = 1$. Then we could consider the transformations

$$T_t(x', x_N) = \left(x', x_N + t\lambda\left(\frac{x_N}{\gamma(x')}\right)\mu(x')\right).$$

This is the first example which illustrates the following general principle: "the transformation T_t of $\overline{D}$ such that $\Omega_t = T_t(\Omega)$ is not unique".

This means that, at least for smooth domains, only the trace $T_t|_{\Gamma_t}$ on Γ_t is important while the displacement of the inner points does not contribute to the definition of Ω_t. Nevertheless this statement is to be used carefully. Here we implicitly assume that the cost function $J(\Omega)$ and the associated constraints $K(\Omega)$ are only a function of the shape of Ω. However in some problems involving singularities at inner points of Ω (e.g. when the solution $y(\Omega)$ of the state equation has a singularity or when some constraints on the domain are active), the situation might require a finer analysis. One such example is the internal displacement of the interior nodes of a triangularization τ_h when the solution $y(\Omega)$ of a partial differential equation is approximated by a piecewise polynomial solution over the triangularized domain Ω_h in the Finite Element Method. Such a displacement does not change the shape of Ω_h but it does change the solution y_h of the problem. When the displacement of the interior nodes is a priori parametrized by the boundary nodes the solution y_h will only depend on the position of the boundary nodes but the interior nodes will contribute through the choice of the specified parametrization.

1.4.4 Polar coordinates

In some examples domains are star-shaped with respect to a point. Since a domain can always be translated we can assume without loss of generality that this point is the origin. Then such domains Ω can be parametrized as follows

$$\Omega = \{x \in \mathbf{R}^N : x = \rho\varsigma, \ \varsigma \in S_{N-1}, \ 0 \le \rho < f(\varsigma)\}$$

where S_{N-1} is the unit sphere in $\mathbf{R}^N$,

$$S_{N-1} = \{x \in \mathbf{R}^N : |x| = 1\},$$

and $f : S_{N-1} \to \mathbf{R}_+$ is a positive continuous mapping from S_{N-1} such that

$$m = \min\{f(\varsigma) : \varsigma \in S_{N-1}\} > 0.$$

Given any $g \in C^0(S_{N-1})$ and a sufficiently small $t \ge 0$ the perturbed domains are defined as

$$\Omega_t = \{x \in \mathbf{R}^N : x = \rho\varsigma, \ \varsigma \in S_{N-1}, \ 0 \le \rho < f(\varsigma) + tg(\varsigma)\}.$$

For example choose $t, 0 \le t \le t_1$, for some $t_1 > 0$,

$$t_1 = \frac{m}{\|g\|_{C^0(S_{N-1})}}.$$

The transformation T_t can be chosen as

$$\begin{cases} T_t(X) = 0, & \text{if } X = 0 \\ T_t(X) = [\rho + t\rho g(\varsigma)/f(\varsigma)]\varsigma, & \text{if } X = \rho\varsigma \ne 0. \end{cases}$$

However as in the previous example T_t is not unique and for any continuous increasing function $\lambda : \mathbf{R}_+ \to \mathbf{R}_+$ such that $\lambda(0) = 0$ and $\lambda(1) = 1$ the transformation

$$\begin{cases} T_t(X) = 0 & \text{if } X = 0 \\ T_t(X) = [\rho + t\lambda(\rho/f(\varsigma))g(\varsigma)]\varsigma, & \text{if } X = \rho\varsigma \ne 0 \end{cases}$$

yields the same domain Ω_t.

1.4.5   Level curves

In some Free Boundary problems (e.g. Plasma Physics in J.P. Zolésio [3]) the free boundary Γ is a level curve of a smooth function u defined over a domain D. Assume that $\overline{D}$ is compact with smooth boundary ∂D. Let $u \in C^2(\overline{D})$ be a positive function on $\overline{D}$ such that

$$\begin{cases} u \geq 0 \text{ in } \overline{D}, \ u = 0 \text{ on } \partial D \\ \exists \text{ a unique } x_u \in D \text{ such that } |\nabla u(x)| > 0, \ \forall x \in D - \{x_u\}. \end{cases}$$

If $m = \max\{u(x) : x \in \overline{D}\}$, then for each t in $[0, m]$ the level set

$$\Gamma_t = u^{-1}(t)$$

is a C^1 manifold in D which is the boundary of the set

$$\Omega_t = \{x \in D : u(x) > t\}.$$

By definition $\Omega_0 = D$, $\forall \, t_1 > t_2$, $\Omega_{t_1} \subset \Omega_{t_2}$ and the domains Ω_t "converge" to the point x_u. Notice that the outward unit normal field on Γ_t is given by

$$n_t(x) = -|\nabla u(x)|^{-1} \nabla u(x), \quad x \in \Gamma_t.$$

This suggests to introduce the velocity field

$$V(x) = |\nabla u(x)|^{-2} \nabla u(x), \quad \forall x \in D - \{x_n\}$$

which is continuous everywhere but at $x = x_u$. If V was continuous everywhere, then for each X, the trajectory $x(t; X)$ of the differential equation

$$\begin{cases} \dfrac{dx}{dt}(t) = V(x(t)) \\ x(0) = X \end{cases}$$

would have the property that

$$u(x(t)) = u(X) + t$$

since formally

$$\frac{d}{dt} u(x(t)) = \nabla u(x(t)) \cdot \frac{dx}{dt}(t) = 1.$$

This means that the map $X \mapsto T_t(X) = x(t; X)$ constructed from (1.36) would map the level curves

$$\Gamma_0 = \{X \in \overline{D} : u(X) = 0\}$$

onto the level curve

$$\Gamma_t = \{x \in \overline{D} : u(x) = t\}$$

and eventually Ω_0 onto Ω_t. Unfortunately it is easy to see that this last property fails on the function $u(x) = 1 - x^2$ defined on the unit disk.

To get around this difficulty, we introduce for some arbitrarily small $\epsilon > 0$ an infinitely differentiable function $\rho_\epsilon : \mathbf{R}^N \to [0,1]$ such that

$$\rho_\epsilon(x) = \begin{cases} 0, & \text{if } |\nabla u(x)| < \epsilon \\ 1, & \text{if } |\nabla u(x)| > 2\epsilon \end{cases}$$

and the velocity

$$V_\epsilon(x) = \rho_\epsilon(x)V(x), \quad x \in \overline{D}.$$

As above define the transformation

$$X \mapsto T_t^\epsilon(X) = x(t; X),$$

where $x(t; X)$ is the solution of the differential equation

$$\begin{cases} \dfrac{dx}{dt}(t) = V_\epsilon(x(t)), & t \geq 0 \\ x(0) = X \in \overline{D}. \end{cases}$$

For $0 \leq t < m - 2\epsilon$, T_t maps Γ_0 onto Γ_t; for $0 \leq s < m - \epsilon$ such that $s + t < m - \epsilon$, T_t maps Γ_s onto Γ_{t+s}. However for $s > m - \epsilon$, T_t is the identity operator. As a result for $0 \leq t < m - 2\epsilon$

$$T_t(\Omega_0) = \Omega_t \text{ and } T_t(\Gamma_0) = \Gamma_t.$$

Of course $\epsilon > 0$ is arbitrary and we can make the construction for t's arbitrarily close to m. This is an example which can be handled by the Velocity (Speed) Method and not by a perturbation of the identity. Here the domains Ω_t are implicitly constrained to stay within the larger domain D. We shall see in §2 how to introduce and characterize such a constraint.

2 Choice of a method of perturbation of a domain

2.1 Unconstrained families of domains

In this section we review and extend the Velocity Method (cf. J.P. Zolésio [1,2]) and prove its equivalence with transformation methods. Under appropriate conditions we show how to construct a family on non-autonomous transformations of $\mathbf{R}^N$ from a family of non-autonomous velocity fields. Conversely we show how to construct the family of non-autonomous velocity fields from a family of non-autonomous transformations of $\mathbf{R}^N$. This construction is applied to various methods based on perturbations of the identity.

2.1.1 Equivalence between velocities and transformations

Let the real number $\tau > 0$ and the map $V : [0, \tau] \times \mathbf{R}^N \to \mathbf{R}^N$ be given. The map V can be viewed as a family $\{V(t) : 0 \le t \le \tau\}$ of non-autonomous velocity fields on $\mathbf{R}^N$ defined by

$$x \mapsto V(t)(x) \overset{\text{def}}{=} V(t, x) : \mathbf{R}^N \mapsto \mathbf{R}^N. \tag{2.1}$$

Assume that

$$(V) \begin{cases} \forall x \in \mathbf{R}^N, \quad V(\cdot, x) \in C^0([0, \tau]; \mathbf{R}^N) \\ \exists c > 0, \forall x, y \in \mathbf{R}^N, \quad \|V(\cdot, y) - V(\cdot, x)\|_{C^0([0,\tau];\mathbf{R}^N)} \le c|y - x|, \end{cases}$$

where $V(\cdot, x)$ is the function $t \mapsto V(t, x)$. Associate with V the solution $x(t; V)$ of the ordinary differential equation

$$\frac{dx}{dt}(t) = V(t, x(t)), \quad t \in [0, \tau], \quad x(0) = X \in \mathbf{R}^N, \tag{2.2}$$

and introduce the homeomorphism

$$X \mapsto T_t(V)(X) = x(t; X) : \mathbf{R}^N \to \mathbf{R}^N \tag{2.3}$$

and the maps

$$(t, X) \mapsto T_V(t, X) \overset{\text{def}}{=} T_t(V)(X) : [0, \tau] \times \mathbf{R}^N \to \mathbf{R}^N, \tag{2.4}$$

$$(t, x) \mapsto T_V^{-1}(t, x) \overset{\text{def}}{=} T_t^{-1}(V)(x) : [0, \tau] \times \mathbf{R}^N \to \mathbf{R}^N. \tag{2.5}$$

Notation 2.1 In the sequel we shall drop the V in $T_V(t, X)$, $T_V^{-1}(t, x)$ and $T_t(V)$ whenever no confusion is possible.

Theorem 2.1 (i) *Under hypothesis* (V) *the map T has the following properties*

$$(T1) \begin{cases} \forall X \in \mathbf{R}^N, \quad T(\cdot, X) \in C^1([0, \tau]; \mathbf{R}^N) \\ \exists c > 0, \forall X, Y \in \mathbf{R}^N, \quad \|T(\cdot, Y) - T(\cdot, X)\|_{C^1([0,\tau];\mathbf{R}^N)} \le c|Y - X|, \end{cases}$$

$(T2) \quad \forall t \in [0, \tau], \quad X \mapsto T_t(X) = T(t, X) : \mathbf{R}^N \to \mathbf{R}^N$ *is bijective,*

$$(T3) \begin{cases} \forall x \in \mathbf{R}^N, \quad T^{-1}(\cdot, x) \in C^0([0, \tau]; \mathbf{R}^N) \\ \exists c > 0, \forall x, y \in \mathbf{R}^N, \quad \|T^{-1}(\cdot, y) - T^{-1}(\cdot, x)\|_{C^0([0,\tau];\mathbf{R}^N)} \le c|y - x|. \end{cases}$$

(ii) *Given a real $\tau > 0$ and a map $T : [0, \tau] \times \mathbf{R}^N \to \mathbf{R}^N$ satisfying hypotheses* (T1), (T2) *and* (T3), *then the map*

$$(t, x) \mapsto V(t, x) = \frac{\partial T}{\partial t}(t, T_t^{-1}(x)) : [0, \tau] \times \mathbf{R}^N \to \mathbf{R}^N, \tag{2.6}$$

satisfies hypothesis (V), *where* T_t^{-1} *is the inverse of* $X \mapsto T_t(X) = T(t, X)$.

A more general version of this theorem for constrained domains (Theorem 2.3) will be given and proved in §2.2.1. The proof of Theorem 2.1 can be found in Delfour-Zolésio [10, Appendix, p. 1435].

This first theorem is an equivalence result which says that we can either start from a family of velocity fields $\{V(t)\}$ on $\mathbf{R}^N$ or a family of transformations $\{T_t\}$ of $\mathbf{R}^N$ provided that the map V, $V(t, x) = V(t)(x)$, satisfies assumption (V) or the map T, $T(t, X) = T_t(X)$, satisfies assumptions (T1) to (T3).

When we start from V, we obtain the *Velocity Method* and the perturbations of an initial domain Ω by the family of homeomorphisms $\{T_t(V)\}$ generates a new family of domains

$$\Omega_t = T_t(V)(\Omega) = \{T_t(V)(X) : X \in \Omega\} \tag{2.7}$$

which will be used in §5 to define shape derivatives. Note that interior (resp. boundary) points of Ω are mapped onto interior (resp. boundary) points of Ω_t.

2.1.2   The special case of perturbations of the identity operator

In examples where we start from T, it is usually possible to satisfy assumptions (T1) to (T3) and construct the corresponding velocity field V defined in (2.6). For instance perturbations of the identity to the first or second order fall in that category:

$$T_t(X) = X + tU(X) + \frac{t^2}{2}A(X) \quad (A = 0 \text{ for the first order}), \quad t \geq 0, \quad X \in \mathbf{R}^N, \tag{2.8}$$

where U and A are transformations of $\mathbf{R}^N$. It turns out that for Lipschitzian transformations U and A, assumptions (T1) to (T3) are satisfied.

Theorem 2.2 *Let U and A be two uniform Lipschitzian transformations of $\mathbf{R}^N$:*

$$\exists c > 0, \; \forall X, Y \in \mathbf{R}^N, \; |U(Y) - U(X)| \leq c|Y - X|, \; |A(Y) - A(X)| \leq c|Y - X|.$$

For $\tau = \min\{1, 1/4c\}$ and T given by (2.8), the map T satisfies hypotheses (T1) to (T3) on $[0, \tau]$. Moreover the associated velocity V is given by

$$(t, x) \mapsto V(t, x) = U(T_t^{-1}(x)) + tA(T_t^{-1}(x)) : [0, \tau] \times \mathbf{R}^N \to \mathbf{R}^N, \tag{2.9}$$

and it satisfies hypothesis (V) *on $[0, \tau]$.*

Remark 2.1   Observe that from (2.8) and (2.9)

$$V(0) = U, \; \dot{V}(0)(x) = \frac{\partial V}{\partial t}(t, x)|_{t=0} = A - [DU]U, \tag{2.10}$$

where DU is the Jacobian matrix of U. The term $\dot{V}(0)$ is an *acceleration* at $t = 0$ which will always be present even when $A = 0$. $\square$

2.2 Constrained families of domains

2.2.1 Equivalence between velocities and transformations

In many applications the family of admissible domains Ω is constrained to subsets of a fixed larger domain or *hold-all* D. To reflect that constraint we consider transformations

$$T : [0,\tau] \times \overline{D} \to \mathbf{R}^N \tag{2.11}$$

with the following properties

(T1_D) $\begin{cases} \forall X \in \overline{D}, \quad T(\cdot, X) \in C^1([0,\tau]; \mathbf{R}^N) \\ \exists c > 0, \forall X, Y \in \overline{D}, \quad \|T(\cdot, Y) - T(\cdot, X)\|_{C^1([0,\tau]; \mathbf{R}^N)} \le c|Y - X|, \end{cases}$

(T2_D) $\forall t \in [0,\tau], \quad X \mapsto T_t(X) = T(t, X) : \overline{D} \to \overline{D}$ is bijective,

(T3_D) $\begin{cases} \forall x \in \overline{D}, \quad T^{-1}(\cdot, x) \in C^0([0,\tau]; \mathbf{R}^N) \\ \exists c > 0, \forall x, y \in \overline{D}, \quad \|T^{-1}(\cdot, y) - T^{-1}(\cdot, x)\|_{C^0([0,\tau]; \mathbf{R}^N)} \le c|y - x|. \end{cases}$

where under hypothesis (T2_D) T^{-1} is defined from the inverse of T_t as

$$(t, x) \mapsto T^{-1}(t, x) = T_t^{-1}(x) : [0,\tau] \times \overline{D} \to \mathbf{R}^N. \tag{2.12}$$

The above three properties are the analogue for $\overline{D}$ of the same three properties obtained for $\mathbf{R}^N$. In fact Theorem 2.1 extends from $\mathbf{R}^N$ to $\overline{D}$ by adding one hypothesis to (V). Specifically we shall consider for $\tau > 0$ velocities

$$V : [0,\tau] \times \overline{D} \to \mathbf{R}^N \tag{2.13}$$

such that

(V1_D) $\begin{cases} \forall x \in \overline{D}, \quad V(\cdot, x) \in C^0([0,\tau]; \mathbf{R}^N) \\ \exists c > 0, \forall x, y \in \overline{D}, \quad \|V(\cdot, y) - V(\cdot, x)\|_{C^0([0,\tau]; \mathbf{R}^N)} \le c|y - x|. \end{cases}$

(V2_D) $\forall x \in \overline{D}, \forall t \in [0,\tau], \quad V(t, x) \in T_{\overline{D}}(x)$ *and* $-V(t, x) \in T_{\overline{D}}(x),$

where $T_{\overline{D}}(x)$ is the Bouligand contingent cone to $\overline{D}$ at the point x in $\overline{D}$ (cf. Aubin-Cellina [1, p. 176]).

The next theorem is a generalization of Theorem 2.1 from $\mathbf{R}^N$ to an arbitrary domain D which shows the equivalence between velocity and transformation viewpoints. Its proof is given in Delfour-Zolésio [13, pp. 7–12].

Theorem 2.3

(i) *Let $\tau > 0$ and V be a family of velocity fields satisfying hypotheses $(V1_D)$ and $(V2_D)$ and consider the family of transformations*

$$(t, X) \mapsto T(t, X) = x(t; X) : [0, \tau] \times \overline{D} \to \mathbf{R}^N \qquad (2.14)$$

where $x(\cdot, X)$ is the solution of

$$\frac{dx}{dt}(t) = V(t, x(t)), \quad 0 \le t \le \tau, \quad x(0) = X. \qquad (2.15)$$

Then the family of transformations T satisfies conditions $(T1_D)$ to $(T3_D)$.

(ii) *Conversely, given a family of transformations T satisfying assumptions $(T1_D)$ to $(T3_D)$, then the family of velocity fields*

$$(t, x) \mapsto V(t, x) = \frac{\partial T}{\partial t}(t, T_t^{-1}(x)) : [0, \tau] \times \overline{D} \to \mathbf{R}^N \qquad (2.16)$$

satisfies conditions $(V1_D)$ and $(V2_D)$, and the transformations constructed from this V coincide with T.

Remark 2.2 Under $(V1_D)$ and $(V2_D)$, $\{T_t : 0 \le t \le \tau\}$ is a family of homeomorphisms of $\overline{D}$ which map the interior $\overset{\circ}{D}$ (resp. the boundary ∂D) of D onto $\overset{\circ}{D}$ (resp. ∂D) (cf. J. Dugundji [1, p. 87–88]). □

Remark 2.3 Assumption $(V2_D)$ is a double viability condition. M. Nagumo [1]'s usual viability condition

$$V(t, x) \in T_{\overline{D}}(x), \quad \forall t \in [0, \tau], \forall x \in \overline{D} \qquad (2.17)$$

is a necessary and sufficient condition for a *viable solution* to (2.15), that is

$$\forall t \in [0, \tau], \forall X \in \overline{D}, \quad x(t; X) \in \overline{D} \text{ (or } T_t(\overline{D}) \subset \overline{D}) \qquad (2.18)$$

(cf. Aubin-Cellina [1, p. 174 and p. 180]). Condition $(V2_D)$

$$\forall t \in [0, \tau], \forall x \in \overline{D}, \quad V(t; x) \in T_{\overline{D}}(x) \text{ and } -V(t, x) \in T_{\overline{D}}(x) \qquad (2.19)$$

is a *strict viability condition* which not only says that T_t maps $\overline{D}$ into $\overline{D}$ but also that

$$\forall t \in [0, \tau], \quad T_t : \overline{D} \to \overline{D} \text{ is a homeomorphism.} \qquad (2.20)$$

In particular it maps interior points onto interior points and boundary points onto boundary points. □

Remark 2.4 Condition $(V2_D)$ is a generalization to arbitrary domains D of the following condition used by J.P. Zolésio [1] in 1979: for all x in ∂D

$$\begin{cases} V(t, x) \cdot n(x) = 0, & \text{if the outward normal } n \text{ exists} \\ 0, & \text{otherwise.} \end{cases}$$

□

2.2.2 Transformation of condition (V2$_D$) into a linear constraint on V

Condition (V2$_D$) is equivalent to

$$\forall t \in [0,\tau], \forall x \in \overline{D}, \quad V(t,x) \in \{-T_D(x)\} \cap T_D(x) \tag{2.21}$$

since $T_{\overline{D}}(x) = T_D(x)$. If $T_D(x)$ was convex, then the above intersection would be a closed linear subspace of $\mathbf{R}^N$. This is true when D is convex. In that case $T_D(x) = C_D(x)$, where $C_D(x)$ is the Clarke tangent cone and

$$L_D(x) = \{-C_D(x)\} \cap C_D(x) \tag{2.22}$$

is a closed linear subspace of $\mathbf{R}^N$. This means that (V2$_D$) reduces to

$$\forall t \in [0,\tau], \forall x \in \overline{D}, \quad V(t,x) \in L_D(x). \tag{2.23}$$

It turns out that for continuous vector fields $V(t,\cdot)$ the equivalence of (V2$_D$) and (2.23) extends to arbitrary domains D.

Theorem 2.4 *Given a velocity field V satisfying (V1$_D$), then condition (V2$_D$) is equivalent to*

$$(\text{V2}_C) \qquad \forall t \in [0,\tau], \forall x \in \overline{D}, \quad V(t,x) \in L_D(x) = \{-C_D(x)\} \cap C_D(x),$$

where $C_D(x)$ is the (closed convex) Clarke tangent cone to $\overline{D}$ at x which is defined by

$$C_D(x) = \left\{ v \in \mathbf{R}^N : \lim_{\substack{h \to 0 \\ y \to_{\overline{D}} x}} \frac{d_D(y + hv)}{h} = 0 \right\}$$

$d_D(y)$ is the minimum distance from y to D, and $\to_{\overline{D}}$ denotes the convergence in $\overline{D}$. Moreover $L_D(x)$ is a closed linear subspace of $\mathbf{R}^N$.

Proof Cf. Delfour-Zolésio [13, pp. 13–14]. $\square$

Notation 2.2 In the sequel it will be convenient to introduce the following spaces and subspaces

$$\mathcal{L} = \{V : [0,\tau] \times \mathbf{R}^N \to \mathbf{R}^N : V \text{ satisfies (V) on } \mathbf{R}^N\} \tag{2.24}$$

and for an arbitrary domain D in $\mathbf{R}^N$

$$\mathcal{L}_D = \{V : [0,\tau] \times \overline{D} \to \mathbf{R}^N : V \text{ satisfies (V1}_D\text{) and (V2}_C\text{) on } \overline{D}\}. \tag{2.25}$$

For any integers $k \geq 0$ and $m \geq 0$ and any compact subset K of $\mathbf{R}^N$ define the following subspaces of $\mathcal{L}$

$$\mathcal{V}_K^{m,k} = C^m([0,\tau], \mathcal{D}^k(K, \mathbf{R}^N)) \cap \mathcal{L}, \tag{2.26}$$

where $\mathcal{D}^k(K, \mathbf{R}^N)$ is the space of k-times continuously differentiable transformations of $\mathbf{R}^N$ with compact support in K. In all cases $\mathcal{V}_K^{m,k} \subset \mathcal{L}_K$. As usual $\mathcal{D}^\infty(K, \mathbf{R}^N)$ will be written $\mathcal{D}(K, \mathbf{R}^N)$. $\square$

3 Shape derivatives by the Velocity (Speed) Method

The object of this section is to study the *Shape gradient* and the *Shape Hessian* by the *Velocity (Speed) Method* (cf. J. Céa [1, 2] and J. P. Zolésio [1, 2]). We extend the results for C^k and Lipschitzian domains to nonsmooth constrained and unconstrained domains and discuss their relationship to various methods based on perturbations of the identity operator.

3.1 Domain functionals

Consider the set $P(D)$ of subsets Ω of a fixed domain D of $\mathbf{R}^N$ (possibly all of $\mathbf{R}^N$) which will play the role of a *hold-all*. A *domain functional* is a function

$$\Omega \mapsto J(\Omega) : \mathcal{A} \subset P(D) \to \mathbf{R}. \tag{3.1}$$

defined on an admissible family $\mathcal{A}$ of subsets of D. The analysis of §2 clarified the connection between methods based on a perturbation of the identity operator and the Velocity (Speed) approach. It has established that we can always construct a family of non-autonomous velocities $\{V(t)\}$ and work within that framework. The key result is the existence of a family of homeomorphisms $\{T_t\}$ which map $\overline{D}$ onto $\overline{D}$, $\overset{\circ}{D}$ onto $\overset{\circ}{D}$ and ∂D onto ∂D under assumptions (V).

3.2 First order semiderivatives and shape gradient

Under the action of a velocity field V in $\mathcal{L}_D$, a domain Ω in the admissible family $\mathcal{A}$ of $P(D)$ is transformed into a new domain

$$\Omega_t(V) = T_t(V)(\Omega) = \{T_t(V)(X) : X \in \Omega\}. \tag{3.2}$$

This provides our first notion of derivative for a domain functional.

Definition 3.1 Given a velocity field V in $\mathcal{L}_D$, J is said to have an *Eulerian semiderivative at Ω in the direction V* if the following limit exists and is finite

$$\lim_{t \searrow 0} \frac{J(\Omega_t(V)) - J(\Omega)}{t}. \tag{3.3}$$

Whenever it exists, the limit will be denoted by $dJ(\Omega; V)$.

This definition is quite general and may include situations where $dJ(\Omega; V)$ is not only a function of $V(0)$ but also of $V(t)$ in a neighbourhood of $t = 0$. This will not occur under some appropriate continuity hypothesis on the map $V \mapsto dJ(\Omega; V)$. This immediately raises the question of the choice of topology and eventually the choice of gradient when we specialize to autonomous vector fields V. We choose to follow the classical framework of the Theory of Distributions (cf. L. Schwartz [1]). Assume that D is an open domain in $\mathbf{R}^N$. Domains Ω in $P(D)$ will be perturbed by velocity fields $V(t)$ with values in $D^k(K, \mathbf{R}^N)$ for some compact subset K of D and integer $k \geq 0$. More precisely we shall consider velocity fields in

$$\vec{\mathcal{V}}_D^{m,k} = \varinjlim_K \{V_K^{m,k} : \forall K \text{ compact in } D\} \tag{3.4}$$

where $\varinjlim$ denotes the inductive limit set with respect to K endowed with its natural inductive limit topology. Four autonomous fields, the above construction reduces to

$$\mathcal{V}_D^k = \begin{cases} D^0(D, \mathbf{R}^N) \cap \text{Lip}(\mathbf{R}^N, \mathbf{R}^N), & k = 0 \\ D^k(D, \mathbf{R}^N), & 1 \leq k \leq \infty \end{cases} \tag{3.5}$$

where $\text{Lip}(\mathbf{R}^N, \mathbf{R}^N)$ denotes the space of uniformly Lipschitzian transformations of $\mathbf{R}^N$. In all cases hypotheses (V1$_D$) and (V2$_D$) are satisfied since for all $t \in [0, \tau], V(t, x) = 0$ for all x in ∂D. When $D = \mathbf{R}^N$ we drop the index D in the above definitions and simply write $\vec{\mathcal{V}}^{m,k}$ and $\mathcal{V}^k$. In the sequel $D^\infty(D, \mathbf{R}^N)$ will be written $D(D, \mathbf{R}^N)$.

Theorem 3.1 *Let Ω be a domain in the fixed open hold-all D. Assume that there exist integers $m \geq 0$ and $k \geq 0$ such that*

$$\forall V \in \vec{\mathcal{V}}_D^{m,k}, \quad dJ(\Omega; V) \text{ exists,} \tag{3.6}$$

and that the map

$$V \mapsto dJ(\Omega; V) : \vec{\mathcal{V}}_D^{m,k} \to \mathbf{R} \tag{3.7}$$

is continuous. Then

$$\forall V \in \vec{\mathcal{V}}_D^{m,k}, \quad dJ(\Omega; V) = dJ(\Omega; V(0)), \tag{3.8}$$

where $dJ(\Omega; V(0))$ is the Eulerian semiderivative for the autonomous vector field equal to $V(0)$.

Proof It is sufficient to prove the theorem for any compact subset K of D. So given V in $\mathcal{V}_K^{m,k}$ construct the sequence

$$V_n(t) = V\left(\frac{t}{n}\right), \quad 0 \leq t \leq \tau, \quad \text{for integers } n \geq 1.$$

By continuity of V, $\{V_n\}$ converges in $\mathcal{V}_K^{m,k}$ to the autonomous field $\tilde{V}(t) = V(0)$. Hence by continuity of (3.7)

$$dJ(\Omega; V_n) \to dJ(\Omega; V(0))$$

and by uniqueness of the limit we obtain (3.8). □

By virtue of this theorem we can now specialize to autonomous vector fields V to further study the properties and the structure of $dJ(\Omega; V)$.

Definition 3.2 Let Ω be a domain in the open hold-all D of $\mathbf{R}^N$.

(i) The functional J is said to be *shape differentiable at* Ω, if the Eulerian semiderivative $dJ(\Omega;V)$ exists for all V in $\mathcal{D}(D,\mathbf{R}^N)$ and the map

$$V \mapsto dJ(\Omega;V) : \mathcal{D}(D,\mathbf{R}^N) \to \mathbf{R} \tag{3.9}$$

is linear and continuous.

(ii) The map (9) defines a vector distribution $G(\Omega)$ which will be referred to as the *shape gradient* of J at Ω.

(iii) When there exists some finite $k \geq 0$ such that $G(\Omega)$ is continuous for the $\mathcal{D}^k(D,\mathbf{R}^N)$-topology, we say that the shape gradient $G(\Omega)$ is of *order* k.

The next theorem gives additional properties of shape differentiable functionals.

Notation 3.1 Associate with a subset A of D and an integer $k \geq 0$ the set

$$L_A^k = \{V \in \mathcal{D}^k(D,\mathbf{R}^N) : \forall x \in A, V(x) \in L_A(x)\}. \tag{3.10}$$

Theorem 3.2 (Hadamard-Zolésio structure theorem) *Let Ω be a domain with boundary Γ in the open hold-all D of $\mathbf{R}^N$ and assume that J has a shape gradient $G(\Omega)$.*

(i) *The support of the shape gradient $G(\Omega)$ is contained in $\Gamma_D \overset{\text{def}}{=} \Gamma \cap D$.*

(ii) *If Ω is open or closed in $\mathbf{R}^N$ and the shape gradient is of order k for some $k \geq 0$, then there exists $[G(\Omega)]$ in $(\mathcal{D}_D^k/L_\Omega^k)'$ such that for all V in $\mathcal{D}_D^k \overset{\text{def}}{=} \mathcal{D}^k(D,\mathbf{R}^N)$*

$$dJ(\Omega;V) = \langle [G(\Omega)], q_L V \rangle_{\mathcal{D}_D^k/L_\Omega^k}, \tag{3.11}$$

where $q_L : \mathcal{D}_D^k \to \mathcal{D}_D^k/L_\Omega^k$ is the canonical quotient surjection. Moreover

$$G(\Omega) = {}^*(q_L)[G(\Omega)], \tag{3.12}$$

where ${}^*(q_L)$ denotes the transpose of the linear map q_L.

Proof Cf. Delfour-Zolésio [13, pp. 17-18]. □

Remark 3.1 When the boundary Γ of Ω is compact and J is shape differentiable at Ω, the distribution $G(\Omega)$ is of finite order. Once this is known, the conclusions of Theorem 3.2 (ii) apply with k equal to the order of $G(\Omega)$. □

The quotient space is very much related to a trace on the boundary Γ and when the boundary Γ is sufficiently smooth we can indeed make that identification.

Corollary *Assume that the hypotheses of Theorem 3.2 are satisfied for an open domain Ω, that the order of $G(\Omega)$ is $k \geq 0$, and that the boundary Γ of Ω is C^{k+1}. Then for all*

x in Γ, $L_\Omega(x)$ is an $(N-1)$- dimensional hyperplane to Ω at x and there exists a unique outward unit normal $n(x)$ which belongs to $C^k(\Gamma; \mathbf{R}^N)$. As a result the kernel of the map

$$V \mapsto \gamma_\Gamma(V) \cdot n : \mathcal{D}^k(D, \mathbf{R}^N) \to \mathcal{D}^k(\Gamma \cap D) \tag{3.13}$$

coincides with L_Ω^k where $\gamma_\Gamma : \mathcal{D}^k(D, \mathbf{R}^N) \to \mathcal{D}^k(\Gamma \cap D, \mathbf{R}^N)$ is the trace of V on $\Gamma \cap D$. Moreover the map $p_L(V)$

$$q_L(V) \mapsto p_L(q_L(V)) = \gamma_\Gamma(V) \cdot n : \mathcal{D}_D^k / L_\Omega^k \to \mathcal{D}^k(\Gamma \cap D) \tag{3.14}$$

is a well-defined isomorphism. In particular there exists a scalar distribution $g(\Gamma)$ in $\mathcal{D}^k(\Gamma \cap D)'$ such that for all V in $\mathcal{D}^k(D, \mathbf{R}^N)$

$$dJ(\Omega; V) = \langle g(\Gamma), \gamma_\Gamma(V) \cdot n \rangle_{\mathcal{D}^k(\Gamma \cap D)} \tag{3.15}$$

and

$$G(\Omega) = {}^*(q_L)[G(\Omega)], \quad [G(\Omega)] = {}^*(p_L)g(\Gamma). \tag{3.16}$$

Proof Cf. Delfour-Zolésio [13, pp. 18–20].   □

Remark 3.2 In 1907, J. Hadamard [1] used velocity fields along the normal to the boundary Γ of a C^∞ domain to compute the derivative of the first eigenvalue of the plate. Theorem 3.2 and its corollary are formalizations to arbitrary shape functionals of that property to open or closed domains with an arbitrary boundary. To our knowledge the first formalization of this structural property to open domains with a C^{k+1} boundary was done by J.P. Zolésio [1] in 1979.   □

Remark 3.3 The space $\mathcal{D}^k(\Gamma \cap D)$ is not simple to characterize. However when Γ is compact and $D = \mathbf{R}^N$, it coincides with $C^k(\Gamma)$.   □

Example 3.1 For any measurable subset Ω of a measurable hold-all D of $\mathbf{R}^N$, consider the volume functional

$$J(\Omega) = \int_\Omega dx. \tag{3.17}$$

For Ω with finite volume and V in $\mathcal{D}^1(D, \mathbf{R}^N)$,

$$dJ(\Omega; V) = \int_\Omega \operatorname{div} V \, dx \tag{3.18}$$

but for a bounded open domain Ω with a C^1 boundary Γ

$$dJ(\Omega; V) = \int_\Gamma V \cdot n \, d\Gamma \tag{3.19}$$

which is continuous on $\mathcal{D}^0(D, \mathbf{R}^N)$. Here the smoothness of the boundary decreases the order of the distribution $G(\Omega)$. This raises the question of the characterization of the family of all domains Ω of D for which the map

$$V \mapsto \int_\Omega \operatorname{div} V \, dx : \mathcal{D}^1(D, \mathbf{R}^N) \to \mathbf{R} \tag{3.20}$$

can be continuously extended to $\mathcal{D}^0(D, \mathbf{R}^N)$. This is the family of *finite perimeter sets* with respect to D. It contains domains Ω whose characteristic function belongs to $BV(D)$, the space of L^1 functions on D with a distributional gradient in the space of (vectorial) Radon measures. They are the sets with finite volume and perimeter which have played a fundamental role in the solution of the Plateau problem. $\qquad\square$

3.3   Second order semiderivative and shape Hessian

We first study the second order Eulerian semiderivative $d^2J(\Omega; V; W)$ of a functional $J(\Omega)$ for two non-autonomous vector field V and W. A first theorem shows that under some natural continuity hypotheses, $d^2J(\Omega; V; W)$ is the sum of two terms: the *canonical term* $d^2J(\Omega; V(0); W(0))$ plus the first order Eulerian semiderivative $dJ(\Omega; \dot{V}(0))$ at Ω in the direction $\dot{V}(0)$ of the time-partial derivative $\partial_t V(t, x)$ at $t = 0$. As in the study of first order Eulerian semiderivatives, this first theorem reduces the study of second order Eulerian semiderivatives to the autonomous case. So we shall specialize to fields V and W in $\mathcal{D}^k(D, \mathbf{R}^N)$ and give the equivalent of the structure theorem 3.2 for the canonical term (cf. Theorem 3.2).

3.3.1   Non-autonomous case

The basic framework introduced in §4.4 and §4.5 has reduced the computation of the Eulerian semiderivative of $J(\Omega)$ to the computation of the derivative

$$j'(0) = dJ(\Omega; V(0)) \tag{3.21}$$

of the function

$$j(t) = J(\Omega_t(V)). \tag{3.22}$$

For $t \geq 0$, we naturally obtain

$$j'(t) = dJ(\Omega_t(V); V(t)). \tag{3.23}$$

This suggests the following definition.

Definition 3.3   Let V and W belong to $\mathcal{L}_D$ and assume that for all $t \in [0, \tau]$, $dJ(\Omega_t(W); V(t))$ exists for $\Omega_t(W) = T_t(W)(\Omega)$. The functional J is said to have a *second order Eulerian semiderivative* at Ω in the directions (V, W) if the following limit exists

$$\lim_{t \searrow 0} \frac{dJ(\Omega_t(W); V(t)) - dJ(\Omega; V(0))}{t}. \tag{3.24}$$

Whenever it exists, it is denoted by $d^2J(\Omega; V; W)$.

Remark 3.4 This last definition is compatible with the second order expansion of $j(t)$ with respect to t around $t = 0$:

$$j(t) \cong j(0) + tj'(0) + \frac{t^2}{2} j''(0), \tag{3.25}$$

where

$$j''(0) = d^2 J(\Omega; V; V). \tag{3.26}$$

□

Remark 3.5 It is easy to construct simple examples (see Example 3.3) with autonomous fields V and W showing that $d^2 J(\Omega; V; W) \neq d^2 J(\Omega; W; V)$ (cf. Delfour-Zolésio [8]). □

The next theorem is the analogue of Theorem 3.1 and provides the canonical structure of the second order Eulerian semiderivative.

Theorem 3.3 *Let Ω be a domain in the fixed open hold-all D of $\mathbf{R}^N$ and let $m \geq 0$ and $\ell \geq 0$ be integers. Assume that*

(i) $\forall V \in \vec{\mathcal{V}}_D^{m+1,\ell}, \forall W \in \vec{\mathcal{V}}_D^{m,\ell}, \quad d^2 J(\Omega; V; W)$ *exists;*

(ii) $\forall W \in \vec{\mathcal{V}}_D^{m,\ell}, \forall t \in [0, \tau]$, J *has a shape gradient at $\Omega_t(W)$ of order ℓ;*

(iii) $\forall U \in \mathcal{V}_D^\ell$, *the map*

$$W \mapsto d^2 J(\Omega; U; W) : \vec{\mathcal{V}}_D^{m,\ell} \to \mathbf{R} \tag{3.27}$$

is continuous.

Then for all V in $\vec{\mathcal{V}}_D^{m+1,\ell}$ and all W in $\vec{\mathcal{V}}_D^{m,\ell}$

$$d^2 J(\Omega; V; W) = d^2 J(\Omega; V(0); W(0)) + dJ(\Omega; \dot{V}(0)), \tag{3.28}$$

where

$$\dot{V}(0)(x) = \lim_{t \searrow 0} [V(t, x) - V(0, x)]/t. \tag{3.29}$$

Proof Cf. Delfour-Zolésio [13, pp. 23–24]. □

This important theorem gives the canonical structure of the second order Eulerian semiderivative: a first term which depends on $V(0)$ and $W(0)$ and a second term which is equal to $dJ(\Omega; \dot{V}(0))$. When V is autonomous the second term disappears and the semiderivative coincides with $d^2 J(\Omega; V; W(0))$ which can be separately studied for autonomous vector fields in $\mathcal{V}_D^\ell$.

3.3.2 Autonomous case

Definition 3.4 Let Ω be a domain in the open hold-all D of $\mathbf{R}^N$.

(i) The functional $J(\Omega)$ is said to be *twice shape differentiable* at Ω if

$$\forall V, \forall W \text{ in } \mathcal{D}(D, \mathbf{R}^N), \quad d^2 J(\Omega; V; W) \text{ exists} \tag{3.30}$$

and the map

$$(V, W) \mapsto d^2 J(\Omega; V; W): \ \mathcal{D}(D, \mathbf{R}^N) \times \mathcal{D}(D, \mathbf{R}^N) \to \mathbf{R} \tag{3.31}$$

is bilinear and continuous. We denote by h the map (12).

(ii) Denote by $H(\Omega)$ the vector distribution in $(\mathcal{D}(D, \mathbf{R}^N) \otimes \mathcal{D}(D, \mathbf{R}^N))'$ associated with h:

$$d^2 J(\Omega; V; W) = \langle H(\Omega), V \otimes W \rangle = h(V, W), \tag{3.32}$$

where $V \otimes W$ is the tensor product of V and W defined as

$$(V \otimes W)_{ij}(x, y) = V_i(x) W_j(y), \ 1 \le i, \ j \le N, \tag{3.33}$$

and $V_i(x)$ (resp. $W_j(y)$) is the i-th (resp. j-th) component of the vector V (resp. W) (cf. L. Schwartz's kernel theorem [2] and Gelfand-Vilenkin [1]). $H(\Omega)$ will be called the *Shape Hessian* of J at Ω.

(iii) When there exists a finite integer $\ell \ge 0$ such that $H(\Omega)$ is continuous for the $\mathcal{D}^\ell(D, \mathbf{R}^N) \otimes \mathcal{D}^\ell(D, \mathbf{R}^N)$-topology, we say that $H(\Omega)$ is of *order ℓ*.

Theorem 3.4 *Let Ω be a domain with boundary Γ in the open hold-all D of $\mathbf{R}^N$ and assume that J is twice shape differentiable.*

(i) *The vector distribution $H(\Omega)$ has support in*

$$(\Gamma \cap D) \times (\Gamma \cap D).$$

(ii) *If Ω is an open or closed domain in D and $H(\Omega)$ is of order $\ell \ge 0$, then there exists a continuous bilinear form*

$$[h] : (\mathcal{D}_D^\ell / \mathcal{D}_\Gamma^\ell) \times (\mathcal{D}_D^\ell / L_\Omega^\ell) \to \mathbf{R} \tag{3.34}$$

such that for all $[V]$ in $\mathcal{D}_D^\ell / \mathcal{D}_\Gamma^\ell$ and $[W]$ in $\mathcal{D}_D^\ell / L_\Omega^\ell$

$$d^2 J(\Omega; V; W) = [h](q_D(V), q_L(W)), \tag{3.35}$$

where $q_D : \mathcal{D}_D^\ell \to \mathcal{D}_D^\ell / \mathcal{D}_\Gamma^\ell$ and $q_L : \mathcal{D}_D^\ell \to \mathcal{D}_D^\ell / L_\Omega^\ell$ are the canonical quotient surjections and

$$\mathcal{D}_\Gamma^\ell = \{ V \in \mathcal{D}^\ell(D, \mathbf{R}^N) : \partial^\alpha V = 0 \text{ on } \Gamma \cap D, \forall \alpha, |\alpha| \le \ell \}. \tag{3.36}$$

Proof Cf. Delfour-Zolésio [13, pp. 25–28]. $\square$

The next and last result is the extension of the structure theorem 3.2 to second order Eulerian semiderivatives. We need the result established in the Corollary to Theorem 3.2. For a domain Ω with a boundary Γ which is $C^{\ell+1}, \ell \geq 0$, the map

$$q_L(W) \mapsto p_L(q_L(W)) = \gamma_L(W) \cdot n : D_\Omega^\ell/L_\Omega^\ell \to D^\ell(\Gamma \cap D) \qquad (3.37)$$

is a well-defined isomorphism. This will be used for the V-component. For the W-component we need the following lemma.

Lemma 3.1 *Assume that the boundary Γ of Ω is $C^{\ell+1}, \ell \geq 0$. Then the map*

$$q_D(V) \mapsto p_D(q_D(V)) = \gamma_\Gamma(V) : D_D^\ell/D_\Gamma^\ell \to D^\ell(\Gamma \cap D, \mathbf{R}^N) \qquad (3.38)$$

is a well-defined isomorphism, where

$$p_D : D_D^\ell \to D_D^\ell/D_\Gamma^\ell \qquad (3.39)$$

is the canonical surjection.

Proof By standard arguments. $\square$

Remark 3.6 When $D = \mathbf{R}^N$ and Γ is compact, $D^\ell(\Gamma \cap D, \mathbf{R}^N) = D^\ell(\Gamma, \mathbf{R}^N)$ coincides with the space of ℓ-times continuously differentiable maps from Γ to $\mathbf{R}^N$. $\square$

Theorem 3.5 *Assume that the hypotheses of Theorem 3.4 (ii) hold and that the boundary Γ of the open domain Ω is $C^{\ell+1}$ for $\ell \geq 0$.*

(i) *The map*

$$\begin{cases} (v, w) \mapsto h_{D \times L}(v, w) = [h](p_D^{-1}v, p_L^{-1}w) \\ : D^\ell(\Gamma_D, \mathbf{R}^N) \times D^\ell(\Gamma_D) \to \mathbf{R} \end{cases} \qquad (3.40)$$

is bilinear and continuous and for all V and W in $D^\ell(D, \mathbf{R}^N)$

$$d^2 J(\Omega; V; W) = h_{D \times L}(\gamma_\Gamma V, ((\gamma_\Gamma W) \cdot n)), \qquad (3.41)$$

where $\Gamma_D = \Gamma \cap D$.

(ii) *This induces a vector distribution $h(\Gamma_D \otimes \Gamma_D)$ on $D^\ell(\Gamma_D, \mathbf{R}^N) \otimes D^\ell(\Gamma_D)$ of order ℓ,*

$$h(\Gamma_D \otimes \Gamma_D) : D^\ell(\Gamma_D, \mathbf{R}^N) \otimes D^\ell(\Gamma_D) \to \mathbf{R}, \qquad (3.42)$$

such that for all V and W in $D^\ell(D, \mathbf{R}^N)$

$$\langle h(\Gamma_D \otimes \Gamma_D), (\gamma_\Gamma V) \otimes ((\gamma_\Gamma W) \cdot n) \rangle = d^2 J(\Omega; V; W), \qquad (3.43)$$

where $(\gamma_\Gamma V) \otimes ((\gamma_\Gamma W) \cdot n)$ is defined as the tensor product

$$((\gamma_\Gamma V) \otimes ((\gamma_\Gamma W) \cdot n))_i(x, y) = (\gamma_\Gamma V_i)(x)((\gamma_\Gamma W) \cdot n)(y), \quad x, y \in \Gamma_D, \qquad (3.44)$$

$V_i(x)$ *is the i-th component of* $V(x)$ *and*

$$(\gamma_\Gamma(W) \cdot n)(y) = (\gamma_\Gamma(W)(y) \cdot n(y), \quad \forall y \in \Gamma_D. \tag{3.45}$$

Remark 3.7 Finally under the hypotheses of Theorems 3.4 and 3.6

$$
\begin{aligned}
d^2 J(\Omega; V; W) &= \langle h(\Gamma_D \otimes \Gamma_D), (\gamma_\Gamma V(0)) \otimes ((\gamma_\Gamma W(0)) \cdot n) \rangle \\
&\quad + \langle (g(\Gamma_D), (\gamma_\Gamma \dot{V}(0)) \cdot n \rangle
\end{aligned}
\tag{3.46}
$$

for all V in $\vec{V}_D^{m+1,\ell}$ and W in $\vec{V}_D^{m,\ell}$. □

Example 3.2 Consider Example 3.1. Recall that for V in $\mathcal{D}^1(D, \mathbf{R}^N)$

$$dJ(\Omega; V) = \int_\Omega \operatorname{div} V \, dx. \tag{3.47}$$

Now for V in $\mathcal{D}^2(D, \mathbf{R}^N)$ and W in $\mathcal{D}^1(D, \mathbf{R}^N)$

$$d^2 J(\Omega; V; W) = \int_\Omega \operatorname{div} [(\operatorname{div} V) W] dx, \tag{3.48}$$

and if Γ is C^1,

$$d^2 J(\Omega; V; W) = \int_\Gamma \operatorname{div} V \, W \cdot n \, d\Gamma \tag{3.49}$$

which is continuous for pairs $(V, W) \in \mathcal{D}^1(D, \mathbf{R}^N) \times \mathcal{D}^0(D, \mathbf{R}^N)$ or $\mathcal{D}^1(\Gamma, \mathbf{R}^N) \times \mathcal{D}^0(\Gamma, \mathbf{R}^N)$. □

Another interesting observation is that the shape Hessian is, in general, not symmetrical as can be seen from the following example in Delfour-Zolésio [8].

Example 3.3 We use the functional (3.47) and expression (3.49) in Example 3.2. Choose the following two vector fields

$$V(x, y) = (1, 0) \quad \text{and} \quad W(x, y) = (\frac{x^2}{2}, 0). \tag{3.50}$$

Then on the open disk Ω with boundary Γ

$$\operatorname{div} V = 0, \quad \text{and} \quad W|_\Gamma = x = \cos\theta,$$

where Γ is parametrized by the angle θ and

$$V \cdot n = n_x = \cos\theta \quad \text{on } \Gamma.$$

As a result $d^2 J(\Omega; V; W) = 0$ and

$$d^2 J(\Omega; W; V) = \int_\Gamma \operatorname{div} W (V \cdot n) d\Gamma = \int_0^{2\pi} \cos^2\theta \, d\theta > 0.$$

□

3.4   Comparison with methods of perturbation of the identity

At this juncture it is instructive to compare first and second order Eulerian semiderivatives obtained by the Velocity (Speed) Method with those obtained by first and second order perturbations of the identity: that is, when the transformations T_t are specified a priori by

$$T_t(X) = X + tU(X) + \frac{t^2}{2}A(X), \quad X \in \mathbf{R}^N, \tag{3.51}$$

where U and A are transformations of $\mathbf{R}^N$ satisfying the hypotheses of Theorem 2.2. The transformation T_t in (3.51) is a *second order* perturbation when $A \neq 0$ and a *first order* perturbation when $A = 0$. According to Theorem 2.2, first and second order Eulerian semiderivatives associated with (3.51) can be equivalently obtained by applying the Velocity (Speed) Method to the non-autonomous velocity fields V_{UA} given by (2.9) in §2.1.2 and

$$dJ(\Omega; V_{UA}) = dJ(\Omega; V_{UA}(0)) = dJ(\Omega; U) \tag{3.52}$$

where we have used Remark 2.1 which says that

$$V_{UA}(0) = U \text{ and } \dot{V}_{UA}(0) = A - [DU]U. \tag{3.53}$$

Similarly if V_{WB} is another velocity field corresponding to

$$T_t(X) = X + tW(X) + \frac{t^2}{2}B(X), \quad X \in \mathbf{R}^N, \tag{3.54}$$

where W and B verify the hypotheses of Theorem 2.2, then

$$d^2J(\Omega; V_{UA}; V_{WB}) = d^2J(\Omega; V_{UA}(0); V_{WB}(0)) + dJ(\Omega; \dot{V}_{UA}(0)) \tag{3.55}$$

and

$$d^2J(\Omega; V_{UA}; V_{WB}) = d^2J(\Omega; U; W) + dJ(\Omega; A - [DU]U). \tag{3.56}$$

Expressions (32) and (35) are to be compared with the following expressions obtained by the Velocity (Speed) Method for two autonomous vector fields U and W

$$dJ(\Omega; U) \text{ and } d^2J(\Omega; U; W). \tag{3.57}$$

For the Shape gradient the two expressions coincide; for the Shape Hessian we recognize the bilinear term in (36) and (37) but the two expressions differ by the term

$$dJ(\Omega; A - [DU]U). \tag{3.58}$$

Even for a first order perturbation $(A = 0)$, we have a quadratic term in U. This situation is analogous to the classical problem of defining second order derivatives on a manifold. The term (3.58) would correspond to the connexion while the bilinear term $d^2J(\Omega; V; W)$ would be the candidate for the *canonical* second order shape derivative. In this context we shall refer to the corresponding distribution $H(\Omega)$ as the *canonical Shape Hessian*. All other second order shape derivatives will be obtained from $H(\Omega)$ by adding the gradient term $G(\Omega)$ acting as the appropriate acceleration field (connexion).

Remark 3.8 The method of perturbation of the identity can be made *more canonical* by using the following family of transformations

$$T_t(X) = X + tU(X) + \frac{t^2}{2}(A + [DU]U) \tag{3.59}$$

which yields

$$dJ(\Omega; U) \text{ for the gradient} \tag{3.60}$$

and

$$d^2 J(\Omega; U; W) + dJ(\Omega; A) \text{ for the Hessian,} \tag{3.61}$$

where for a first order perturbation $(A = 0)$ the second term disappears. □

Remark 3.9 When Ω^* is an appropriately smooth domain which minimizes a twice shape differentiable functional $J(\Omega)$ without constraints on Ω, the classical necessary conditions would be (at least formally)

$$dJ(\Omega^*; V) = 0, \quad \forall V, \tag{3.62}$$

$$d^2 J(\Omega^*; W; W) \geq 0, \quad \forall W, \tag{3.63}$$

or equivalently for "smooth" velocity fields V and W

$$dJ(\Omega^*; V(0)) = 0, \quad \forall V \tag{3.64}$$

$$d^2 J(\Omega^*; W(0); W(0)) + dJ(\Omega^*; \dot{V}(0)) \geq 0, \quad \forall W. \tag{3.65}$$

But in view of (3.64), condition (3.65) reduces to the following condition on the canonical Shape Hessian

$$d^2 J(\Omega^*; W(0); W(0)) \geq 0, \quad \forall W. \tag{3.66}$$

□

4 Elements of shape calculus

4.1 Derivative of a domain integral

In this section we quote a number of useful formulae from J.P. Zolésio [1, 2]. Other classical computations can be found in the book by Sokolowski-Zolésio [1]. Fix the open domain D with smooth boundary ∂D. The simplest examples of shape functionals are given by integrals over the domain Ω in D or its boundary $\Gamma = \partial \Omega$. We assume that ∂D and Γ are smooth manifolds.

Let φ be a fixed function in $W^{1,1}(D)$ and $\tau > 0$ a real parameter such that

$$\forall t \in [0, \tau], \quad J_t \geq 0.$$

Assume that $\{V(t) : 0 \leq t \leq \tau\}$ is a family of vector fields in $C^0([0, \tau]; C^1(\overline{D}, \mathbf{R}^N))$ such that

$$V(t) \cdot \nu = 0 \quad \text{on } \partial D,$$

where ν is the unit outward normal to D. We consider

$$J(\Omega) = \int_\Omega \varphi \, dx. \tag{4.1}$$

By the classical change of variable theorem we have

$$J(\Omega_t) = \int_{\Omega_t(V)} \varphi \, dx = \int_\Omega \varphi \circ T_t \, J_t \, dx, \tag{4.2}$$

where $T_t = T_t(V)$ is the transformation of $\overline{D}$ associated with the velocity vector field V,

$$J_t(X) = \det DT_t(X),$$

and $DT_t(X)$ is the Jacobian matrix of transformation T_t at X. In the sequel we shall use the notation

$$J_t = \det DT_t. \tag{4.3}$$

The following result is easy to check.

Proposition 4.1 *The map*

$$t \mapsto \varphi \circ T_t : [0, \tau] \to L^1(D)$$

is differentiable and

$$\frac{\partial}{\partial t} \varphi(T_t(X)) = \nabla \varphi(T_t(X)) \cdot V(t, T_t(X));$$

or in compact form

$$\frac{d}{dt} \varphi \circ T_t = (\nabla \varphi \circ T_t) \cdot (V(t) \circ T_t) \quad in \;\; L^1(\overline{D}). \tag{4.4}$$

Similarly the map

$$t \mapsto J_t : [0, \tau] \to C^0(\overline{D})$$

is differentiable and

$$\frac{\partial}{\partial t} J_t(X) = \operatorname{div} V(t, T_t(X)) J_t(X);$$

or in compact form

$$\frac{dJ_t}{dt} = [\operatorname{div} V(t)] \circ T_t J_t \quad in \;\; C^0(\overline{D}). \tag{4.5}$$

Indeed it is easy to check that

$$\frac{d}{dt} DT_t(X) = DV(t, T_t(X)) DT_t(X)$$

and

$$\frac{d}{dt} \det DT_t(X) = \operatorname{tr} DV(t, T_t(X)) \det DT_t(X).$$

Finally

$$\frac{d}{dt} \det DT_t(X) = \operatorname{div} V(t, T_t(X)) \det DT_t(X),$$

and (4.5) follows directly by definition of $J_t(X)$.

From (4.2), (4.4), (4.5) we get

$$dJ(\Omega; V) = \frac{\partial}{\partial t} J(\Omega_t(V))|_{t=0} = \int_\Omega \{\nabla\varphi \cdot V(0) + \varphi \operatorname{div} V(0)\} dx$$

that is

$$dJ(\Omega; V) = \int_\Omega \operatorname{div} (\varphi V(0)) dx$$

and as Ω is smooth, by Stokes' theorem

$$dJ(\Omega; V) = \int_\Gamma \varphi V(0) \cdot n \, d\Gamma. \tag{4.6}$$

4.2 Derivative of a boundary integral

Let now ψ be given in $H^2(D)$ and consider the shape functional

$$J(\Omega) = \int_\Gamma \psi \, d\Gamma. \tag{4.7}$$

One can easily verify from the material introduced in §1.4 the change of variable formula on the boundary Γ_t:

$$\int_{\Gamma_t} \psi \, d\Gamma_t = \int_\Gamma \psi \circ T_t \, \omega_t \, d\Gamma, \tag{4.8}$$

where the density ω_t is obtained as

$$\omega_t = |M(DT_t)n|_{\mathbf{R}^N}, \tag{4.9}$$

where n is the outward normal field on Γ and $M(DT_t)$ is the cofactor matrix of DT_t, that is,

$$M(DT_t) = J_t \, {}^*(DT_t)^{-1}, \tag{4.10}$$

${}^*A$ denotes the transpose of a matrix A. In particular

$${}^*(DT_t)^{-1} = {}^*((DT_t)^{-1}) = ({}^*(DT_t))^{-1}.$$

It can easily be verified from (4.9) that $t \mapsto \omega_t$ is differentiable in $C^0(\Gamma)$ and that

$$\omega' = \lim_{t \searrow 0} \frac{1}{t}(\omega_t - \omega_0) \quad (\text{limit in } C^0(\Gamma) \text{ norm})$$

is itself linear and continuous with respect ot $V(0)$ (in $C^k(\overline{D}, \mathbf{R}^N)$ topology). Then formula (3.15) in §3 (Corollary to Theorem 3.2) applies and $dJ(\Omega; V)$ just depends on the normal component of the normal field

$$v = V(0) \cdot n \quad \text{on } \Gamma. \tag{4.11}$$

The idea is now to suppose that the field $V(0)$ itself is proportional to an extension N_0 of the normal field n. From §4.1 we know that if we assume Γ is of class C^2 then the normal field n belongs to $C^1(\Gamma)$ and then using the expression (1.36) or (1.38) in §1.4 for N_0, when Γ is C^2, we obtain $N_0 \in C^1(\overline{D}, \mathbf{R}^N)$ and also we can assume N_0 to be unitary, i.e. $\|N_0(x)\| = 1$ for all x in a neighborhood of Γ in $\overline{D}$.

Proposition 4.2 *Let Γ be C^2 and let N_0 be an extension of the normal field n, $N_0 \in C^1(\overline{D}, \mathbf{R}^N)$, $N_0|_\Gamma = n$ and such that $\|N_0\| = 1$ in a neighborhood of Γ. Then*

$$^*DN_0\, N_0 = 0 \tag{4.12}$$

in this neighborhood.

In fact it can easily be verified from (4.9) that

$$\omega' = \operatorname{div} V(0) - [DV(0)]n \cdot n \quad \text{on } \Gamma \tag{4.13}$$

in an element of $C^1(\Gamma)$ if $V \in C^0([0,\tau]; C^2(\overline{D}, \mathbf{R}^n))$. When V is chosen proportional to the normal field, that is $V = vN_0$ with N_0 as above we get a simplification in the expression (4.13) of ω':

Proposition 4.3 *Let $V \in C^2(\overline{D}, \mathbf{R}^N)$ be such that $V = vN_0$ with $\frac{\partial v}{\partial n} = 0$ on Γ. Then*

$$\omega' = v(\operatorname{div}(N_0))|_\Gamma. \tag{4.14}$$

Proposition 4.4

$$(\operatorname{div}(N_0))|_\Gamma = H \quad \text{the mean curvature of the surface } \Gamma. \tag{4.15}$$

Using (4.14) and (4.15) we obtain the

Corollary *Let Γ be of class C^2, $\Omega \subset \overline{D}$, and $V \in C^1(\overline{D}, \mathbf{R}^N)$ in the form*

$$\begin{cases} V(x) = v(x)N_0(x) \text{ for } x \text{ in a neighborhood of } \Gamma, \\[2mm] \text{where } N_0 \text{ is a unitary extension of } n, N_0 \in C^1(\overline{D}) \\[2mm] (\text{unitary in a neighborhood of } \Gamma), v \in C^1(\overline{D}) \\[2mm] \text{with } \frac{\partial v}{\partial n} = 0 \text{ on } \Gamma. \end{cases} \tag{4.16}$$

Then

$$\omega'(0) = v\,H \ \text{ on } \ \Gamma, \tag{4.17}$$

where H is the main curvature of the surface Γ.

From (4.8) and (4.16) we obtain the

Proposition 4.5 *Let $\psi \in H^2(D)$, Γ of class C^2 and $V \in C^1(\overline{D}, \mathbf{R}^N)$ and consider the functional*

$$J(\Omega) = \int_\Gamma \psi \, d\Gamma. \tag{4.18}$$

Then J is shape differentiable and we have

$$dJ(\Omega; V) = \int_\Gamma \left(\frac{\partial \psi}{\partial n} + H\psi \right) v \, d\Gamma, \tag{4.19}$$

where

$$v = V \cdot n \ \text{ on } \ \Gamma. \tag{4.20}$$

4.3 Shape derivative of the volume

Consider

$$J(\Omega) = \int_\Omega dx \tag{4.21}$$

(we can simply write $|\Omega|$). This shape functional is used as a constraint on the domain in several examples of shape optimization problems. From (4.5) we get

$$dJ(\Omega; V) = \int_\Omega \operatorname{div} V(0) dx. \tag{4.22}$$

A sufficient condition on the field $V(0)$, for the volume to be preserved is div $V(0) = 0$, but if Γ is of class C^1, then we get

$$dJ(\Omega; V) = \int_\Gamma V(0) \cdot n \, d\Gamma. \tag{4.23}$$

4.4 Shape derivative of the length of a boundary Γ

We consider

$$J(\Omega) = \int_\Gamma d\Gamma. \tag{4.24}$$

Assuming Γ of class C^2 and $V(0)$ in $C^1(\overline{D}, \mathbf{R}^N)$ we get from (4.19), (4.20)

$$dJ(\Omega; V) = \int_\Gamma H \, v \, d\Gamma. \tag{4.25}$$

Then the condition for the length of Γ to be preserved is that $v = V(0) \cdot n$ be orthogonal (in $L^2(\Gamma)$) to the main curvature H.

5 Shape gradients via a Min formulation

5.1 An illustrative example and a shape variational principle

Let Ω be a bounded open domain in $\mathbf{R}^N$ with a smooth boundary Γ. Let $y = y(\Omega)$ be the solution of the Dirichlet problem

$$-\Delta y = f \quad \text{in } \Omega, \tag{5.1}$$

$$y = 0 \quad \text{on } \Gamma, \tag{5.2}$$

where f is a fixed function in $H^1(\mathbf{R}^N)$. The solution of (5.1) is the minimizing element in $H_0^1(\Omega)$ of the energy functional

$$E(\Omega, \varphi) = \int_\Omega \left[\frac{1}{2} |\nabla \varphi|^2 - f\varphi \right] dx. \tag{5.3}$$

Introduce the domain functional

$$J(\Omega) = \inf_{\varphi \in H_0^1(\Omega)} E(\Omega, \varphi). \tag{5.4}$$

We want to show that

$$dJ(\Omega; V) = -\frac{1}{2} \int_\Gamma \left| \frac{\partial y}{\partial n} \right|^2 V \cdot n \, d\Gamma. \tag{5.5}$$

The motivation behind this example is that it is the prototype of a *Free Boundary* problem which can be obtained from the following *Shape Variational* principle:

$$dJ(\Omega; V) = 0 \quad \text{for all } V. \tag{5.6}$$

It yields the extra boundary condition

$$\frac{\partial y}{\partial n} = 0 \quad \text{on } \Gamma. \tag{5.7}$$

Equations (5.1), (5.2) and (5.7) characterize the Free Boundary problem. It is the simplest example of a large family of problems where the domain functional is the minimum of a natural internal energy. They correspond to the so-called "compliance problems" in Elasticity Theory. They also occur in Fracture Theory and Image Segmentation. The first order variation of this shape functional yields the extra boundary condition which is typical of a Free Boundary problem.

5.2 Function space parametrization

To compute the first order derivative of $J(\Omega)$ we perturb the domain Ω by a velocity field V which generates the family of transformations $\{T_t : 0 \leq t \leq \tau\}$ of $\mathbf{R}^N$ and the family of domains $\{\Omega_t = T_t(\Omega) : 0 \leq t \leq \tau\}$. At t

$$J(\Omega_t) = \inf_{\varphi \in H_0^1(\Omega_t)} E(\Omega_t, \varphi) \tag{5.8}$$

and the minimizing element $y_t = y(\Omega_t)$ is the solution of the Dirichlet problem

$$- \Delta y_t = f \ \text{in} \ \Omega_t \tag{5.9}$$

$$y_t = 0 \ \text{on} \ \Gamma_t \ (\text{boundary of } \Omega_t). \tag{5.10}$$

We want to compute the derivative

$$dj(0) = \lim_{t \searrow 0} \frac{j(t) - j(0)}{t} \tag{5.11}$$

of the function

$$j(t) = J(\Omega_t). \tag{5.12}$$

We are looking for a theorem which would give the derivative of a Min with respect to a parameter $t > 0$ at $t = 0$. The difficulty here is that the function space $H^1(\Omega_t)$ depends on the parameter t. To get around this difficulty and obtain a Min with respect to a function space which is independent of t, we introduce the following parametrization

$$H_0^1(\Omega_t) = \{\varphi \circ T_t^{-1} : \varphi \in H_0^1(\Omega)\}. \tag{5.13}$$

Notice that since T_t is a homeomorphism, it transforms the open domain Ω onto the open domain Ω_t and sends the boundary Γ of Ω onto the boundary Γ_t of Ω_t. In particular when V is sufficiently smooth for all φ in $H_0^1(\Omega)$, $\varphi \circ T_t^{-1} \in H_0^1(\Omega_t)$ and conversely for all ψ in $H_0^1(\Omega_t)$, $\psi \circ T_t \in H_0^1(\Omega)$. This parametrization does not affect the value of the minimum $J(\Omega_t)$ but changes the functional E

$$J(\Omega_t) = \inf_{\varphi \in H_0^1(\Omega)} E(T_t(\Omega), \varphi \circ T_t^{-1}). \tag{5.14}$$

This type of parametrization seems to be unique to *Shape Analysis*. It amounts to introducing the new energy functional

$$\tilde{E}(t, \varphi) = E(T_t(\Omega), \varphi \circ T_t^{-1}), \quad \varphi \in H_0^1(\Omega). \tag{5.15}$$

Our objective is to compute the limit (5.11) where

$$j(t) = \inf_{\varphi \in H_0^1(\Omega)} \tilde{E}(t, \varphi). \tag{5.16}$$

This will be done in the next section. Before closing it is interesting to characterize the minimizing element y^t in $H_0^1(\Omega)$ of

$$\tilde{E}(t, \varphi) = \int_{\Omega_t} \left[\frac{1}{2} |\nabla(\varphi \circ T_t^{-1})|^2 - f(\varphi \circ T_t^{-1}) \right] dx \tag{5.17}$$

which is the solution of the variational equation

$$\begin{cases} y^t \in H_0^1(\Omega), \ \text{and} \ \forall \varphi \in H_0^1(\Omega) \\ \\ \int_{\Omega_t} \{\nabla(y^t \circ T_t^{-1}) \cdot \nabla(\varphi \circ T_t^{-1}) - f(\varphi \circ T_t^{-1})\} dx = 0. \end{cases} \tag{5.18}$$

This expression is to be compared with the characterization of the minimizing element y_t of $E(\Omega_t, \varphi)$ on $H_0^1(\Omega_t)$:

$$\begin{cases} y_t \in H_0^1(\Omega), \text{ and } \forall \varphi \in H_0^1(\Omega_t) \\[2mm] \displaystyle\int_{\Omega_t} \{\nabla y_t \cdot \nabla \varphi - f\varphi\} dx = 0. \end{cases} \tag{5.19}$$

It is easy to verify that

$$y_t = y^t \circ T_t^{-1} \quad \text{and} \quad y^t = y_t \circ T_t. \tag{5.20}$$

So y^t is the solution y_t of (5.9)-(5.10) transported back onto the fixed domain Ω by the change of variable induced by T_t.

In view of the above considerations we can rewrite expression (5.17) on the fixed domain Ω as

$$\tilde{E}(t, \varphi) = \int_\Omega \left\{ \frac{1}{2}[A(t)\nabla\varphi] \cdot \nabla\varphi - (f \circ T_t)\varphi J_t \right\} dx, \tag{5.21}$$

where for t in $[0, \tau]$ small

$$DT_t = \text{Jacobian matrix of } T_t \tag{5.22}$$

$$J_t = \det DT_t \ (\det DT_t = |\det DT_t| \text{ for } t \geq 0 \text{ small}) \tag{5.23}$$

$$A(t) = J_t[DT_t]^{-1} \, ^*[DT_t]^{-1}, \tag{5.24}$$

where $^*[\ ]$ denotes the transposed matrix. With this change of variable y^t is now characterized by the variational equation

$$\begin{cases} y^t \in H_0^1(\Omega), \ \forall \varphi \in H_0^1(\Omega) \\[2mm] \displaystyle\int_\Omega [A(t)\nabla y^t \cdot \nabla\varphi - J_t(f \circ T_t)\varphi] dx = 0. \end{cases} \tag{5.25}$$

5.3   Differentiability of a Min with respect to a parameter

Consider a functional

$$G : [0, \tau] \times X \to \mathbf{R} \tag{5.26}$$

for some $\tau > 0$ and some set X. For each t in $[0, \tau]$ define

$$g(t) = \inf\{G(t, x) : x \in X\} \tag{5.27}$$

and the set

$$X(t) = \{x \in X : G(t, x) = g(t)\}. \tag{5.28}$$

We wish to study the existence and characterization of the limit

$$dg(0) = \lim_{t \searrow 0} \frac{g(t) - g(0)}{t} \tag{5.29}$$

when $X(t)$ is not empty for $0 \leq t \leq \tau$.

When $X(t) = \{x^t\}, 0 \leq t \leq \tau$, and the derivative

$$\dot{x} = \lim_{\substack{t>0 \\ t \to 0}} \frac{x^t - x^0}{t} \tag{5.30}$$

of x is known, then it is easy to obtain $dg(0)$ under appropriate differentiability of the functional G with respect to t and x. When $\dot{x}$ is not readily available or when the sets $X(t)$ are not singletons, this direct approach fails or becomes very intricate. In this section we present a theorem which gives an explicit expression for $dg(0)$, the derivative of the Min of the functional G with respect to t at $t = 0$. Its originality is that the differentiability of x^t is replaced by a continuity hypothesis on the set-valued function and the existence of the partial derivative of the functional G with respect to the parameter t. In other words, this technique does not require a priori knowledge of the derivative $\dot{x}$ of the minimizing elements x^t with respect to t.

Theorem 5.1 *Let $\tau > 0$ be a real number, X an arbitrary set and $G : [0,\tau] \times X \to \mathbf{R}$ a well-defined functional. Assume that the following conditions are satisfied:*

(H1) *for all $t \in [0,\tau]$, $X(t) \neq \emptyset$;*

(H2) *for all x in $X(0)$, $\partial_t G(t, x)$ exists everywhere in $[0,\tau]$;*

(H3) *there exists a topology $\mathcal{T}_X$ on X such that for any sequence $\{t_n\} \subset \,]0,\tau]$, converging to $t_0 = 0$, $\exists x_0 \in X(0)$, $\exists$ a subsequence $\{t_{n_k}\}$ of $\{t_n\}$, and for each $k \geq 1$, $\exists x_{n_k} \in X(t_{n_k})$ such that*

    (i) *$x_{n_k} \to x_0$ in the $\mathcal{T}_X$-topology; and*

    (ii)

$$\liminf_{\substack{t \searrow 0 \\ k \to \infty}} \partial_t G(t, x_{n_k}) \geq \partial_t G(0, x^0);$$

(H4) *for all x in $X(0)$, the map $t \mapsto \partial_t G(t, x)$ is upper semicontinuous at $t = 0$.*

Then there exists $x^0 \in X(0)$ such that

$$dg(0) = \lim_{t \searrow 0} \frac{g(t) - g(0)}{t} = \inf_{x \in X(0)} \partial_t G(0, x) = \partial_t G(0, x^0). \tag{5.31}$$

Remark 5.1 In the literature condition H3(i) is known as sequential semicontinuity for set-valued functions. When $X(0)$ is a singleton $\{x_0\}$ we automatically have $dg(0) = \partial_t G(0, x_0)$. □

Remark 5.2 This theorem and in particular the last part of property (5.31) extends for instance an older result by B. Lemaire [1, Thm 2.1, p. 38] where sequential compactness of the set X was assumed. It also completes and extends Theorem 1 in Delfour-Zolésio [2] and J.P. Zolésio [3]. □

Proof of Theorem 5.1 (i) We first establish upper and lower bounds to the differential quotient

$$\frac{\Delta(t)}{t}, \quad \Delta(t) \overset{\text{def}}{=} g(t) - g(0).$$

Choose arbitrary x_0 in $X(0)$ and x_t in $X(t)$. Then by definition

$$\begin{aligned} G(t, x_t) &\leq G(t, x_0) \\ -G(0, x_t) &\leq -G(0, x_0). \end{aligned}$$

Add up the above two inequalities to obtain

$$G(t, x_t) - G(0, x_t) \leq \Delta(t) \leq G(t, x_0) - G(0, x_0).$$

By hypothesis (H2), there exist θ_t, $0 < \theta_t < 1$, and α_t, $0 < \alpha_t < 1$, such that

$$\begin{aligned} G(t, x_t) - G(0, x_t) &= t \partial_t G(\theta_t t, x_t) \\ G(t, x_0) - G(0, x_0) &= t \partial_t G(\alpha_t t, x_0) \end{aligned}$$

and by dividing by $t > 0$

$$\partial_t G(\theta_t t, x_t) \leq \frac{\Delta(t)}{t} \leq \partial_t G(\alpha_t t, x_0). \tag{5.32}$$

(ii) Define

$$\underline{dg}(0) = \liminf_{t \searrow 0} \frac{\Delta(t)}{t}, \quad \overline{dg}(0) = \limsup_{t \searrow 0} \frac{\Delta(t)}{t}.$$

There exists a sequence $\{t_n : 0 < t_n \leq \tau\}$, $t_n \to 0$, such that

$$\lim_{n \to \infty} \frac{\Delta(t_n)}{t_n} = \underline{dg}(0).$$

By hypothesis (H3), $\exists x^0 \in X(0)$, $\exists$ a subsequence $\{t_{n_k}\}$ of $\{t_n\}$, for each $k \geq 1$, $\exists x_{n_k} \in X(t_{n_k})$ such that $x_{n_k} \to x^0$ in $\mathcal{T}_X$ and

$$\liminf_{\substack{t \searrow 0 \\ k \to \infty}} \partial_t G(t, x_{n_k}) \geq \partial_t G(0, x^0).$$

So from the first part of the estimate (5.32) for $t = t_{n_k}$

$$\partial_t G(\theta_{t_{n_k}} t_{n_k}, x_{n_k}) \leq \frac{\Delta(t_{n_k})}{t_{n_k}}$$

and

$$\partial_t G(0, x^0) \leq \liminf_{k \to \infty} \partial_t G(\theta_{t_{n_k}} t_{n_k}, x_{n_k}) \leq \lim_{k \to \infty} \frac{\Delta(t_{n_k})}{t_{n_k}} = \underline{dg}(0).$$

Therefore

$$\exists x^0 \in X(0), \quad \partial_t G(0, x^0) \leq \underline{dg}(0)$$

and

$$\inf_{z \in X(0)} \partial_t G(0, x) \leq \partial_t G(0, x^0) \leq \underline{dg}(0). \tag{5.33}$$

From the second part of (5.32) and assumption (H4) we also obtain

$$\forall x \in X(0), \quad \partial_t G(0, x) \geq \overline{dg}(0)$$

$$\overline{dg}(0) \leq \inf_{z \in X(0)} \partial_t G(0, x) \tag{5.34}$$

and necessarily

$$\inf_{z \in X(0)} \partial_t G(0, x) = \underline{dg}(0) = \overline{dg}(0) = \inf_{z \in X(0)} \partial_t G(0, x).$$

In particular from (5.33) and (5.34)

$$\partial_t G(0, x^0) = dg(0) = \inf_{z \in X(0)} \partial_t G(0, x)$$

and x^0 is a minimizing point of $\partial_t G(0, \cdot)$.   □

5.4   Application of the theorem

Our example has a unique minimizing point y^t for $t \geq 0$ small. Here $X = H_0^1(\Omega)$, $X(t) = \{y^t\}$ and it is sufficient to establish the continuity of the map $t \mapsto y^t$ at $t = 0$ for an appropriate topology on $H_0^1(\Omega)$.

We now proceed to the verification of hypotheses H1 to H4. Assume that $V \in C^0([0, \tau]; \mathcal{D}^2(\mathbf{R}^N, \mathbf{R}^N))$ and that $f \in H^1(\mathbf{R}^N)$. Choose $\tau > 0$ small enough such that

$$J_t = |J_t|, \quad 0 \leq t \leq \tau, \tag{5.35}$$

and that there exist constants $0 < \alpha < \beta$ such that

$$\forall \xi \in \mathbf{R}^N, \quad \alpha |\xi|^2 \leq A(t)\xi \cdot \xi \leq \beta |\xi|^2, \quad \text{and} \quad \alpha \leq J_t \leq \beta. \tag{5.36}$$

Since the bilinear form associated with (5.25) is coercive, there exists a unique solution y^t to (5.25) and

$$\forall t \in [0, \tau], \quad X(t) = \{y^t\} \neq \emptyset. \tag{5.37}$$

So hypothesis H1 is verified. To check H2 we use expression (5.21) and compute

$$\partial_t \tilde{E}(t, \varphi) = \int_\Omega \left\{ \frac{1}{2}[A'(t)\nabla\varphi] \cdot \nabla\varphi - [\operatorname{div} V_t(f \circ T_t) + J_t \nabla f \cdot V_t]\varphi \right\} dx \tag{5.38}$$

where

$$V_t(X) = V(t, T_t(X)), \quad DV_t(X) = DV(t, T_t(X)), \tag{5.39}$$

and

$$A'(t) = (\text{div } V_t)I - {}^*[DV_t] - [DV_t]. \tag{5.40}$$

By hypotheses on V and f, $\partial_t \tilde{E}(t, \varphi)$ exists everywhere in $[0, \tau]$ for all φ in $H_0^1(\Omega)$ and hypothesis H2 is verified.

To check hypothesis H3(i) we first show that $\{y^t\}$ is bounded in $H_0^1(\Omega)$. From (5.36)

$$\begin{aligned}
\alpha \|\nabla y^t\|_{L^2(\Omega)}^2 &\leq \int_\Omega A(t) \nabla y^t \cdot \nabla y^t \, dx = \int_\Omega J_t(f \circ T_t) y^t \, dx \\
&\leq \|J_t \, f \circ T_t\|_{L^2(\Omega)} \|y^t\|_{L^2(\Omega)}.
\end{aligned} \tag{5.41}$$

By using the norm

$$\|\varphi\|_{H_0^1(\Omega)} = \|\nabla \varphi\|_{L^2(\Omega)}$$

and the continuous injection of $H_0^1(\Omega)$ into $L^2(\Omega)$

$$\exists c > 0, \quad \|\varphi\|_{L^2(\Omega)} \leq c \|\varphi\|_{H_0^1(\Omega)}.$$

So from (5.41)

$$\|y^t\|_{H_0^1(\Omega)} \leq \frac{c}{\alpha} \|J_t \, f \circ T_t\|_{L^2(\Omega)}. \tag{5.42}$$

But $J_t \to 1$ as $t \to 0$ and $f \circ T_t \to f$ in $L^2(\Omega)$ by the following lemma.

Lemma 5.1 *Assume that $V \in C^0([0, \tau]; \mathcal{D}^1(\mathbf{R}^N, \mathbf{R}^N))$ satisfies hypothesis* (V) *and that $f \in L^2(\mathbf{R}^N)$. Then*

$$\lim_{t \searrow 0} f \circ T_t = f \quad \text{and} \quad \lim_{t \searrow 0} f \circ T_t^{-1} = f \text{ in } L^2(\mathbf{R}^N). \tag{5.43}$$

So by (5.42) y^t is bounded:

$$\exists c > 0, \quad \sup_{t \in [0, \tau]} \|y^t\|_{H_0^1(\Omega)} \leq c. \tag{5.44}$$

The next step is to prove the continuity by substracting (5.25) at $t > 0$ from (5.25) at $t = 0$:

$$\int_\Omega \nabla y^t \cdot \nabla \varphi \, dx + \int_\Omega (A(t) - I) \nabla y^t \cdot \nabla \varphi \, dx = \int_\Omega f \varphi \, dx + \int_\Omega [J_t(f \circ T_t) - f] \varphi \, dx$$

$$\int_\Omega \nabla y \cdot \nabla \varphi \, dx = \int_\Omega f \varphi \, dx.$$

Substract and set $\varphi = y^t - y$

$$\int_\Omega |\nabla(y^t - y)|^2 dx \;=\; -\int_\Omega (A(t) - I)\nabla y^t \cdot \nabla(y^t - y) + [J_t(f \circ T_t) - f](y^t - y)dx$$

$$\leq |A(t) - I| \, \|\nabla y^t\|_{L^2(\Omega)} \, \|\nabla(y^t - y)\|_{L^2(\Omega)}$$

$$+\|J_t f \circ T_t - f\|_{L^2(\Omega)} \, \|y^t - y\|_{L^2(\Omega)}$$

and

$$\|y^t - y\|_{H_0^1(\Omega)} \leq c\{|A(t) - I| + \|J_t f \circ T_t - f\|_{L^2(\Omega)}\}.$$

But $A(t) - I \to 0$ and $J_t f \circ T_t \to f$ in $L^2(\Omega)$ and finally $y^t \to y$ in $H_0^1(\Omega)$. So hypothesis H3(i) is verified for $H_0^1(\Omega)$-strong.

For hypothesis H3(ii) we have

$$\partial_t G(t, \varphi) = \int_\Omega \left\{ \frac{1}{2}[A'(t)\nabla\varphi] \cdot \nabla\varphi - [\mathrm{div}\, V_t(f \circ T_t) + J_t \nabla f \circ V_t]\varphi \right\} dx$$

and for φ in $H_0^1(\Omega)$, f in $H^1(\mathbf{R}^N)$ and V in $C^0([0,\tau]; \mathcal{D}^2(\mathbf{R}^N, \mathbf{R}^N))$

$$\partial_t G(t, \varphi) - \partial_t G(0, y) = \int_\Omega \{ \frac{1}{2}(A'(t) - A'(0))\nabla\varphi \cdot \nabla\varphi$$

$$- [\mathrm{div}\, V_t(f \circ T_t) + J_t \nabla f \cdot V_t - \mathrm{div}\, V(0)f + \nabla f \cdot V(0)]\varphi\}dx$$

$$+ \int_\Omega \frac{1}{2}\{A'(0)\nabla\varphi \cdot \nabla\varphi - A'(0)\nabla y \cdot \nabla y\}dx.$$

As φ goes to y in $H_0^1(\Omega)$ and $t \to 0$, the first term converges to zero since φ is bounded and

$$A'(t) \to A'(0)$$
$$\mathrm{div}\, V_t(f \circ T_t) \to \mathrm{div}\, V(0)f \text{ in } L^2(\Omega)$$
$$J_t \nabla f \cdot V_t \to \nabla f \cdot V(0) \text{ in } L^2(\Omega).$$

The second term is continuous with respect to φ in $H_0^1(\Omega)$ and goes to zero as $\varphi \to y$ in $H_0^1(\Omega)$. So hypothesis H3(ii) is verified.

Finally H4 is satisfied since

$$t \mapsto \partial_t \tilde{E}(t, y) = \int_\Omega \left\{ \frac{1}{2}A'(t)\nabla y \cdot \nabla y - [\mathrm{div}\, V_t f + \nabla f \cdot V_t]y \right\} dx \tag{5.45}$$

is continuous in $[0, \tau]$. So all the hypotheses of Theorem 2.1 are verified and for V in $C^0([0,\tau]; \mathcal{D}^2(\mathbf{R}^N, \mathbf{R}^N))$ and f in $H^1(\mathbf{R}^N)$,

$$dJ(\Omega; V) = \int_\Omega \left\{ \frac{1}{2}A'(0)\nabla y \cdot \nabla y - [\mathrm{div}\, V(0)f + \nabla f \cdot V(0)]y \right\} dx. \tag{5.46}$$

For V autonomous we see that expression (5.46) is continuous with respect to $\mathcal{D}^1(\mathbf{R}^N, \mathbf{R}^N)$ and the shape gradient is of order 1. We know by Hadamard's structure theorem that for Ω open and a C^2 boundary Γ

$$\exists g(\Gamma) \in \mathcal{D}^1(\Gamma), \quad dJ(\Omega; V) = \langle g(\Gamma), V \rangle_{\mathcal{D}^1(\Gamma)}.$$

The characterization of $g(\Gamma)$ will be given in the next section.

We complete this section with the proof of Lemma 5.1.

Proof of Lemma 5.1 (i) By density of $\mathcal{D}^1(\mathbf{R}^N)$ in $L^2(\mathbf{R}^N)$, for all $\epsilon > 0$ there exists f_ϵ such that

$$\|f - f_\epsilon\|_{L^2} < \frac{\epsilon}{\max\{J_t^{-1} : 0 \le t \le \tau\}} \quad (\le \epsilon\alpha \le \epsilon).$$

Hence

$$\|f \circ T_t - f\| \le \|f_\epsilon \circ T_t - f_\epsilon\| + \|f \circ T_t - f_\epsilon \circ T_t\| + \|f - f_\epsilon\|. \tag{5.47}$$

The last term is less that ϵ and the middle term can be rewritten after a change of variable

$$\int_{\mathbf{R}^N} |f \circ T_t - f_\epsilon \circ T_t|^2 dx = \int_{\mathbf{R}^N} |f - f_\epsilon|^2 J_t^{-1} dx \le \epsilon^2.$$

A function f_ϵ in $\mathcal{D}^1(\mathbf{R}^N)$ has a compact support and is uniformly Lipschitz continuous, that is

$$\exists c > 0, \forall x, y \in \mathbf{R}^N, \quad |f_\epsilon(y) - f_\epsilon(x)| \le c|y - x|.$$

Thus for all X in $\mathbf{R}^N$

$$|f_\epsilon(T_t(X)) - f_\epsilon(X)| \le c|T_t(X) - X|. \tag{5.48}$$

Now

$$T_t(X) = X + \int_0^t V(s, X) ds + \int_0^t [V(s, T_s(X)) - V(s, X)] ds.$$

Since $V \in C^0([0, \tau]; \mathcal{D}^1(\mathbf{R}^N, \mathbf{R}^N))$ is also uniformly Lipschitz continuous by hypothesis (V)

$$\exists c > 0, \forall (X, Y), \forall s \in [0, \tau], |V(s, Y) - V(s, X)| \le c|Y - X|$$

and for all t in $[0, \tau]$

$$|T_t(X) - X| \le t \, \|V(\cdot, X)\|_{C^0([0,\tau];\mathbf{R}^N)} + \int_0^t c|T_s(X) - X| ds.$$

It is now easy to verify that there exists a constant $c > 0$ such that

$$\max_{s \in [0,\tau]} |T_s(X) - X| \le ct \, \|V(\cdot, X)\|_{C^0([0,\tau];\mathbf{R}^N)}. \tag{5.49}$$

Finally in view of (5.48) and (5.49) the term

$$\int_{\mathbf{R}^N} |f_\epsilon(T_t(X)) - f_\epsilon(X)|^2 dX \le ct^2 \int_{\mathbf{R}^N} \|V(\cdot, X)\|_{C^0}^2 dX$$

$$\int_{\mathbf{R}^N} |f_\epsilon(T_t(X)) - f_\epsilon(X)|^2 dX = \int_{K_\epsilon} |f_\epsilon(T_t(X)) - f_\epsilon(X)|^2 dX$$

$$\le ct^2 \int_{K_\epsilon} \|V(\cdot, X)\|_{C^0}^2 dx \le c't^2.$$

So for t small enough the right-hand side of (5.47) is less than 3ϵ and this completes the proof of the first part of (5.44).

(ii) The second part of (5.44) can be obtained by the following change of variable

$$\int_{\mathbf{R}^N} |f \circ T_t^{-1} - f|^2 dx = \int_{\mathbf{R}^N} |f - f \circ T_t|^2 J_t^{-1} dX$$

and the fact that $\beta^{-1} < J_t^{-1} < \alpha^{-1}$. This completes the proof of the lemma. □

5.5 Domain and boundary expressions of the shape gradient

Expression (5.46) for the shape gradient is a volume (or domain) integral and it is easily verified that the map

$$V \mapsto dJ(\Omega;V) : \vec{\mathcal{V}}^{0,1} \to \mathbf{R} \tag{5.50}$$

is linear and continuous (cf. §3.2). So by the Corollary to Theorem 3.2, we know that for a domain Ω with a C^2 boundary Γ there exists a scalar distribution $g(\Gamma)$ in $\mathcal{D}(\Gamma)'$ such that

$$dJ(\Omega;V) = \langle g(\Gamma), \ V(0) \cdot n \rangle_{\mathcal{D}^1(\Gamma)}. \tag{5.51}$$

The next objective is to further characterize the boundary expression. Recall that we have assumed that $f \in H^1(\mathbf{R}^N)$. So for a C^2 boundary Γ the solution $y = y(\Omega)$ of the problem

$$-\Delta y = f \text{ in } \Omega$$
$$y = 0 \text{ on } \Gamma$$

belongs to $H^2(\Omega)$. For velocity fields V in $C^0([0,\tau]; \mathcal{D}^1(\mathbf{R}^N, \mathbf{R}^N))$ satisfying hypothesis (V) the transported solution y^t in $H^2(\Omega)$ is the solution of the system

$$-\text{div}\,(A(t)\nabla y^t) = J_t f \circ T_t \text{ in } \Omega$$
$$y^t = 0 \text{ on } \Gamma. \tag{5.52}$$

Knowing that for all t in $[0,\tau]$, $y^t \in H^2(\Omega) \cap H_0^1(\Omega)$, we can repeat the computation of $\partial_t \tilde{E}(t,\varphi)$ for φ in $H^2(\Omega) \cap H_0^1(\Omega)$ instead of $H_0^1(\Omega)$. With this extra smoothness we can use the formula of §4

$$\frac{d}{dt}\int_{\Omega_t} F(t, T_t^{-1}x)dx|_{t=0} = \int_\Gamma F(0,x)V(0) \cdot n \ d\Gamma + \int_\Omega \frac{\partial F}{\partial t}(0,x)dx, \tag{5.53}$$

for a sufficiently smooth function $F : [0,\tau] \times \Omega \to \mathbf{R}$. Prior to applying this formula to expression (5.17) in §5.2, notice that for φ in $H^2(\Omega)$,

$$\dot{\varphi} = \frac{d}{dt}\varphi \circ T_1^{-1}|_{t=0} = -\nabla\varphi \cdot V(0) \in H^1(\Omega) \tag{5.54}$$

but it generally does not belong to $H_0^1(\Omega)$. Then

$$\partial_t \tilde{E}(t,\varphi)|_{t=0} = \int_\Gamma \left\{\frac{1}{2}|\nabla\varphi|^2 - f\varphi\right\}V(0) \cdot n \ d\Gamma$$
$$+ \int_\Omega \{\nabla\varphi \cdot \nabla\dot{\varphi} - f\dot{\varphi}\}dx. \tag{5.55}$$

Substitute $\varphi = y$ in (5.55)

$$\partial_t \tilde{E}(t,y)|_{t=0} = \int_\Gamma \left\{ \frac{1}{2} |\nabla y|^2 - fy \right\} V(0) \cdot n \, d\Gamma$$
$$- \int_\Omega \{ \nabla y \cdot \nabla(\nabla y \cdot V(0)) - f(\nabla y \cdot V(0)) \} dx. \tag{5.56}$$

But $y \in H^2(\Omega) \cap H_0^1(\Omega)$ and

$$\int_\Omega \nabla y \cdot \nabla(\nabla y \cdot V(0)) dx = - \int_\Omega \Delta y \, \nabla y \cdot V(0) dx + \int_\Omega \frac{\partial y}{\partial n} \nabla y \cdot V(0) d\Gamma$$

and

$$\partial_t \tilde{E}(t,y)|_{t=0} = \int_\Gamma \left\{ \left[\frac{1}{2} |\nabla y|^2 - fy \right] V(0) \cdot n - \frac{\partial y}{\partial n} \nabla y \cdot V(0) \right\} d\Gamma. \tag{5.57}$$

But

$$y = 0 \text{ on } \Gamma \Rightarrow \nabla y = \frac{\partial y}{\partial n} n \text{ on } \Gamma,$$

and finally

$$\partial_t \tilde{E}(t,y)|_{t=0} = - \int_\Gamma \frac{1}{2} \left| \frac{\partial y}{\partial n} \right|^2 V(0) \cdot n \, d\Gamma. \tag{5.58}$$

This is the boundary expression which is continuous for $V(0)$ in $\mathcal{D}^0(\mathbf{R}^N, \mathbf{R}^N)$. It has been obtained via a parametrization of the function space appearing in the Min formulation. Thus

$$dJ(\Omega; V) = - \int_\Omega \frac{1}{2} \left| \frac{\partial y}{\partial n} \right|^2 V(0) \cdot n \, d\Gamma \tag{5.59}$$

as predicted in §5.1.

6   Shape gradients via a Min Max formulation and Function Space Embedding

6.1   An illustrative example

Let Ω be a bounded open domain in $\mathbf{R}^n$ with a smooth boundary Γ. Let $y = y(\Omega)$ be the solution of the Neumann problem

$$-\Delta y + y = f \text{ in } \Omega,$$
$$\frac{\partial y}{\partial n} = 0 \text{ on } \Gamma, \tag{6.1}$$

where f is a fixed function in $H^1(\mathbf{R}^n)$. Associate with $y(\Omega)$ the cost function

$$J(\Omega) = \frac{1}{2} \int_\Omega |y(\Omega) - y_d|^2 dx \tag{6.2}$$

where y_d is a fixed function in $H^1(\mathbf{R}^n)$.

The solution of (6.1) coincides with the minimizing element of the following variational problem

$$\inf\{E(\Omega, \varphi) : \varphi \in H^1(\Omega)\}, \tag{6.3}$$

where

$$E(\Omega, \varphi) = \frac{1}{2}\int_\Omega \{|\nabla\varphi|^2 + \varphi^2 - 2f\varphi\}dx. \tag{6.4}$$

The minimizing element y of (6.4) is the solution in $H^1(\Omega)$ of Euler's equation

$$dE(\Omega, y; \varphi) = 0, \quad \forall \varphi \in H^1(\Omega), \tag{6.5}$$

where

$$dE(\Omega, y; \varphi) = \int_\Omega [\nabla y \cdot \nabla\varphi + y\varphi - f\varphi]dx. \tag{6.6}$$

Equation (6.5) is the *variational equation* for y.

The cost function $J(\Omega)$ is a function of the domain Ω and the solution of (6.1) which will be called the *state*. It is convenient to introduce the *cost functional*

$$F(\Omega, \varphi) = \frac{1}{2}\int_\Omega |\varphi - y_d|^2 dx \tag{6.7}$$

which clearly expresses the dependence on Ω and φ. To sum up we consider the cost function

$$J(\Omega) = F(\Omega, y(\Omega)), \tag{6.8}$$

where $y = y(\Omega)$ is the solution of

$$y \in H^1(\Omega), \ dE(\Omega, y; \varphi) = 0, \quad \forall \varphi \in H^1(\Omega). \tag{6.9}$$

We wish to find an expression for the shape derivative $dJ(\Omega; V)$.

6.2 The Min Max formulation

The basic idea is the same as the one encountered in Control Theory. Equation (6.1) (or in its variational form equation (6.9)) is considered as a constraint in the minimization problem. We construct a Lagrangian functional by introducing a "multiplier function" ψ :

$$G(\Omega, \varphi, \psi) = F(\Omega, \varphi) + dE(\Omega, \varphi; \psi). \tag{6.10}$$

Then the cost function is given by

$$J(\Omega) = \min_{\varphi \in H^1(\Omega)} \ \sup_{\psi \in H^1(\Omega)} G(\Omega, \varphi, \psi) \tag{6.11}$$

since

$$\sup_{\psi \in H^1(\Omega)} G(\Omega, \varphi, \psi) = \begin{cases} F(\Omega, y(\Omega)), & \text{if } \varphi = y(\Omega) \\ +\infty, & \text{if } \varphi \neq y(\Omega). \end{cases} \tag{6.12}$$

In our example the Lagrangian G is convex and continuous with respect to the variable φ and concave and continuous with respect to the variable ψ. Moreover the space $H^1(\Omega)$ is convex and closed. So the functional G has a saddle point if and only if the saddle point equations have a solution (y, p) (cf. Ekeland-Temam [1]):

$$p \in H^1(\Omega), \quad dG(\Omega, y, p; 0, \psi) = 0, \quad \forall \psi \in H^1(\Omega) \tag{6.13}$$

$$y \in H^1(\Omega), \quad dG(\Omega, y, p; \varphi, 0) = 0, \quad \forall \varphi \in H^1(\Omega). \tag{6.14}$$

They are completely equivalent to

$$y \in H^1(\Omega), \quad dE(\Omega, y; \psi) = 0, \quad \forall \psi \in H^1(\Omega) \tag{6.15}$$

$$p \in H^1(\Omega), \quad dF(\Omega, y; \varphi) + d^2 E(\Omega, y; p; \varphi) = 0, \quad \forall \varphi \in H^1(\Omega) \tag{6.16}$$

or

$$-\Delta y + y = f \text{ in } \Omega, \quad \frac{\partial y}{\partial n} = 0 \text{ on } \Gamma \tag{6.15a}$$

$$-\Delta p + p + y - y_d = 0 \text{ in } \Omega, \quad \frac{\partial p}{\partial n} = 0 \text{ on } \Gamma. \tag{6.16a}$$

System (6.15)–(6.16) has a unique solution in $H^1(\Omega) \times H^1(\Omega)$ which coincides with the unique saddle point of $G(\Omega, \varphi, \psi)$ in $H^1(\Omega) \times H^1(\Omega)$.

6.3   Function Space Parametrization

We have shown that the cost function $J(\Omega)$ can be expressed as a Min Max of a functional G with a unique saddle point (y, p) which is completely characterized by the variational equations (6.15)–(6.16). The same result holds when Ω is transformed into a domain $\Omega_t = T_t(\Omega)$ under the action of the velocity field V for $t \geq 0$:

$$J(\Omega_t) = \min_{\varphi \in H^1(\Omega_t)} \sup_{\psi \in H^1(\Omega_t)} G(\Omega_t, \varphi, \psi) \tag{6.17}$$

where the saddle point (y_t, p_t) is completely characterized by

$$y_t \in H^1(\Omega_t), \quad dE(\Omega_t, y_t; \psi) = 0, \quad \forall \psi \in H^1(\Omega_t), \tag{6.18}$$

$$p_t \in H^1(\Omega_t), \quad dF(\Omega_t, y_t; \varphi) + d^2 E(\Omega_t, y_t; p_t, \varphi) = 0, \quad \forall \varphi \in H^1(\Omega_t). \tag{6.19}$$

We are looking for a theorem which will give an expression for the derivative of a Min Sup with respect to a parameter $t \geq 0$. However in (6.17) the space $H^1(\Omega_t)$ depends on the parameter t. To get around this difficulty and obtain a Min Sup expression for $J(\Omega_t)$ over spaces which are independent of $t \geq 0$, we introduce the following parametrization

$$H^1(\Omega_t) = \{\varphi \circ T_t^{-1} : \varphi \in H^1(\Omega)\} \tag{6.20}$$

since T_t and T_t^{-1} are diffeomorphisms. This parametrization does not affect the value of the saddle point $J(\Omega_t)$ but will change the parametrization of the functional G:

$$J(\Omega_t) = \inf_{\varphi \in H^1(\Omega)} \sup_{\psi \in H^1(\Omega)} G(\Omega_t, \varphi \circ T_t^{-1}, \psi \circ T_t^{-1}). \tag{6.21}$$

This parametrization is apparently unique to *Shape Analysis*. It amounts to introducing the new Lagrangian functional

$$\tilde{G}(t, \varphi, \psi) = G(T_t(\Omega), \varphi \circ T_t^{-1}, \psi \circ T_t^{-1}), \quad \varphi \text{ and } \psi \text{ in } H^1(\Omega). \tag{6.22}$$

Our next objective is to find an expression for the limit

$$dg(0) = \lim \frac{g(t) - g(0)}{t}, \tag{6.23}$$

where

$$g(t) = J(\Omega_t) = \inf_{\varphi \in H^1(\Omega)} \sup_{\psi \in H^1(\Omega)} \tilde{G}(t, \varphi, \psi). \tag{6.24}$$

This will be done in the next section.

Before closing, it is useful to look at the expression for $\tilde{G}$ and the resulting saddle point (y^t, p^t) in $H^1(\Omega) \times H^1(\Omega)$. By definition $\tilde{G}$ is given by the expression

$$\tilde{G}(t, \varphi, \psi) = \frac{1}{2} \int_{\Omega_t} |\varphi \circ T_t^{-1} - y_d|^2 dx$$
$$+ \int_{\Omega_t} [\nabla(\varphi \circ T_t^{-1}) \cdot \nabla(\psi \circ T_t^{-1}) + (\varphi \circ T_t^{-1})(\psi \circ T_t^{-1}) - f(\psi \circ T_t^{-1})] dx \tag{6.25}$$

and its saddle point is a solution of the variational equations

$$y^t \in H^1(\Omega), \text{ and for all } \psi \text{ in } H^1(\Omega)$$
$$\int_{\Omega_t} [\nabla(y^t \circ T_t^{-1}) \cdot \nabla(\psi \circ T_t^{-1}) + (y^t \circ T_t^{-1})(\psi \circ T_t^{-1}) - f(\psi \circ T_t^{-1})] dx = 0 \tag{6.26}$$

$$p^t \in H^1(\Omega), \text{ and for all } \varphi \text{ in } H^1(\Omega)$$
$$\int_{\Omega_t} [(y^t \circ T_t^{-1} - y_d)(\varphi \circ T_t^{-1}) + \nabla(p^t \circ T_t^{-1}) \cdot \nabla(\varphi \circ T_t^{-1}) + (p^t \circ T_t^{-1})(\varphi \circ T_t^{-1})] dx = 0. \tag{6.27}$$

It is readily seen that $(y^t \circ T_t^{-1}, p^t \circ T_t^{-1})$ coincides with the saddle point (y_t, p_t) in $H^1(\Omega_t) \times H^1(\Omega_t)$:

$$y_t = y^t \circ T_t^{-1}, \quad p_t = p^t \circ T_t^{-1},$$

or equivalently

$$y^t = y_t \circ T_t, \quad p^t = p_t \circ T_t. \tag{6.28}$$

The solutions (y^t, p^t) can easily be interpreted as the solutions (y_t, p_t) on Ω_t transported back to the fixed domain Ω by the transformation T_t.

In view of this observation, we can rewrite expressions (6.25) to (6.27) on the fixed domain Ω by using the coordinate transformation T_t. Expression (6.25) becomes

$$\tilde{G}(t, \varphi, \psi) = \frac{1}{2} \int_\Omega |\varphi - y_d \circ T_t|^2 J_t \, dx \\ + \int_\Omega \{A(t)\nabla\varphi \cdot \nabla\psi + J_t[\varphi\psi - (f \circ T_t)\psi]\} \, dx, \tag{6.29}$$

where for $t > 0$ small

$$
\begin{aligned}
DT_t &= \quad \text{Jacobian matrix of } T_t & (6.30)\\
J_t &= \quad \det DT_t \text{ (since det } DT_t = |\det DT_t| \text{ for } t \geq 0 \text{ small)} & (6.31)\\
A(t) &= \quad J_t[DT_t]^{-1} \, {}^*[DT]^{-1}. & (6.32)
\end{aligned}
$$

Similarly the variational equations (6.26)-(6.27) reduce to

$$y_t \in H^1(\Omega), \text{ and } \forall\psi \in H^1(\Omega)$$
$$\int_\Omega \{A(t)\nabla y^t \cdot \nabla\psi + J_t[y^t\psi - (f \circ T_t)\psi]\}dx = 0 \tag{6.33}$$

$$p^t \in H^1(\Omega) \text{ and } \forall\varphi \in H^1(\Omega)$$
$$\int_\Omega \{A(t)\nabla p^t \cdot \nabla\varphi + J_t[p^t\varphi + (y^t - y_d \circ T_t)\varphi]\}dx = 0. \tag{6.34}$$

6.4   Differentiability of a saddle point with respect to a parameter

Consider a functional

$$G : [0, \tau] \times X \times Y \to \mathbf{R} \tag{6.35}$$

for some $\tau > 0$ and sets X and Y. For each t in $[0, \tau]$ define

$$g(t) = \inf_{x \in X} \sup_{y \in Y} G(t, x, y) \tag{6.36}$$

and the sets

$$X(t) = \left\{ x^t \in X : \sup_{y \in Y} G(t, x^t, y) = g(t) \right\} \tag{6.37}$$

$$Y(t, x) = \left\{ y^t \in Y : G(t, x, y^t) = \sup_{y \in Y} G(t, x, y) \right\}. \tag{6.38}$$

Similarly define

$$h(t) = \sup_{y \in Y} \inf_{x \in X} G(t, x, y) \tag{6.39}$$

and the sets

$$Y(t) = \left\{ y^t \in Y : \inf_{x \in X} G(t, x, y^t) = h(t) \right\} \tag{6.40}$$

$$X(t, y) = \left\{ x^t \in X : G(t, x^t, y) = \inf_{x \in X} G(t, x, y) \right\}. \tag{6.41}$$

In general we always have the inequality

$$h(t) \leq g(t). \tag{6.42}$$

To complete the set of notations, we introduce the set of *saddle points*

$$S(t) = \{(x, y) \in X \times Y : g(t) = G(t, x, y) = h(t)\} \tag{6.43}$$

which may be empty.

Our objective is to find realistic conditions under which the limit

$$dg(0) = \lim_{t \searrow 0} \frac{g(t) - g(0)}{t} \tag{6.44}$$

exists. A case of special interest is when G has a saddle point for all t in $[0, \tau]$. It can be viewed as an extension of Theorem 5.1 in §5.3 on the differentiability of a Min with respect to a parameter. It is used when the functional to be minimized is a function of the state which is itself a function of the domain through the boundary value problem. In that case the saddle point equations coincide with the "state equation" and the "adjoint state equation" as illustrated in the previous section. The main advantage of this approach is to avoid the problem of the existence and characterization of the derivative of the state x^t with respect to t. In a Control problem this would be the directional derivative of the state with respect to the control variable. In particular it is not necessary to invoque any Implicit Function Theorem with possibly restrictive differentiability conditions. It will be sufficient to check two continuity conditions for the set-valued maps $X(\cdot)$ and $Y(\cdot)$. To complete this discussion we recall

Lemma 6.1 *Fix t in $[0, \tau]$, Then*

$$\forall (x^t, y^t) \in X(t) \times Y(t), \quad h(t) \leq G(t, x^t, y^t) \leq g(t) \tag{6.45}$$

and if $h(t) = g(t)$

$$X(t) \times Y(t) = S(t). \tag{6.46}$$

Proof (i) If $X(t) \times Y(t) = \emptyset$ there is nothing to prove. If there exist $x^t \in X(t)$ and $y^t \in Y(t)$, then by definition

$$h(t) = \inf_{x \in X} G(t, x, y^t) \leq G(t, x^t, y^t) \leq \sup_{y \in Y} G(t, x^t, y^t) = g(t). \tag{6.47}$$

(ii) If $h(t) = g(t)$, then in view of (6.47), $X(t) \times Y(t) \subset S(t)$. Conversely if there exists $(x^t, y^t) \in S(t)$, then $h(t) = G(t, x^t, y^t) = g(t)$ and by definitions (6.37) and (6.40) of $X(t)$ and $Y(t), (x^t, y^t) \in X(t) \times Y(t)$.   □

It is important to keep in mind that identity (6.46) is always true when

$$h(t) = \sup_{y \in Y} \inf_{x \in X} G(t, x, y) = \inf_{x \in X} \sup_{y \in Y} G(t, x, y) = g(t)$$

but that $S(t)$ may be empty.

Theorem 6.1 (Correa-Seeger [1]) *Let the real number $\tau > 0$, the sets X and Y and the functional $G : [0, \tau] \times X \times Y \to \mathbf{R}$ be given. Assume that the following hypotheses hold:*

(H1) $S(t) \neq \emptyset$, $0 \leq t \leq \tau$;

(H2) *for all (x, y) in $[\cup\{X(t) : 0 \leq t \leq \tau\} \times Y(0)] \cup [X(0) \times \cup\{Y(t) : 0 \leq t \leq \tau\}]$ the partial derivative $\partial_t G(t, x, y)$ exists everywhere in $[0, \tau]$;*

(H3) *there exists a topology $\mathcal{T}_X$ on X such that for any sequence $\{t_n : 0 < t_n \leq \tau\}$, $t_n \to t_0 = 0$, $\exists x^0 \in X(0)$, $\exists$ a subsequence $\{t_{n_k}\}$ of $\{t_n\}$, and for each $k \geq 1$, $\exists x_{n_k} \in X(t_{n_k})$ such that*

   (i) *$x_{n_k} \to x_0$ in the $\mathcal{T}_X$-topology; and*

   (ii) *for all y in $Y(0)$,*

$$\liminf_{\substack{t \searrow 0 \\ k \to \infty}} \partial_t G(t, x_{n_k}, y) \geq \partial_t G(0, x^0, y); \tag{6.48}$$

(H4) *there exists a topology $\mathcal{T}_Y$ on Y such that for any sequence $\{t_n : 0 < t_n \leq \tau\}$, $t_n \to t_0 = 0$, $\exists y^0 \in Y(0)$, $\exists$ a subsequence $\{t_{n_k}\}$ of $\{t_n\}$, and for each $k \geq 1$, $\exists y_{n_k} \in Y(t_{n_k})$ such that*

   (i) *$y_{n_k} \to y^0$ in the $\mathcal{T}_Y$-topology; and*

   (ii) *for all x in $X(0)$,*

$$\limsup_{\substack{t \searrow 0 \\ k \to \infty}} \partial_t G(t, x, y_{n_k}) \geq \partial_t G(0, x, y^0). \tag{6.49}$$

Then there exists $(x^0, y^0) \in X(0) \times Y(0)$ such that

$$\begin{aligned} dg(0) &= \inf_{x \in X(0)} \sup_{y \in Y(0)} \partial_t G(0, x, y) = \partial_t G(0, x^0, y^0) \\ &= \sup_{y \in Y(0)} \inf_{x \in X(0)} \partial_t G(0, x, y). \end{aligned} \tag{6.50}$$

Thus (x^0, y^0) is a saddle point of $\partial_t G(0, x, y)$ on $X(0) \times Y(0)$.

Proof (i) We first establish upper and lower bounds to the differential quotient

$$\frac{\Delta(t)}{t}, \quad \Delta(t) \stackrel{\text{def}}{=} g(t) - g(0).$$

Choose arbitrary x_0 in $X(0)$, x_t in $X(t)$, y_0 in $Y(0)$ and y_t in $Y(t)$. Then by definition

$$\begin{aligned}
G(t, x_t, y_0) &\le & G(t, x_t, y_t) &\le & G(t, x_0, y_t) \\
-G(0, x_t, y_0) &\le & -G(0, x_0, y_0) &\le & -G(0, x_0, y_t).
\end{aligned}$$

Add up the above two chains of inequalities to obtain

$$G(t, x_t, y_0) - G(0, x_t, y_0) \le \Delta(t) \le G(t, x_0, y_t) - G(0, x_0, y_t).$$

By hypothesis (H2), there exist θ_t, $0 < \theta_t < 1$, and α_t, $0 < \alpha_t < 1$, such that

$$\begin{aligned}
G(t, x_t, y_0) - G(0, x_t, y_0) &= t \partial_t G(\theta_t t, x_t, y_0) \\
G(t, x_0, y_t) - G(0, x_0, y_t) &= t \partial_t G(\alpha_t t, x_0, y_t)
\end{aligned}$$

and by dividing by $t > 0$

$$\partial_t G(\theta_t t, x_t, y_0) \le \frac{\Delta(t)}{t} \le \partial_t G(\alpha_t t, x_0, y_t). \tag{6.51}$$

(ii) Define

$$\underline{dg}(0) = \liminf_{t \searrow 0} \frac{\Delta(t)}{t}, \quad \overline{dg}(0) = \limsup_{t \searrow 0} \frac{\Delta(t)}{t}.$$

There exists a sequence $\{t_n : 0 < t_n \le \tau\}$, $t_n \to 0$, such that

$$\lim_{n \to \infty} \frac{\Delta(t_n)}{t_n} = \underline{dg}(0).$$

By hypothesis (H3), $\exists x^0 \in X(0)$, $\exists$ a subsequence $\{t_{n_k}\}$ of $\{t_n\}$, for each $k \ge 1$, $\exists x_{n_k} \in X(t_{n_k})$ such that $x_{n_k} \to x^0$ in $\mathcal{T}_X$ and

$$\forall y \in Y(0), \quad \liminf_{\substack{t \searrow 0 \\ k \to \infty}} \partial_t G(t, x_{n_k}, y) \ge \partial_t G(0, x^0, y).$$

So from the first part of the estimate (6.51) for any $y \in Y(0)$ and $t = t_{n_k}$,

$$\partial_t G(\theta_{t_{n_k}} t_{n_k}, x_{n_k}, y) \le \frac{\Delta(t_{n_k})}{t_{n_k}}$$

and

$$\partial_t G(0, x^0, y) \le \liminf_{k \to \infty} \partial_t G(\theta_{t_{n_k}} t_{n_k}, x_{n_k}, y) \le \lim_{k \to \infty} \frac{\Delta(t_{n_k})}{t_{n_k}} = \underline{dg}(0).$$

Therefore

$$\exists x^0 \in X(0), \forall y \in Y(0), \quad \partial_t G(0, x^0, y) \le \underline{dg}(0)$$

and

$$\inf_{x \in X(0)} \sup_{y \in Y(0)} \partial_t G(0, x, y) \le \sup_{y \in Y(0)} \partial_t G(0, x^0, y) \le \underline{dg}(0). \tag{6.52}$$

By a dual argument and assumption (H4) we also obtain

$$\exists y^0 \in Y(0), \forall x \in X(0), \quad \partial_t G(0, x, y^0) \le \overline{dg}(0)$$

$$\overline{dg}(0) \le \inf_{x \in X(0)} \partial_t G(0, x, y^0) \le \sup_{y \in Y(0)} \inf_{x \in X(0)} \partial_t G(0, x, y) \tag{6.53}$$

and necessarily

$$\inf_{x \in X(0)} \sup_{y \in Y(0)} \partial_t G(0, x, y) = \underline{dg}(0) = \overline{dg}(0) = \sup_{y \in Y(0)} \inf_{x \in X(0)} \partial_t G(0, x, y). \tag{6.54}$$

In particular from (6.52) and (6.53)

$$\sup_{y \in Y(0)} \partial_t G(0, x^0, y) = dg(0) = \inf_{x \in X(0)} \partial_t G(0, x, y^0) \tag{6.55}$$

and (x^0, y^0) is a saddle point of $\partial_t G(0, \cdot, \cdot)$. □

Remark 6.1 In the applications, this formulation of the theorem presents some definite technical advantages over its original version.

(i) Equation (6.50) establishes the existence of a saddle point of $\partial_t G(0, \cdot, \cdot)$ with respect to $X(0) \times Y(0)$.

(ii) Another important feature is the use of subsequences in hypotheses (H3) and (H4). This makes it possible to work with weak topologies in reflexive Banach spaces and use the eventual boundedness of the sets of saddle points.

(iii) Finally hypothesis (H2) and conditions (6.48) and (6.49) in (H3) and (H4) need only be verified on the family of saddle points at $t = 0$. For instance the first part of hypotheses (H3) and (H4) could be verified in $H^1(\Omega) \times H^1(\Omega)$. Yet, if the saddle points are smoother, say in $H^2(\Omega) \times H^2(\Omega)$, this extra smoothness can be used to verify (H2) and (6.48) and (6.49) in (H3) and (H4). □

6.5   Application of the theorem

Our example has a unique saddle point (y^t, p^t) for $t \ge 0$ small and we can use the corollary to Theorem 6.1. The set-valued maps X and Y reduce to ordinary functions

$$t \mapsto X(t) = y^t, \ t \mapsto Y(t) = p^t \tag{6.56}$$

and it is sufficient to show their continuity at $t = 0$ in $H^1(\Omega)$. So we now proceed to the verification of hypotheses H1 to H4.

Assume that V belongs to $\mathcal{V}^1$, that is $\mathcal{D}^1(\mathbf{R}^N, \mathbf{R}^N)$, and that f and y belong to $H^1(\mathbf{R}^N)$. Choose $\tau > 0$ small enough such that

$$J_t = \det DT_t = |\det DT_t| = |J_t|, \quad 0 \le t \le \tau, \tag{6.57}$$

and that there exist constants $0 < \alpha < \beta$ such that

$$\forall \xi \in \mathbf{R}^N, \quad \alpha|\xi|^2 \le A(t)\xi \cdot \xi \le \beta|\xi|^2, \quad \text{and } \alpha \le J_t \le \beta. \tag{6.58}$$

Since the bilinear forms associated with (6.33) and (6.34) are coercive, there exists a unique solution pair (y^t, p^t) of the system (6.33)-(6.34). Hence

$$\forall t \in [0,\tau], \quad X(t) = \{y^t\} \ne \emptyset, \quad Y(t) = \{p^t\} \ne \emptyset. \tag{6.59}$$

So hypothesis H1 is satisfied. To check H2 we use expression (6.29) and compute for φ and ψ in $H^1(\Omega)$

$$\begin{aligned}
\partial_t \tilde{G}(t, \varphi, \psi) &= \int_\Omega \left\{ \frac{1}{2}(\varphi - y_d \circ T_t)^2 \operatorname{div} V_t - (\varphi - y_d \circ T_t)\nabla y_d \cdot V_t J_t \right\} dx \\
&\quad + \int_\Omega \{A'(t)\nabla\varphi \cdot \nabla\psi + \operatorname{div} V_t(\varphi\psi - f \circ T_t\psi) - J_t\nabla f \cdot V_t \psi\} dx,
\end{aligned} \tag{6.60}$$

where

$$V_t(X) = V(T_t(X)), \quad A'(t) = (\operatorname{div} V_t)I - {}^*DV_t - DV_t, \tag{6.61}$$

I is the identity matrix on $\mathbf{R}^N$ and DV_t is the Jacobian matrix of V_t. By choice of V in $\mathcal{D}^1(\mathbf{R}^N, \mathbf{R}^N)$, $t \mapsto V_t$ and $t \mapsto DV_t$ are continuous on $[0,\tau]$. Moreover f and y_d belong to $H^1(\mathbf{R}^N)$. As a result expression (6.60) is well-defined and $\partial_t\tilde{G}(t, \varphi, \psi)$ exists everywhere in $[0,\tau]$ for all φ and ψ in $H^1(\Omega)$. This can be proved in many ways. For instance we establish (6.60) for f and y_d in $\mathcal{D}(\mathbf{R}^N)$. Then we show that the affine map $(f, y_d) \mapsto \partial_t\tilde{G}(\cdot, \varphi, \psi)$ is continuous from $H^1(\mathbf{R}^N) \times H^1(\mathbf{R}^N)$ to $C^1([0,\tau])$. So it extends to all (f, y_d) in $H^1(\mathbf{R}^N) \times H^1(\mathbf{R}^N)$ by uniform continuity and density of $\mathcal{D}(\mathbf{R}^N)$ in $H^1(\mathbf{R}^N)$. Hypothesis H2 is satisfied.

To ckeck hypotheses H3(i) and H4(i), we first show that for any sequence $\{t_n\} \subset [0,\tau]$, $t_n \to 0$, there exists a subsequence of $\{y^{t_n}\}$, still denoted $\{y^{t_n}\}$, such that

$$y^{t_n} \rightharpoonup y^0 = y \text{ in } H^1(\Omega) - \text{weak}$$
$$y^{t_n} \rightharpoonup p^0 = p \text{ in } H^1(\Omega) - \text{weak},$$

where (y, p) is the solution of system (6.15)-(6.16) or (6.15a)-(6.16a). By choice of τ satisfying condition (6.58), there exists a constant $c > 0$ such that

$$\alpha \|y^t\|_{H^1(\Omega)} \le \beta c \|f\|_{L^2(\mathbf{R}^N)}$$
$$\alpha \|p^t\|_{H^1(\Omega)} \le \beta c \|y^t - y_d\|_{L^2(\Omega)}.$$

So the pair $\{y^t, p^t\}$ is bounded in $H^1(\Omega) \times H^1(\Omega)$ and there exists a subsequence $\{y^{t_n}, p^{t_n}\}$ and a pair (z, q) in $H^1(\Omega) \times H^1(\Omega)$ such that

$$y^{t_n} \rightharpoonup z \text{ in } H^1(\Omega) - \text{weak, and } p^{t_n} \rightharpoonup q \text{ in } H^1(\Omega) - \text{weak}.$$

The pair (z, q) can be characterized by going to the limit in the variational equations (6.33)-(6.34)

$$\int_{\Omega} \{A(t_n)\nabla y^{t_n} \cdot \nabla \psi + J_{t_n}[y^{t_n}\psi - (f \circ T_{t_n})\psi]\} dx = 0$$

$$\int_{\Omega} \{A(t_n)\nabla p^{t_n} \cdot \nabla \varphi + J_{t_n}[p^{t_n}\varphi + (y^{t_n} - y_d \circ T_{t_n})\varphi]\} dx = 0.$$

So we proceed as in §5.4 and use Lemma 5.1 to obtain

$$\int_{\Omega} \{\nabla z \cdot \nabla \psi + z\psi - f\psi\} dx = 0, \ \forall \psi,$$

$$\int_{\Omega} \{\nabla q \cdot \nabla \varphi + q\varphi + (z - y_d)\varphi\} dx = 0, \ \forall \varphi.$$

By uniqueness, $(z, q) = (y, p)$. We now proceed as in §5.4 and prove that

$$y^{t_n} \to y \ \text{in} \ H^1(\Omega) - \text{strong}, \ p^{t_n} \to p \ \text{in} \ H^1(\Omega) - \text{strong}$$

by the same argument. So hypotheses H3(i) and H4(i) are satisfied for the strong topology of $H^1(\Omega)$. Finally hypotheses H3(ii) and H4(ii) are readily verified in view of the strong continuity of $(t, \varphi) \mapsto \partial_t \tilde{E}(t, \varphi, \psi)$ and $(t, \psi) \mapsto \partial_t \tilde{E}(t, \varphi, \psi)$. In fact it would have been sufficient to check hypothesis H3 with $H^1(\Omega)$-strong and H4 with $H^1(\Omega)$-weak.

So all hypotheses of Theorem 3.1 are satisfied and

$$dJ(\Omega; V) = \int_{\Omega} \left\{ \frac{1}{2}(y - y_d)^2 \operatorname{div} V - (y - y_d)\nabla y_d \cdot V \right\} dx$$
$$+ \int_{\Omega} \{A'(0)\nabla y \cdot \nabla p + \operatorname{div} V(0)(yp - fp) - \nabla f \cdot V(0)p\} dx, \tag{6.62}$$

where (y, p) is the solution of equations (6.15a)-(6.16a) or in variational form

$$y \in H^1(\Omega), \quad \forall \psi \in H^1(\Omega)$$
$$\int_{\Omega} \{\nabla y \cdot \nabla \psi + y\psi - f\psi\} dx = 0, \tag{6.63}$$

$$p \in H^1(\Omega), \quad \forall \varphi \in H^1(\Omega)$$
$$\int_{\Omega} \{\nabla p \cdot \nabla \varphi + p\varphi + (y - y_d)\varphi\} dx. \tag{6.64}$$

6.6   Domain and boundary expressions for the shape gradient

Expression (6.62) for the shape gradient is a volume or domain integral. For y_d and f in $H^1(\mathbf{R}^N)$ it is readily verified that the map

$$V \mapsto dJ(\Omega; V) : \mathcal{D}^1(\mathbf{R}^N, \mathbf{R}^N) \to \mathbf{R} \tag{6.65}$$

is linear and continuous. So by the Corollary to Theorem 3.2 in §3.1, we know that for a domain Ω with a C^2 boundary Γ there exists a scalar distribution $g(\Gamma)$ in $\mathcal{D}^1(\Gamma)'$ such that

$$dJ(\Omega;V) = \langle g(\Gamma), V \cdot n \rangle. \tag{6.66}$$

We now further characterize this boundary expression. In view of the hypotheses on f, y_d and Ω, the pair (y,p) is the solution in $H^2(\Omega) \times H^2(\Omega)$ of the system

$$-\Delta y = f \text{ in } \Omega \tag{6.67}$$

$$y = 0 \text{ on } \Gamma$$

$$-\Delta p + (y - y_d) = 0 \text{ in } \Omega \tag{6.68}$$

$$p = 0 \text{ on } \Gamma.$$

Similarly for V in $\mathcal{D}^1(\mathbf{R}^N, \mathbf{R}^N)$ the system

$$\begin{aligned} -\text{div}\,[A(t)\nabla y^t] &= J_t f \circ T_t \text{ in } \Omega \\ y^t &= 0 \text{ on } \Gamma \end{aligned} \tag{6.69}$$

$$\begin{aligned} -\text{div}\,[A(t)\nabla p^t] + (y^t - y_d \circ T_t)J_t &= 0 \text{ in } \Omega \\ p^t &= 0 \text{ on } \Gamma \end{aligned} \tag{6.70}$$

has a unique solution in $H^2(\Omega) \times H^2(\Omega)$ instead of $H^1(\Omega) \times H^1(\Omega)$. With this extra smoothness we can use the formula obtained in §4

$$\begin{aligned} \frac{d}{dt}\int_\Omega F(t, T_t^{-1}(x))dx|_{t=0} &= \int_\Gamma F(0,x)V(0) \cdot n \, d\Gamma \\ &+ \int_\Omega \frac{\partial F}{\partial t}(0,x) \, dx \end{aligned} \tag{6.71}$$

for a sufficiently smooth function $F : [0,\tau] \times \Omega \to \mathbf{R}$. We easily obtain

$$\begin{aligned} \partial_t \tilde{G}(0,\varphi,\psi) &= \int_\Gamma \left\{ \frac{1}{2}(\varphi - y_d)^2 + \nabla\varphi \cdot \nabla\psi + \varphi\psi - f\psi \right\} d\Gamma \\ &+ \int_\Omega \left\{ (\varphi - y_d)\dot{\varphi} + \nabla\psi \cdot \nabla\dot{\varphi} + \psi\dot{\varphi} \right\} dx \\ &+ \int_\Omega \left\{ \nabla\varphi \cdot \nabla\dot{\psi} + \varphi\dot{\psi} - f\dot{\psi} \right\} dx, \end{aligned} \tag{6.72}$$

where

$$\dot{\varphi} = \frac{d}{dt}\varphi \circ T_t^{-1}|_{t=0} = -\nabla\varphi \cdot V(0) \tag{6.73}$$

and

$$\dot{\psi} = \frac{d}{dt}\psi \circ T_t^{-1}|_{t=0} = -\nabla\psi \cdot V(0). \tag{6.74}$$

Now substitute for (φ, ψ) the solution (y, p) of (6.63)-(6.64)

$$
\begin{aligned}
\partial_t \tilde{G}(0, y, p) &= \int_\Gamma \left\{ \frac{1}{2}(y - y_d)^2 + \nabla y \cdot \nabla p + yp - fp \right\} V \cdot n \, d\Gamma \\
&+ \int_\Omega \{ (y - y_d)(-\nabla y \cdot V) + \nabla p \cdot \nabla(-\nabla y \cdot V) + p(-\nabla p \cdot V) \} \, dx \quad (6.75) \\
&+ \int_\Omega \{ \nabla y \cdot \nabla(-\nabla p \cdot V) + y(-\nabla p \cdot V) - f(-\nabla p \cdot V) \} \, dx.
\end{aligned}
$$

We recognize that the second term is equation (6.64) with $\varphi = -\nabla y \cdot V$ and that the third term is equation (6.63) with $\psi = -\nabla p \cdot V$. So they are both zero and finally

$$
dJ(\Omega; V) = \int_\Gamma \left\{ \frac{1}{2}(y - y_d)^2 + \nabla y \cdot \nabla p + yp - fp \right\} V \cdot n \, d\Gamma. \quad (6.76)
$$

It must be emphasized that this last expression has been obtained under the assumption that both y and p belong to $H^2(\Omega)$. We shall see later that the shape gradient can also be obtained by our technique for the finite element approximations of y and p. However, for piecewise linear elements, formula (6.76) fails since the finite element solutions y_h and p_h belong to $H^1(\Omega_h)$ but not to $H^2(\Omega_h)$. However, the domain formula (6.62) will remain true. The crucial point is that for the continuous problem a smooth boundary plus f and y_d in $H^1(\mathbf{R}^N)$ put the solution (y, p) in $H^2(\Omega) \times H^2(\Omega)$. However, the smoothness of the finite element solution (y_h, p_h) cannot be improved.

7 Shape gradients using multipliers and Function Space Embedding

7.1 The non-homogeneous Dirichlet problem

Let Ω be a bounded open domain in $\mathbf{R}^N$ with a sufficiently smooth boundary Γ. Let $y = y(\Omega)$ be the solution of the non-homogeneous Dirichlet problem

$$
-\Delta y = f \quad \text{in } \Omega, \quad (7.1)
$$

$$
y = g \quad \text{on } \Gamma, \quad (7.2)
$$

where f and g are fixed functions in $H^{\frac{1}{2}+\epsilon}(\mathbf{R}^N)$ and $H^{2+\epsilon}(\mathbf{R}^N)$, respectively, for some arbitrary fixed $\epsilon > 0$.

Associate with the solution of (7.1)-(7.2) the cost function

$$
J(\Omega) = \frac{1}{2} \int_\Omega |y(\Omega) - y_d|^2 dx \quad (7.3)
$$

for some fixed function y_d in $H^{\frac{1}{2}+\epsilon}(\mathbf{R}^N)$ and some arbitrary fixed $\epsilon > 0$. We want to compute the derivative of $J(\Omega)$ with respect to Ω subject to the state equation system

(7.1)-(7.2). Our objective is to transform this problem into finding the saddle point of a volume Lagrangian functional. This technique can be applied to other boundary value problems with Dirichlet conditions.

7.2 A saddle point formulation of the state equation

When $g = 0$ problem (7.1)-(7.2) is equivalent to a variational problem on $H_0^1(\Omega)$. When $g \neq 0$ the extra constraint $\phi = g$ makes the Sobolev space dependent on g. To get around this difficulty we introduce a Lagrange multiplier and the new functional

$$L(\phi, \psi, \mu) = \int_\Omega (\Delta\phi + f)\psi \, dx + \int_\Gamma (\phi - g)\mu \, d\Gamma \tag{7.4}$$

for all $\psi \in H^2(\Omega)$ and $\mu \in H^{\frac{1}{2}}(\Gamma)$. This is a convex-concave functional with a unique saddle point $(\hat\phi, \hat\psi, \hat\mu)$ which is completely characterized by the equations

$$\Delta\hat\phi + f = 0 \quad \text{in } \Omega, \tag{7.5}$$

$$\hat\phi - g = 0 \quad \text{in } \Gamma, \tag{7.6}$$

$$\int_\Omega \Delta \phi\hat\psi \, dx + \int_\Gamma \phi\hat\mu \, d\Gamma = 0, \quad \forall\phi \in H^2(\Omega). \tag{7.7}$$

The last equation characterizes $\hat\psi$ and $\hat\mu$:

$$\Delta\hat\psi = 0 \quad \text{in } \Omega, \tag{7.8}$$

$$\hat\psi = 0 \quad \text{on } \Gamma, \tag{7.9}$$

$$\hat\mu = \frac{\partial\hat\psi}{\partial n} \quad \text{on } \Gamma. \tag{7.10}$$

The proof of this can be found in Ekeland-Temam [1, Prop. 1.6]. Of course, this implies that the saddle point is unique and given by

$$(\hat\phi, \hat\psi, \hat\mu) = (y, 0, 0). \tag{7.11}$$

The purpose of the above computation was to find out the form of the multiplier $\hat\mu$

$$\hat\mu = \frac{\partial\hat\psi}{\partial n} \quad \text{on } \Gamma, \tag{7.12}$$

in order to rewrite the previous functional as a function of two variables instead of three:

$$L(\Omega, \phi, \psi) = \int_\Omega (\Delta\phi + f)\psi \, dx + \int_\Gamma (\phi - g)\frac{\partial\psi}{\partial n} \, d\Gamma, \tag{7.13}$$

for (ϕ, ψ) in $H^2(\Omega) \times H^2(\Omega)$. It is also advantageous for shape problems to get rid of boundary integrals whenever it is possible. So noting that

$$\int_\Gamma (\phi - g)\frac{\partial \psi}{\partial n}d\Gamma = \int_\Omega \operatorname{div}[(\phi - g)\nabla \psi]dx, \tag{7.14}$$

we finally use the functional

$$L(\Omega, \phi, \psi) = \int_\Omega \{(\Delta \phi + f)\psi + (\phi - g)\Delta \psi + \nabla(\phi - g) \cdot \nabla \psi\}dx \tag{7.15}$$

on $H^2(\Omega) \times H^2(\Omega)$. It is readily seen that it has a unique saddle point $(\hat{\phi}, \hat{\psi})$ in $H^2(\Omega) \times H^2(\Omega)$ which is completely characterized by the saddle point equations:

$$\Delta\hat{\phi} + f \;\; = \;\; 0 \text{ in } \Omega, \tag{7.16}$$

$$\hat{\phi} \;\; = \;\; g \text{ on } \Gamma, \tag{7.17}$$

$$\Delta\hat{\psi} \;\; = \;\; 0 \text{ in } \Omega, \tag{7.18}$$

$$\hat{\psi} \;\; = \;\; 0 \text{ on } \Gamma. \tag{7.19}$$

7.3   Min Max expression of the cost function

Now repeat the above constructions taking into account the cost function. First introduce the cost functional

$$F(\Omega, \phi) = \frac{1}{2}\int_\Omega |\phi - y_d|^2 dx \tag{7.20}$$

and the new Lagrangian functional

$$G(\Omega, \phi, \psi) = F(\Omega, \phi) + L(\Omega, \phi, \psi).$$

Then it is easy to verify that

$$J(\Omega) = \mathop{\mathrm{Min}}_{\phi \in H^2(\Omega)} \mathop{\mathrm{Max}}_{\psi \in H^2(\Omega)} \; G(\Omega, \phi, \psi). \tag{7.21}$$

The Lagrangian $G(\Omega, \phi, \psi)$ is given by the expression

$$\begin{aligned} G(\Omega, \phi, \psi) \;\; &= \frac{1}{2}\int_\Omega |\phi - y_d|^2 dx \\ &+ \int_\Omega \{(\Delta \phi + f)\psi + (\phi - g)\Delta \psi + \nabla(\phi - g) \cdot \nabla \psi\}dx \end{aligned} \tag{7.22}$$

on $H^2(\Omega) \times H^2(\Omega)$. It is readily seen that it has a unique saddle point $(\hat{\phi}, \hat{\psi})$ which is completely characterized by the following saddle point equations:

$$\Delta\hat{\phi} + f = 0 \text{ in } \Omega, \tag{7.23}$$

$$\hat{\phi} = g \text{ on } \Gamma \tag{7.24}$$

$$\int_\Omega \{(\hat{\phi} - y_d)\phi + \Delta\phi \, \hat{\psi} + \phi \, \Delta\hat{\psi} + \nabla\phi \cdot \nabla\hat{\psi}\}dx = 0, \quad \forall \phi \in H^2(\Omega). \tag{7.25}$$

But the last equation is equivalent to

$$\int_\Omega [(\hat{\phi} - y_d) + \Delta\hat{\psi}]\phi \, dx + \int_\Gamma \frac{\partial\phi}{\partial n}\hat{\psi} \, d\Gamma = 0, \quad \forall\phi \in H^2(\Omega) \tag{7.26}$$

or

$$\Delta\hat{\psi} + (\hat{\phi} - y_d) = 0 \text{ in } \Omega, \tag{7.27}$$

$$\hat{\psi} = 0 \text{ on } \Gamma, \tag{7.28}$$

by using the theorem on the surjectivity of the trace. In the sequel, we shall use the notation (y, p) for the saddle point $(\hat{\phi}, \hat{\psi})$. As a result, we have

$$J(\Omega) = \underset{\phi \in H^2(\Omega)}{\text{Min}} \underset{\psi \in H^2(\Omega)}{\text{Max}} G(\Omega, \phi, \psi). \tag{7.29}$$

We shall now use the above Lagrangian formulation combined with the Velocity method to compute the Shape gradient of $J(\Omega)$. Given a velocity field V in $\mathcal{D}^1(\mathbf{R}^N, \mathbf{R}^N)$ and the parametrized domains $\Omega_t = T_t(\Omega)$,

$$J(\Omega_t) = \underset{\phi \in H^2(\Omega_t)}{\text{Min}} \underset{\psi \in H^2(\Omega_t)}{\text{Max}} G(\Omega_t, \phi, \psi). \tag{7.30}$$

There are two methods to get rid of the time dependence in the underlying function spaces:

- the *Function Space Parametrization*

- the *Function Space Embedding*.

In the first case, we parametrize the functions in $H^2(\Omega_t)$ by elements of $H^2(\Omega)$ through the transformation

$$\phi \mapsto \phi \circ T_t^{-1} = H^2(\Omega) \to H^2(\Omega_t), \tag{7.31}$$

where "$\circ$" denotes the composition of the two maps and we introduce the *Parametrized Lagrangian*,

$$\tilde{G}(t, \phi, \psi) = G(T_t(\Omega), \phi \circ T_t, \psi \circ T_t^{-1}) \tag{7.32}$$

on $H^2(\Omega) \times H^2(\Omega)$. In the Function Space Embedding Method, we introduce a large enough domain, D, which contains all the transformations $\{\Omega_t : 0 \le t \le \bar{t}\}$ of Ω for some small $\bar{t} > 0$.

In this section, we use the Function Space Embedding Method with $D = \mathbf{R}^N$ and

$$J(\Omega_t) = \underset{\Phi \in H^2(\mathbf{R}^N)}{\text{Min}} \underset{\Psi \in H^2(\mathbf{R}^N)}{\text{Max}} G(\Omega_t, \Phi, \Psi). \tag{7.33}$$

As can be expected, the price to pay for the use of this method, is the fact that the set of saddle points

$$S(t) = X(t) \times Y(t) \subset H^2(\mathbf{R}^N) \times H^2(\mathbf{R}^N) \tag{7.34}$$

is not a singleton any more since

$$X(t) = \{\Phi \in H^2(\mathbf{R}^N) : \Phi|_{\Omega_t} = y_t\} \tag{7.35}$$

$$Y(t) = \{\Psi \in H^2(\mathbf{R}^N) : \Psi|_{\Omega_t} = p_t\} \tag{7.36}$$

where (y_t, p_t) is the unique solution in $H^2(\Omega_t) \times H^2(\Omega_t)$ to the previous saddle point equations on Ω_t,

$$\Delta y_t + f = 0 \text{ in } \Omega_t, \quad y_t = g \text{ in } \Gamma_t, \tag{7.37}$$

$$\Delta p_t + (y_t - y_d) = 0 \text{ in } \Omega_t, \quad p_t = 0 \text{ on } \Gamma_t. \tag{7.38}$$

We are now ready to apply the theorem of Correa-Seeger [1] which says that under appropriate hypotheses (to be checked in the next section)

$$dJ(\Omega; V) = \underset{\Phi \in X(0)}{\text{Min}} \underset{\Psi \in Y(0)}{\text{Max}} \partial_t G(\Omega_t, \Phi, \Psi). \tag{7.39}$$

Since we have already characterized $X(0)$ and $Y(0)$, we only need to compute the partial derivative of

$$G(\Omega_t, \Phi, \Psi) = \int_{\Omega_t} \{\frac{1}{2}|\Phi - y_d|^2 + (\Delta\Phi + f)\Psi + (\Phi - g)\Delta\Psi + \nabla(\Phi - g) \cdot \nabla\Psi\}dx. \tag{7.40}$$

If we assume that Ω_t is sufficiently smooth, then

$$f, y_d \in H^{\frac{1}{2}+\epsilon}(\mathbf{R}^N) \text{ and } g \in H^{2+\epsilon}(\mathbf{R}^N) \Rightarrow y \text{ and } p \in H^{\frac{5}{2}+\epsilon}(\Omega) \tag{7.41}$$

and we can choose to consider our saddle points $S(t)$ in $H^{\frac{5}{2}+\epsilon}(\mathbf{R}^N) \times H^{\frac{5}{2}+\epsilon}(\mathbf{R}^N)$ rather than $H^2(\mathbf{R}^N) \times H^2(\mathbf{R}^N)$. If Φ and Ψ belong to $H^{\frac{5}{2}+\epsilon}(\mathbf{R}^N)$, then

$$\partial_t G(\Omega_t, \Phi, \Psi) = \int_{\Gamma_t} \{\frac{1}{2}(\Phi - y_d)^2 + (\Delta\Phi + f)\Psi + (\Phi - g)\Delta\Psi + \nabla(\Phi - g) \cdot \nabla\Psi\}V \cdot n_t d\Gamma_t. \tag{7.42}$$

This expression is an integral over the boundary Γ which will not depend on Φ and Ψ outside of $\bar{\Omega}$. As a result the Min and the Max can be dropped in expression (7.39) which reduces to

$$dJ(\Omega; V) = \int_{\Gamma} \{\frac{1}{2}(y - y_d)^2 + (\Delta y + f)p + (y - g)\Delta p + \nabla(y - g) \cdot \nabla p\}V \cdot n \, d\Gamma. \tag{7.43}$$

But

$$p = 0 \text{ and } y - g = 0 \Rightarrow \nabla p = \frac{\partial p}{\partial n}n, \quad \nabla(y - g) = \frac{\partial}{\partial n}(y - g)n \text{ on } \Gamma \tag{7.44}$$

and finally

$$dJ(\Omega; V) = \int_{\Gamma} \{\frac{1}{2}(g - y_d)^2 + \frac{\partial}{\partial n}(y - g)\frac{\partial p}{\partial n}\}V \cdot n \, d\Gamma. \tag{7.45}$$

7.4   Verification of the hypotheses of Theorem 6.1

As we have seen the computation of the Shape gradient is both quick and easy. We now turn to the step by step verification of the hypotheses of the Theorem 6.1. Many of the constructions given below are "canonical" and can be repeated for different problems in different contexts.

Let y_d and $f \in H^1(\mathbf{R}^N)$ and $g \in H^{\frac{5}{2}}(\mathbf{R}^N)$ so that

$$X = Y = H^3(\mathbf{R}^N). \tag{7.46}$$

The saddle points $S(t) = X(t) \times Y(t)$ are given by

$$X(t) = \{\Phi \in X : \Phi|_{\Omega_t} = y_t\} \tag{7.47}$$

$$Y(t) = \{\Psi \in Y : \Psi|_{\Omega_t} = p_t\}. \tag{7.48}$$

The sets $X(t)$ and $Y(t)$ are not empty since it is always possible to construct a continuous linear extension

$$\Pi^m : H^m(\Omega) \to H^m(\mathbf{R}^N) \tag{7.49}$$

for each $m \geq 1$. For instance with $m = 1$ and a boundary Γ which is $W^{1,\infty}$, see Agmon-Douglis-Nirenberg [1,2] and for $m > 1$ V.M. Babić [1] (cf. also J. Nečas [1]). Using this Π^m, we then define the following extension

$$\Pi_t^m : H^m(\Omega_t) \to H^m(\mathbf{R}^N) \tag{7.50}$$

$$\Pi_t^m(\phi) = [\Pi^m(\phi \circ T_t)] \circ T_t^{-1}. \tag{7.51}$$

In the sequel m is fixed and equal to 3, so we shall drop the superscript m and define the extensions

$$Y_t = \Pi_t y_t, \quad P_t = \Pi_t p_t \tag{7.52}$$

of y_t and p_t, respectively. Hence,

$$Y_t \in X(t) \text{ and } P_t \in Y(t) \Rightarrow S(t) \neq \emptyset. \tag{7.53}$$

So condition (H1) is verified. Condition (H2) follows from the hypotheses on f, y_d and g. To check conditions (H3) and (H4), we need two general theorems which can be used in various contexts and problems.

Theorem 7.1   *For $V \in \mathcal{D}^1(\mathbf{R}^N, \mathbf{R}^N)$ and $\Phi \in L^2(\mathbf{R}^N)$,*

$$\lim_{t \searrow 0} \Phi \circ T_t = \Phi \text{ and } \lim_{t \searrow 0} \Phi \circ T_t^{-1} = \Phi \text{ in } L^2(\mathbf{R}^N). \tag{7.54}$$

Proof   (i) The space $\mathcal{D}(\mathbf{R}^N)$ of continuous functions with compact support in $\mathbf{R}^N$ is dense in $L^2(\mathbf{R}^N)$. So given $\epsilon > 0$, there exists Φ_ϵ in $\mathcal{D}(\mathbf{R}^N)$ such that

$$\|\Phi - \Phi_\epsilon\|_{L^2}^2 < \frac{\epsilon^2}{\max\{J_t^{-1} : 0 \leq t \leq \tau\}}.$$

Hence,

$$||\Phi \circ T_t - \Phi|| \le ||\Phi_\epsilon \circ T_t - \Phi_\epsilon|| + ||\Phi \circ T_t - \Phi_\epsilon \circ T_t|| + ||\Phi - \Phi_\epsilon||. \qquad (7.55)$$

But

$$\forall t \in [0,\tau], \quad \int_{\mathbf{R}^N} |\Phi \circ T_t - \Phi_\epsilon \circ T_t|^2 dx = \int_{\mathbf{R}^N} |\Phi - \Phi_\epsilon|^2 J_t^{-1} dx \le \epsilon^2.$$

So the last two terms in (7.55) are less than 2ϵ. It remains to evaluate the first term for a fixed function Φ_ϵ with compact support K in $\mathbf{R}^N$. Recall that, since $\Phi_\epsilon = 0$ on the boundary ∂K of K, $T_t(K) = K$ for all t in $[0,\tau]$ (use N. Nagumo [1]'s theorem twice as in the proof of Theorem 2.4(i)). Moreover, by compactness of K, Φ_ϵ is uniformly continuous on $\mathbf{R}^N$ and

$$\exists \delta > 0, \forall x, y \in \mathbf{R}^N, \quad |x - y| < \delta \implies |\Phi_\epsilon(y) - \Phi_\epsilon(x)| < \frac{\epsilon}{m(K)^{\frac{1}{2}}}.$$

But, T_t is also uniformly continuous on K and

$$\exists \eta > 0, \forall t, \; 0 \le t < \eta, \; \forall x \in K, \; |T_t x - x| < \delta.$$

By construction,

$$\text{supp}(\Phi_\epsilon \circ T_t) = T_t(\text{supp } \Phi_\epsilon) \subset K$$

and

$$\Phi_\epsilon = 0 \text{ and } \Phi_\epsilon \circ T_t = 0 \text{ outside of } K.$$

Finally,

$$\int_{\mathbf{R}^N} |\Phi_\epsilon(T_t x) - \Phi_\epsilon(x)|^2 dx = \int_K |\Phi_\epsilon(T_t x) - \Phi_\epsilon(x)|^2 dx \le \epsilon^2$$

and this implies that

$$\forall \epsilon > 0, \; \exists \eta > 0, \; \forall 0 \le t \le \eta, \; ||\Phi \circ T_t - \Phi||_{L^2(\mathbf{R}^N)} \le 3\epsilon.$$

(ii) For the second part of (7.54) we make a change of variable and use the result of part (i)

$$\int_{\mathbf{R}^N} |\Phi \circ T_t^{-1} - \Phi|^2 dx = \int_{\mathbf{R}^N} |\Phi - \Phi \circ T_t|^2 J_t dx \le \epsilon^2.$$

This completes the proof.   □

Corollary   *Under the assumptions of Theorem 4.1 for $m \ge 1$, V in $\mathcal{D}^m(\mathbf{R}^N, \mathbf{R}^N)$ and $\Phi \in H^m(\mathbf{R}^N)$,*

$$\lim_{t \searrow 0} \Phi \circ T_t = \Phi \quad and \quad \lim_{t \searrow 0} \Phi \circ T_t^{-1} = \Phi \quad in \quad H^m(\mathbf{R}^N). \qquad (7.56)$$

Remark 7.1   In fact, for $m \ge 1$ and $V \in \mathcal{D}^m(\mathbf{R}^N, \mathbf{R}^N)$ the transformation

$$S(t)\Phi = \Phi \circ T_t, \quad \forall \Phi \in H^m(\mathbf{R}^N), \; \forall t, \; 0 \le t \le \tau, \qquad (7.57)$$

defines a strongly continuous semigroup of class C_0 on $H^m(\mathbf{R}^N)$ with infinitesimal generator

$$\mathcal{A}\Phi = \nabla\Phi \cdot V, \ \mathcal{D}(\mathcal{A}) = \{\Phi \in H^m(\mathbf{R}^N) : \nabla\Phi \cdot V \in H^m(\mathbf{R}^N)\}.$$

$\square$

Theorem 7.2 *Under the assumptions of Theorem 4.1,*

$$y^t \to y^0 \ in \ H^m(\Omega) - strong \ (resp. \ weak) \tag{7.58}$$

implies that

$$Y_t \to Y_0 \ in \ H^m(\mathbf{R}^N) - strong \ (resp. \ weak).$$

Proof The strong case is obvious. We prove the weak case for $m = 0$. By definition,

$$Y_t = (\Pi y^t) \circ T_t^{-1}$$

and for all Φ in $L^2(\mathbf{R}^N)$, we consider

$$\int_{\mathbf{R}^N} Y_t \Phi \, dx = \int_{\mathbf{R}^N} (\Pi y^t) \circ T_t^{-1} \Phi \, dx = \int_{\mathbf{R}^N} \Pi y^t \Phi \circ T_t J_t \, dx.$$

We have shown in Theorem 4.1 that

$$\Phi \circ T_t \to \Phi \ in \ L^2(\mathbf{R}^N) \ strong.$$

In addition, $J_t \to 1$ and by linearity and continuity of Π,

$$\Pi y^t \to \Pi y \ in \ L^2(\mathbf{R}^N) \ weak.$$

Hence,

$$\forall \Phi \in L^2(\mathbf{R}^N), \ \int_{\mathbf{R}^N} Y_t \, \Phi \, dx \to \int_{\mathbf{R}^N} \Pi y \, \Phi \, dx = \int_{\mathbf{R}^N} Y_0 \, \Phi \, dx.$$

This proves the weak convergence. $\square$

To verify condition (H3), we transform (y_t, p_t) on Ω_t to $(y^t, p^t) = (y_t \circ T_t, p_t \circ T_t)$ on Ω. The pair (y^t, p^t) is the transported pair of solutions from Ω_t to Ω. It is the unique solution in $H^1(\Omega) \times H^1(\Omega)$ of the system

$$-\text{div}[A(t)\nabla y^t] = J_t f \circ T_t \ in \ \Omega, \ y^t = g \circ T_t \ on \ \Gamma, \tag{7.59}$$

$$-\text{div}[A(t)\nabla p^t] = J_t(y^t - y_d \circ T_t) \ in \ \Omega, \ p^t = 0 \ on \ \Gamma, \tag{7.60}$$

where

$$A(t) = J_t[DT_t]^{-1} {}^*[DT_t]^{-1}, \quad J_t = |\det DT_t|, \tag{7.61}$$

DT_t is the Jacobian matrix of T_t and ${}^*[DT_t]^{-1}$ is the transpose of $[DT_t]^{-1}$.

For sufficiently smooth domains Ω and vector fields V, the pair $\{y^t, p^t\}$ is bounded in $H^1(\Omega) \times H^1(\Omega)$ as t goes to zero. Since $H^1(\Omega)$ is a Hilbert space, we can extract weakly convergent subsequences to some $(\bar{y}, \bar{p})$ in $H^1(\Omega) \times H^1(\Omega)$. However, by linearity of the equation with respect to (y^t, p^t) and continuity of the coefficients with respect to t, the limit point $(\bar{y}, \bar{p})$ will coincide with (y^0, p^0), since the system has a unique solution at $t = 0$. Then we go back to the equation for y^t and y and show that the convergence is strong in $H^1(\Omega)$. Finally by using the regularity of the data and the classical regularity theorems we show that $(y^t, p^t) \rightarrow (y, p)$ in $H^3(\Omega) \times H^3(\Omega)$.

For the verification of condition (H4), we go back to expression (7.42) which can be rewritten as a volume integral

$$\partial_t G(\Omega_t, \Phi, \Psi) = \int_{\Omega_t} \operatorname{div}\{[\frac{1}{2}(\Phi - y_d)^2 + (\Delta\Phi + f)\Psi + (\Phi - g)\Delta\Psi + \nabla(\Phi - g)\cdot\nabla\Psi]V\}dx \quad (7.62)$$

for $(\Phi, \Psi) \in H^3(\mathbf{R}^N) \times H^3(\mathbf{R}^N)$. Now introduce the map

$$\begin{cases} (\Phi, \Psi) \mapsto F(\Phi, \Psi) = [\frac{1}{2}(\Phi - y_d)^2 + (\Delta\Phi + f)\Psi + (\Phi - g)\Delta\Psi + \nabla(\Phi - g)\cdot\nabla\Psi]V \\ : H^3(\mathbf{R}^N) \times H^3(\mathbf{R}^N) \rightarrow (H^1(\mathbf{R}^N))^N. \end{cases}$$

It is bilinear and continuous. Finally, the map

$$(t, F) \mapsto \int_{\Gamma_t} F \circ n_t \, d\Gamma = \int_{\Omega_t} F \, dx = \int_\Omega (\operatorname{div} F) \circ T_t J_t^{-1} dx \; : [0, \tau] \times H^1(\mathbf{R}^N) \rightarrow \mathbf{R} \quad (7.63)$$

is continuous. Then

$$(t, \Phi, \Psi) \mapsto \partial_t G(\Omega_t, \Phi, \Psi) = \int_{\Gamma_t} F(\Phi, \Psi) \cdot n_t \, d\Gamma_t \qquad (7.64)$$

is continuous and condition (H4) is verified. This completes the verification of the four conditions of Theorem 6.1.

7.5   Shape Hessian

For the interested reader detailed computations of the shape Hessian for this problem can be found in Delfour-Zolésio [12], [13].

References

S. Agmon, A. Douglis and L. Nirenberg

[1] Estimates near the boundary for solutions of elliptic partial differential equations satisfying general boundary conditions, I., *Comm. Pure Appl. Math.* **12** (1959), 623–727.

[2] Estimates near the boundary for solutions of elliptic partial differential equations satisfying general boundary conditions, II., *Comm. Pure Appl. Math.* **17** (1964), 35–92.

G. Arumugam and O. Pironneau

[1] On the problems of riblets as a drag reduction device, *Optimal Control Appl. Methods* **10** (1989), 93–112.

[2] Sur le problème des "riblets", Rapport de recherche R87027, Publications du Laboratoire d'analyse numérique, Université Pierre et Marie Curie, Paris, 1987.

J.P. Aubin and A. Cellina

[1] *Differential Inclusions*, Springer-Verlag, Berlin, 1984.

J.P. Aubin and H. Frankowska

[1] *Set-Valued Analysis*, Birkhäuser, Basel, Berlin, 1990.

V.M. Babič

[1] Sur le prolongement des fonctions, *Uspekhi Mat. Nauk* **8** (1953), 111–113 (in Russian).

A. Bern

[1] Contribution à la modélisation par éléments finis des surfaces libres en régime permanent, Thesis, École Nationale Supérieure des Mines de Paris, CEMEF, Sophia Antipolis, 1987.

A. Bern, J.L. Chenot, Y. Demay and J.P. Zolésio

[1] Numerical computation of the free boundary in non-Newtonain stationary flows, in: *Proc. Sixth Internat. Symp. on Finite Element Methods in Flow Problems (June 1986)*, Publications INRIA, Rocquencourt, 383–390.

P. Cannarsa and H.M. Soner

[1] On the singularities of the viscosity solutions to Hamilton-Jacobi-Bellman equations, *Indiana Univ. Math. J.* **36** (1987), 501–524.

J. Céa

[1] Problems of shape optimal design, in: *Optimization of Distributed Parameter Structures*, vol. II (E.J. Haug and J. Céa, eds.), Sijthoff and Noordhoff, Alphen aan den Rijn, 1981, 1005–1048.

[2] Numerical methods of shape optimal design, in: *Optimization of Distributed Parameter Structures*, vol. II (E.J. Haug and J. Céa, eds.), Sijthoff and Noordhoff, Alphen aan den Rijn, 1981, 1049–1087.

[3] Conception optimale ou identification de formes: calcul rapide de la dérivée directionnelle de la fonction coût, *RAIRO Modél. Math. Anal. Numér.* **20** (1986), 371–402.

K.-T. Cheng and N. Olhoff

[1] Regularized formulation for optimal design of axisymmetric plates, *Internat. J. Solids and Structures* **18** (1982), 153–169.

R. Correa and A. Seeger

[1] Directional derivatives of a minimax function, *Nonlinear Anal.* **9** (1985), 13–22.

M.C. Delfour and J. Morgan

[1] Differentiability of Min Max and saddle points under relaxed assumptions, in: *Boundary Control and Boundary Variations* (J.-P. Zolésio, ed.), Springer-Verlag, Berlin, Heidelberg, 1991, 1–12.

[2] A complement to the differentiability of saddle points and Min Max, *Optimal Control and Applications* (to appear).

[3] Derivative of min max and saddle points with respect to a parameter, Report no. 1784, Centre de recherches mathématiques, Université de Montréal, October 1991.

M.C. Delfour, G. Payre and J.P. Zolésio

[1] Shape optimal design of a radiating fin, in: *System Modelling and Optimization* (P. Thoft-Christensen, ed.), Springer-Verlag, Berlin, Heidelberg, 1984, 810–818.

[2] An optimal triangulation for second order elliptic problems, *Comput. Methods Appl. Mech. Engrg.* **50** (1985), 231–261.

M.C. Delfour and J.P. Zolésio

[1] Dérivation d'un MinMax et application à la dérivation par rapport au contrôle d'une observation non différentiable de l'état, *C.R. Acad. Sci. Paris Sér. I Math.* **302** (1986), 571–574.

[2] Shape sensitivity analysis via MinMax differentiability, *SIAM J. Control Optim.* **26** (1988), 834–862.

[3] Differentiability of a MinMax and application to optimal control and design problems, Part I, in: *Control Problems for Systems Described as Partial Differential Equations and Applications* (I. Lasiecka and R. Triggiani, eds.), Springer-Verlag, New York, 1987, 204–219.

[4] Differentiability of a MinMax and application to optimal control and design problems, Part II, in: *Control Problems for Systems Described as Partial Differential Equations and Applications* (I. Lasiecka and R. Triggiani, eds.), Springer-Verlag, New York, 1987, 220–229.

[5] Further developments in shape sensitivity analysis via a penalization method, in: *Boundary Control and Boundary Variations* (J.P. Zolésio, ed.), Springer-Verlag, Berlin, Heidelberg, New York, Tokyo, 1988, 153–191.

[6] Shape sensitivity analysis via a penalization method, *Ann. Mat. Pura Appl. (4)* **151** (1988), 179–212.

[7] Analyse des problèmes de forme par la dérivation des Min Max, in: *Analyse non linéaire* (H. Attouch et al., eds.), Série Analyse Non Linéaire, *Ann. Inst. H. Poincaré*, Special volume in honor of J.-J. Moreau, Gauthier-Villars, Bordas, Paris, 1989, 211–228.

[8] Anatomy of the shape Hessian, *Ann. Mat. Pura Appl. (4)* **158** (1991), 315–339.

[9] Computation of the shape Hessian by a Lagrangian method, in: *Fifth Symp. on Control of Distributed Parameter Systems* (A. El Jai and M. Amouroux, eds.), Pergamon Press, Oxford, New York, 1989, 85–90.

[10] Shape Hessian by the Velocity Method: a Lagrangian approach, in: *Stabilization of Flexible Structures* (J.P. Zolésio, ed.), Springer Verlag, Berlin, Heidelberg, New York, 1990, 255–279.

[11] Shape derivatives for non-smooth domains, in: *Optimal Control of Partial Differential Equations* (K.-H. Hoffmann and W. Krabs, eds.), Springer Verlag, Berlin, Heidelberg, New York, 1991, 38–55.

[12] Velocity Method and Lagrangian formulation for the computation of the shape Hessian, *SIAM J. Control Optim.* **29** (1991), 1414–1442.

[13] Structure of shape derivatives for nonsmooth domains, *J. Funct. Anal.* **104** (1992), 1–33.

[14] Adjoint state in the control of variational inequalities, in: *Proc. Free Boundary Conference (Montreal 1990)* (J. Chadam and H. Rasmussen, eds.), Longman-Pitman, (to appear).

J. Dugundji

[1] *Topology*, Allyn and Bacon, Boston, 1966.

I. Ekeland and R. Temam

[1] *Analyse convexe et problèmes variationnels*, Dunod, Paris, 1974.

M. Fortin and R. Glowinski

[1] *Augmented Lagrangian Methods: Applications to the Numerical Solution of Boundary-Value Problems*, North-Holland, Amsterdam, 1983.

N. Fujii

[1] Domain optimization problems with a boundary value problem as a constraint, in: *Control of Distributed Parameter Systems*, Pergamon Press, Oxford, New York, 1986, 5–9.

[2] Second variation and its application in a domain optimization problem, in: *Control of Distributed Parameter Systems*, Pergamon Press, Oxford, New York, 1986, 431–436.

M. Guelfand and N.Y. Vilenkin

[1] *Les distributions, Applications de l'analyse harmonique* (French translation by G. Rideau), Dunod, Paris, 1967.

J. Hadamard

[1] Mémoire sur le problème d'analyse relatif à l'équilibre des plaques élastiques encastrées, in: *Œuvres de J. Hadamard*, vol. II (original reference: *Mém. Sav. Etrang.* **33** (1907), mémoire couronné par l'Académie des Sciences), C.N.R.S., Paris, 1968, 515–641.

P. Laurence

[1] Une nouvelle formule pour dériver par rapport aux lignes de niveau et applications, *C.R. Acad. Sci. Paris Sér. I Math.* **306** (1988), 455–460.

F. Murat and J. Simon

[1] Sur le contrôle par un domaine géométrique, Rapport 76015, Université Pierre et Marie Curie, Paris, 1976.

M. Nagumo

[1] Über die Lage der Integralkurven gewöhnlicher Differentialgleichungen, *Proc. Phys. Math. Soc. Japan* **24** (1942), 551–559.

J. Nečas

[1] *Les méthodes directes en théorie des équations elliptiques*, Masson (Paris) et Academia (Praha), 1967.

O. Pironneau

[1] *Optimal Design for Elliptic Systems*, Springer-Verlag, New York, 1984.

L. Schwartz

[1] *Théorie des distributions*, Hermann, Paris, 1966.

[2] Théorie des noyaux, in: *Proc. Internat. Congress of Mathematicians*, vol. I, 1950, 220–230.

J. Simon

[1] Second variations for domain optimization problems, in: *Control of Distributed Parameter Systems (Proc. 4th Internat. Conf. in Vorau)*, Birkhäuser Verlag, (to appear).

J. Sokolowski and J.P. Zolésio

[1] *Introduction to Shape Optimization: Shape Sensitivity Analysis*, Series in Computational Mathematics, vol. 16, Springer-Verlag, Berlin, Heidelberg, New York, 1992.

J.P. Zolésio

[1] Identification de domaines par la déformation, Thèse de doctorat d'état, Université de Nice, 1979.

[2] The material derivative (or speed) method for shape optimization, in: *Optimization of Distributed Parameter Structures*, vol. II (E.J. Haug and J. Céa, eds.), Sijthoff and Nordhoff, Alphen aan den Rijn, 1981, 1089–1151.

[3] Domain variational formulation for free boundary problems, in: *Optimization of Distributed Parameter Structures*, vol. II (E.J. Haug and J. Céa, eds.), Sijthoff and Nordhoff, Alphen aan den Rijn, 1981, 1152–1194.

Some Free Boundary Problems
With Industrial Applications

Antonio FASANO

Dipartimento di Matematica U. Dini
Università di Firenze
Viale Morgagni 67a
I-50134 Firenze
Italy

Abstract

These lecture notes deal with two topics each one representing a free boundary problem for a partial differential equation, arising from the analysis of interesting phenomena in different areas. The first problem models the steady states of the so-called electrochemical machining process (governed by the Laplace equation), and is characterized by free boundary conditions of Cauchy type. The second subject is a Bingham flow in a 1-D geometry; the p.d.e. is parabolic and the free boundary conditions are of nonstandard type.

Preface

First of all I wish to thank Professor Daigneault and Professor Delfour for their kind invitation to lecture at the 29th Session of the *Séminaire de mathématiques supérieures* and for the excellent way it was organized.

I have chosen two topics very different from each other, in order to give an idea of the diversity of the subjects one can find in the vast area of free boundary problems.

The first one is a steady state problem, apparently the simplest free boundary problem one can formulate for the Laplace equation, which happens to have interesting links with the other "soul" of the Session, i.e., shape optimization. The problem is introduced as the mathematical model of the process of electrochemical machining of metals with a threshold current, but it also models other processes in different areas.

It seemed to me a splendid opportunity to present a theory developed in an old paper (A. Beurling, 1957), which, far from being obsolete, represents – still today – a rare instance of a treatment of classical solutions of a free boundary problem in more than one space dimension. This part of the course is complemented by some more recent material on the same subject.

113

M. C. Delfour and G. Sabidussi (eds.), Shape Optimization and Free Boundaries, 113–142.
© 1992 *Kluwer Academic Publishers.*

The second subject is an evolution problem in the dynamics of Bingham fluids. Confining ourselves to the one-dimensional case, we shall deal with a free boundary problem for the heat equation, which, although not resembling any of the usual parabolic free boundary problems, is somehow related to the Stefan problem.

I like to thank Dr. E. Comparini for her help in preparing these notes.

I have greatly enjoyed teaching at the *Séminaire* and attending the lectures of the Colleagues.

A. Fasano

Part I

A free boundary problem for the Laplace equation

1 The electrochemical machining problem with a threshold current

Electrochemical machining (ECM) is a very efficient method for shaping metals. It is based on an electrochemical process in which the workpiece and the tool are the anode and the cathode, respectively. The machining of the anode occurs as a consequence of the current flow through the electrolyte, which is continuously replaced, so that the material coming from the anode is removed from the system. Fig. 1.1 is a schematic representation of a 2-D geometry (the electrolytic drill).

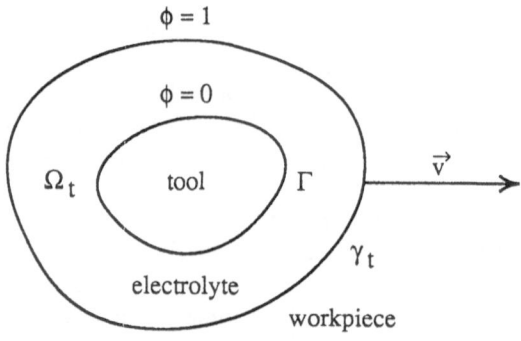

Figure 1.1

The machining rate of the anode is a function of the local value of the electric field. Therefore, denoting by γ_t the anode surface at time t, and by ϕ the electric potential in the electrolyte, we have some empirical relationship between the normal velocity of the points of γ_t and the normal derivative of ϕ:

$$v_n = f\left(\frac{\partial \phi}{\partial n}\right) \qquad \text{on } \gamma_t$$

($\dfrac{\partial}{\partial n}$ denoting the outer normal derivative).

In most practical cases the function f is characterized by the presence of a threshold. In other words, no machining occurs when $\left|\dfrac{\partial \phi}{\partial n}\right|$ on γ_t is below some value $\lambda > 0$ (in that case $f(s) \equiv 0$ for $s \le \lambda$ and is increasing for $s > \lambda$).

In the following we shall use nondimensional variables, so that $\phi = 1$ on $\gamma_t, \phi = 0$ on the cathode surface Γ. Neglecting charge effects, we assume ϕ to be harmonic in the region Ω occupied by the electrolyte.

At this point the evolution problem can be stated as follows.

Given the cathode surface Γ and the initial configuration γ_0 of the anode surface, find the pair (γ_t, ϕ) such that

$$\Delta \phi = 0 \quad \text{in } \Omega_t, \tag{1}$$
$$\phi|_\Gamma = 0, \ t > 0, \tag{2}$$
$$\phi|_{\gamma_t} = 1, \ t > 0, \tag{3}$$
$$v_n = f(\frac{\partial \phi}{\partial n}|_{\gamma_t}), \ t > 0, \tag{4}$$
$$\gamma_{t=0} = \gamma_0, \tag{5}$$

Ω_t being the domain bounded by Γ and γ_t.

In the scheme above it is understood that ϕ possesses enough regularity, so that all equations hold in the classical sense.

Remark 1.1 In the general case also Γ can be a moving surface, but its motion is prescribed.

Remark 1.2 If $f(\frac{\partial \phi}{\partial n}) = k\frac{\partial \phi}{\partial n}$ and if $S(x,t) = 0$ is the equation of the free boundary, then from $\nabla S.\vec{v} + \frac{\partial S}{\partial t} = 0$ and $\vec{v}.\vec{n} = k\nabla\phi.\vec{n}$, since $\vec{n} = \nabla S/|\nabla S|$ the free boundary condition takes the form

$$k\nabla S.\nabla\phi + \frac{\partial S}{\partial t} = 0 \tag{4'}$$

(or if $S = t - \tau(x), \nabla\tau.\nabla\phi = 1$). Hence (1)–(4) is nothing but a one phase Stefan problem with zero specific heat. Thus there is a strong analogy with the Hele-Shaw problem and this explains the possibility of reducing the problem to an elliptic variational inequality to be satisfied for each t. This point of view has been developed in [Elli 80], [Elli 83]. For more information on the Stefan and Hele-Shaw problems see e.g. [ElOc 82], [Fasa 89].

In the case considered in the Remark above there cannot be any nontrivial equilibrium. Here we are interested in the case in which the function $f(s)$ has a threshold $\lambda > 0$ and in the associated steady state problem, in which the electric field on the anode equals the threshold value λ. Such a problem consists in finding a pair (γ, ϕ) satisfying the equations

$$\Delta\phi = 0 \quad \text{in} \quad \Omega, \tag{6}$$
$$\phi|_\Gamma = 0, \tag{7}$$
$$\phi|_\gamma = 1, \tag{8}$$
$$\frac{\partial\phi}{\partial n}\bigg|_\gamma = \lambda. \tag{9}$$

It is reasonable to expect that (6)–(9) is the asymptotic form of the evolution problem for $t \to \infty$.

The literature on the ECM or related problems is quite large. Let us mention the book [Mcge 74] and the papers [McRa 74], [Mack 86], [Mcge 85], [Roge 80], besides the ones which will be explicitly considered in the sequel.

Here we want to concentrate mainly on a generalization of the steady state problem (6)–(9) in two dimensions, referring to some theorems from a fundamental paper [Beur 57] about existence and uniqueness, as well as some more specific results about the case of a convex cathode [Tepp 74], [Tepp 75]. Our exposition follows in part the pattern of [Fasa 89, Part II].

The variational formulation of the same problem (fully investigated in [AlCa 81]) will be described shortly in the last section.

The evolution problem with threshold current is open. Some a-priori results on the asymptotic behaviour of (possibly existing) solutions have been discussed in [LaSh 87].

Remark 1.3 It is interesting to remark that problem (6)–(9) can receive different physical interpretations (see e.g. [Carl 18], [Frie 34], [Acke 77]). For instance the same scheme can be used in order to solve the following problem: let Ω be a heat conducting medium, whose temperature is prescribed as a constant on the boundaries Γ and γ; determine γ so that the heat flux through γ is exactly λ.

The next section will be devoted to the study of some very simple case.

Throughout the paper we use the abbreviations f.b.(p.) [c.] for free boundary (problem) [condition].

2   Some simple solution

Let us consider the evolution problem with no threshold current in two space dimensions with

$$f(\frac{\partial \phi}{\partial n}) = k\frac{\partial \phi}{\partial n}.$$

If $y = a(x,t)$, $y = c(x,t)$ are the equations of the anode and of the cathode, respectively, and in (4') we set $S(x,y,t) = y - a(x,t)$, we derive the f.b.c.

$$\frac{\partial a}{\partial t} + k\frac{\partial a}{\partial x}\frac{\partial \phi}{\partial x} = k\frac{\partial \phi}{\partial y} \qquad \text{on}\ \ y = a(x,t). \tag{10}$$

Of course the function $a(x,t)$ is unknown, while $c(x,t)$ is prescribed.

We can obtain a simple solution e.g. in the case of *planar electrodes*: $y = c(t)$, $y = a(t)$. In this case the potential ϕ is a linear function of y:

$$\phi(y,t) = \frac{y - c(t)}{a(t) - c(t)} \tag{11}$$

and (10) becomes

$$[a(t) - c(t)]\dot{a}(t) = k, \tag{12}$$

i.e., an o.d.e. for $a(t)$, to be integrated starting for some given initial value $a(0)$.

In the situation of Fig. 2.1 we have $\dot{c} = -V < 0$, and introducing the unknown $s(t) = c(t) - a(t) > 0$, (12) takes the form

$$\dot{s}(t) = \frac{k}{s(t)} - V, \qquad s(0) = s_0, \tag{13}$$

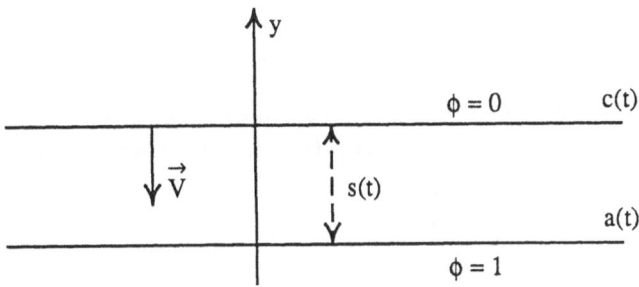

Figure 2.1

whose solution is given implicitly by

$$- s(t) + \frac{k}{V} \log \frac{(k/V) - s_0}{(k/V) - s(t)} = Vt. \tag{14}$$

The stationary solution $s^* = k/V$ gives the asymptotic distance between the electrodes.

In the case $f(\frac{\partial \phi}{\partial n}) = [k \frac{\partial \phi}{\partial n} - \lambda]_+$ the f.b.c. changes to

$$\frac{\partial a}{\partial t} + \left(k \frac{\partial \phi}{\partial x} \frac{\partial a}{\partial x} - k \frac{\partial \phi}{\partial y} - \lambda \sqrt{1 + (\frac{\partial a}{\partial x})^2} \right)_+ = 0. \tag{15}$$

In the one-dimensional case (11) is still valid and the f.b.c. is

$$\dot{a} + \left(\frac{k}{c - a} - \lambda \right)_+ = 0,$$

implying that $a =$ constant as long as $s \geq s_\lambda = k/\lambda$. If $s < s_\lambda$ the anode moves. When $\dot{c} = -V < 0$ and $s(0) < s_\lambda$, the differential equation satisfied by $s(t)$ is obtained replacing V with $V + \lambda$ in (13). The same substitution in (14) gives the solution, whose asymptotic value is now $s_\infty = \frac{k}{V + \lambda} < s_\lambda$.

Notice that s_∞ reduces to s_λ when $V = 0$.

For more information on special solutions see [Mack 86] ($\lambda = 0$) and [LaSh 87] ($\lambda > 0$).

3 Statement of the free boundary problem to be studied

Let us now state the f.b.p. we are going to deal with for most of the course. The basic reference is [Beur 57].

Let R be a region in $\mathbb{R}^2$ external to a bounded domain, whose boundary is a simple closed (or Jordan) curve Γ, which is prescribed and fixed.

Problem (P) *Given a positive continuous function Q defined in R, find a bounded open annulus $\omega \subset R$, bounded by Γ and by a free boundary γ, and a function $V \in C^2(\omega) \cap C(\overline{\omega})$ such that*

$$
\begin{aligned}
\Delta V &= 0 \quad \text{in } \omega, & (16) \\
V|_\Gamma &= 0, & (17) \\
V|_\gamma &= 1, & (18) \\
|\nabla V| &= Q \quad \text{on } \gamma. & (19)
\end{aligned}
$$

The free boundary γ we are looking for is another simple closed curve not touching Γ.

The pair (ω, V), or simply ω, will be called a solution of the problem.

Remark 3.1 Since ∇V is not necessarily defined on γ, condition (19) is meant in the sense

$$\liminf |\nabla V|/Q = \limsup |\nabla V|/Q = 1 \qquad \text{as} \quad P \to \gamma, \quad P \in \omega.$$

Remark 3.2 If ∇V exists on γ, then $|\nabla V| = \dfrac{\partial V}{\partial n} > 0$ on γ (n = exterior normal vector) and therefore (18), (19) are the Cauchy data.

Remark 3.3 Q need not be bounded in the vicinity of Γ.

4 Reduction of Problem (P) to a standard form

Let us now associate the complex variable $z = x + iy$ to the points $(x, y) \in \mathbb{R}^2$. If $z_1 = f(z)$ is a conformal mapping and $V_1(z_1) = V(z)$, (16)–(18) remain unchanged for V_1, while (19) becomes

$$|\nabla V_1| = Q_1(z_1) \equiv Q[z(z_1)] \left| \frac{dz}{dz_1} \right|. \tag{20}$$

A convenient transformation is the one which reduces R to the half strip $0 < x < 2\pi$, $y > 0$ in which the points on the lines $x = 0$, $x = 2\pi$ are identified.[1] In other words the new functions V_1, Q_1 are 2π-periodic in x.

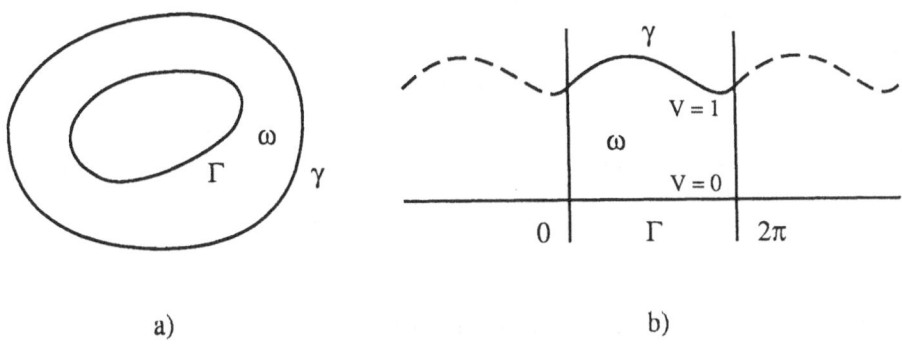

a)   b)

Figure 4.1 a,b

---

[1]First map Γ into the unit circle, then use the transformation $z = e^{-iz_1}$.

In the transformed plane a bounded annulus ω has the periodic structure represented in Fig. 4.1b), with Γ on the real axis and the outer boundary γ being a periodic curve. The set of such domains (in which the Dirichlet problem is solvable) will be denoted by $\mathcal{A}$.

Our program is now the following: first we study Problem (P) in the latter formulation, then we discuss how to revert to the original variables.

For the moment we shall omit the subscripts to V and Q, resuming the standard notation when necessary.

An interesting information can be obtained from the one-dimensional problem, i.e. when $Q = Q(y)$ and V is a linear function of y:

Proposition 4.1 *The one-dimensional problem has as many solutions as the roots of the equation*

$$yQ(y) = 1. \tag{21}$$

The proof is elementary. Such a result shows that for the problem considered we may have nonexistence or nonuniqueness (we can have even a continuum of solutions).

5   Auxiliary results

For ω given in $\mathcal{A}$ we denote by V_ω the solution of the Dirichlet problem (16)–(18) in ω (we refer to the x-periodic formulation).

The key point in order to decide whether ω is a solution or not is to compare $|\nabla V_\omega|$ and Q at the boundary γ. Such a comparison is made simpler if we introduce a pair of harmonic functions associated to $|\nabla V_\omega|$ and to Q.

Therefore we define

$$p_\omega = \log|\nabla V_\omega|. \tag{22}$$

It is well known that p_ω is harmonic in $\omega \cup \Gamma$. Moreover p_ω can be continued beyond Γ by reflection, so that

$$\left.\frac{\partial p_\omega}{\partial y}\right|_\Gamma = 0. \tag{23}$$

Next we define the function q_ω as the harmonic function in ω with boundary conditions

$$q_\omega = \log Q(x, y) \qquad \text{on } \gamma, \tag{24}$$

$$\frac{\partial q_\omega}{\partial y} = 0 \qquad \text{on } \Gamma. \tag{25}$$

Moreover q_ω is continued outside ω in R as $\log Q(x, y)$.

We also define the following three subsets of Q:

$$A(Q) = \{\omega \in \mathcal{A} : \liminf Q^{-1}|\nabla V_\omega| \geq 1\},$$

$$B(Q) = \{\omega \in \mathcal{A} : \limsup Q^{-1}|\nabla V_\omega| \leq 1\},$$

$$B_0(Q) = \{\omega \in \mathcal{A} : \limsup Q^{-1}|\nabla V_\omega| < 1\},$$

where the limits are taken for $z \in \omega$ tending to γ.

If there is no ambiguity we shall write A, B, B_0.

Remark 5.1 The set of the solutions of (P) coincides with $A \cap B$.

Lemma 5.2 *If $\omega \in A$ then $p_\omega \geq q_\omega$ in ω. If $\omega \in B$ then $p_\omega \leq q_\omega$ in ω. The converse statements are also true.*

Proof It is an easy consequence of the maximum principle. $\qquad \square$

Lemma 5.3 *If $\{\omega_n\}$ is an increasing sequence of domains in A (or B) having a limit $\omega \in \mathcal{A}$, then $\omega \in A$ (or B).*

Proof From the uniform convergence of the sequences $\{q_{\omega_n}\}, \{V_{\omega_n}\}, \{p_{\omega_n}\}$ to their respective limits $q_\omega, V_\omega, p_\omega$ and from Lemma 1 we deduce that if $\omega_n \in A$ then $p_{\omega_n} \geq q_{\omega_n}$, implying $p_\omega \geq q_\omega$. In turn this implies $\omega \in A$. $\qquad \square$

Lemma 5.4 *Let $\omega_1 \in A, \omega_2 \in B_0$ and $\omega_1 \subset \omega_2$. Then ω_2 contains the outer boundary of ω_1.*

Proof Let us suppose that the outer boundaries γ_1, γ_2 have a point z_0 in common (Fig. 5.1). We set $V_1 = V_{\omega_1}, V_2 = V_{\omega_2}$.

By the maximum principle we know that

$$V_2 < V_1 \quad \text{in} \quad \omega_1 \tag{26}$$

(note that $\omega_1 \neq \omega_2$ since $A \cap B_0 = \emptyset$).

Let χ_0 be the line passing through z_0 and tangent to the vector field ∇V_1 (such a line exists because $\liminf\limits_{z \to z_0} |\nabla V_1| > 0)$.

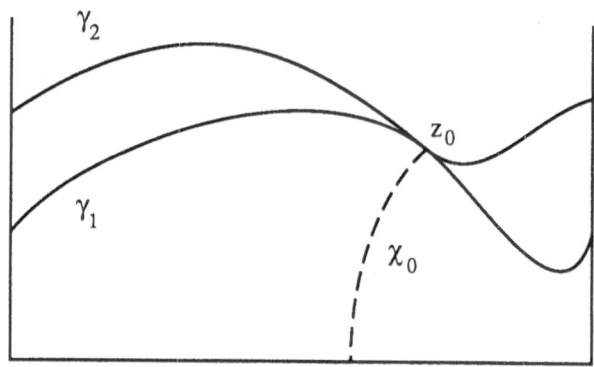

Figure 5.1

Performing the integral

$$\int_{z}^{\frown} z_0 |\nabla V_1||dz|$$

along χ_0 we obtain $1 - V_1(z)$. The integral

$$\int_{z}^{\frown} z_0 |\nabla V_2||dz|$$

along the same path is not less than $1 - V_2(z)$, since $|\nabla V_2|$ is not less than the tangential derivative of V_2 along χ_0. On the other hand from the assumptions on ω_1, ω_2 it follows that if z is sufficiently close to z_0

$$|\nabla V_2| < (1 - \epsilon)Q < |\nabla V_1|$$

along χ_0, for some $\epsilon > 0$. Hence

$$1 - V_1(z) > 1 - V_2(z),$$

i.e., $V_2(z) > V_1(z)$, contradicting (5.5).

Definition 5.5 Given $\omega_1, \omega_2 \in \mathcal{A}$, the *extended union* of ω_1, ω_2 is the smallest set in $\mathcal{A}$ which contains $\omega_1 \cup \omega_2$.

Fig. 5.2 shows an example in which the extended union is larger than $\omega_1 \cup \omega_2$. The points in the dashed area lie in the extended union, but not in the union.

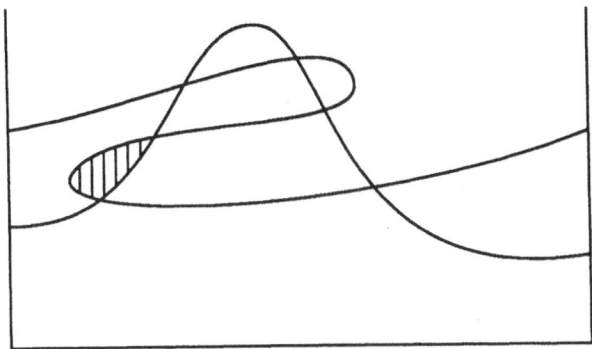

Figure 5.2

Definition 5.6 The *reduced intersection* of ω_1, ω_2 is the largest set in $\mathcal{A}$ contained in $\omega_1 \cap \omega_2$.

Fig. 5.3 shows an example in which the reduced intersection is smaller than $\omega_1 \cap \omega_2$. The points in the dashed area do not belong to the reduced intersection, although they are in $\omega_1 \cap \omega_2$.

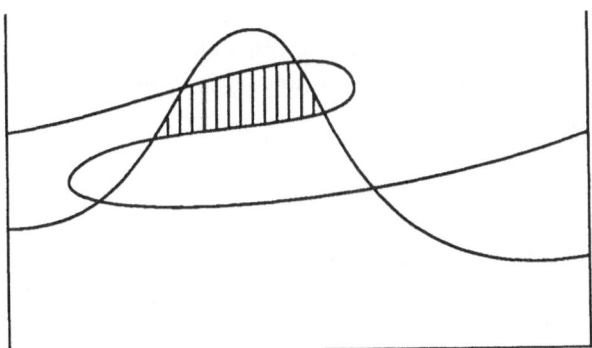

Figure 5.3

In the following we shall use the symbols $\cup_e, \cap_r$.

Lemma 5.7 *If $\omega_1, \omega_2 \in A$ so does their extended union. If $\omega_1, \omega_2 \in B$ so does their reduced intersection.*

Proof Let $\omega = \omega_1 \cup_e \omega_2$ and set $V_\omega = V, V_{\omega_1} = V_i, i = 1, 2$. Since $V \le V_i$ in $\omega_i, i = 1, 2$, the inequality $\liminf |\nabla V|/|\nabla V_i| \ge 1$ on $\gamma \cap \gamma_1$ and $\gamma \cap \gamma_2$ can be demonstrated easily if such boundaries are smooth or by means of approximations if they are not. Thus ω belongs to A if ω_1, ω_2 belong to A.

The proof of the second part of the lemma is symmetric. □

6 Existence

The aim of this section is to show some criteria for the existence of a solution. Although such criteria are not explicitly expressed as sufficient conditions on the data, they will play an important role in providing existence theorems for some specific case.

Theorem 6.1 *Assume that there exist two domains $\Omega \in A, \Omega_0 \in B_0$ such that $\Omega \subset \Omega_0$. Then (P) has at least one solution ω^* such that $\Omega \subset \omega^* \subset \Omega_0$.*

Proof Consider the set

$$S = \{\omega \in A : \Omega \subset \omega \subset \Omega_0\},$$

which is not void since it contains Ω. We show that

(i) S contains an element ω^* which is maximal, i.e., $\omega \in S \Rightarrow \omega \subset \omega^*$, and its outer boundary γ^* is contained in Ω_0,

(ii) ω^* is locally maximal in A, in the sense that there exists a neighborhood N of γ^* such that for any $\omega \in A$ with outer boundary in N and $\omega \ne \omega^*$ one has $\omega^* \supset \omega$,

(iii) as a consequence of (ii) ω^* is a solution of (P).

Proof of (i). Consider the union U of all $\omega \in S$ and choose a sequence $\{D_n\}$ of domains of S (with $D_1 = \Omega$), such that each point $z \in U$ is contained in some D_n. Next define the sequence $\{\omega_n\}$:

$$\omega_1 = D_1, \qquad \omega_{n+1} = \omega_n \cup_e D_{n+1}, \qquad n > 1.$$

By Lemma 5.7 each ω_n is in S, by Lemma 5.3 the limit ω^* of $\{\omega_n\}$ is also in S, and by Lemma 5.4 its outer boundary γ^* is contained in Ω_0. The maximality of ω^* follows from the fact that $\omega^* \supset U$.

Proof of (ii). By Lemma 5.4 $\gamma^* \subset \Omega_0$. If we take $N \subset \Omega_0$, then $\omega \in A$ with outer boundary in N is trivially an element of S.

Proof of (iii) We want to prove that any $\omega^* \in A$ satisfying (ii) is a solution to (P).

We denote by V, p, q the functions related with ω^* and we suppose that ω^* is not a solution. Since $\omega^* \in A$ we would have $p > q$ in ω^*, i.e.,

$$|\nabla V| > e^q. \tag{27}$$

For $z \in \omega^*$ let $\chi(z)$ be the arc ending in z and starting from Γ and being the shortest in the metric $e^q|dz|$. We introduce the distance

$$\delta(z) = \int_{\chi(z)} e^q|dz|. \tag{28}$$

From (27) it follows that

$$\delta(z) < V(z) \tag{29}$$

and consequently

$$k = \min_{z \in \gamma^*} \delta(z) < 1. \tag{30}$$

The set $D = \{z \in R : 0 < \delta(z) < k\} \in A$ is contained in ω^* and has at least one boundary point z_0 on γ^*.

Now we define the function V_D such that

$$|\nabla V_D| = e^q/k, \qquad V_D|_\Gamma = 0. \tag{31}$$

We remark that integrating $|\nabla V_D|$ along lines connecting a point $z \in D$ with Γ, the minimum is obtained by definition on $\chi(z)$, which is therefore the line of the vector field ∇V_D passing through z. Hence $V_D(z) = \delta(z)$ and $V_D = 1$ on the boundary γ_D of D.

Setting $\Delta V_D = \xi$ it is easy to compute that $\Delta p_D = 0$ is equivalent to

$$V_{D_x}\xi_x + V_{D_y}\xi_y = \xi g(x, y), \tag{32}$$

where g contains some derivatives of V_D up to the second order. Thus ξ solves a first order p.d.e. with the Cauchy data

$$\xi = 0 \quad \text{on} \quad \Gamma. \tag{33}$$

Equation (33) follows from $V_D|_\Gamma = 0$ and from $\left.\dfrac{\partial p_D}{\partial y}\right|_\Gamma = \left.\dfrac{1}{k}\dfrac{\partial q}{\partial y}\right|_\Gamma = 0$.

Hence $\xi \equiv 0$ i.e., V_D is harmonic. Moreover $D \in A(e^q)$.

Now we translate γ_D by an amount ϵ in the y-direction and we consider the solution $V_D^{(\epsilon)}$ of (16)–(18) in the enlarged domain D_ϵ. Obviously z_0 is an inner point of D_ϵ. We have $V_D^{(\epsilon)} < V_D$ in D and

$$0 < V_D(x, \epsilon) - V_D^{(\epsilon)}(x, \epsilon) < m(\epsilon), \tag{34}$$

where $m(\epsilon) \to 0$ as $\epsilon \to 0$, linearly in ϵ. In order to compare $|\nabla V_D|, |\nabla V_D^{(\epsilon)}|$ on the respective boundaries $\gamma_D, \gamma_D^{(\epsilon)}$, we introduce the function $\overline{V}_D(x, y) = V_D^{(\epsilon)}(x, y + \epsilon)$ and we consider the difference

$$W = V_D - \overline{V}_D \leq 0 \quad \text{in} \quad D.$$

The function Z, harmonic in D with boundary values

$$Z|_{\gamma_D} = 0, \qquad Z|_\Gamma = m(\epsilon)$$

is a barrier for W on γ_D, implying that

$$|\nabla W| = |\nabla V_D| - |\nabla \overline{V}_D| \leq |\nabla Z| = O(\epsilon) \quad \text{in} \quad D \tag{35}$$

From (35) it is easy to infer that

$$\liminf_{\substack{z \to \gamma_D^{(\epsilon)} \\ z \in D_\epsilon}} \frac{|\nabla V_D^{(\epsilon)}|}{e^q} \geq \liminf_{\substack{z \to \gamma_D^{(\epsilon)} \\ z \in D_\epsilon}} \frac{|\nabla V_D|}{e^q} - O(\epsilon) = \frac{1}{k} - O(\epsilon)$$

Therefore we can choose ϵ so small that $D_\epsilon \in A(e^q)$.

From Lemma 5.7 we conclude that $\omega^{**} = \omega^* \cup_\epsilon D_\epsilon \in A(e^q)$.

The points of the boundary γ^{**} of ω^{**} either lie on γ^* or are external to ω^*. Thus $e^q = Q$ everywhere on γ^{**}. Therefore we have found a set $\omega^{**} \supset \omega^*, \omega^{**} \neq \omega^*$ with γ^{**} lying in a prescribed neighborhood of γ^* (reduce ϵ if needed) and belonging to $A(Q)$. Hence we have reached a contradiction to the fact that ω^* is locally maximal in A. □

Now we prove that under an additional assumption on Q the fact that the set $B(Q)$ is not void is a characteristic condition for existence.

It is obvious that if a solution exists then $B(Q) \neq \emptyset$, since every solution belongs to $B(Q)$. So the nontrivial part of our claim is the sufficiency of such a condition. More precisely we show the following

Theorem 6.2 *Assume that*

$$\lim_{y \downarrow o} y Q(x, y) = 0 \tag{36}$$

uniformly w.r.t.x. Let $B(Q) \neq \emptyset$. Then each $\Omega \in B(Q)$ contains a solution.

Proof Let us assume first that $B_0 \neq \emptyset$ and consider $\Omega \in B_0$ with the associated solution V_Ω of (16)–(18).

We note that $|\nabla V_\Omega|$ is bounded in a neighborhood of Γ and bounded away from zero.

Now we choose $s > 0$ so small that the level curve $V_\Omega = s$, henceforth denoted by γ_s, lies in such a neighborhood. The inequalities $k_1 \leq \left|\frac{\partial V_\Omega}{\partial y}\right|$, $k_1 \leq |\nabla V_\Omega| \leq k_2$ valid here imply that on γ_s we have $s/k_2 \leq y \leq s/k_1$ and therefore $|\nabla V_\Omega/s|y \geq k_1/k_2$. Recalling (36), we conclude that a constant c exists such that for s sufficiently small we have

$$|\nabla V_\Omega/s| > Q(x,y) \qquad \text{on } \gamma_s. \tag{37}$$

If $\Omega_s \in \mathcal{A}$ is the domain bounded by γ_s, the function $V_s = V_\Omega/s$ is the solution of (16)–(18) in Ω_s and because of (37) Ω_s belongs to $A(Q)$. Existence now follows from Theorem 6.1.

If Ω is in B but not in B_0, we replace Q by λQ, λ being a real parameter greater than 1. So $\Omega \in B_0(\lambda Q)$ and we can apply the above argument to obtain a solution ω_λ, which is going to be the maximal domain in the set $S_\lambda = \{\omega \in \mathcal{A} : \omega \in A(\lambda Q) \subset A(Q), \Omega_s \subset \omega \subset \Omega\}$.

It is not difficult to see that we can select ω_λ increasing as $\lambda \downarrow 1$. Indeed if $1 < \mu < \lambda$ we have $\omega_\mu \in B(\mu Q) \subset B_0(\lambda Q)$ and from the previous result ω_λ can be chosen so that $\omega_\lambda \subset \omega_\mu$. Hence ω_λ has a limit $\omega \subset \Omega$.

Setting $p_{\omega_\lambda} = p_\lambda$ and $q_{\omega_\lambda} = q_\lambda$, from Lemma 1, Section 2 we have $p_\lambda = q_\lambda$. The value of q_λ on the free boundary γ_λ is $\log Q + \log \lambda$. Therefore q_λ converges to $q = q_\omega$. Moreover V_λ and ∇V_λ converge uniformly in compact subsets of ω to V and to ∇V respectively, i.e., $\omega \in A(Q) \cap B(Q)$, thus being a solution. □

Exercise 6.3 Interpret the results of Theorems 6.1 and 6.2 for the one-dimensional solution considered in Proposition 4.1.

Still referring to the same one-dimensional solution, it is clear that uniqueness requires some additional constraint on Q. The following theorem is concerned with non-uniqueness cases. The question of uniqueness will be treated in the next section.

Theorem 6.4 *Let (36) be valid and assume that ω_1, ω_2 are solutions, none contained in the other. Then they both contain a third solution ω_3.*

Proof From the fact that $\omega_1, \omega_2 \in A \cap B$ and from Lemma 5.7 it follows that $\Omega = \omega_1 \cap_r \omega_2 \in B$. From Theorem 6.2 we conclude that there exists a solution $\omega_3 \subset \Omega$. □

7   Uniqueness

Theorem 7.1 *Let Q be twice continuously differentiable in R (satisfying (28)) and such that*

$$(i) \quad \Delta \log Q \geq 0 \qquad in \quad R,$$

$$(ii) \quad \liminf_{y \downarrow 0} \frac{\partial Q}{\partial y} \geq 0 \quad for \ any \quad x.$$

Then (P) cannot have more than one solution.

Proof   For any $\omega \subset \mathcal{A}$ the difference $W = q_\omega - \log Q$ is such that $\Delta W \leq 0$ in ω, $W = 0$ on γ_ω, $\limsup_{y \downarrow 0} \dfrac{\partial W}{\partial y} \leq 0$.

Therefore W cannot have a minimum in ω nor on Γ and $q_\omega \geq \log Q$ in ω.

If we now consider two domains $\omega_1 \subset \omega_2$ with boundaries γ_1, γ_2, the associated functions q_1, q_2 are such that $q_2|_{\gamma_1} \geq \log Q|_{\gamma_1} = q_1|_{\gamma_1}$. Hence $q_2 \geq q_1$ in ω_1, that is to say that q_ω is nondecreasing for ω increasing.

At this point we assume that (P) has more than one solution. Then from Theorem 6.4 we know that we can find two distinct solutions ω_1, ω_2, such that $\omega_1 \subset \omega_2$.

Since $p_\omega = q_\omega$ for every solution ω, on Γ we have

$$|\nabla V_1| = e^{q_1} \leq e^{q_1} = |\nabla V_2|. \tag{38}$$

However it is elementary to show that $|\nabla V_\omega|$ on Γ is strictly decreasing for ω increasing, thus contradicting (38).   □

8   Reverting to the original variables

The results above have been proved for the standard form of problem (P). Therefore the conditions imposed on Q have to be re-interpreted in the framework of the original variables and data. Some of these conditions will obviously be influenced by the original shape of Γ.

The most important conditions on Q are the ones in Theorem 7.1. We claim that the assumption

$$\nabla \log Q \geq 0 \qquad \text{(Theorem 7.1, (i))}$$

can be referred directly to the original data, irrespective of the curve Γ.

This follows from the fact that the quantity $Q^{-2}\Delta\log Q$ is invariant under conformal mappings.

Condition (36) can be easily expressed in terms of the original data.

It is more difficult to interpret condition (ii) of Theorem 7.1. This time the shape of Γ plays an important role. For simplicity we shall assume from now on that Γ is analytic and that Q is differentiable up to Γ. Such restrictions can be removed by a simple approximation argument. Here we denote by Γ, Q the original data and we recall the symbols z_1, Q_1 of Section 4.

Proposition 8.1 *If* $\left.\dfrac{\partial Q}{\partial n}\right|_\Gamma \geq 0$ *(n denoting the normal to Γ pointing in R) and if Γ is convex, then* $\left.\dfrac{\partial Q_1}{\partial y_1}\right|_{y_1=0} \geq 0.$

Proof For any given annulus ω (in the original x-y plane) we consider the solution V of (16)–(18) and its harmonic conjugate U. Then we consider the conformal mapping generated by

$$f(z) = U + iV, \tag{39}$$

which maps ω into a rectangle $0 < V < 1, 0 < U < U_0$ (we can normalize U so that $U = 0$ on a point of Γ).

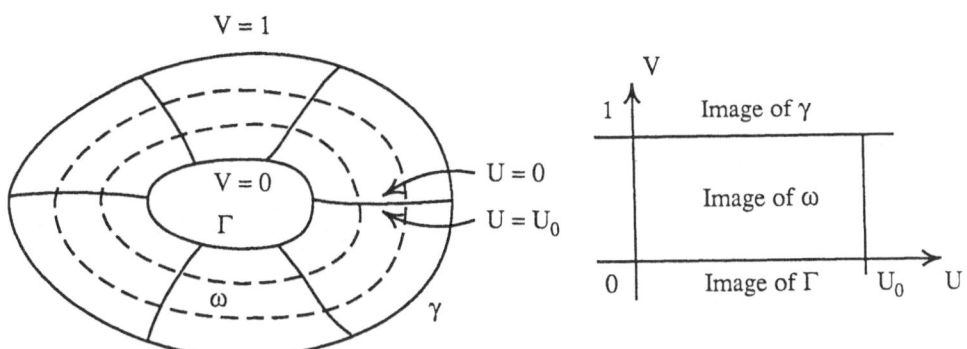

Figure 8.1

We represent $f'(z)$, which is also analytic, in the form

$$f'(z) = \exp[\rho(z) + i\theta(z)], \tag{40}$$

the exponent being analytic as well.

We recall that θ is the angle by which the tangent to a curve is rotated in the positive sense by effect of the mapping.

If (s, η) is the orthogonal coordinate system of the level lines of V, U (so that $\left.\frac{\partial Q}{\partial \eta}\right|_{s=0} = \left.\frac{\partial Q}{\partial n}\right|_{\Gamma}$), we can write down the Cauchy-Riemann equations for $\rho + i\theta$ in the form

$$\frac{\partial \rho}{\partial s} = \frac{\partial \theta}{\partial \eta}, \qquad \frac{\partial \rho}{\partial \eta} = -\frac{\partial \theta}{\partial s}. \tag{41}$$

Assume that s increases counterclockwise. Let us evaluate $\frac{\partial \theta}{\partial s}$ on Γ, noting that the tangent to Γ after the mapping goes over the U axis. It is therefore clear that the convexity of Γ is equivalent to $\frac{\partial \theta}{\partial s} \geq 0$, the inequality being strict if Γ is strictly convex.

Thus from (41)

$$\left.\frac{\partial \rho}{\partial \eta}\right|_{\Gamma} \leq 0 \tag{42}$$

(< 0 for Γ strictly convex).

From (20) we have

$$Q(x, y) = \tilde{Q}(U, V)|f'(z)| = \tilde{Q}(U, V)e^{\rho}.$$

We have $\left.\frac{\partial U}{\partial \eta}\right|_{\Gamma} = 0$ and consequently

$$\left.\frac{\partial Q}{\partial n}\right|_{\Gamma} = \left.\frac{\partial \tilde{Q}}{\partial V}\right|_{V=0} \left.\frac{\partial V}{\partial \eta}\right|_{\Gamma} e^{\rho} + \tilde{Q}e^{\rho} \left.\frac{\partial \rho}{\partial \eta}\right|_{\Gamma}. \tag{43}$$

Since $\left.\frac{\partial V}{\partial \eta}\right|_{\Gamma} > 0$, if we assume that $\left.\frac{\partial Q}{\partial n}\right|_{\Gamma} \geq 0$, then $\left.\frac{\partial \tilde{Q}}{\partial V}\right|_{V=0} \geq 0$.

Furthermore it is quite obvious that $\left.\frac{\partial \tilde{Q}}{\partial V}\right|_{V=0}$ and $\left.\frac{\partial Q_1}{\partial y_1}\right|_{y_1=0}$ have the same sign and the Proposition is proved.

$\square$

Corollary 8.2 *If* Γ *is convex and the original function* Q *is such that* $\Delta \log Q \geq 0$ *in* R *and* $\left.\dfrac{\partial Q}{\partial n}\right|_{\Gamma} \geq 0$, *then the solution of* (P) *is unique.*

Remark 8.3 In particular if $Q \equiv$ constant and Γ is convex the solution is unique.

The following section will be devoted to the study of the latter case.

9 Existence and qualitative properties for Γ convex, $Q =$ constant [Tepp 74]

In this section we analyse the following special case:

$$\Gamma \quad \text{convex}, \qquad Q \equiv \lambda > 0, \quad \text{constant}.$$

Thus we have a one parameter family of problems (P_λ) for which uniqueness has been proved.

We denote by ω_λ the unique solution of (P_λ). We want to show that ω_λ exists, that its free boundary γ_λ is also convex, and that ω_λ depends monotonically on λ and invades R as $\lambda \downarrow 0$.

For simplicity we assume that Γ is analytic, although this restriction can be easily removed.

Theorem 9.1 (Existence) *For each* $\lambda > 0$, (P_λ) *has a solution* ω_λ. *Moreover, the distance of the points of* ω_λ *from* Γ *is less than* λ^{-1}.

Proof According to Theorem 6.2 all we have to prove in order to get existence is that $B(\lambda) \neq \emptyset$.

For any z in R we consider its distance $d(z)$ from Γ (Fig. 9.1). It is not difficult to see that $d(z)$ is subharmonic, i.e. $\Delta d \geq 0$ in R. In the frame of reference in which y' is a ray orthogonal to Γ on a point P of the y' axis $\dfrac{\partial^2 d}{\partial y'^2} = 0$. It is easy to check that $\dfrac{\partial^2 d}{\partial x'^2} \geq 0$ if Γ is convex. Since Δ is invariant under rotation $\Delta d \geq 0$ is demonstrated.

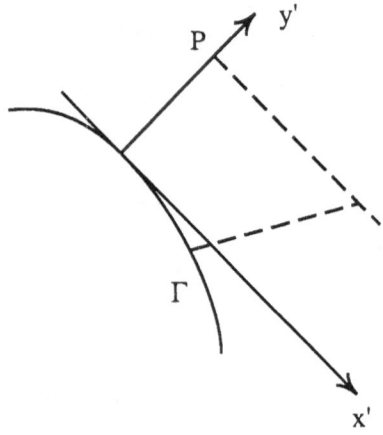

Figure 9.1

Now we define the set (Fig. 9.2)

$$\omega_\lambda^* = \{z \in R : d(z) < \lambda^{-1}\} \in \mathcal{A}. \tag{44}$$

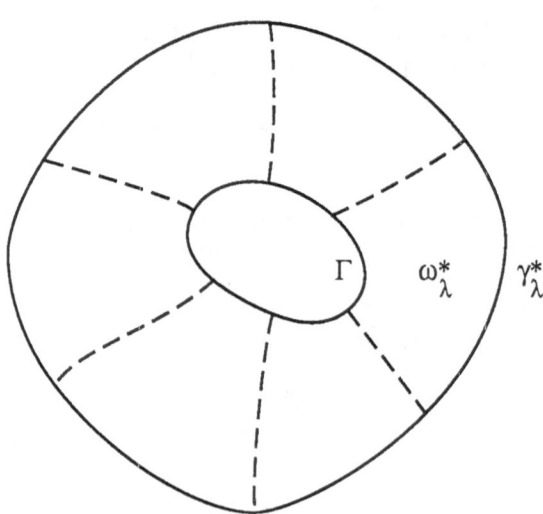

Figure 9.2

The rays normal to Γ are also normal to the outer boundary γ_λ^* of ω_λ^*. Let V_λ^* be the solution of (16)–(18) in ω_λ^*, and let $U(z) = \lambda d(z)$.

U and V_λ^* are both zero on Γ and both 1 on γ_λ^*.

Moreover V_λ^* is harmonic and U is subharmonic, hence

$$V_\lambda^* \geq U \qquad \text{in} \quad \omega_\lambda^*$$

and on $\gamma_\lambda^*, 0 < \dfrac{\partial V_\lambda^*}{\partial n} \leq \dfrac{\partial U}{\partial n}$ (n outer normal).

But the direction of n is also the direction of a normal ray to Γ and obviously $\dfrac{\partial d(z)}{\partial n} = 1$, so $\left.\dfrac{\partial U}{\partial n}\right|_{\gamma_\lambda^*} = \lambda$, implying that $\omega_\lambda^* \in B(\lambda)$. Thus $B(\lambda)$ is nonvoid and (P_λ) has a solution ω_λ contained in ω_λ^*.

Theorem 9.2 (Convexity) *The free boundary γ_λ is strictly convex.*

Proof Let us consider the analytic function $\rho + i\theta$ introduced in (40), with ρ and θ satisfying (41) in ω_λ. We have seen that the convexity of Γ implies $\dfrac{\partial \rho}{\partial \eta} \leq 0$ on Γ. Since ρ is harmonic in ω_λ, this excludes that it can have a minimum on Γ. Therefore the minimum of ρ has to be taken on the free boundary γ_λ. Here

$$\lambda = |\nabla V| = |f'| = e^\rho,$$

i.e. $\rho = \log \lambda$. Thus ρ equals its minimum at every point of γ_λ and therefore

$$\left.\frac{\partial \rho}{\partial \eta}\right|_{\gamma_\lambda} \leq 0, \tag{45}$$

which means $\left.\dfrac{\partial \theta}{\partial s}\right|_{\gamma_\lambda} \geq 0$. By the same argument explained in the previous section this implies that γ_λ is convex.

Actually γ_λ is strictly convex, since if it contains a segment the inequality (45) holds in the strict sense on it, thus leading to a contradiction. $\qquad \square$

Theorem 9.3 (Monotonicity) *If $\lambda_1 > \lambda_2$, then $\omega_{\lambda_1} \subset \omega_{\lambda_2}$.*

Proof We know that $\omega_{\lambda_2} \in B(\lambda_1)$ and consequently ω_{λ_2} contains a solution to (P_{λ_1}) (Theorem 6.2). Owing to the uniqueness theorem such a solution coincides with ω_{λ_1}. $\qquad \square$

Theorem 9.4 *Given any $z_0 \in R$ there exists λ so small that $z_0 \in \omega_\lambda$.*

Proof Take a disk E such that z_0 and Γ are contained in E. Let D be the annulus $E \cap R$ and V_D be the associated solution of (16)–(18). Set

$$0 < \delta = \min_{\gamma_D} |\nabla V_D|,$$

γ_D being the boundary of E. Clearly $D \in A(\delta)$ and for some $\delta_0 < \delta$, $D \subset \omega_{\delta_0}^*$ (see (44) for the definition of the sets ω^*). We know that $\omega_{\delta_0}^* \in B(\delta_0) \subset B_0(\delta)$.

Hence we can apply Theorem 6.1, concluding that (P_δ) has a solution ω_δ such that $D \subset \omega_\delta \subset \omega_{\delta_0}^*$. Since $z_0 \in D$ the theorem is proved. $\square$

10   An associated minimum problem [AlCa81]

The problem discussed so far is a particular case of the following variational problem studied in [AlCa81] (see also [Dani72]).

Let $\Omega \subset \mathbb{R}^n$ be an open connected (possibly unbounded) set, whose boundary $\partial\Omega$ is locally a Lipschitz graph (for the two dimensional problem we had a milder condition). For the sake of simplicity assume that the Dirichlet data are prescribed on $\partial\Omega$ as the trace of a non-negative function $u^0 \in L^1_{loc}(\Omega)$ with $\nabla u^0 \in L^2(\Omega)$ (in the previous case we had $u^0 = 0$). The function Q is assumed to be non-negative and measurable.

Let K be the convex set

$$K = \{v \in L^1_{loc}(\Omega) : \nabla v \in L^2(\Omega), \quad v = u^0 \text{ on } \partial\Omega\}.$$

The problem consists in finding an absolute minimum of the functional

$$J(v) = \int_\Omega [|\nabla v|^2 + \chi(v)Q^2]dx,$$

in K, $\chi(v)$ being the characteristic function of the positivity set of v.

This problem is non-standard since the term χQ^2 makes the domain $\{v > 0\}$ appear in its formulation (in other words the variational formulation does not eliminate the f.b. completely). An accurate analysis is performed in [AlCa81].

Here we report the existence proof, which is surprisingly simple if compared with the classical theory exposed in the previous sections.

Theorem 10.1 *If $J(u^0) < \infty$ then the minimum problem has a solution.*

Proof Assume first Ω is bounded. Let $\{u_k\}$ be a minimizing sequence for J. Since the L^2 norms of ∇u_k are uniformly bounded, the L^2 norms of $u_k - u^0$ are also uniformly bounded. Therefore for a subsequence $\{u_{k'}\}$ and for some $u \in K$

$$\nabla u_{k'} \to \nabla u, \quad \text{weakly in } L^2(\Omega),$$
$$u_{k'} \to u, \quad \text{a.e. in } \Omega.$$

Moreover

$$\chi(u_{k'}) \to \gamma, \quad \text{weakly star in} \quad L^\infty,$$

for some $\gamma \in L^\infty, 0 \le \gamma \le 1$, such that $\gamma = 1$ a.e. in $\{u > 0\}$.

Now

$$\int_\Omega |\nabla u|^2 dx \le \liminf \int_\Omega |\nabla u_{k'}|^2 dx$$

as an easy consequence of Mazur's lemma [Yosi 80].

Since

$$\int_\Omega (|\nabla u|^2 + \gamma Q^2) dx \ge J(u),$$

we conclude that

$$J(u) \le \lim J(u_{k'}),$$

i.e., that u minimizes J in K. $\qquad\qquad\qquad\qquad\qquad\qquad\qquad\qquad$ $\square$

In the same paper some regularity results are obtained. In two dimensions it is shown that the f.b. is analytic if Q is analytic (use is made of previous results of [KiNi 77]).

A two-phase extension has been studied in [ACF 84].

Part II

A free boundary problem for a Bingham flow

1   Introduction. Bingham fluids (visco-plastic flows)

A Bingham fluid is a non-Newtonian fluid characterized by the presence of a threshold value τ_0 for the stress, such that when the stress τ is less than τ_0 the fluid behaves like a rigid body, while for $\tau > \tau_0$ the relationship between stress and strain rate is linear and in a laminar flow we can write

$$\tau = \tau_0 + \eta\dot{\gamma}, \tag{46}$$

where $\dot{\gamma}$ is the strain rate and η plays the role of the viscosity. For a general treatment see [DuLi76].

In other words, the dynamics of a Bingham fluid obeys the Navier-Stokes equation in the region $\{\tau > \tau_0\}$, while on the boundary with the rigid core (the f.b.) we have $\tau = \tau_0$, i.e., zero strain rate, and another f.b.c. resulting from the balance of momentum.

Let us consider the specific example of an incompressible Bingham fluid flowing between two parallel plates. Let x be the coordinate along the direction of motion and y the coordinate in the direction perpendicular to the plates.

In the representative xy-plane the velocity has the form $\vec{v} = (v(y,t),0)$. Therefore the equation of motion in the viscous region is

$$\rho\frac{\partial v}{\partial t} = -\frac{\partial p}{\partial x} + \eta\frac{\partial^2 v}{\partial y^2}, \tag{47}$$

from which we deduce that $\dfrac{\partial p}{\partial x}$ does not depend on x.

On the other hand, since the y-component of v is identically zero, p minus the hydrostatic pressure cannot depend on y, i.e.,

$$-\frac{\partial p}{\partial x} = f(t), \tag{48}$$

which we assume to be given.

Therefore, if $y = \pm s(t)$ denotes the equation of the boundary of the rigid core, the velocity satisfies the parabolic equation

$$\rho\frac{\partial v}{\partial t} - \eta\frac{\partial^2 v}{\partial y^2} = f(t) \tag{49}$$

in the regions $-L < y < -s(t)$, $s(t) < y < L$, where $2L$ denotes the thickness of the layer. By symmetry we can consider the upper half layer and we impose the no-slip condition

$$v(L,t) = 0 \tag{50}$$

and some initial condition

$$v(y,0) = v_0(y) \tag{51}$$

such that $v_0(y)$ =constant for $0 < y < s_0$, $v_0(L) = 0$, v_0' continuous and non-positive. Of course s_0 gives the initial position of the f.b.

The no deformation condition of the f.b. is

$$\left.\frac{\partial v}{\partial y}\right|_{x=s(t)} = 0. \tag{52}$$

In order to derive the second condition we write the equation of motion for a portion of the rigid core lying between two unit squares parallel to the plates:

$$s(t)\rho\frac{\partial v}{\partial t} = s(t)f(t) - \tau_0, \tag{53}$$

where $2s(t)\rho$ represents the mass of the body considered, $2s(t)f(t)$ is the driving force due to the pressure gradient, and $2\tau_0$ is the drag force due to the viscosity.

Remark For $f = f_0$, constant, and $f_0 > \tau_0/L$ the stationary solution is $s \equiv s_\infty = \tau_0/f_0, v(y) = (f_0/2\eta)(x^2 - L^2) - (f_0/\eta)(x - L)$.

For a more detailed description of the physical problem (and some short notes on fluid dynamics) see [Eijn 90].

2   Other formulations of the problem

The f.b.c.'s of the Bingham flow, i.e.,

$$\left.\frac{\partial v}{\partial y}\right|_{x=s(t)} = 0, \qquad \left.\frac{\partial v}{\partial t}\right|_{x=s(t)} = \frac{1}{\rho}f(t) - \frac{\tau_0}{\rho s(t)},$$

are neither of Cauchy (v and $\frac{\partial v}{\partial y}$ prescribed) nor of Stefan type (v given and $\frac{\partial v}{\partial y}$ related to $\dot{s}$).

However, setting $w = \frac{\partial v}{\partial y}$, the problem becomes

$$\rho\frac{\partial w}{\partial t} - \eta\frac{\partial^2 w}{\partial y^2} = 0,$$

$$w_y(L, t) = -\frac{1}{\eta}f(t),$$

(CP) $w(y, 0) = v_0'(y),$ $(w = 0 \quad \text{for } 0 < y < s_0),$

$$w(s(t), t) = 0,$$

$$\eta w_y(s(t), t) = -\tau_0/s(t),$$

i.e., a f.b.p. with Cauchy data on the f.b.

It is possible to get a Stefan type problem with the transformation $z = v_t = \frac{\eta}{\rho}w_y + \frac{1}{\rho}f$:

$$\rho z_t - \eta z_{yy} = \dot{f},$$

$$z(L, t) = 0,$$

$$z(y, 0) = \frac{\eta}{\rho}v_0'' + \frac{1}{\rho}f(0),$$

(SP)

$$z(s(t), t) = \left(f(t) - \frac{\tau_0}{s(t)}\right)/\rho,$$

$$z_y(s(t), t) = \frac{1}{\eta}\frac{\tau_0}{s(t)}\dot{s}(t).$$

In the latter form the problem falls in the class studied in [FaPr 77], where existence and uniqueness of a classical solution in a sufficiently small time interval are proved under very mild assumptions on the data.

At this point a large digression on the equivalence of different classes of free boundary problems in one space dimension would be necessary. Such an analysis reveals interesting features and turns out to be particularly delicate in those cases in which the free boundary exhibits a singularity (blowing up solutions). For a detailed discussion on the relationship between two standard parabolic f.b.p. problems with particular reference to singularities (namely, the Stefan problem and the oxygen diffusion-consumption problem) we refer to [FPHO 89], [FPHO 90].

Here we recall that in one-phase problems the most typical situation in which singularities can occur is when the phase is receding. Therefore, generally speaking the study of problems with receding phases (a fact which can occur in Bingham flows) is expected to present some difficulty.

3 Analysis of problem (SP)

In this section we refer to the results obtained in [Comp 90] about global existence of a classical solution as well as on its asymptotic behaviour.

We sketch the proof of the following theorem, omitting the precise statement of the assumptions (including some restrictions on the derivatives of $v_0(y)$), for which we refer to the quoted paper.

Theorem 3.1 *Under suitable assumptions on the initial data* $s_0, v_0(y)$ *and on the driving term* $f(t)$ *problem (SP) has one unique global classical solution. Moreover, if* $f_0 = \lim_{t \to \infty} f(t)$, *then* $s(t)$ *tends to* $s_\infty = \tau_0/f_0$ *as* $t \to \infty$ *and*

$$\int_0^\infty (s_\infty/s(t) - f(t)/f_0)dt < \infty. \tag{54}$$

We have already said that the theory developed in [FaPr 77a] guarantees local existence and uniqueness. Therefore the only fact to be estabilished is that the solution can be continued to infinity.

For simplicity we consider the case $f(t) = f_0$. The proof is based on a series of a-priori estimates on v, v_y, v_t and on the following monotone dependence Lemma.

Lemma 3.2 *Let* (v_1, s_1, T_1) *and* (v_2, s_2, T_2) *be the solutions of the Bingham flow model corresponding to the respective data* $(v_1(y,0), s_1(0)), (v_2(y,0), s_2(0))$. *Suppose that* $s_1(0) > s_2(0)$ *and* $v_{1y}(y,0) \geq v_{2y}(y,0)$, *then*

$$s_1(t) > s_2(t), \quad 0 < t < \bar{T} = \min(T_1, T_2), \tag{55}$$
$$v_1(y,t) \leq v_2(y,t), \quad s_1(t) \leq y \leq L, \ 0 \leq t \leq \bar{T}, \tag{56}$$
$$v_{1y}(y,t) \geq v_{2y}(y,t), \quad s_1(t) \leq y \leq L, \ 0 \leq t \leq \bar{T}. \tag{57}$$

If $s_1(0) = s_2(0)$ *the inequality in (3.2) is valid in the weak sense* $s_1(t) \geq s_2(t)$.

For the sake of brevity we omit the proof of the above results and we outline the main ideas in the proof of the theorem.

Thanks to Lemma 3.2 and under suitable assumptions on $v_0(y)$ one can see that if $s_0 > s_\infty$ or $s_0 < s_\infty$ then the respective inequalities $s(t) > s_\infty, s(t) < s_\infty$ are satisfied for all t such that the solution exists.

Next, an argument based on the maximum principle applied to the function $z_y(y,t)$ shows that the free boundary is monotone (provided a sign restriction and a compatibility condition are imposed on $v'''(y)$).

Finally it is possible to obtain a uniform bound for $\dot{s}(t)$ constructing appropriate barriers, thus leading to the conclusion that the solution is continuable beyond any finite time.

The information on the asymptotic behaviour can be deduced using Green's identity.

$$\iint_D (\phi Lu - uL^*\phi)dy\ dt = \int_{\partial D}[(\phi u_y - u\phi_y)d\tau + u\phi\ dy],$$

where L denotes the heat operator $\rho\dfrac{\partial}{\partial t} - \eta\dfrac{\partial^2}{\partial y^2}$, and L^* its adjoint.

Setting $u = z(y,t)$ and $\phi = y - L$, one obtains the integral relationship:

$$\frac{\eta f_0}{\rho}\int_0^t(\frac{s_\infty}{s(\tau)} - 1)d\tau = \rho\int_{s(t)}^L (L - y)z(y,t)dy$$
$$- \rho\int_{s_0}^L (L - y)z_0(y,t)dy + \frac{f_0}{2}[(L - s_0)^2 - (L - s(t))^2],$$

from which (54) follows, letting t tend to ∞ and using the fact that $z(y,t)$ tends to zero uniformly as $t \to \infty$, so that the limit of the right hand side is easily computed.

4 Further references

Beside the book [DuLi 76] we mention the clear statement of the one-dimensional problem of [Rubi 71]. In the series of papers [MoMi 65, 66, 67] a number of interesting results have been obtained about the (multidimensional) steady-state problem. In [Kim 86, 87] a large analysis of the evolution problem in variational form is carried out.

References

[AlCa81]   H.W. Alt, L.A. Caffarelli, Existence and regularity for a minimum problem with free boundary, *J. Reine Angew. Math* **325** (1981), 105–144.

[ACF84]   H.W. Alt, L.A. Caffarelli, A. Friedman, Variational problems with two phases and their free boundaries, *Trans. Amer. Math. Soc.* **282** (1984) 431–461.

[Acke77]   A. Acker, Heat flow inequalities with applications to heat flow optimization problems, *SIAM J. Math. Anal.* **8** (1977), 604–618.

[Beur57]   A. Beurling, On free boundary problems for the Laplace equation. Seminars on Analytic Functions, I, *Advanced Study Seminars* (1957), 248–263.

[Carl18]   T. Carleman, Über ein Minimalproblem der mathematischen Physik, *Math. Z.* **1** (1918), 208–212.

[Comp90]   E. Comparini, A one-dimensional Bingham flow. To appear in *J. Math. Anal. Appl.*

[Dani72]   I.I. Daniliuk, On integral functionals with a variable domain of integration, *Proc. Steklov Inst. Math.* **118** (1972), Engl. Transl. AMS (1976).

[DuLi76]   G. Duvaut, J.L. Lions, *Inequalities in Mechanics and Physics*, Springer-Verlag, Berlin 1976.

[Eijn90]   S.J.L. van Eijndhoven, Mathematical models based on free boundary problems (based on a course by A. Fasano), Opleiding Wiskunde voor de Industrie, Eindhoven, Report **90–01** (1990).

[Elli80]   C.M. Elliot, On a variational inequality formulation of an electrochemical machining moving boundary problem and its approximation by the finite element method, *J. Inst. Math. Appl.* **25** (1980), 121–131.

[Elli83]   C.M. Elliot, A variational inequality formulation of a steady state electrochemical machining free boundary problem, in [FaPr83], Vol II, *Research Notes Math.* **79** (1983), Pitman, London, 505–512.

[ElOc82]   C.M. Elliot, J.R. Ockendon, Weak and variational methods for moving boundary problems, *Research Notes Math.* **59** (1982), Pitman, London (1982).

[Fasa89]   A. Fasano, Free boundary problems and their applications, SASIAM (Bari, Italy) Report.

[FHPO89]   A. Fasano, S.D. Howison, M. Primicerio, J.R. Ockendon, On the singularities of the one-dimensional Stefan problems with supercooling, in *Mathematical Models for Phase Change Problems* (J.F. Rodrigues ed.), Birkhäuser, Basel, 1989.

[FHPO90]   A. Fasano, S.D. Howison, M. Primicerio, J.R. Ockendon, Some remarks on the regularization of supercooled one-phase Stefan problems in one dimension, *Quart. Appl. Math.* **98** (1990), 153–168.

[FaPr77]   A. Fasano, M. Primicerio, General free-boundary problems for the heat equation I, *J. Math. Anal. Appl.* **57** (1977), 694–723.

[Frie34]   K. Friedrichs, Über ein Minimumproblem für Potentialströmungen mit freiem Rande, *Math. Ann.* **109** (1934), 60–82.

[Kim86]   J.U. Kim, On the Cauchy problem associated with the motion of a Bingham fluid in the plane, *Trans. Amer. Math. Soc.* **298** (1986), 371–400.

[Kim87]   J.U. Kim, On the initial-boundary value problem for a Bingham fluid in a three dimensional domain, *Trans. Amer. Math. Soc.* **304** (1987), 751–770.

[KiNi77]   D. Kinderlehrer, L. Nirenberg, Regularity in free boundary problems, *Ann. Scuola Norm. Sup. Pisa (4)* **4** (1987), 373–391.

[LaSh87]   A.A. Lacey, M. Shillor, Electrochemical and electro-discharge machining with a threshold current, *IMA J. Appl. Math.* **39** (1987), 121–142.

[Mack86]   A.G. MacKie, The mathematics of electrochemical machining, *J. Math. Anal. Appl.* **117** (1986), 548–560.

[Mcge74]   J.A. McGeough, *Principles of Electrochemical Machining*, Chapman & Hall, London 1974.

[Mcge85]   J.A. McGeough, Unsolved moving boundary problem in electro-chemical machining, in *Free Boundary Problems: Applications and Theory*, Vol III (A. Bossavit et al., eds.), *Research Notes Math* **120** (1985), 152–156.

[McRa74]   J.A. McGeough, H. Rasmussen, On the derivation of the quasi-steady model in electrochemical machining, *J. Inst. Math. Appl.* **13** (1974), 13–21.

[MoMi65]   P.P. Mosolov, V.P. Miasnikov, Variational methods in the theory of the fluidity of a viscous-plastic medium, *J. Appl. Math. Mech.* **29** (1965), 545–577.

[MoMi66]   P.P. Mosolov, V.P. Miasnikov, On stagnant flow regions of a viscous-plastic medium in pipes, *J. Appl. Math. Mech.* **30** (1966), 841–854.

[MoMi67]   P.P. Mosolov, V.P. Miasnikov, On qualitative singularities of the flow of a viscous-plastic medium in pipes, *J. Appl. Math. Mech.* **31** (1967), 609–613.

[Roge80]   J.C.W. Rogers, Relation of the one-phase Stefan problem to the seepage of liquids and electrochemical machining, in *Free Boundary Problems*, Vol I (E. Magenes, ed.), Ist. Naz. Alta Mat., Roma (1980), 333–382.

[Rubi71]   L.I. Rubinstein, *The Stefan Problem*, Transl. Math. Monographs. **27** Amer. Math. Soc., Providence 1971.

[Tepp74]   D.E. Tepper, Free boundary problem, *SIAM J. Math. Anal.* **5** (1974), 841–846.

[Tepp75]   D.E. Tepper, On a free boundary problem, the starlike case, *SIAM J. Math Anal.* **6** (1975), 503–505.

[Yosi80]   K. Yosida, *Functional Analysis*, 6th edition, Springer Verlag, Berlin 1980.

Problèmes de surfaces libres en mécanique des fluides

Michel FORTIN

Département de mathématiques, statistique et actuariat
Université Laval
Québec, Québec G1K 7P4
Canada

Résumé

Ce cours porte sur divers aspects des problèmes de surfaces libres que l'on rencontre en mécanique des fluides. Il comporte deux parties.

Dans la première partie, plus théorique, on cherche à poser les problèmes de surface libre dans un cadre d'optimisation par rapport à la forme du domaine dans lequel l'écoulement s'effectue. Notre succès ne sera que partiel mais les résultats obtenus permettent de mieux saisir les similitudes et les différences entre les problèmes d'élasticité et les problèmes de fluides.

Dans la seconde partie on traitera de questions numériques. On y présentera la formulation eulérienne-lagrangienne des équations de Navier-Stokes et on l'utilisera pour la construction d'une méthode de type Newton pour le calcul du déplacement d'une frontière libre. On considérera aussi quelques approximations de la méthode ainsi développée. On présentera également un algorithme de type quasi-Newton appuyé sur l'utilisation de la méthode des résidus minimaux généralisé (GMRES) pour la résolution des sous-problèmes linéarisés.

1 Surfaces libres et optimisation de forme

1.1 Généralités

1.1.1 Description des problèmes considérés

Nous considérons comme à la Figure 1.1, deux fluides incompressibles de densité différente en contact selon une surface S. Nous notons Ω_1 et Ω_2 les sous-domaines occupés respectivement par chacun de ces fluides et nous supposons qu'ils sont mis en mouvement par des forces f_1 et f_2.

M. C. Delfour and G. Sabidussi (eds.), Shape Optimization and Free Boundaries, 143–172.

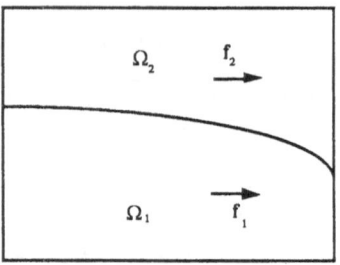

Figure 1.1

Les équations de Navier-Stokes s'écrivent alors

$$\left.\begin{aligned}\rho_1\left(\frac{\partial \boldsymbol{u}_1}{\partial t} + \boldsymbol{u}_1\cdot\operatorname{\mathbf{grad}}\boldsymbol{u}_1 - \nu\operatorname{div}\boldsymbol{\varepsilon}(\boldsymbol{u}_1)\right) + \operatorname{\mathbf{grad}}p_1 = \rho_1\boldsymbol{f}_1, \\ \rho_2\left(\frac{\partial \boldsymbol{u}_2}{\partial t} + \boldsymbol{u}_2\cdot\operatorname{\mathbf{grad}}\boldsymbol{u}_2 - \nu\operatorname{div}\boldsymbol{\varepsilon}(\boldsymbol{u}_2)\right) + \operatorname{\mathbf{grad}}p_2 = \rho_2\boldsymbol{f}_2,\end{aligned}\right\} \tag{1.1}$$

alors que sur l'interface S, on doit respecter les conditions de raccord

$$\boldsymbol{u}_1|_S = \boldsymbol{u}_2|_S \tag{1.2}$$

$$p_1 + \nu\varepsilon_{n^1n^1}(\boldsymbol{u}_1) = p_2 + \nu\varepsilon_{n^2n^2}(\boldsymbol{u}_2) \tag{1.3}$$

$$\varepsilon_{t^1n^1}(\boldsymbol{u}_1) = \varepsilon_{t^2n^2}(\boldsymbol{u}_2). \tag{1.4}$$

On a noté ici,

$$\varepsilon_{n^in^i}(\boldsymbol{u}_i) = \sum_{k,l}\varepsilon_{kl}(\boldsymbol{u}_i)n_k^i n_l^i \tag{1.5}$$

et

$$\varepsilon_{t^in^i}(\boldsymbol{u}_i) = \sum_{k,l}\varepsilon_{kl}(\boldsymbol{u}_i)t_k^i n_l^i, \tag{1.6}$$

n_k^i et t_l^i étant respectivement les composantes de la normale et de la tangente à S, la normale étant orientée vers l'extérieur de Ω_i.

Un cas particulier important de ce problème est celui où le fluide dans Ω_2 est supposé de densité et de viscosité négligeables. La seconde équation de (1.1) se réduit alors à

$$\operatorname{\mathbf{grad}}p_2 = 0 \tag{1.7}$$

d'où $p_2 = p_{ext}$ est constante dans Ω_2. La condition (1.2) disparait et (1.3)–(1.4) se réduisent, notant $\boldsymbol{u}$ et p au lieu de $\boldsymbol{u}_1$ et p_1, à

$$p + \varepsilon_{nn}(\boldsymbol{u}) = p_{ext}, \tag{1.8}$$

$$\varepsilon_{tn}(\boldsymbol{u}) = 0. \tag{1.9}$$

On dit maintenant que l'on a affaire à un problème de surface libre. Nous voulons pouvoir calculer de façon efficace et précise la position de cette surface libre dans l'écoulement d'un fluide visqueux incompressible en faisant un minimum d'hypothèses simplificatrices.

Le problème traité peut être stationnaire ou évolutif. Dans le cas stationnaire, on doit adjoindre à (1.8) et (1.9) la condition,

$$\boldsymbol{u} \cdot \boldsymbol{n} = 0 \tag{1.10}$$

alors que dans le cas évolutif on a, notant $\boldsymbol{x}$ la position d'un point de la surface libre,

$$\frac{d}{dt}(\boldsymbol{x} \cdot \boldsymbol{n}) = \boldsymbol{u} \cdot \boldsymbol{n} \tag{1.11}$$

Nous n'aborderons pas la question de l'existence d'une solution. On sait que les équations de Navier-Stokes ont des solutions turbulentes pour les faibles valeurs de la viscosité. On doit donc s'attendre à des difficultés dans le cas d'écoulements peu visqueux. De fait, il est plausible que les solutions physiquement stables soient instationnaires dans beaucoup de problèmes pratiques. Il sera donc nécessaire de pouvoir calculer des solutions de ce type.

1.1.2 Optimisation de forme

Le problème de surface libre que nous venons de décrire peut se résumer à trouver la forme d'un domaine. Il est naturel de se demander si ce domaine est optimal par rapport à une énergie bien choisie ou à quelque autre critère. Dans le cas des écoulements irrotationnels des fluides non visqueux, pour lesquels il existe un potentiel ϕ tel que $\boldsymbol{u} = \mathbf{grad}\,\phi$, la réponse est affirmative. On sait alors (Zolesio [1991]) que le domaine Ω minimise

$$\frac{1}{2} \int_{\Omega} |\mathbf{grad}\,\phi|^2 d\boldsymbol{x} + \int_{\Omega} \Xi\,d\boldsymbol{x} + \sigma \int_{S} ds, \tag{1.12}$$

c'est-à-dire la somme des énergies cinétique, potentielle et de tension superficielle. On a noté par Ξ un potentiel donné dont est supposé découler l'ensemble des forces extérieures. Dans la pratique, on aura souvent $\Xi = g x_3$ où g est la gravité et x_3 la coordonnée verticale. On rencontre aussi des cas où le potentiel Ξ est engendré par un phénomène électromagnétique. (Bourgeois-Chevalier-Picasso-Rappaz-Touzani [1989])

Dans le cas d'un problème hydrostatique, i.e. $\boldsymbol{u} = 0$, la condition d'optimalité de (1.12) se réduit à

$$\sigma H + \Xi = Cte, \tag{1.13}$$

où H est la courbure de S. La condition (1.13) est un problème de surface minimale classique. Nous retrouverons plus loin un critère analogue à (1.12) dans le cas plus général des équations d'Euler. Nous n'aurons plus dans ce cas minimum mais un équilibre.

Pour ce qui est des fluides visqueux, la situation est moins claire. Considérons par exemple le problème de Stokes stationnaire,

$$\left.\begin{array}{l} - \text{div } \boldsymbol{\varepsilon}(\boldsymbol{u}) + \text{ grad } p = \boldsymbol{f} \\ \text{div } \boldsymbol{u} = 0, \\ p + \varepsilon_{nn}(\boldsymbol{u}) = p_{ext}, \ \ \varepsilon_{tn}(\boldsymbol{u}) = 0, \ \ \text{sur } S \end{array}\right\} \tag{1.14}$$

qui est, comme on le sait, équivalent dans le cas d'un domaine Ω connu, à la minimisation de la fonctionnelle

$$\frac{1}{2}\int_\Omega |\boldsymbol{\varepsilon}(v)|^2 dx - \int_\Omega \boldsymbol{f}\cdot v dx \tag{1.15}$$

sur les fonctions à divergence nulle. Par analogie avec le problème scalaire correspondant, on pourrait espérer retrouver la condition de surface libre $\boldsymbol{u}\cdot\boldsymbol{n} = 0$ en minimisant (1.15) par rapport à la forme de Ω, en utilisant les techniques décrites par Delfour-Zolésio [1991a]. Les calculs montrent cependant que ce n'est pas le cas: la minimisation conduit à des conditions sur la surface n'ayant aucun sens physique. Pour chercher à mieux comprendre et à expliciter éventuellement l'origine des conditions de surface libre dans le cas visqueux, nous allons d'abord considérer le problème de l'élasticité pour lequel la situation est bien maîtrisée.

1.2 Problèmes de frontières libres en élasticité

1.2.1 Notations

Nous considérons le problème de la déformation d'un corps élastique soumis à l'action de forces extérieures volumiques ou de surface. Nous nous plaçons dans le cadre des déformations finies sans aucune hypothèse de "petits déplacements".

Soit donc B le "corps matériel" ou corps de référence et soit Ω_t le corps tel que déformé au temps t. (Figure 1.2)

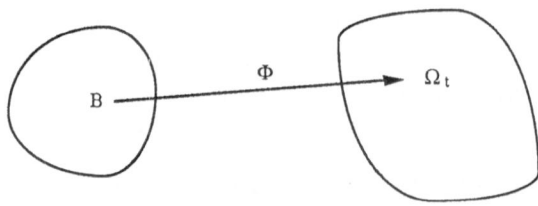

Figure 1.2

Nous noterons respectivement X les coordonnées de B dites coordonnées matérielles ou lagrangiennes et x celles de Ω_t dites coordonnées spatiales ou eulériennes. Nous noterons $\mathbf{GRAD}$, DIV, $\mathbf{grad}$ et div les opérateurs gradient et divergence dans chacun de ces systèmes de référence. La déformation de B en Ω_t est décrite par la relation

$$x = \boldsymbol{\Phi}(X, t). \tag{2.1}$$

On associe classiquement à $\boldsymbol{\Phi}$ les quantités suivantes:

$$F = \frac{\partial x}{\partial X} = \mathbf{GRAD}\,\boldsymbol{\Phi} \tag{2.2}$$

sera appelé tenseur de déformation,

$$U(X, t) = u(x, t) = \frac{\partial \boldsymbol{\Phi}}{\partial t} \tag{2.3}$$

sera la vitesse et

$$A(X, t) = a(x, t) = \frac{\partial^2 \boldsymbol{\Phi}}{\partial t^2} = \frac{\partial U(X, t)}{\partial t} = \frac{du(x, t)}{dt} = \frac{\partial u}{\partial t} + v \cdot \mathbf{grad}\,u \tag{2.4}$$

sera l'accélération.

La Figure 1.3 résume les conditions aux limites du problème que nous considérons.

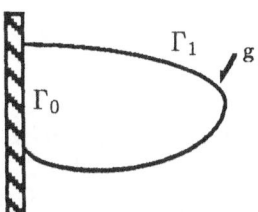

Figure 1.3

Sur la partie Γ_0 de la frontière on suppose le déplacement $\boldsymbol{\Phi}$ connu alors que sur la partie complémentaire Γ_1, des forces de surface, éventuellement nulles, sont imposées.

1.2.2 Élasticité, cas stationnaire

Pour décrire le comportement du matériau, on postule l'existence d'une énergie interne $W(F)$, fonction donc du tenseur de déformation en coordonnées lagrangiennes. Ce type de matériau est dit hyperélastique. On introduit la fonctionnelle (Marsden-Hughes [1983])

$$E(\boldsymbol{\Phi}) = \int_B \rho_{ref} W(F) dX - \int_B \rho_{ref} f \cdot \boldsymbol{\Phi} dX \tag{2.5}$$

qui est donc implicitement une fonction du domaine Ω_t.

Soit alors η une variation de Φ. Cette variation devra être choisie dans l'espace tangent à la variété des Φ admissibles, par exemple dans le cas incompressible la variété des Φ préservant le volume. On change alors formellement Φ en $\Phi + \epsilon\eta$ et F en $F + \epsilon\,\mathbf{GRAD}\,\eta$. La dérivée de $E(\Phi)$ dans la direction η s'écrit,

$$dE(\Phi) \cdot \eta = \int_B \rho_{ref}\left(\frac{\partial W}{\partial F} : \mathbf{GRAD}\,\eta - f \cdot \eta\right)dX \tag{2.6}$$

où nous avons noté par ":" la double contraction des tenseurs, c'est-à-dire en coordonnées cartésiennes

$$P : Q = \sum_{i,j} P_{ij}Q_{ij}.$$

Définissons alors par

$$P = \rho_{ref}\frac{\partial W}{\partial F} \tag{2.7}$$

le tenseur de Piola-Kirchhoff et en coordonnées eulériennes, le tenseur de Cauchy σ par

$$P = JF^{-1}\sigma \tag{2.8}$$

où $J = det F$. La transformation de P en σ s'appelle la transformation de Piola et les propriétés fondamentales en sont:

$$\int_B P : \mathbf{GRAD}\,\eta\,dX = \int_{\Omega_t} \sigma : \mathbf{grad}\,\eta\,dx \tag{2.9}$$

et

$$\int_{\partial B} \eta \cdot P \cdot N dS = \int_{\partial\Omega_t} \eta \cdot \sigma \cdot n\,ds. \tag{2.10}$$

Par (2.6), dire que Φ minimise l'énergie est alors équivalent à

$$\int_B (P : \mathbf{grad}\,\eta - \rho_{ref}f \cdot \eta)dX = 0, \quad \forall\,\eta, \tag{2.11}$$

ce qui s'interprète en coordonnées lagrangiennes sous la forme,

$$\left.\begin{array}{l} \mathbf{DIV}\,P + \rho_{ref}f = 0, \\ P \cdot N = 0 \text{ sur } \Gamma_1 \end{array}\right\} \tag{2.12}$$

et en coordonnées eulériennes par,

$$\left.\begin{array}{l} \mathbf{div}\,\sigma + \rho f = 0, \\ \sigma \cdot n = 0 \text{ sur } \Gamma_1. \end{array}\right\} \tag{2.13}$$

On vérifie facilement que l'introduction de la variation η est précisément équivalente à une variation du domaine par un champ de vitesse au sens de l'optimisation de domaine.

Notons que dans la perturbation $\boldsymbol{\Phi} \to \boldsymbol{\Phi} + \epsilon\boldsymbol{\eta}$ on a

$$\left.\begin{array}{l} \boldsymbol{F} \to \boldsymbol{F} + \epsilon\, \mathbf{GRAD}\,\eta \\[2mm] \dot{\boldsymbol{F}} = \dfrac{\partial \boldsymbol{F}}{\partial t} \to \dot{\boldsymbol{F}} + \epsilon\, \mathbf{GRAD}\, V, \end{array}\right\} \tag{2.14}$$

où $V = \dfrac{\partial \eta}{\partial t}$.

La condition de frontière libre $\boldsymbol{P} \cdot \boldsymbol{N} = 0$ ou $\boldsymbol{\sigma} \cdot \boldsymbol{n} = 0$ est donc ici une condition d'optimalité par rapport à la frontière du domaine, car l'énergie W dépend directement de la déformation.

1.2.3 Élasticité, cas évolutif

Les considérations précédentes s'étendent au cas évolutif dans le cadre de la mécanique hamiltonienne. Soit donc $[t_1, t_2]$ l'intervalle de temps dans lequel on désire décrire l'évolution d'un corps élastique. On introduit le lagrangien,

$$\mathrm{L}(\boldsymbol{\Phi}, \dot{\boldsymbol{\Phi}}) = \int_B (\tfrac{1}{2}\rho_{ref}|U|^2 - \rho_{ref}W + \rho_{ref}\boldsymbol{f} \cdot \boldsymbol{\Phi})d\boldsymbol{X} \tag{2.15}$$

et on considère

$$L(\boldsymbol{\Phi}, \dot{\boldsymbol{\Phi}}) = \int_{t_1}^{t_2} \mathrm{L}(\boldsymbol{\Phi}, \dot{\boldsymbol{\Phi}})dt + \int_B \rho_{ref}\boldsymbol{u}_0 \cdot \boldsymbol{\Phi}d\boldsymbol{X}.$$

On va caractériser les points d'équilibre de L par rapport aux variations de $\boldsymbol{\Phi}$ de la forme

$$\left.\begin{array}{l} \boldsymbol{\Phi}(\boldsymbol{X};t) \to \boldsymbol{\Phi}(\boldsymbol{X};t) + \epsilon\boldsymbol{\eta}(\boldsymbol{X};t) \\[2mm] \boldsymbol{\eta}(\boldsymbol{X};t_2) = 0. \end{array}\right\} \tag{2.16}$$

Dérivant en effet $L(\boldsymbol{\Phi}, \dot{\boldsymbol{\Phi}})$ par rapport à ϵ, on obtient, posant toujours

$$\boldsymbol{P} = \rho_{ref}\partial W/\partial \boldsymbol{F},$$

$$\int_{t_1}^{t_2}\int_B (\rho_{ref}U \cdot \dot{\boldsymbol{\eta}} - \boldsymbol{P} : \mathbf{GRAD}\,\eta + \rho_{ref}\boldsymbol{f} \cdot \boldsymbol{\eta})d\boldsymbol{X}\,dt + \int_B \rho_{ref}\boldsymbol{u}_0 \cdot \boldsymbol{\eta}d\boldsymbol{X} = 0, \tag{2.17}$$

d'où, en intégrant par parties,

$$\left.\begin{array}{l} \rho_{ref}\dot{U} - \mathbf{DIV}\,\boldsymbol{P} = \rho_{ref}\boldsymbol{f}, \\[2mm] \boldsymbol{P} \cdot \boldsymbol{N} = 0 \text{ sur } \Gamma_1, \\[2mm] U(0) = \boldsymbol{u}_0, \end{array}\right\} \tag{2.18}$$

$\boldsymbol{\Phi}$ étant toujours supposé connu sur Γ_0.

Pour écrire le problème dans un cadre hamiltonien, il existe plusieurs façons de procéder. La plus classique consiste à introduire une variable $\boldsymbol{Y}$ que l'on veut égale à U

et à imposer cette égalité au moyen d'un multiplicateur de Lagrange. On écrit alors les conditions d'équilibre de

$$L(\boldsymbol{\Phi}, \boldsymbol{Y}, \boldsymbol{\Lambda}) = \int_{t_1}^{t_2} \left(\mathbb{L}(\boldsymbol{\Phi}, \boldsymbol{Y}) - \int_B \rho_{ref} \boldsymbol{\Lambda} \cdot (\boldsymbol{Y} - \boldsymbol{V}) dX \right) dt. \qquad (2.19)$$

Nous prenons dorénavant $\boldsymbol{u}_0 = 0$ pour alléger. La condition d'optimalité en $\boldsymbol{Y}$ donne immédiatement $\boldsymbol{\Lambda} = \boldsymbol{Y}$ d'où éliminant $\boldsymbol{Y}$, on a à chercher l'équilibre de l'hamiltonien

$$\mathcal{H}(\boldsymbol{\Lambda}, \boldsymbol{\Phi}) = \int_{t_1}^{t_2} \int_B (\rho_{ref} \boldsymbol{\Lambda} \cdot \boldsymbol{V} - \frac{\rho_{ref}}{2} |\boldsymbol{\Lambda}|^2 - \rho_{ref} W + \rho_{ref} \boldsymbol{f} \cdot \boldsymbol{\Phi}) dX dt. \qquad (2.20)$$

La variation de $\mathcal{H}$ par rapport à $\boldsymbol{\Lambda}$ nous donne,

$$\boldsymbol{\Lambda} = \boldsymbol{U} = \frac{\partial \boldsymbol{\Phi}}{\partial t} \qquad (2.21)$$

et la variation par rapport à $\boldsymbol{\Phi}$,

$$\int_{t_1}^{t_2} \int_B (\rho_{ref} \boldsymbol{\Lambda} \cdot \dot{\boldsymbol{\eta}} - \boldsymbol{P} : \mathbf{GRAD} \boldsymbol{\eta} + \rho_{ref} \boldsymbol{f} \cdot \boldsymbol{\eta}) dX \, dt = 0, \qquad (2.22)$$

c'est-à-dire (2.18). On a donc écrit le problème sous la forme classique,

$$\left. \begin{array}{rcl} \dfrac{\partial \boldsymbol{\Phi}}{\partial t} &=& \dfrac{\partial \mathcal{H}}{\partial \boldsymbol{\Lambda}}, \\[2mm] \dfrac{\partial \boldsymbol{\Lambda}}{\partial t} &=& -\dfrac{\partial \mathcal{H}}{\partial \boldsymbol{\Phi}}. \end{array} \right\} \qquad (2.23)$$

On aurait pu aussi introduire une variable S que l'on voudrait représenter $\boldsymbol{F}(\boldsymbol{X}, t)$ et chercher un point d'équilibre de:

$$\int_{t_1}^{t_2} \int_B (\frac{\rho_{ref}}{2} |\boldsymbol{Y}|^2 \quad -\rho_{ref} W(S) + \rho_{ref} \boldsymbol{f} \cdot \boldsymbol{\Phi}$$
$$-\rho_{ref} \boldsymbol{\Lambda}_1 \cdot (\boldsymbol{Y} - \boldsymbol{V}) + \rho_{ref} \boldsymbol{\Lambda}_2 : (S - \boldsymbol{F})) dX \, dt. \qquad (2.24)$$

L'optimalité en $\boldsymbol{Y}$ donne toujours $\boldsymbol{Y} = \boldsymbol{\Lambda}_1$. L'optimalité en S est plus subtile. On a en effet,

$$\sup_S \int_{t_1}^{t_2} \int_B \rho_{ref} (\boldsymbol{\Lambda}_2 : S - W(S)) dX \, dt = \int_{t_1}^{t_2} \int_B \rho_{ref} W^*(\boldsymbol{\Lambda}_2) dX \, dt, \qquad (2.25)$$

où W^* est la fonction conjuguée de W (que l'on suppose convexe) par rapport au produit scalaire,

$$\langle \boldsymbol{S}, \boldsymbol{T} \rangle = \int_B \rho \boldsymbol{S} : \boldsymbol{T} \, dX.$$

Le problème se ramène donc à optimiser,

$$\int_{t_1}^{t_2} \int_B \rho_{ref} (\boldsymbol{\Lambda}_1 \cdot \boldsymbol{V} - \boldsymbol{\Lambda}_2 : \boldsymbol{F} - \frac{1}{2} |\boldsymbol{\Lambda}_1|^2 + W^*(\boldsymbol{\Lambda}_2) + \boldsymbol{f} \cdot \boldsymbol{\Phi}) dX \, dt. \qquad (2.26)$$

Dérivant, on trouve:

$$\langle \mathbf{A_1}, \mathbf{V} \rangle = \langle \mathbf{U}, \mathbf{V} \rangle, \quad \forall \mathbf{V}, \tag{2.27}$$

$$\langle \mathbf{F}, \mathbf{T} \rangle = \langle \frac{\partial W^*}{\partial \mathbf{A_2}}, \mathbf{T} \rangle, \quad \forall \mathbf{T}, \tag{2.28}$$

$$\langle \mathbf{A_1}, \dot{\boldsymbol{\eta}} \rangle - \langle \mathbf{A_2}, \mathbf{GRAD} \, \boldsymbol{\eta} \rangle + \langle \mathbf{f}, \boldsymbol{\eta} \rangle = 0, \quad \forall \boldsymbol{\eta}. \tag{2.29}$$

L'équation (2.28) s'inverse par un résultat classique d'analyse convexe (ou si l'on préfère par la propriété bien connue de la transformation de Legendre), pour donner,

$$\langle \mathbf{A_2}, \mathbf{T} \rangle = \langle \frac{\partial W}{\partial \mathbf{F}}, \mathbf{T} \rangle$$

ou encore

$$\rho_{ref} \mathbf{A_2} = \rho_{ref} \frac{\partial W}{\partial \mathbf{F}} = \mathbf{P}. \tag{2.30}$$

$\rho_{ref} \mathbf{A_2}$ est donc le tenseur de Piola-Kirchhoff. Si on ne retrouve plus ainsi la forme classique (2.23), on voit cependant mieux le rôle du tenseur $\mathbf{P}$.

1.3 Problèmes à frontière libre en mécanique des fluides

1.3.1 Équations d'Euler

Nous sommes maintenant en mesure de retrouver les équations d'Euler et la condition de frontière libre qui leur est associée. On considère simplement, utilisant les mêmes notations qu'à la section précédente, l'énergie cinétique et le lagrangien,

$$\int_{t_1}^{t_2} \int_B \rho_{ref} \left(\frac{1}{2} |\mathbf{U}|^2 + \mathbf{f} \cdot \boldsymbol{\Phi} \right) d\mathbf{X} \, dt + \int_B \rho_{ref} \mathbf{u_0} \cdot \boldsymbol{\Phi} d\mathbf{X} \tag{3.1}$$

dont on prend les variations par rapport à une perturbation préservant le volume (du moins dans le cas des fluides incompressibles que nous considérons d'abord). On sait alors (cf. Ebin-Marsden [1970]), que la perturbation $\boldsymbol{\eta}$ vérifie,

$$\text{div} \left(\boldsymbol{\eta} \circ \boldsymbol{\Phi}^{-1} \right) = 0, \tag{3.2}$$

c'est-à-dire que l'image de $\boldsymbol{\eta}$ en coordonnées eulériennes est à divergence nulle.

En dérivant (3.1), on obtient,

$$\int_{t_1}^{t_2} \int_B \rho_{ref} (\mathbf{U} \cdot \dot{\boldsymbol{\eta}} + \mathbf{f} \cdot \boldsymbol{\eta}) d\mathbf{X} \, dt + \int_b \rho_{ref} \mathbf{u_0} \cdot \boldsymbol{\eta} d\mathbf{X} = 0, \tag{3.3}$$

pour tout $\boldsymbol{\eta}$ vérifiant (3.2). Après intégration par parties, ceci devient, en coordonnées spatiales,

$$\left.\begin{aligned}
&\int_{t_1}^{t_2} \int_{\Omega_t} \rho \left(\frac{\partial \boldsymbol{u}}{\partial t} + \boldsymbol{u} \cdot \ \mathbf{grad}\ \boldsymbol{u} - \boldsymbol{f} \right) \cdot \boldsymbol{v} \, d\boldsymbol{x} \, dt, \\
&\int_B \rho(\boldsymbol{u}_0 - \boldsymbol{u}(0)) \cdot \boldsymbol{v}(0) d\boldsymbol{X} = 0, \quad \forall \boldsymbol{v}, \ \mathrm{div}\ \boldsymbol{v} = 0.
\end{aligned}\right\} \tag{3.4}$$

Par un résultat classique, il existe donc une fonction p telle que,

$$\int_{t_1}^{t_2} \int_{\Omega_t} \left(\rho \left(\frac{\partial \boldsymbol{u}}{\partial t} + \boldsymbol{u} \cdot \ \mathbf{grad}\ \boldsymbol{u} - \boldsymbol{f} \right) \cdot \boldsymbol{v} - p \ \mathrm{div}\ \boldsymbol{v} \right) d\boldsymbol{x} \, dt = 0, \quad \forall \boldsymbol{v} \tag{3.5}$$

et on obtient ainsi les équations d'Euler,

$$\left.\begin{aligned}
&\rho \left(\frac{\partial \boldsymbol{u}}{\partial t} + \boldsymbol{u} \cdot \ \mathbf{grad}\ \boldsymbol{u} \right) + \ \mathbf{grad}\ p = \rho \boldsymbol{f}, \\
&\mathrm{div}\ \boldsymbol{u} = 0, \\
&\boldsymbol{u}(0) = \boldsymbol{u}_0
\end{aligned}\right\} \tag{3.6}$$

et sur la partie libre de la frontière,

$$p = 0. \tag{3.7}$$

Ce résultat est essentiellement instationnaire; il exprime un équilibre dynamique.

Remarque   On peut aussi considérer (cf. Marsden-Hughes [1983]) le cas d'un fluide compressible. Dans ce cas, il existe comme dans le cas de l'élasticité une énergie emmagasinée que l'on suppose ne dépendre que de $J = det\boldsymbol{F}$, i.e. $W = h(J)$. On ne peut plus ici supposer la densité constante. On pose comme à la section 2,

$$\boldsymbol{P} = \rho_{ref} \frac{\partial W}{\partial \boldsymbol{F}} = \rho h'(J) J \boldsymbol{F}^{-1}, \tag{3.8}$$

ou encore

$$J \boldsymbol{F}^{-1} \boldsymbol{\sigma} = \rho h'(J) J \boldsymbol{F}^{-1}.$$

On en déduit pour $\boldsymbol{\sigma}$ la relation,

$$\boldsymbol{\sigma} = \rho h' \boldsymbol{I} = -p(\rho) \boldsymbol{I}. \tag{3.9}$$

On retrouve donc l'équation d'Euler,

$$\rho \left(\frac{\partial \boldsymbol{u}}{\partial t} + \boldsymbol{u} \cdot \ \mathbf{grad}\ \boldsymbol{u} \right) + \ \mathbf{grad}\ p = \rho \boldsymbol{f}, \tag{3.10}$$

que l'on doit compléter par l'équation de conservation de la masse,

$$\frac{\partial \rho}{\partial t} + \mathrm{div}\ (\rho \boldsymbol{u}) = 0. \tag{3.11}$$

1.3.2 Équations de Navier-Stokes

Nous aimerions maintenant étendre les résultats précédents au cas des fluides visqueux. Nous nous placerons d'emblée dans le cas incompressible et nous supposerons dorénavant la densité constante et égale à 1. On aura toujours dans ce cadre $J = det F = 1$. Si on se reporte au cas de l'élasticité, on voit que nous devons parvenir à écrire le tenseur des contraintes comme la dérivée d'une énergie interne appropriée. Contrairement à un matériau élastique, un fluide n'accumule pas d'énergie sous l'effet d'une déformation. Il dissipe plutôt en chaleur de l'énergie cinétique par frottement visqueux. Dans le cas le plus simple, adiabatique, on peut supposer que cette énergie dissipée est entraînée par le fluide et n'interagit plus avec ce dernier. On peut alors écrire pour l'énergie W l'équation,

$$\dot{W} = \boldsymbol{\sigma} : \mathbf{grad} \; \mathbf{u}. \tag{3.12}$$

Cherchons donc quelle expression générale pourrait servir à définir $\boldsymbol{\sigma}$ et $\boldsymbol{P}$. Dans le cas d'un fluide newtonien incompressible à densité constante, on a en coordonnées spatiales,

$$\dot{W} = \mu |\mathbf{grad} \; \mathbf{u}|^2 \tag{3.13}$$

et la relation

$$\boldsymbol{\sigma} = 2\mu \mathbf{grad} \; \mathbf{u} = \frac{\partial \dot{W}}{\partial \mathbf{grad} \; \mathbf{u}} = 2\mu (\boldsymbol{F}^{-t} \mathbf{GRAD} \; U). \tag{3.14}$$

On peut traduire la relation (3.13) dans le cadre lagrangien par

$$\dot{W} = \mu |(\boldsymbol{F}^{-t} \mathbf{GRAD} \; U)|^2 = 2\mu |\boldsymbol{F}^{-t} \dot{\boldsymbol{F}}|^2 \tag{3.15}$$

et il s'ensuit (rappelons que $J = 1$),

$$
\begin{aligned}
\frac{\partial \dot{W}}{\partial \dot{\boldsymbol{F}}} = \frac{\partial \dot{W}}{\partial \mathbf{GRAD} \; U} &= 2\mu \boldsymbol{F}^{-t} \mathbf{GRAD} \; U : \boldsymbol{F}^{-t} \\
&= \boldsymbol{F}^{-1}(2\mu \mathbf{grad} \; \mathbf{u}) : I = \boldsymbol{F}^{-1} \boldsymbol{\sigma} : I = P.
\end{aligned}
\tag{3.16}
$$

On supposera donc dans le cas général une énergie interne $W(\dot{\boldsymbol{F}})$ vérifiant

$$\frac{\partial \dot{W}}{\partial \dot{\boldsymbol{F}}} = \boldsymbol{P}. \tag{3.17}$$

On peut vérifier que cette définition reste cohérente dans le cas des fluides newtoniens généralisés où l'on a par exemple,

$$\boldsymbol{\sigma} = k |\boldsymbol{\varepsilon}(\boldsymbol{u})|^{p-2} \boldsymbol{\varepsilon}(\boldsymbol{u}). \tag{3.18}$$

Dans le cas d'un fluide viscoélastique on devrait sans doute considérer une énergie dépendant de $\boldsymbol{F}$ et de $\dot{\boldsymbol{F}}$. Comme nous l'avons vu à la section précédente, on peut traiter le cas compressible en introduisant une énergie supplémentaire et récupérable.

Revenons cependant à notre question de départ. Nous introduisons comme en élasticité le lagrangien,

$$L(\boldsymbol{\Phi}, \dot{\boldsymbol{\Phi}}) = \int_B \left(\frac{1}{2} |U|^2 + W(\dot{\boldsymbol{F}}) + \boldsymbol{f} \cdot \boldsymbol{n} \right) dX, \tag{3.19}$$

$$L(\boldsymbol{\phi}, \dot{\boldsymbol{\phi}}) = \int_{t_1}^{t_2} L(\boldsymbol{\Phi}, \dot{\boldsymbol{\Phi}}) dt + \int_B \boldsymbol{u}_0 \cdot \boldsymbol{\Phi} dX \tag{3.20}$$

et on introduit une variation de la forme $\boldsymbol{\Phi} \to \boldsymbol{\Phi} + \epsilon \boldsymbol{\eta}(\boldsymbol{X}, t)$. Dérivant, on obtient (formellement toujours...):

$$\int_{t_1}^{t_2} \int_B \left(U \cdot \dot{\boldsymbol{\eta}} + \frac{\partial W}{\partial \dot{\boldsymbol{F}}} : \mathbf{GRAD} \, \dot{\boldsymbol{\eta}} + \boldsymbol{f} \cdot \boldsymbol{\eta} \right) dX \, dt + \int_B \boldsymbol{u}_0 \cdot \boldsymbol{\eta} dX = 0. \tag{3.21}$$

Posant $S = \partial W / \partial \dot{\boldsymbol{F}}$ et intégrant par parties en temps, il vient:

$$\begin{aligned}
\int_{t_1}^{t_2} \int_B \ & \left(-\dot{U} \cdot \boldsymbol{\eta} - \dot{S} \cdot \mathbf{GRAD} \, \boldsymbol{\eta} + \boldsymbol{f} \cdot \boldsymbol{\eta} \right) dX \, dt \\
& + \int_B (\boldsymbol{u}_0 - U(0)) \cdot \boldsymbol{\eta}(0) dX - \int_B S : \mathbf{GRAD} \, \boldsymbol{\eta}(0) dX = 0,
\end{aligned} \tag{3.22}$$

pour tout $\boldsymbol{\eta}$ à divergence nulle. En écrivant $P = \dot{S}$, on obtient

$$\left. \begin{aligned}
& \dot{U} + \mathrm{DIV} \, \boldsymbol{P} = \boldsymbol{f}, \\
& U(0) = \boldsymbol{U}_0, \\
& \mathrm{DIV} \, S(0) = 0.
\end{aligned} \right\} \tag{3.23}$$

Passant en coordonnées spatiales, on obtient en introduisant comme dans le cas des équations d'Euler une pression,

$$\left. \begin{aligned}
& \frac{\partial \boldsymbol{u}}{\partial t} + \boldsymbol{u} \cdot \mathrm{grad} \, \boldsymbol{u} + \mathrm{div} \, \boldsymbol{\sigma} + \mathrm{grad} \, p = \boldsymbol{f}, \\
& \mathrm{div} \, \boldsymbol{u} = 0, \\
& p\boldsymbol{I} + \boldsymbol{\sigma} \cdot \boldsymbol{n} = 0 \ \text{sur} \ \Gamma_1, \\
& \boldsymbol{u}(0) = \boldsymbol{u}_0, \\
& \mathrm{div} \, \boldsymbol{s}(0) = 0,
\end{aligned} \right\} \tag{3.24}$$

où le tenseur $\boldsymbol{s}$ vérifie,

$$\mathcal{L}\boldsymbol{s} = \boldsymbol{\sigma}, \tag{3.25}$$

$\mathcal{L}$ étant la dérivée de Lie contravariante (Marsden-Hughes [1983]). Ce tenseur $\boldsymbol{s}$ ou son équivalent en coordonnées lagrangiennes S représente l'histoire des contraintes. Cependant cette histoire n'intervient dans les équations que par son taux de variation. La condition supplémentaire $\mathrm{DIV} \, S(0) = 0$ est ainsi sans influence sur le comportement du fluide.

Une remarque s'impose en ce qui concerne le lagrangien (3.19): même dans le cas d'un écoulement stationnaire ($\partial \boldsymbol{u} / \partial t = 0$), le terme d'énergie cinétique, $\frac{1}{2} \int_B |U|^2 d\boldsymbol{x}$ ne sera pas nul. On peut toutefois le négliger, obtenant ainsi l'approximation de Stokes, valable pour les écoulements à faible vitesse.

On peut également introduire un hamiltonien de la forme (2.26) tenant compte de la dépendance de W en $\dot{F}$.

$$\int_{t_1}^{t_2} \int_B \left(Y \cdot V - A_2 : \dot{F} - \frac{1}{2}|Y|^2 + W^*(A_2) + f \cdot \Phi \right) dX \, dt \qquad (3.26)$$

d'où en dérivant

$$\int_B Y \cdot V \, dX = \int_B U \cdot V \, dX, \quad \forall V \qquad (3.27)$$

$$\dot{F} = \frac{\partial W^*}{\partial A_2}, \quad A_2 = \frac{\partial W}{\partial \dot{F}}(\dot{F}) \qquad (3.28)$$

$$\int_{t_1}^{t_2} \int_B \left(Y \cdot \dot{\eta} + \int A_2 : \mathbf{GRAD} \, \dot{\eta} - f \cdot \eta \right) dX \, dt = 0. \qquad (3.29)$$

On retrouve bien les équations de Navier-Stokes et $A_2 = S$.

On pourrait aussi considérer le problème dans le cadre des structures de Poisson en utilisant un crochet de Poisson bien défini. Pour les fluides visqueux et visco-élastiques, cette approche est développé dans Beris-Edwards [1990]. Guénette [1990] a aussi développé cette approche pour les fluides orientés (cristaux liquides en phase nématique. Dans le cadre élastique la formulation hamiltonienne et la formulation en crochet de Poisson sont équivalentes. Dans le contexte des fluides visqueux, le passage d'une approche à l'autre reste à faire.

1.3.3 Conclusion

Un certain nombre de problèmes de mécanique impliquent la détermination de la forme d'un domaine. Dans le cas de l'élasticité statique, la forme du domaine minimise une énergie. Dans le cas dynamique on a plutôt un équilibre. En mécanique des fluides, sauf sous des hypothèses très restrictives (écoulement irrotationnel), le processus est toujours dynamique.

S'il ne s'agit pas d'un problème de minimisation par rapport à Ω, du type

$$\inf_{\Omega} J(\Omega), \qquad (3.30)$$

on peut cependant l'écrire sous la forme d'une équation

$$F(u, \Omega) = 0. \qquad (3.31)$$

Posant,

$$V = \{v | v \in (H^1(\Omega))^2, \operatorname{div} v = 0\},$$
$$V_0 = \{v | v \in V, v \cdot n|_{\Gamma} = 0\},$$

la formulation variationnelle habituelle des équations de Navier-Stokes stationnaire, pour une frontière donnée, consiste à chercher $u \in V_0$ vérifiant

$$\int u \cdot \operatorname{grad} u \cdot v \, dx + \nu \int_\Omega \operatorname{grad} u \cdot \operatorname{grad} v dx - \int_\Omega f \cdot v dx = 0 \quad \forall v \in V_0. \quad (3.32)$$

La formulation (3.32) contient implicitement la condition naturelle

$$\nu \left(\frac{\partial u \cdot n}{\partial t} + \frac{\partial u \cdot t}{\partial n} \right) = 0.$$

Si on demande de plus que (3.32) soit vérifiée pour tout $v \in V$ le problème est surdéterminé et ne pourra avoir éventuellement une solution que pour un domaine bien particulier. Les fonctions tests non nulles au bord définissent donc des conditions sur la forme du domaine.

Nous allons à la Section 2 nous placer dans ce cadre et chercher à construire des méthodes de type Newton pour la résolution des problèmes de la forme (3.31).

2   Formulations eulériennes lagrangiennes

Le calcul d'une surface libre peut se faire selon plusieurs techniques.

- En coordonnées lagrangiennes: on fixe alors le domaine et on suit les "particules".

- En coordonnées spatiales: le domaine de calcul se déforme en fonction du temps.

La première approche est attrayante mais mal adaptée à la mécanique des fluides car même pour un domaine fixe, les particules se déplacent.

L'approche eulérienne-lagrangienne que nous allons introduire permet de fixer le domaine sans avoir à suivre les particules. Elles nous servira aussi à construire une méthode de type Newton pour le calcul des surfaces libres stationnaires ou non. Notre présentation ne saurait être exhaustive et nous tenons à signaler l'article très complet de Cuvelier-Schulkes [1990] pour une présentation générale des problèmes de surface libre.

2.1   Description de la méthode

On considère toujours comme en élasticité, un corps matériel B et une déformation $\Phi(x, t)$ de B. Nous noterons $\Omega_t = \Phi(B, t)$ l'image de B au temps t par la transformation Φ. De plus nous introduisons un domaine $\hat{\Omega}$ et une transformation $\lambda(\hat{x}, t)$ de $\hat{\Omega}$. (Figure 2.1)

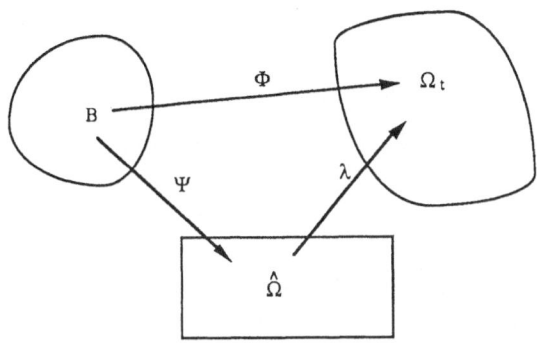

Figure 2.1

On définit ainsi un cadre intermédiaire entre la formulation lagrangienne et la formulation eulérienne (Hughes-Liu-Zimmermann [1981], Donéa [1983]). Nous nous inspirerons pour notre présentation de Soulaimani-Fortin-Dhatt-Ouellet [1991]. On a donc alors pour le mouvement $\boldsymbol{\Phi}$

$$\left. \begin{aligned} U(X,t) &= u(x,t) = \frac{\partial \boldsymbol{\Phi}}{\partial t}(X,t), \\ a(x,t) &= \frac{\partial U}{\partial t} = \frac{d}{dt}u(x,t) = \frac{\partial u}{\partial t} + u \cdot \operatorname{grad} u \end{aligned} \right\} \tag{2.1}$$

et pour $\lambda(\hat{x}, t)$

$$\left. \begin{aligned} \hat{w}(\hat{x},t) &= \frac{\partial \lambda}{\partial t}(\hat{x},t), \\ w(x,t) &= \hat{w}(\hat{x},t), \end{aligned} \right\} \tag{2.2}$$

c'est-à-dire $w = \hat{w} \circ \lambda^{-1}$. On définit aussi le tenseur de déformation

$$\hat{F}_{ij} = \frac{\partial x_i}{\partial \hat{x}_j} \tag{2.3}$$

On a alors, $\hat{x} = \lambda^{-1}(\boldsymbol{\Phi}(X,t)) =: \psi(X,t)$, ou encore

$$x = \lambda(\psi(X,t),t) = \boldsymbol{\Phi}(X,t). \tag{2.4}$$

Dérivant (2.4) par rapport à t, nous obtenons,

$$u(x,t) = w(x,t) + \hat{F}(\hat{x},t)V^{\psi}(\hat{x},t), \tag{2.5}$$

où V^{ψ} est la vitesse relative,

$$V^{\psi} = \frac{\partial}{\partial t}\psi(X,t).$$

Notant $\hat{\boldsymbol{u}}(\hat{\boldsymbol{x}}, t) = \boldsymbol{u}(\boldsymbol{x}, t)$, nous avons donc,

$$\hat{\boldsymbol{u}}(\hat{\boldsymbol{x}}, t) = \hat{\boldsymbol{w}}(\hat{\boldsymbol{x}}, t) + \hat{\boldsymbol{F}} V^\psi(\hat{\boldsymbol{x}}, t).$$

Nous écrivons de même,

$$\hat{\boldsymbol{a}}(\hat{\boldsymbol{x}}, t) = \boldsymbol{a}(\boldsymbol{x}, t) = \frac{d}{dt} \boldsymbol{u}(\boldsymbol{x}, t) \tag{2.6}$$

où $\boldsymbol{u}(\boldsymbol{x}, t) = \hat{\boldsymbol{u}}(\lambda^{-1}(\boldsymbol{x}, t), t) = \hat{\boldsymbol{u}}(\psi(\boldsymbol{X}, t), t)$ et on obtient donc,

$$\boldsymbol{a}(\boldsymbol{x}, t) = \hat{\boldsymbol{a}}(\hat{\boldsymbol{x}}, t) = \frac{\partial \hat{\boldsymbol{u}}}{\partial t} + \hat{\boldsymbol{F}}^{-1}(\hat{\boldsymbol{u}} - \hat{\boldsymbol{w}}) \cdot \widehat{\mathbf{grad}}\, \hat{\boldsymbol{u}}(\hat{\boldsymbol{x}}, t). \tag{2.7}$$

Si $\hat{J} = det \hat{\boldsymbol{F}}$, on peut écrire l'équation de conservation de la quantité de mouvement dans $\hat{\Omega}$ sous la forme

$$\hat{J}\hat{\boldsymbol{a}}(\hat{\boldsymbol{x}}, t) = \hat{J}\hat{\boldsymbol{f}} + \widehat{\mathrm{div}}\, \hat{\boldsymbol{P}}$$

où $\hat{\boldsymbol{P}} = \hat{J}\hat{\boldsymbol{F}}^{-1}\sigma$ est un "tenseur de Piola-Kirchhoff" différent, cependant, de celui utilisé en description matérielle, que nous avons considéré dans la première partie. Par (2.7), on a donc,

$$\hat{J}\left[\frac{\partial \hat{\boldsymbol{u}}}{\partial t} + \hat{\boldsymbol{F}}^{-1}(\hat{\boldsymbol{u}} - \hat{\boldsymbol{w}}) \cdot \widehat{\mathbf{grad}}\, \hat{\boldsymbol{u}}\right] = \hat{J}\hat{\boldsymbol{f}} + \widehat{\mathrm{div}}\, \hat{\boldsymbol{P}}, \tag{2.8}$$

$$\widehat{\mathbf{grad}}\, \hat{\boldsymbol{u}}(\hat{\boldsymbol{x}}, t) : \hat{\boldsymbol{F}}^{-1}(\hat{\boldsymbol{x}}, t) = 0, \tag{2.9}$$

cette dernière équation traduisant la condition de divergence nulle. Pour un fluide newtonien, on a de plus, en coordonnées eulériennes, c'est-à-dire dans Ω_t,

$$\sigma^D(\boldsymbol{x}, t) = 2\mu\boldsymbol{\varepsilon}(\boldsymbol{u}). \tag{2.10}$$

Pour obtenir une forme variationnelle dans $\hat{\Omega}$, on définit donc les formes bilinéaires,

$$\left.\begin{aligned}
\hat{a}(\hat{\boldsymbol{u}}, \hat{\boldsymbol{v}}) &= 2\mu \int_{\hat{\Omega}} \hat{\boldsymbol{F}}^{-t} \widehat{\mathbf{grad}}\, \hat{\boldsymbol{v}} : [\hat{\boldsymbol{F}}^{-t} \widehat{\mathbf{grad}}\, \hat{\boldsymbol{u}} + (\hat{\boldsymbol{F}}^{-t} \widehat{\mathbf{grad}}\, \hat{\boldsymbol{u}})^t]\hat{J} d\hat{\boldsymbol{x}} \\
\hat{b}(\hat{\boldsymbol{v}}, \hat{q}) &= -\int_{\hat{\Omega}} \hat{q}(\widehat{\mathbf{grad}}\, \hat{\boldsymbol{v}} : \hat{\boldsymbol{F}}^{-1}) d\hat{\boldsymbol{x}} \\
\hat{c}(\hat{\boldsymbol{z}}, \hat{\boldsymbol{u}}, \hat{\boldsymbol{v}}) &= \int_{\hat{\Omega}} \hat{J}(\hat{\boldsymbol{F}}^{-1}\hat{\boldsymbol{z}}) \cdot \widehat{\mathbf{grad}}\, \hat{\boldsymbol{u}} \cdot \hat{\boldsymbol{v}} d\hat{\boldsymbol{x}},
\end{aligned}\right\} \tag{2.11}$$

d'où la formulation variationnelle des équations de Navier-Stokes dans $\hat{\Omega}$,

$$\left.\begin{aligned}
\int_{\hat{\Omega}} \hat{J}\frac{\partial \hat{\boldsymbol{u}}}{\partial t} \cdot \hat{\boldsymbol{v}}\, d\hat{\boldsymbol{x}} + \hat{a}(\hat{\boldsymbol{u}}, \hat{\boldsymbol{v}}) &+ \hat{c}(\hat{\boldsymbol{u}} - \hat{\boldsymbol{w}}, \hat{\boldsymbol{u}}, \hat{\boldsymbol{v}}) + \hat{b}(\hat{\boldsymbol{v}}, \hat{p}) \\
&= \int_{\hat{\Omega}} \hat{J}\hat{\boldsymbol{f}} \cdot \hat{\boldsymbol{v}}\, d\hat{\boldsymbol{x}}, \quad \forall \hat{\boldsymbol{v}}, \\
\hat{b}(\hat{\boldsymbol{u}}, \hat{q}) &= 0, \quad \forall \hat{q},
\end{aligned}\right\} \tag{2.12}$$

pour des fonctions tests choisies selon les conditions aux limites à imposer. Nous reviendrons plus loin sur ce point.

2.2 Forme variationnelle approchée

Le principe de l'utilisation de la formulation dans $\hat{\Omega}$ est de trouver un mouvement simple (e.g., vertical) tel que $\boldsymbol{u} \cdot \boldsymbol{n} = \boldsymbol{w} \cdot \boldsymbol{n}$ sur la surface libre. Afin de simplifier la construction des méthodes numériques, nous allons considérer dans ce qui suit une perturbation de la forme,

$$\boldsymbol{x} = \boldsymbol{\lambda}(\hat{\boldsymbol{x}}, t) = \hat{\boldsymbol{x}} + \Delta t \hat{\boldsymbol{w}}(\hat{\boldsymbol{x}}) = \hat{\boldsymbol{x}} + \widehat{\delta \boldsymbol{x}}. \tag{2.13}$$

On a alors pour Δt petit,

$$\left. \begin{aligned} \hat{\boldsymbol{F}}(\hat{\boldsymbol{x}}, t) &= \boldsymbol{I} + \widehat{\text{grad}\,\delta \hat{\boldsymbol{x}}} = \boldsymbol{I} + \Delta t\,\widehat{\text{grad}\,\hat{\boldsymbol{w}}}, \\ \hat{\boldsymbol{F}}^{-1}(\hat{\boldsymbol{x}}, t) &\approx \boldsymbol{I} - \widehat{\text{grad}\,\delta \hat{\boldsymbol{x}}} = \boldsymbol{I} + \Delta t\,\widehat{\text{grad}\,\hat{\boldsymbol{w}}}, \\ \hat{J}(\hat{\boldsymbol{x}}, t) &\approx 1 + \widehat{\text{div}\,\delta \hat{\boldsymbol{x}}} = 1 + \Delta t\,\widehat{\text{div}\,\hat{\boldsymbol{w}}}. \end{aligned} \right\} \tag{2.14}$$

Reportant ces expressions dans (2.11), il vient, en négligeant les termes d'ordre deux en $\delta \hat{\boldsymbol{x}}$,

$$\left. \begin{aligned} \hat{a}(\hat{\boldsymbol{u}}, \hat{\boldsymbol{v}}) &\approx 2\mu \int_{\hat{\Omega}} (1 + \widehat{\text{div}\,\delta \hat{\boldsymbol{x}}})[(\boldsymbol{I} - \widehat{\text{grad}\,\delta \hat{\boldsymbol{x}}})^t \widehat{\text{grad}\,\hat{\boldsymbol{u}}} \\ &\quad + (\widehat{\text{grad}\,\hat{\boldsymbol{u}}})^t (\boldsymbol{I} - \widehat{\text{grad}\,\delta \hat{\boldsymbol{x}}}) : (\boldsymbol{I} - \widehat{\text{grad}\,\delta \hat{\boldsymbol{x}}})^t \widehat{\text{grad}\,\hat{\boldsymbol{v}}}\,d\hat{\boldsymbol{x}} \\ &\approx 2\mu \int_{\hat{\Omega}} \left(\widehat{\text{grad}\,\hat{\boldsymbol{u}}} + (\widehat{\text{grad}\,\hat{\boldsymbol{u}}})^t \right) : \widehat{\text{grad}\,\hat{\boldsymbol{v}}}\,d\hat{\boldsymbol{x}} \\ &\quad + 2\mu \int_{\hat{\Omega}} \widehat{\text{div}\,\delta \boldsymbol{x}} \left(\widehat{\text{grad}\,\hat{\boldsymbol{u}}} + (\widehat{\text{grad}\,\hat{\boldsymbol{u}}})^t \right) : \widehat{\text{grad}\,\hat{\boldsymbol{v}}}\,d\hat{\boldsymbol{x}} \\ &\quad - 2\mu \int_{\hat{\Omega}} \left((\widehat{\text{grad}\,\delta \hat{\boldsymbol{x}}})^t \widehat{\text{grad}\,\hat{\boldsymbol{u}}} + (\widehat{\text{grad}\,\hat{\boldsymbol{u}}})^t \widehat{\text{grad}\,\delta \hat{\boldsymbol{x}}} \right) : \widehat{\text{grad}\,\hat{\boldsymbol{v}}}\,d\hat{\boldsymbol{x}} \\ &\quad - 2\mu \int_{\hat{\Omega}} (\widehat{\text{grad}\,\hat{\boldsymbol{u}}} + (\widehat{\text{grad}\,\hat{\boldsymbol{u}}})^t) : (\widehat{\text{grad}\,\delta \hat{\boldsymbol{x}}})^t \widehat{\text{grad}\,\hat{\boldsymbol{v}}}\,d\hat{\boldsymbol{x}} \\ &=: a(\hat{\boldsymbol{u}}, \hat{\boldsymbol{v}}) + \Delta t\,\delta a(\hat{\boldsymbol{w}}, \hat{\boldsymbol{u}}, \hat{\boldsymbol{v}}). \end{aligned} \right\} \tag{2.15}$$

De même on a, en négligeant toujours les termes d'ordre deux,

$$\left. \begin{aligned} \hat{b}(\hat{\boldsymbol{v}}, \hat{q}) &\approx - \int_{\hat{\Omega}} \hat{q}\,\widehat{\text{grad}\,\hat{\boldsymbol{v}}} : (\boldsymbol{I} - \widehat{\text{grad}\,\delta \hat{\boldsymbol{x}}})\,d\hat{\boldsymbol{x}} \\ &= - \int_{\hat{\Omega}} \hat{q}(\widehat{\text{grad}\,\hat{\boldsymbol{v}}} : \boldsymbol{I})d\hat{\boldsymbol{x}} + \int_{\hat{\Omega}} \hat{q}\,\widehat{\text{grad}\,\hat{\boldsymbol{v}}} : \widehat{\text{grad}\,\delta \hat{\boldsymbol{x}}}\,d\hat{\boldsymbol{x}} \\ &= - \int_{\hat{\Omega}} \hat{q}\,\widehat{\text{div}\,\hat{\boldsymbol{v}}}\,d\hat{\boldsymbol{x}} + \int_{\hat{\Omega}} \hat{q}\,\widehat{\text{grad}\,\hat{\boldsymbol{v}}} : \widehat{\text{grad}\,\delta \hat{\boldsymbol{x}}}\,d\hat{\boldsymbol{x}} \\ &=: b(\hat{\boldsymbol{v}}, \hat{q}) + \Delta t\,\delta b(\hat{\boldsymbol{w}}, \hat{\boldsymbol{v}}, \hat{q}), \end{aligned} \right\} \tag{2.16}$$

$$\hat{c}(\hat{\boldsymbol{u}} - \hat{\boldsymbol{w}}, \hat{\boldsymbol{u}}, \hat{\boldsymbol{v}}) \ \approx \int_{\hat{\Omega}} (1 + \widehat{\mathrm{div}\,\delta\hat{\boldsymbol{x}}})(\hat{\boldsymbol{u}} - \hat{\boldsymbol{w}}) \cdot \widehat{\mathrm{grad}\,\hat{\boldsymbol{u}} \cdot \hat{\boldsymbol{v}}}\,d\hat{\boldsymbol{x}}$$

$$- \int_{\hat{\Omega}} (\boldsymbol{I} - \widehat{\mathrm{grad}\,\delta\hat{\boldsymbol{x}}})(\hat{\boldsymbol{u}} - \hat{\boldsymbol{w}}) \cdot \widehat{\mathrm{grad}\,\hat{\boldsymbol{u}} \cdot \hat{\boldsymbol{v}}}\,d\hat{\boldsymbol{x}}$$

$$= \int_{\hat{\Omega}} (\hat{\boldsymbol{u}} - \hat{\boldsymbol{w}}) \cdot \widehat{\mathrm{grad}\,\hat{\boldsymbol{u}} \cdot \hat{\boldsymbol{v}}}\,d\hat{\boldsymbol{x}}$$

$$+ \int_{\hat{\Omega}} \widehat{\mathrm{div}\,\delta\hat{\boldsymbol{x}}}(\hat{\boldsymbol{u}} - \hat{\boldsymbol{w}}) \cdot \widehat{\mathrm{grad}\,\hat{\boldsymbol{u}} \cdot \hat{\boldsymbol{v}}}\,d\hat{\boldsymbol{x}} \qquad (2.17)$$

$$- \int_{\hat{\Omega}} \widehat{\mathrm{grad}\,\delta\hat{\boldsymbol{x}}}(\hat{\boldsymbol{u}} - \hat{\boldsymbol{w}}) \cdot \widehat{\mathrm{grad}\,\hat{\boldsymbol{u}} \cdot \hat{\boldsymbol{v}}}\,d\hat{\boldsymbol{x}}$$

$$=: c(\hat{\boldsymbol{u}} - \hat{\boldsymbol{w}}, \hat{\boldsymbol{u}}, \hat{\boldsymbol{v}}) + \Delta t\,\delta c(\hat{\boldsymbol{w}}, \hat{\boldsymbol{u}} - \hat{\boldsymbol{w}}, \hat{\boldsymbol{u}}, \hat{\boldsymbol{v}})$$

et de même pour les autres termes.

Le problème s'écrit maintenant sous la forme:

$$\int_{\hat{\Omega}} (1 + \Delta t\,\mathrm{div}\,\hat{\boldsymbol{w}}) \frac{\partial \hat{\boldsymbol{u}}}{\partial t} \cdot \hat{\boldsymbol{v}}\,d\hat{\boldsymbol{x}} + \int_{\hat{\Omega}} (\hat{\boldsymbol{u}} - \hat{\boldsymbol{w}}) \cdot (\widehat{\mathrm{grad}\,\hat{\boldsymbol{u}}}) \cdot \hat{\boldsymbol{v}}\,d\hat{\boldsymbol{x}}$$

$$+ 2\mu \int_{\hat{\Omega}} \left(\widehat{\mathrm{grad}\,\hat{\boldsymbol{u}}} + (\widehat{\mathrm{grad}\,\hat{\boldsymbol{u}}})^t \right) : \widehat{\mathrm{grad}\,\hat{\boldsymbol{v}}}\,d\hat{\boldsymbol{x}} - \int_{\hat{\Omega}} \hat{p}\,\widehat{\mathrm{div}}\,\hat{\boldsymbol{v}}\,d\hat{\boldsymbol{x}}$$

$$+ \Delta t(\delta a(\hat{\boldsymbol{w}}, \hat{\boldsymbol{u}}, \hat{\boldsymbol{v}}) + \delta b(\hat{\boldsymbol{w}}, \hat{\boldsymbol{v}}, \hat{q}) + \delta c(\hat{\boldsymbol{w}}, \hat{\boldsymbol{u}} - \hat{\boldsymbol{w}}, \hat{\boldsymbol{u}}, \hat{\boldsymbol{v}})) \qquad (2.18)$$

$$= \int_{\hat{\Omega}} (1 + \Delta t\,\mathrm{div}\,\hat{\boldsymbol{w}})\hat{\boldsymbol{f}} \cdot \hat{\boldsymbol{v}}\,d\hat{\boldsymbol{x}},$$

$$- \int_{\hat{\Omega}} \hat{q}\,\widehat{\mathrm{div}}\,\hat{\boldsymbol{u}}\,d\hat{\boldsymbol{x}} + \Delta t\,\delta b(\hat{\boldsymbol{w}}, \hat{\boldsymbol{u}}, \hat{q}) = \int_{\hat{\Omega}} \hat{g}\hat{q}\,d\hat{\boldsymbol{x}}. \qquad (2.19)$$

Ces équations doivent être lues, pour l'instant, comme des équations en $\hat{\boldsymbol{u}}$ et $\hat{p}$, en supposant $\hat{\boldsymbol{w}}$ connu. Pour un choix approprié des espaces dans lesquelles on les pose, elles contiennent implicitement les conditions naturelles qui impliquent que les contraintes sont nulles sur la frontière libre. L'approximation ci-dessus n'est valable que sur un petit intervalle. Il faut répéter son application en utilisant comme configuration de référence le domaine tel que connu en un certain temps t_0.

2.2.1   Conditions aux limites

Il nous faut aussi des conditions aux limites sur $\hat{\boldsymbol{u}}$. Le choix de ces conditions dépendra du contexte dans lequel on utilisera la formulation précédente.

Dans le cas d'une frontière mobile *imposée* avec une condition de non-glissement à la paroi, la condition serait

$$\hat{\boldsymbol{u}} = \hat{\boldsymbol{w}} \quad \mathrm{sur} \ \ |_{\Gamma}, \qquad (2.20)$$

c'est-à-dire une condition de type Dirichlet.

Dans le cas du calcul d'une surface libre, deux options s'offrent à nous.

- On peut imposer sur la frontière les conditions de contraintes nulles, qui sont des conditions naturelles pour la formulation variationnelle, et la condition supplémentaire

$$\boldsymbol{u} \cdot \boldsymbol{n} = \boldsymbol{w} \cdot \boldsymbol{n} \quad \text{sur} \quad \Gamma_0, \tag{2.21}$$

ce qui impose une condition sur le choix de $\hat{\boldsymbol{w}}(\hat{\boldsymbol{x}}, t)$ et donc de Ω_t.

- On peut imposer (2.21) et $\sigma_{nt} = 0$, en considérant la condition supplémentaire $\sigma_{nn} = 0$ comme déterminant la frontière.

La seconde possibilité sera abordée à la Section 2.4. Pour l'instant nous allons expliciter davantage la première option. Notons que (2.21) s'écrit sur Ω_t et non sur $\hat{\Omega}$. Pour fixer les idées, nous allons considérer en détail le cas où la frontière libre est décrite simplement par sa hauteur $h(\hat{x}_1)$ et où on a $\hat{\boldsymbol{w}}(\hat{\boldsymbol{x}}, t) = (0, w(\hat{x}_1))$ (Figure 2.2). (Nous présenterons simplement le cas bidimensionnel, mais le passage au cas tridimensionnel est immédiat.) Dans ce cas, nous avons

$$\frac{\partial h(\hat{x}_1)}{\partial t} = w(\hat{x}_1). \tag{2.22}$$

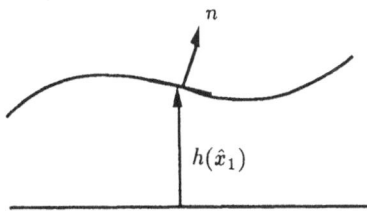

Figure 2.2

La normale (non unitaire) à la surface est maintenant $\left(-\dfrac{\partial h}{\partial \hat{x}_1}, 1\right)$ et la condition (2.21) s'écrit

$$-\hat{u}_1 \frac{\partial h}{\partial \hat{x}_1} + \hat{u}_2 = w \tag{2.23}$$

ou encore, utilisant (2.22),

$$\frac{\partial h(\hat{x}_1)}{\partial t} + \hat{u}_1 \frac{\partial h}{\partial \hat{x}_1} = \hat{u}_2. \tag{2.24}$$

On pourrait procéder de la même façon pour d'autres paramétrisations de la surface. Revenant au cas défini par (2.13), et notant $h_0(\hat{x}_1)$ la hauteur de la surface libre au temps t_0, l'équation (2.24) devient alors:

$$w(\hat{x}_1) + \hat{u}_1 \frac{\partial h_0}{\partial \hat{x}_1} + \Delta t \, \hat{u}_1 \frac{\partial w}{\partial \hat{x}_1} = \hat{u}_2. \tag{2.25}$$

Les équations (2.18), (2.19) et (2.25) forment ainsi un système non linéaire en $\hat{\boldsymbol{u}}$ et w qui doit être vérifié pour toute valeur de Δt. Pour l'utiliser dans la pratique, nous allons devoir en déduire un schéma d'approximation en temps puis utiliser une linéarisation de type Newton.

2.2.2 Problèmes stationnaires

Les considérations précédentes s'appliquent également aux problèmes stationnaires. Le domaine étant inconnu, il nous sera utile d'écrire notre problème dans un domaine de référence $\hat{\Omega}$. Les équations à résoudre sont dans ce cas les équations (2.12) avec $\hat{\boldsymbol{w}} = 0$, c'est-à-dire,

$$\left.\begin{array}{l} \hat{a}(\hat{\boldsymbol{u}}, \hat{\boldsymbol{v}}) + \hat{c}(\hat{\boldsymbol{u}}, \hat{\boldsymbol{u}}, \hat{\boldsymbol{v}}) + \hat{b}(\hat{\boldsymbol{v}}, \hat{p}) = \displaystyle\int_{\hat{\Omega}} \hat{J}\hat{\boldsymbol{f}} \cdot \hat{\boldsymbol{v}} \, d\hat{\boldsymbol{x}}, \quad \forall \hat{\boldsymbol{v}}, \\[3mm] \hat{b}(\hat{\boldsymbol{u}}, \hat{q}) = 0, \quad \forall \hat{q}. \end{array}\right\} \tag{2.26}$$

La condition aux limites (2.21) devient pour sa part

$$\boldsymbol{u} \cdot \boldsymbol{n} = 0. \tag{2.27}$$

Nous allons nous inspirer des développements précédents pour construire une linéarisation de ce problème non linéaire. Nous nous plaçons donc dans le contexte d'une itération, convergeant éventuellement vers la solution recherchée. Supposons connu, à l'itération n, $\hat{\Omega} - \Omega_n$ et cherchons $\Omega_t = \Omega_{n+1}$ défini de nouveau par (2.13). On supposera de plus que $\hat{\boldsymbol{u}}_{n+1}$ s'écrit sous la forme $\hat{\boldsymbol{u}}_n + \delta\hat{\boldsymbol{u}}$. On développe alors comme précédemment en négligeant tous les termes du second ordre pour obtenir:

$$\begin{aligned} \int_{\hat{\Omega}} \hat{\boldsymbol{u}}_n \cdot (\widehat{\operatorname{grad}} \, \hat{\boldsymbol{u}}_n) \cdot \hat{\boldsymbol{v}} \, d\hat{\boldsymbol{x}} \quad &+ \int_{\hat{\Omega}} \delta\hat{\boldsymbol{u}} \cdot (\widehat{\operatorname{grad}} \, \hat{\boldsymbol{u}}_n) \cdot \hat{\boldsymbol{v}} \, d\hat{\boldsymbol{x}} \\[2mm] &+ \int_{\hat{\Omega}} \hat{\boldsymbol{u}}_n \cdot (\widehat{\operatorname{grad}} \, \delta\hat{\boldsymbol{u}}) \cdot \hat{\boldsymbol{v}} \, d\hat{\boldsymbol{x}} \\[2mm] &+ 2\mu \int_{\hat{\Omega}} \left(\widehat{\operatorname{grad}} \, \hat{\boldsymbol{u}}_n + (\widehat{\operatorname{grad}} \, \hat{\boldsymbol{u}}_n)^t \right) : \widehat{\operatorname{grad}} \, \hat{\boldsymbol{v}} \, d\hat{\boldsymbol{x}} \\[2mm] &+ 2\mu \int_{\hat{\Omega}} \left(\widehat{\operatorname{grad}} \, \delta\hat{\boldsymbol{u}} + (\widehat{\operatorname{grad}} \, \delta\hat{\boldsymbol{u}})^t \right) : \widehat{\operatorname{grad}} \, \hat{\boldsymbol{v}} \, d\hat{\boldsymbol{x}} \\[2mm] &- \int_{\hat{\Omega}} \hat{p}_n \, \widehat{\operatorname{div}} \, \hat{\boldsymbol{v}} \, d\hat{\boldsymbol{x}} - \int_{\hat{\Omega}} \delta\hat{p} \, \widehat{\operatorname{div}} \, \hat{\boldsymbol{v}} \, d\hat{\boldsymbol{x}} \\[2mm] &+ \delta a(\delta\hat{\boldsymbol{x}}, \hat{\boldsymbol{u}}_n, \hat{\boldsymbol{v}}) + \delta b(\delta\hat{\boldsymbol{x}}, \hat{\boldsymbol{v}}, \hat{p}_n) \\[2mm] &+ \delta c(\delta\hat{\boldsymbol{x}}, \hat{\boldsymbol{u}}_n, \hat{\boldsymbol{u}}_n, \hat{\boldsymbol{v}}) \\[2mm] &= \int_{\hat{\Omega}} (1 + \operatorname{div} \delta\hat{\boldsymbol{x}}) \hat{\boldsymbol{f}} \cdot \hat{\boldsymbol{v}} \, d\hat{\boldsymbol{x}}, \end{aligned} \tag{2.28}$$

$$\int_{\hat{\Omega}} \hat{q} \, \widehat{\operatorname{div}} \, \hat{\boldsymbol{u}}_n \, d\hat{\boldsymbol{x}} + \int_{\hat{\Omega}} \hat{q} \, \widehat{\operatorname{div}} \, \delta\hat{\boldsymbol{u}} \, d\hat{\boldsymbol{x}} + \int_{\hat{\Omega}} \hat{q} \, \widehat{\operatorname{grad}} \, \hat{\boldsymbol{u}}_n : \widehat{\operatorname{grad}} \, \delta\hat{\boldsymbol{x}} \, d\hat{\boldsymbol{x}}. \tag{2.29}$$

La condition de frontière libre (2.25) devient

$$\hat{u}_1 \frac{\partial h_0}{\partial \hat{x}_1} + \delta \hat{u}_1 \frac{\partial h_0}{\partial \hat{x}_1} + \hat{u}_1 \frac{\partial \delta \hat{x}}{\partial \hat{x}_1} = \hat{u}_2 + \delta \hat{u}_2. \tag{2.30}$$

Le système (2.28)–(2.29)–(2.30) permet de déterminer $\delta \hat{u}, \delta \hat{p}$ et $\delta \hat{x}$, à condition de le compléter par une paramétrisation de $\delta \hat{x}$ à l'intérieur du domaine ou une équation de la forme

$$\int_{\hat{\Omega}} \widehat{\mathbf{grad}}\, \delta \hat{x} \cdot \mathbf{grad}\, \hat{w} = 0 \quad \forall \hat{w} \in H_0^1, \tag{2.31}$$

c'est-à-dire $-\Delta \delta \hat{x} = 0$. La solution $\delta \hat{u}, \delta \hat{p}, \delta \hat{x}$ est ainsi complètement déterminée. On procède alors à la mise à jour du domaine et on poursuit l'itération jusqu'à convergence. Le développement que nous avons effectué permet en fait de construire une variante de la méthode de Newton pour le problème de surface libre. On trouvera des idées analogues, par exemple, dans Saito-Scriven [1982] ou dans Kruyt-Cuvelier-Segal-Van der Zanden [1988].

2.3 Discrétisation en temps

A la section précédente, nous avons introduit une forme variationnelle approchée, valable au voisinage d'une configuration donnée. Il nous reste à en déduire un schéma numérique. Supposons donc connus au temps t_n une solution (u_n, p_n) et un domaine Ω_n. Nous cherchons à calculer au temps $t_{n+1} = t_n + \Delta t$ le nouveau domaine Ω_{n+1} et la solution (u_{n+1}, p_{n+1}) qui lui est associée. Dans le cadre de la section précédente, plusieurs approches s'offrent à nous, en particulier quant au choix du domaine de référence $\hat{\Omega}$ et au type de schéma utilisé. Pour ce dernier point, nous devrons suivre quelques lignes directrices.

- La première est que la condition d'incompressibilité et le caractère implicite qu'elle confère aux équations rendent presque nécessaire l'utilisation d'un schéma implicite.

- La seconde vient de la nature potentiellement instationnaire des solutions recherchées. L'expérience (Fortin-Fortin-Gervais [1991]) montre que la détection correcte d'une bifurcation de Hopf, vers une solution périodique, exige l'emploi d'un schéma en temps qui soit au minimum d'ordre deux.

2.3.1 Le cas $\hat{\Omega} = \Omega_n$

La technique la plus simple est évidemment de choisir $\hat{\Omega} = \Omega_n$ et d'utiliser (2.18), (2.19) et, dans le cas où $\hat{w} = (0, w)$, (2.25) en remplaçant dans (2.18) et (2.25) les dérivées en temps par une formule de différence rétrograde. On obtient ainsi un schéma où les inconnues sont $\hat{u}_{n+1}, \hat{p}_{n+1}$ et $\hat{w}_{n+1} = (0, w_{n+1})$. On suppose la frontière libre décrite par sa hauteur $h(\hat{x}_1)$ et la relation

$$h_{n+1} - h_n = \Delta t \hat{w}_{n+1}.$$

On a alors

$$
\int_{\hat{\Omega}} \left(1 + \Delta t \operatorname{div} \hat{w}_{n+1}\right) \frac{1}{\Delta t} \left(\frac{3}{2}\hat{u}_{n+1} - 2\hat{u}_n + \frac{1}{2}\hat{u}_{n-1}\right) \cdot \hat{v} \, d\hat{x}
$$

$$
+ \int_{\hat{\Omega}} (\hat{u}_{n+1} - \hat{w}_{n+1}) \cdot (\widehat{\operatorname{grad}} \, \hat{u}_{n+1}) \cdot \hat{v} \, d\hat{x}
$$

$$
+ 2\mu \int_{\hat{\Omega}} \left(\widehat{\operatorname{grad}} \, \hat{u}_{n+1} + (\widehat{\operatorname{grad}} \, \hat{u}_{n+1})^t\right) : \widehat{\operatorname{grad}} \, \hat{v} \, d\hat{x}
$$

$$
- \int_{\hat{\Omega}} \hat{p}_{n+1} \widehat{\operatorname{div}} \, \hat{v} \, d\hat{x} \qquad\qquad (3.1)
$$

$$
+ \Delta t \, \delta a(\hat{w}_{n+1}, \hat{u}_{n+1}, \hat{v}) + \Delta t \, \delta b(\hat{w}_{n+1}, \hat{v}, \hat{p}_{n+1})
$$

$$
+ \Delta t \, \delta c(\hat{w}_{n+1}, \hat{u}_{n+1} - \hat{w}_{n+1}, \hat{u}_{n+1}, \hat{v}))
$$

$$
= \int_{\hat{\Omega}} \left(1 + \Delta t \operatorname{div} \hat{w}_{n+1}\right)\hat{f} \cdot \hat{v} \, d\hat{x},
$$

$$
- \int_{\hat{\Omega}} \hat{q} \, \widehat{\operatorname{div}} \, \hat{u}_{n+1} d\hat{x} + \Delta t \, \delta b(\hat{w}_{n+1}, \hat{u}_{n+1}, \hat{q}) = \int_{\hat{\Omega}} \hat{g}\hat{q} \, d\hat{x}, \qquad\qquad (3.2)
$$

$$
\left(\frac{3}{2}w_{n+1} - \frac{1}{2}w_n\right) + \hat{u}_{1,n+1}\frac{\partial h_0}{\partial \hat{x}_1} + \Delta t \, \hat{u}_{1,n+1}\frac{\partial w_{n+1}}{\partial \hat{x}_1} = \hat{u}_{2,n+1}. \qquad\qquad (3.3)
$$

Nous avons utilisé pour cette dernière équation la relation

$$
\frac{3}{2}h_{n+1} - 2h_n + \frac{1}{2}h_{n-1} = \Delta t\left(\frac{3}{2}w_{n+1} - \frac{1}{2}w_n\right).
$$

Nous obtenons ainsi un système non linéaire que l'on peut résoudre, par exemple, par une linéarisation de type Newton. On peut, ce faisant, négliger un certain nombre de termes si l'on considère que les termes en Δt constituent déjà une linéarisation.

Dans cette procédure, le domaine reste fixe tant que l'on n'a pas obtenu $(\hat{u}_{n+1}, \hat{p}_{n+1})$. On effectue ensuite la mise à jour de $\hat{\Omega}$ et on recommence le processus au pas de temps suivant.

2.3.2   Cas $\hat{\Omega} = \Omega_{n+1}$

On peut considérer une autre approche consistant à prendre comme domaine de référence le domaine inconnu Ω_{n+1}. Supposons temporairement $\hat{w}_{n+1}$ connu et discrétisons (2.18) et (2.19) par le même schéma qu'à la Section 3.1, en prenant cette fois $\hat{\Omega} = \Omega_{n+1}$. On n'a plus de termes en Δt et l'on obtient:

$$
\int_{\hat{\Omega}} \frac{1}{\Delta t} \left(\frac{3}{2} \hat{u}_{n+1} - 2\hat{u}_n + \frac{1}{2} \hat{u}_{n-1} \right) \cdot \hat{v} \, d\hat{x}
$$

$$
+ \int_{\hat{\Omega}} (\hat{u}_{n+1} - \hat{w}_{n+1}) \cdot (\widehat{\mathrm{grad}\,} \hat{u}_{n+1}) \cdot \hat{v} \, d\hat{x}
$$

$$
+ 2\mu \int_{\hat{\Omega}} \left(\widehat{\mathrm{grad}\,} \hat{u}_{n+1} + (\widehat{\mathrm{grad}\,} \hat{u}_{n+1})^t \right) : \widehat{\mathrm{grad}\,} \hat{v} \, d\hat{x} \tag{3.4}
$$

$$
- \int_{\hat{\Omega}} \hat{p}_{n+1} \widehat{\mathrm{div}\,} \hat{v} \, d\hat{x} = \int_{\hat{\Omega}} \hat{f} \cdot \hat{v} \, d\hat{x},
$$

$$
- \int_{\hat{\Omega}} \hat{q} \, \widehat{\mathrm{div}\,} \hat{u}_{n+1} d\hat{x} = \int_{\hat{\Omega}} \hat{g} \hat{q} \, d\hat{x}. \tag{3.5}
$$

Résolvant (3.4)–(3.5), on obtient une solution $(\hat{u}_{n+1}, \hat{p}_{n+1})$ que l'on peut considérer comme une fonction de w. On peut alors considérer (2.20) ou sa forme (2.25) comme une équation en w,

$$
\left(\frac{3}{2} w_{n+1} - \frac{1}{2} w_n \right) + \hat{u}_{1,n+1} \frac{\partial h_0}{\partial \hat{x}_1} = \hat{u}_{2,n+1}. \tag{3.6}
$$

On peut alors itérer selon plusieurs tactiques.

- On considère le problème en $(\hat{u}_{n+1}, \hat{p}_{n+1}, w_{n+1})$ comme un problème stationnaire et on applique la technique de la Section 2.2.2. Ceci nécessite un nouveau développement au premier ordre et fait réapparaître les termes supplémentaires de la méthode précédente.

- On se donne w_0 au moyen duquel on calcule un estimé de $\hat{\Omega}$. On résout alors simultanément (3.4), (3.5) et (3.6). Le w_{n+1} calculé n'est pas, en général, égal à w_0. On effectue donc une itération de type point fixe en prenant cette nouvelle valeur de w pour recalculer $\hat{\Omega}$.

- On se donne w_0 au moyen duquel on calcule un estimé de $\hat{\Omega}$. On résout alors (3.4), (3.5). Connaissant $\hat{u}_{n+1}$ et p_{n+1}, on résout (3.6) en w_{n+1} et on procède encore par point fixe. Cette méthode est "moins implicite" que la précédente.

- On considère dans (3.6) $\hat{u}_{n+1}$ et p_{n+1} comme des fonctions implicites de w_{n+1} et on cherche une linéarisation de type Newton du problème en w ainsi défini. Ceci oblige à calculer la matrice jacobienne de l'équation et donc à dériver $\hat{u}_{n+1}$ et p_{n+1} par rapport à w. Ce calcul n'est pas simple mais peut se faire en dérivant les équations (3.4) et (3.5) par rapport au domaine. Le tout ressemble fort au calcul du hessien de forme d'une fonctionnelle (Delfour-Zolésio [1991b]). Nous verrons à la Section 4 comment éviter le calcul de ces dérivées par l'emploi de formules de différences.

2.3.3   L'algorithme suisse

On peut considérer des variantes plus simples. Nous avons déjà signalé à la section 2.2.1 la possibilité d'utiliser la condition $\sigma_{n,n} = p + 2\mu\varepsilon_{n,n} = 0$ pour déterminer la frontière libre. Ce type d'approche a été développé par Bourgeois-Chevalier-Picasso-Rappaz-Touzani [1989] dans un cas où la pression contient un potentiel électromagnétique connu, engendrant des forces dominant les autres forces en présence.

Nous allons présenter l'algorithme dans le cas stationnaire seulement. On cherche à résoudre un problème de la forme

$$\left. \begin{aligned} &-2\mu \ \mathbf{DIV} \ \mathbf{grad} \ \boldsymbol{u} + \rho\boldsymbol{u} \cdot \ \mathbf{grad} \ \boldsymbol{u} + \ \mathbf{grad} \ \tilde{p} = \boldsymbol{f}, \\ &\tilde{p} = p - \rho x_2 - \phi \\ &\mathrm{div} \ \boldsymbol{u} = 0, \\ &\boldsymbol{u} \cdot n = 0, \quad \varepsilon_{tn} = 0 \end{aligned} \right\} \tag{3.7}$$

où ρx_2 représente les forces gravitationnelles et ϕ un potentiel connu, en général d'origine électromagnétique. Par ailleurs, la condition de surface libre s'écrit, supposant une tension superficielle constante σ et notant H la courbure de la surface

$$\sigma H + 2\mu\varepsilon_{nn} + p - \rho x_2 - \phi(x_2) = C_0, \tag{3.8}$$

la pression étant définie à une constante près. On suppose, de plus, la surface paramétrisée (Figure 2.3),

$$\begin{aligned} x_1 &= x_0 + r(\theta)\cos\theta \\ x_2 &= r(\theta)\sin\theta. \end{aligned} \tag{3.9}$$

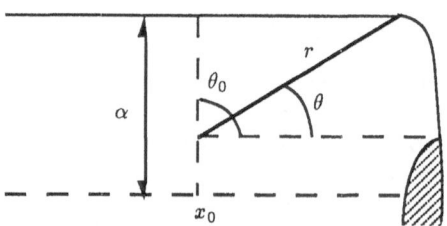

Figure 2.3

La condition (3.8) s'écrit alors:

$$\sigma \frac{r^2(\theta) + 2(r'(\theta))^2 - r(\theta)r''(\theta)}{(r^2(\theta) + (r'(\theta))^2)^{\frac{3}{2}}} + 2\mu\varepsilon_{nn} + p - \rho r(\theta)\sin\theta - \phi(r,\theta) = C_0 \tag{3.10}$$

ce que l'on peut lire comme une équation différentielle en $r(\theta)$, que l'on peut aussi écrire, en abrégé

$$F(r) + s_n - \phi(r) - C_0 = 0.$$

Il faut alors des conditions aux limites sur r. Nous pouvons prendre, par exemple,

$$r'(0) = 0 \tag{3.11}$$

et, la hauteur du ménisque étant α (Figure 2.3),

$$r(\theta_0) = \frac{\alpha}{\sin \theta_0}, \tag{3.12}$$

$$r'(\theta_0) = \frac{-\alpha \cos \theta_0}{\sin^2 \theta_0}. \tag{3.13}$$

L'algorithme consistera alors en une méthode de point fixe. La surface étant donnée, on résoudra (3.7), puis u et p étant connus, on cherchera $r(\theta)$ et donc une nouvelle position de la surface en fixant les forces visqueuses. Cette dernière étape se fera par une itération de type Newton. On va donc chercher $r(\theta)$ sous la forme $r_0(\theta) + \xi$ et chercher la solution de

$$S(\xi) = F(r_0 + \xi) - \phi(r_0 + \xi) + s_n - C_0 = 0,$$

en supposant s_n fixé. Développant (3.9) au premier ordre, il vient

$$\left. \begin{aligned} DF(r_0) \cdot \xi + D\phi(r_0) \cdot \xi &= -S(r_0), \\ \xi'(0) = \xi'(\theta_0) = \xi(\theta_0) &= 0, \end{aligned} \right\} \tag{3.14}$$

où l'on a,

$$\begin{aligned} DF(s) \cdot \xi = {} & \sigma \frac{2s\xi + 4s'\xi' - s''\xi - s\xi''}{(s^2 + (s')^2)^3} \\ & - 3\sigma \frac{(s^2 + 2(s')^2 - ss')(\xi s + s'\xi')}{(s^2 + (s')^2)^{\frac{5}{2}}}. \end{aligned} \tag{3.15}$$

La frontière est donc ajustée, u et p étant connus, à partir de considérations purement géométriques, l'effet d'une variation de la frontière sur la solution étant négligé. A noter le rôle important joué par la tension superficielle.

2.4 Méthodes de gradient conjugué et de résidu minimal

2.4.1 Linéarisation des problèmes non linéaires

Une façon standard de résoudre un problème non linéaire est d'utiliser la méthode de Newton.

$$F(x_n - \delta x) = 0 \Rightarrow F'(x_n)\delta x = F(x_n) \tag{4.1}$$

$$x_{n+1} = x_n - \delta x. \tag{4.2}$$

Nous devons alors être en mesure d'évaluer $F(x)$, ce qui est inévitable, mais aussi $F'(x)$ ce qui est coûteux (et pénible dans le cas de la dérivée par rapport à un domaine). A chaque itération de la méthode de Newton, nous avons à résoudre un système linéaire, dont la matrice est $F'(x_n)$, matrice qui changera d'une itération à l'autre, ce qui entraînera un coût d'assemblage élevé. Cependant, si on résout les problèmes linéarisés par une méthode de type gradient conjugué, on constate qu'il n'est pas nécessaire de connaître $F'(x_n)$ mais uniquement $F'(x_n) \cdot w_i$, $(i \leq 1 \leq m - 1)$. Or on peut approcher cette quantité par

$$\frac{F(x_n + hw_i) - F(x_n)}{h} \tag{4.3}$$

pour h assez (mais pas trop!) petit.

L'algorithme que nous allons présenter, basé sur l'algorithme GMRES de Saad-Schulz [1986], consiste en gros en une approximation de la méthode de Newton dont chaque étape est d'éviter le calcul explicite de $F'(x_n)$.

2.4.2   Méthodes de type gradient conjugué

Les méthodes de type gradient conjugué préconditionné, ou de résidu conjugué sont devenues très populaires pour la résolution numérique des problèmes linéaires. Elles ont montré leur efficacité dans de nombreuses situations. Nous allons ici en rappeler rapidement le principe.

On cherche à résoudre un système linéaire $Ax = b$ dans un espace de dimension N. Supposons connues m directions, $(w_1, w_2, \ldots, w_m)$ et cherchons la "meilleure solution" dans l'espace qu'elles engendrent. L'approche de Galerkin consiste, si A est définie positive à minimiser la norme du résidu en norme A^{-1}, c'est-à-dire

$$\begin{aligned}
\|Ax - b\|_{A^{-1}} &= \langle A^{-1}(Ax - b), Ax - b \rangle \\
&= \langle Ax, x \rangle - 2\langle b, x \rangle + cte.
\end{aligned} \tag{4.4}$$

Il est facile de voir que si on écrit $u_m = \sum_{j=1}^{m} a_j w_j$ cette minimisation est équivalent à la résolution du système

$$\langle Aw_j, w_i \rangle \alpha_j = \langle f, w_i \rangle \quad i = 1, m \tag{4.5}$$

Le système (2.2) sera très simple à résoudre si la matrice $\langle Aw_j, w_i \rangle$ est triangulaire supérieure ou a la forme de Hessenberg. Pour arriver à cette propriété, nous allons poser

$$w_1 = r_0 = Ax_0 - b, \qquad w_{i+1} = Aw_i \quad i \geq 1. \tag{4.6}$$

On obtient ainsi l'espace de Krilov

$$(r_0, Ar_0, A^2 r_0, \ldots, A^{m-1} r_0) \tag{4.7}$$

auquel on applique le procédé de Gram-Schmidt modifié de façon à rendre la base orthonormale.

On obtient aussi $\langle Av_j, v_i \rangle = 0$, $j < i - 1$ et le système linéaire (4.5) est triangulaire. Cette méthode est connue sous le nom de méthode d'Arnoldi.

On peut minimiser sur le même espace de Krilov, la norme euclidienne du résidu. A noter qu'on n'utilise pas l'espace de Krilov de $A^t A$. On obtient ainsi une méthode résidu minimal. Au point de vue pratique on a un décalage d'une ligne dans la matrice formée dans la méthode d'Arnoldi. Dans GMRES, Saad et Schultz considèrent un système rectangulaire résolu lui-même par moindres carrés. Cette méthode s'est avérée plus stable et plus robuste que l'algorithme standard. Nous renvoyons le lecteur à Saad-Schulz [1986] pour plus de détails. Il existe des versions de la méthode disponibles dans le domaine public.

2.4.3 Calcul d'une surface libre (vitesse normale nulle)

Considérons pour fixer les idées le problème de Stokes dans un domaine Ω dont une partie de la frontière Γ_0 est libre alors que sur la partie complémentaire Γ_1, on supposera des conditions de Dirichlet. Le problème est donc,

$$\left.\begin{array}{l} -\Delta u + \operatorname{grad} p = f \\ \operatorname{div} u = 0 \\ u \cdot n = 0, \quad \varepsilon_{tn} = 0 \quad \text{sur } \Gamma_0 \\ u = 0 \quad \text{sur } \Gamma_1 \end{array}\right\} \tag{4.8}$$

On cherche à modifier Ω de façon à vérifier la condition supplémentaire

$$p + \varepsilon_{nn}(u) = 0 \quad \text{sur } \Gamma_0. \tag{4.9}$$

A noter que dans une discrétisation, on a accès à (4.9) via la formulation variationnelle. On a donc à résoudre

$$\begin{aligned} F(u, \Omega) &= \int_\Omega \varepsilon(u) : \varepsilon(v) \, dx - \int_\Omega p \operatorname{div} v \, dx \\ &- \int_\Omega f \cdot v \, dx = 0, \quad \forall v \in (H^1(\Omega))^2, u \in V \end{aligned} \tag{4.10}$$

où $V = \{v | v \cdot n = 0 \text{ sur } \Gamma_0\}$. Nous pouvons alors résoudre en prenant des fonctions tests dans V. Il restera uniquement à vérifier (4.10) pour des fonctions de composante normale non nulle au bord, ce qui est formellement équivalent à (4.9). Cette condition porte en fait sur Ω car la solution à Ω fixé est unique.

Dans une méthode de gradient conjugué préconditionné on cherche à résoudre $F(x) = 0$ et on écrit $x_{n+1} = x_n + \rho_n S^{-1} F(x_n)$. La variable est ici la forme du domaine. Dans notre contexte, il nous faut convertir (4.9), qui est le résidu de notre problème, en une variation de Ω, c'est-à-dire en un champ de vecteurs.

Supposant résolu le problème (4.8), (i.e. (4.10) avec $v \in V$), posons

$$\begin{aligned} \int_\Omega \varepsilon(w) : \varepsilon(v) \, dx - \int_\Omega \pi \operatorname{div} v \, dx = \\ \int_\Omega \varepsilon(u) : \varepsilon(v) dx - \int_\Omega p \operatorname{div} v \, dx - \int_\Omega f \cdot v \, dx = 0, \quad \forall v, \end{aligned} \tag{4.11}$$

$$\int_\Omega \operatorname{div} \boldsymbol{w} \, q \, d\boldsymbol{x} = 0, \quad \forall q. \tag{4.12}$$

On a alors,

$$\left. \begin{array}{l} -\Delta \boldsymbol{w} + \operatorname{\mathbf{grad}} \pi = 0, \\ \pi + \varepsilon_{nn}(w) = p + \varepsilon_{nn}(u), \quad \varepsilon_{tn}(w) = 0 \quad \text{sur } \Gamma_0, \\ \boldsymbol{w} = 0 \quad \text{sur } \Gamma_1. \end{array} \right\} \tag{4.13}$$

Le champ $\boldsymbol{w}$ est bien défini dans Ω. Il est nul si le problème est résolu. Il est à divergence nulle et il préserve le volume. La variable $\boldsymbol{w}$ ainsi définie est effectivement utilisable pour construire un déplacement. Nous définirons $S^{-1} F(\boldsymbol{u}, \Omega) = \boldsymbol{w}$ et nous fournissons donc $\boldsymbol{w}$ à GMRES comme préconditionnement. Les dérivées sont approchées, comme nous l'avons décrit plus haut par des quotients différentiels. Dans l'application de l'orthogonalisation de Gram-Schmidt, à l'intérieur de GMRES, il sera important d'employer le produit scalaire de $H^1(\Omega)$. A noter qu'une méthode semblable, où Ω était modifié par $\boldsymbol{x}_{n+1} = \boldsymbol{x}_n - \rho \boldsymbol{w}_n$ avec ρ fixé petit a été appliquée par Tanguy-Fortin-Choplin [1984] avec succès, mais évidemment avec une convergence plus lente.

2.4.4 Calcul d'une surface libre (contrainte normale nulle)

Le problème précédent peut aussi être attaqué en résolvant dans un domaine Ω donné en imposant sur la frontière libre la condition de contrainte nulle et en cherchant à retrouver par la forme du domaine la condition $\boldsymbol{u} \cdot \boldsymbol{n} = 0$.

La valeur de $\boldsymbol{u} \cdot \boldsymbol{n}$ devient alors le résidu de l'équation. Elle est directement utilisable pour le déplacement de la frontière et peut donc être fournie directement à GMRES.

Cette façon de procéder est donc apparemment plus simple que la précédente car on n'a pas de problème auxiliaire à résoudre. Cependant, il semble que le conditionnement du problème soit moins bon, ou à tout le moins que la répartition des valeurs propres du système soit moins bonne, la convergence étant moins rapide qu'avec la méthode précédente.

Bibliographie

Beris, A.N., Edwards, B.J. [1990], Poisson bracket formulation of incompressible flow equations in continuum mechanics, *J. Rheol.* **34**, 55–78.

Bourgeois, J., Chevalier, P.A., Picasso, M., Rappaz, J., Touzani, R. [1989], An MHD problem in the aluminium industry, *Proceedings of the 7th International Conference on Finite Elements in Flow Problems*, Huntsville, AL.

Cuvelier, C., Schulkes, R.M.S. [1990], Some numerical methods for the computation of capillary free boundaries governed by the Navier-Stokes equations, *SIAM Rev.* **32**, 355–423.

Delfour, M.C., Zolésio, J.P. [1991a], Velocity method and Lagrangian formulation for the computation of the shape Hessian, *SIAM J. Control Optim.* **29**, 1414–1442.

Delfour, M.C., Zolésio, J.P. [1991b], Anatomy of the shape hessian, *Ann. Mat. Pura Appl. (4)* (to appear).

Donea, J. [1983], Arbitrary Lagrangian-eulerian finite element methods, *Comput. Methods Transient Anal.* **1**, 473–516.

Ebin, D.G., Marsden, J. [1970], Groups of diffeomorphisms and the motion of an incompressible fluid, *Ann. Math.* **92**, 102–163.

Fortin, A., Fortin, M., Gervais, J.J. [1991], Complex transition to chaotic flow in a periodic array of cylinders, *J. Theoret. Comput. Fluid Dynamics* (to appear).

Guénette, R. [1990], *Études théoriques et numériques du modèle de Leslie-Ericksen pour les cristaux liquides*. Thèse, Unviersité Laval, Québec.

Hughes, T.J.R., Liu, W.K., Zimmermann, T.K. [1981], Lagrangian-eulerian finite element formulation for incompressible viscous flows, *Comput. Methods Appl. Mech. Engrg.* **29**, 329–349.

Kruyt, N.P., Cuvelier, C., Segal, A., Van der Zanden, J. [1988], A total linearization method for solving viscous free boundary flow problems by the finite element method, *Internat. J. Numer. Methods Engrg.* **8**, 351–363.

Marsden, J., Hughes, T.J.R. [1983], *Mathematical Foundations of Elasticity*, Prentice Hall, Englewood Cliffs, N.J.

Saad, Y., Schultz, M.H. [1986], A generalized minimum residual method for solving nonsymmetric linear systems, *SIAM J. Sci. Statist. Comput.* **7**, 856–869.

Saito, H., Scriven, L.E. [1982], Study of coating flow by the finite element method, *J. Comput. Phys.* **42**, 53–76.

Soulaimani, A., Fortin, M., Dhatt, G., Ouellet, Y. [1991], Finite element simulation of two and three dimensional free surface flows, *Comput. Methods Appl. Mech. Engrg.* **86**, 265–296.

Tanguy, P., Fortin M., Choplin, L. [1984], Finite element simulation of dip coating, *Internat. J. Numer. Methods Fluids* **4**, 441–457.

Zolésio, J.P. [1992], Introduction to shape optimization problems and free boundary problems, *this volume.*

Numerical Structural Optimization via a Relaxed Formulation

Robert V. KOHN[1]

Courant Institute of Mathematical Sciences
New York University
251 Mercer Street
New York, NY 10012
USA

Abstract

The basic problem of structural optimization is to choose the shape or composition of a structure so as to optimize some feature of its elastic behavior. The use of a relaxed formulation involving composite materials puts the theory on a sound mathematical basis. It also leads to better designs than the traditional formulation, because the relaxed formulation has fewer local minima. These lectures explain the relaxed viewpoint as it applies to shape optimization of two-dimensional structures for minimum compliance in plane stress. They draw heavily from recent joint work with G. Allaire.

Lecture 1

Introduction to structural optimization

Summary This lecture presents some typical model problems of structural optimization. It discusses the "standard" approach, then describes the "new" approach based on a relaxed formulation. Mathematical details are postponed until later.

The basic problem of structural optimization is to choose the shape or composition of a structure so as to optimize some feature of its elastic behavior. These lectures will concentrate primarily on *shape optimization*.

Figures 1.1 and 1.2 show two types of structures whose shape one might wish to optimize. In Figure 1.1 the structure serves to transmit a load from the left-hand side to the right-hand side. The total force on each side must be the same, so the force per unit length is small on the left and large on the right. The upper and lower boundaries are traction-free. Such a structure is commonly called a "planar fillet".

---

[1]Support is gratefully acknowledged from NSF Grant DMS-9102829, ARO contract DAAL03-89-K-0039, and AFOSR Grant 90-0090.

M. C. Delfour and G. Sabidussi (eds.), Shape Optimization and Free Boundaries, 173–210.
© 1992 *Kluwer Academic Publishers.*

Figure 1.2 gives another type of structure. It serves to transmit a load from the top to a pair of supports on a rigid surface. This time the traction is specified on the top, and the displacement must vanish where the supports meet the "ground". The other parts of the boundary are traction-free. Such a structure might be called a "planar bridge".

These examples are cited only to fix ideas. Our discussion actually applies to *any* elastic structure, including three-dimensional ones as well as two-dimensional problems of plane stress or plane strain. (We shall make special use of two-dimensionality in Lectures 3 and 5; also, we consider only boundary loads, ignoring body loads such as gravity which are shape-dependent. These hypotheses simplify matters, but they are not essential.)

A typical goal of shape optimization is to *minimize weight subject to one or more constraints* on the elastic behavior. The weight is of course proportional to the total amount of elastic material. Commonly considered constraints are

(i) the work done by the load ("compliance")

(ii) the maximum stress produced by the load

(iii) the average displacement over some part of the structure.

One might impose constraints involving the responses to several different loads.

Shape optimization can be viewed as a problem of optimal control, in which the "equation of state" is the elliptic system of elastostatics, and the "control variable" is the shape of the domain on which this PDE is to be solved.

From a slightly different perspective, shape optimization can be viewed as an extreme case of a material distribution problem. In the context of Figure 1.1, for example, consider a rectangular structure with the same loads, composed of two distinct elastic materials (Figure 1.3). Assume that material 1 is more rigid than 2 elastically, but also more expensive per unit area; the "material distribution problem" seeks an arrangement of the two materials that minimized the total cost, subject to constraints on the elastic response. In the limit as material 2 becomes elastically degenerate (i.e. its bulk and shear moduli both tend to zero), the region occupied by this material behaves as a union of holes. Hence the material distribution problem becomes, in this limit, a shape optimization problem.

We have presented the problem as one of constrained optimization. It is standard to observe that there is also an unconstrained formulation, obtained by adding a Lagrange multiplier times the constraint to the objective. For example, minimizing weight subject to a compliance constraint is the same as the unconstrained optimization

$$\min_{(designs)} \quad \text{weight} \ + \lambda \cdot \ \text{compliance.} \tag{1.1}$$

For given $\lambda > 0$, the design which solves (1.1) will have minimum weight *for its compliance*. As λ varies, one gets designs which optimize weight for different values of the compliance.

Shape optimization has been studied for many years by both mathematicians and engineers. It would be impossible to give a comprehensive review in the space of a few pages. Let us nevertheless say a word about what has been achieved.

The standard approach

Numerical optimization

For the numerical solution of a shape optimization problem, one typically starts by guessing an initial design. Some parts of the boundary are designated as being "free", i.e. subject to optimization (see Figure 1.4). They should be parametrized in a way that enforces some degree of smoothness, e.g. using splines. One then discretizes the elasticity problem, using finite elements or a boundary integral method. After discretization, the optimal design problem becomes a large nonlinear programming problem. It can be solved using a version of steepest descent or some more sophisticated method.

This approach has the advantage of being conceptually straightforward: no difficult PDE results are required for its execution. Moreover, it usually achieves some incremental improvement without introducing radical changes in the design. There are number of highly sophisticated and efficient implementations, and the method is widely used.

This approach has some drawbacks, however. The most significant is the problem of *local minima*. In most implementations the topology of the design is constrained by that of the initial guess. Thus one obtains at best a local optimum among designs that are grossly similar to the initial one. A second disadvantage, at least when finite elements are used, is the need for regridding. (Boundary element methods avoid this, but they lead to dense rather than sparse linear systems.)

We also note that the continuum limit of this approach is not well understood. In particular, there is no theory to explain why calculations of the type just described should converge in the limit as the mesh size tends to zero. This might be a defect of the theory rather than the method. My personal belief is that, for two dimensional problems at least, there should exist an "optimal design with N holes" for each fixed N. Perhaps the standard method computes this optimum, with N determined by the initial design.

Theory

The "standard approach" to the theory is more or less parallel to the numerical approach just described. One begins by guessing the gross form of the optimal design (i.e. its topology). Taking the first variation with respect to changes of geometry leads to an overdetermined boundary condition at each free boundary. One then seeks the optimal design as a solution of this free boundary problem.

This method has led to explicit solutions in some special cases, usually obtained via complex variable methods and conformal mapping.

As a mathematical theory, however, this approach is incomplete at best. Since the existence of an optimal design is uncertain, the existence of a solution of the associated free boundary problem is not assured. If the free boundary problem does indeed have a solution, one must still ask whether it represents a minimum of the design problem or a saddle point. (The second variation is not easy to evaluate.) Even if these difficulties could be resolved, we would still be left with the problem of topology optimization: at best, the standard approach could give us a list of critical points in design space, one (or more) for each choice of topology.

The relaxed formulation

The new "relaxed formulation" improves on the standard approach in two crucial respects: it yields a "more global" optimum, and it places the entire endeavor of structural optimization on a sound mathematical basis.

The main idea is to consider designs made from *composite materials* obtained by perforation. Such designs are limiting cases of structures made using many small holes. Thus it is natural to include them as legitimate designs, if we do not intend to constrain the topology or penalize the number of holes.

To be more explicit: in the classical formulation of shape optimization, we have at each point x either a given elastic material or nothing at all. In the relaxed formulation we consider, in addition to those alternatives, perforated composites with volume fraction $\theta(x)$ of holes, $0 \le \theta(x) \le 1$, and Hooke's law $A_{ijkl}(x)$. Notice that θ and A are not assumed to be constant: the choice of composite is expected to vary with position in the optimal design. The underlying design problem is not being changed, provided that we consider only Hooke's laws $A(x)$ which can actually be realized by perforation from the originally given material. The relaxed problem "looks different", however: it appears to have more options at its disposal. The classical formulation corresponds to forcing $\theta(x)$ to take the values 0 or 1 almost everywhere. In a sense, the process of relaxation turns a discrete problem $(\theta(x) = 0 \text{ or } 1)$ into a continuous one $(0 \le \theta(x) \le 1)$.

The relaxed formulation is useful both for the direct numerical solution of optimal design problems and for proving the existence of optimal designs.

Numerical optimization

A conceptually simple approach to computing relaxed optimal designs is as follows. One starts with a given region from which material is to be removed. The equations of elasticity are discretized using finite elements, with Hooke's law taken to be piecewise constant (varying from element to element). A "design" is an assignment to each element of a number θ, representing the volume fraction of holes, and a Hooke's law A, representing the elastic behavior of some perforated composite with volume fraction θ. (See Figure 1.5.) The resulting nonlinear optimization problem is once again to be solved by a version of steepest descent or some more sophisticated method.

To specify the algorithm completely, it is obviously necessary that the set of all admissible pairs (θ, A) be given explicitly. One approach, explored in recent work of Bendsoe, Kikuchi, and Suzuki, is to permit only composites of some specified and relatively simple form – e.g. with a spatially periodic microstructure consisting of rectangular holes. We like to call this a "partial relaxation". A different idea is to characterize in advance the set G_θ of all Hooke's laws obtainable by perforation from the originally given material with volume fraction θ of holes. One could then permit any pair (θ, A) with $A \in G_\theta$; we like to call this the "full relaxation". The set G_θ is unfortunately not known explicitly, so we do not know the full relaxation of a general design problem. However, we have partial information about G_θ, which permits us to consider the "full relaxation" of problems involving compliance. This will be explained in Lecture 3.

The relaxed formulation has several advantages over the standard one. First, it has the appearance at the discrete level of an expansion of the design space. (Physically speaking it amounts to the introduction of subgrid structure.) As a result, the "optimum" obtained this way is almost certain to be better than that obtained from a standard method with a similar mesh. Perhaps more important, the apparent expansion of the design space has the effect of destroying local optima. While the relaxed formulation can have local optima, this phenomenon is much less pronounced than in the standard approach: the character of the optimal design seems not to depend very much on the numerical discretization or the initial guess. Finally, the relaxed formulation is easier to implement numerically, since there is no need for regridding as the optimization proceeds.

There is, of course, one feature that could be viewed as a major disadvantage: the relaxed formulation yields an optimal design which is made from composite materials. Such a design may be impossible to manufacture, or it might be impractical for other reasons. Even so, the relaxed optimum provides a benchmark for the output of a more standard approach. It can also be useful in choosing a good initial guess. (Recent work of Kikuchi and Suzuki suggests that for their formulation of the relaxed problem, the optimal design makes relatively little use of composites. This would seem a major advantage, though an understanding of this phenomenon is presently lacking.)

Theory

We have been cavalier about the use of the term "perforated composite". The proper mathematical tool is the theory of G-convergence. It applies most readily to the material distribution problem, i.e. the optimal arrangement of two nondegenerate elastic materials in a given region. The existence of an optimal design is immediate from the compactness theorem for G-convergence. To formulate the first-order conditions of optimality in a general setting would require knowledge of the set G_θ of Hooke's laws obtained by mixing the two component materials in specified volume fractions. However, as we shall explain in Lecture 3, compliance optimization requires less information – specifically, an optimal lower bound on complementary energy. The numerical method described above (based on the "full relaxation" not just a "partial relaxation") can be interpreted as finding the optimal material distribution, in the limit as the weaker material becomes totally degenerate. Thus, unlike the standard approach the relaxed formulation has a well-defined continuum limit.

Lecture 2

Introduction to elastostatics and composite materials

Summary   This lecture gives a fast introduction to linear elasticity, including the principles of minimum elastic and complementary energy. It also gives an introduction to homogenization, in the setting of spatial periodicity. It closes with a discussion of layered composites.

The equation of elastotatics are

$$\operatorname{div} \sigma = 0 \text{ in } \Omega$$

$$\sigma = Ae(u), \quad e(u) = \frac{1}{2}(\nabla u + \nabla u^T) \tag{2.1}$$

$$\sigma \cdot n = f \text{ at } \partial\Omega.$$

(We focus for simplicity on the case of a traction boundary condition, with no body load.) Here

$\Omega \subset R^d$ is the domain occupied by the body ($d = 2$ and 3 are the cases of physical interest).

$u : \Omega \to R^d$ is the linearized elastic displacement. (The physical point at $x \in \Omega$ goes to $x + \delta u(x), \delta \approx 0$.)

$e(u)$ is the linear strain tensor. (It represents the displacement gradient, neglecting the infinitesimal rotation $\omega(u) = \frac{1}{2}(\nabla u - \nabla u^T)$.)

σ is the stress tensor, a symmetric-matrix valued function. (For any unit vector n, $\sigma \cdot n$ gives the vector of force per unit area on the plane perpendicular to n.)

> A is Hooke's law, a symmetric linear map from symmetric
> tensors to symmetric tensors. (The elastic energy quadratic
> form $\langle A\xi, \xi \rangle$ should be positive definite.)
>
> f is the boundary load, i.e., the force per unit area on $\partial\Omega$.

An elastic body may be homogeneous ($A = $ constant) or inhomogeneous ($A = A(x)$), and it may be locally isotropic or anisotropic. If it is both isotropic and homogeneous, then Hooke's law has the form

$$A\xi = \kappa(\text{tr } \xi)I + 2\mu(\xi - \frac{1}{d}(\text{tr } \xi)I).$$

The elastic constants κ and μ are called the bulk and shear moduli, respectively.

Because we have chosen to discuss the traction boundary value problem, there is a consistency condition on the load f. It must be orthogonal to any infinitesimal rigid motion, i.e.

$$\int_{\partial\Omega} \langle f, v \rangle = 0 \quad \text{for } v \in R$$

with

$$R = \{v : e(v) \equiv 0\}.$$

Every element of R is the sum of a skew-symmetric linear map and a constant.

An easy integration by parts shows that the solution of (2.1) satisfies

$$\int_\Omega \langle Ae(u), e(u) \rangle = \int_{\partial\Omega} \langle u, f \rangle.$$

Thus the internal elastic energy is equal to the work done by the load. This quantity, called *compliance*, represents one measure of the structure's rigidity under the given load.

Variational principles

The Principle of Minimum Elastic Energy asserts that the elastic displacement u solves

$$\min_v \int_\Omega \frac{1}{2}\langle Ae(v), e(v) \rangle - \int_{\partial\Omega} \langle v, f \rangle. \tag{2.2}$$

The proof is standard, making use of Korn's inequality in the form

$$\min_{\gamma \in R} ||v - \gamma||_{H^1(\Omega)} \leq C||e(v)||_{L^2(\Omega)}.$$

The Principle of Minimum Complementary Energy asserts that the stress σ solves

$$\min_{\substack{\text{div } \tau = 0 \\ \tau \cdot n = f}} \int_\Omega \frac{1}{2}\langle A^{-1}\tau, \tau \rangle. \tag{2.3}$$

This is (up to sign) the convex dual of (2.2). Here is a formal derivation of (2.3) from (2.2) via a saddle point principle:

$$
\begin{aligned}
\min_v (2.2) &= \min_v \max_\tau \int_\Omega [\langle \tau, e(v) \rangle - \frac{1}{2}\langle A^{-1}\tau, \tau \rangle] - \int_{\partial\Omega} \langle v, f \rangle \\
&= \min_v \max_\tau \int_\Omega [\langle -\operatorname{div}\tau, v \rangle - \frac{1}{2}\langle A^{-1}\tau, \tau \rangle] + \int_{\partial\Omega} \langle (\tau \cdot n - f), v \rangle \\
&= \max_\tau \min_v \int_\Omega [\langle -\operatorname{div}\tau, v \rangle - \frac{1}{2}\langle A^{-1}\tau, \tau \rangle] + \int_{\partial\Omega} \langle (\tau \cdot n - f), v \rangle \\
&= \max_{\substack{\operatorname{div}\tau=0 \\ \tau\cdot n=f}} -\frac{1}{2}\int_\Omega \langle A^{-1}\tau, \tau \rangle.
\end{aligned}
$$

A rigorous justification of this argument is not difficult.

We note that $\int_\Omega \langle A^{-1}\sigma, \sigma \rangle = \int_\Omega \langle Ae(u), e(u) \rangle$. Hence the extreme value of the complementary energy variational principle (2.3) is precisely half the compliance.

Structures made of two materials

We say a structure is made of two materials with Hooke's laws A_1 and A_2 respectively if $A = A(x)$ has the form

$$
A(x) = \begin{cases} A_1 & \text{in material 1} \\ A_2 & \text{in material 2} \end{cases}
$$

The transmission boundary conditions are easily deduced from the weak formulation of (2.1): u and $\sigma \cdot n$ are continuous at the material interface. This corresponds physically to the hypothesis that the materials are perfectly bonded.

In the limit as $A_1 \to 0$, the region occupied by material 1 can be viewed as a hole (or a union of holes), with a traction-free boundary.

Elastic composites

A composite material is really an idealization, representing the limiting behavior of a mixture with fine scale structure as the ratio

$$
\epsilon = \frac{\text{length scale of microstructure}}{\text{length scale of macroscopic structure}}
$$

tends to zero. For simplicity, we shall focus on composites made by mixing *two* elastic materials, with *spatially periodic* microstructure. Such a composite is determined by its cell of periodicity, in which A_1 and A_2 can be arranged in any way whatsoever. This cell is then rescaled to length scale ϵ and repeated periodically in space, in such a way as to fill out the physical domain Ω. (See Figure 2.1.)

The mathematical description of such a composite is as follows. Hooke's law has the form $A_\epsilon(x) = \tilde{A}(x/\epsilon)$, where

$$\tilde{A}(y) = A_1\chi_1(y) + A_2\chi_2(y). \tag{2.4}$$

Here $\chi_i(y)$ is a spatially periodic "characteristic function", equal to 1 in the ith phase and zero elsewhere. For each $\epsilon > 0$, consider the solution u_ϵ of the associated boundary-value problem:

$$\text{div}(A_\epsilon(x)e(u_\epsilon)) = 0, \quad [A_\epsilon e(u_\epsilon)] \cdot n = f.$$

The basic theorem of homogenization asserts that in the limit as $\epsilon \to 0$, the structure behaves like an elastic material with some new Hooke's law A_*, the "effective Hooke's law" of the composite. In other words, $u_\epsilon \to u_*$, where u_* solves

$$\text{div}(A_*e(u_*)) = 0, \quad [A_*e(u_*)] \cdot n = 0.$$

The effective Hooke's law A_* depends only on the microstructure (2.4), not on Ω or f. It is characterized by the variational principle

$$\langle A_*\xi, \xi \rangle = \min_{\phi \text{ per}} \int \langle \tilde{A}(y)[\xi + e(\phi)], \; \xi + e(\phi) \rangle, \tag{2.5}$$

in which ϕ ranges over periodic displacement fields and $\int$ denotes spatial averaging. There is also a variational characterization of A_*^{-1}, essentially the convex dual of (2.5):

$$\langle A_*^{-1}\eta, \eta \rangle = \min_{\substack{\text{div } \tau = 0 \\ \int \tau = \eta}} \int \langle \tilde{A}(y)^{-1}\tau, \tau \rangle. \tag{2.6}$$

The effective Hooke's law A_* is usually anisotropic, even if the components A_1 and A_2 are isotropic. For example, a periodic array of spherical conclusions of A_1 in a matrix of A_2 yields an effective tensor A_* with cubic symmetry.

The preceding discussion extends easily to the case of a composite with piecewise periodic structure. Then Hooke's law takes the form $A_\epsilon(x) = \tilde{A}(x; x/\epsilon)$, with

$$\tilde{A}(x; y) = A_1\chi_1(x; y) + A_2\chi_2(x; y)$$

piecewise constant in x and periodic in y. (See Figure 2.2.) The effective Hooke's law $A_*(x)$ is piecewise constant, and is locally determined by the periodic theory.

The spatially periodic or piecewise periodic theory is convenient, but one might wonder whether it is sufficiently general. An apparently more general theory is provided by the notion of G-convergence, which places no hypothesis on the fine-scale structure and requires no separation of length scales. In fact, however, the Hooke's laws associated with piecewise periodic composites are dense (in $L^p, 1 < p < \infty$) in the space of all Hooke's laws obtainable by G-convergence. Thus we lose no generality by focussing on the spatially periodic setting.

Layered composites

For most microstructures one cannot write down a formula for A_*; rather, one must evaluate it numerically. This can be done by solving finitely many elasticity problems with periodic boundary conditions, corresponding to the characterization (2.5) with various choices of ξ.

One important case *can* be done explicitly, however. This is the example of a layered composite. Specifically, suppose that A_* arises by mixing A_1 and A_2 in layers orthogonal to a unit vector n, using volume fractions θ_1 and $\theta_2 = 1 - \theta_1$, respectively (see Figure 2.3). We shall present a simple, explicit formula for A_* in Lecture 4. In preparation for that calculation, let us show here that the associated minimization (2.5) amounts to a problem of linear algebra, not one of PDE.

Proposition *For a layered microstructure, the optimal $e(\phi)$ in (2.5) is piecewise constant, of the form*

$$e(\phi) = \begin{cases} \theta_2 a \odot n & \text{in material 1} \\ -\theta_1 a \odot n & \text{in material 2} \end{cases} \tag{2.7}$$

for some vector $a \in R^d$. Hence

$$\begin{aligned} \langle A_* \xi, \xi \rangle = \inf_{a \in R^d} \{ &\theta_1 \langle A_1(\xi + \theta_2 a \odot n), (\xi + \theta_2 a \odot n) \rangle \\ &+ \theta_2 \langle A_2(\xi - \theta_1 a \odot n), (\xi - \theta_1 a \odot n) \rangle \}. \end{aligned} \tag{2.8}$$

Proof We are using the notation $a \odot n = \frac{1}{2}(a \otimes n + n \otimes a)$. Notice that (2.8) represents $\langle A_* \xi, \xi \rangle$ as the minimum of a quadratic expression involving only the vector a as an unknown. Evaluating such a minimum is clearly just a matter of linear algebra.

We may view the layered microstructure as being spatially periodic, by working in a coordinate system in which the layer normal n is one of the basis vectors.

It is easy to see that the right-hand side of (2.7) is the strain of a periodic displacement field ϕ. Indeed, we may take the jth component ϕ_j to have the form $\phi_j(x) = \psi_j(x \cdot n)$, where $\psi_j(t)$ is periodic with period 1 and

$$\psi_j'(t) = \begin{cases} \theta_2 a_j & 0 \leq t \leq \theta_1 \\ -\theta_1 a_j & \theta_1 \leq t \leq \theta_1. \end{cases}$$

Thus (2.8) amounts to the minimization of (2.5) over a *subspace* of the original family of test fields. If the solution of (2.8) should happen to solve the optimality condition for (2.5), then it must actually give the solution of (2.5) as well as (2.8).

The optimality condition for (2.8) is most easily obtained by rewriting (2.8) in the form

$$\inf_{\substack{\xi_2 - \xi_1 \in V(n) \\ \theta_1 \xi_1 + \theta_2 \xi_2 = \xi}} \theta_1 \langle A_1 \xi_1, \xi_1 \rangle + \theta_2 \langle A_2 \xi_2, \xi_2 \rangle, \tag{2.9}$$

where we have set

$$V(n) = \{a \odot n : a \in R^n\}.$$

The optimality condition is

$$\begin{aligned} \dot{\xi}_1 - \dot{\xi}_2 \in V(n) \\ \\ \theta_1 \dot{\xi}_1 + \theta_2 \dot{\xi}_2 = 0 \end{aligned} \Rightarrow \theta_1 \langle A_1 \xi_1, \dot{\xi}_1 \rangle + \theta_2 \langle A_2 \xi_2, \dot{\xi}_2 \rangle = 0.$$

This is equivalent to

$$A_1 \xi_1 - A_2 \xi_2 \perp V(n). \tag{2.10}$$

A simple calculation shows that

$$\tau \perp V(n) \Leftrightarrow \tau \cdot n = 0$$

for any symmetric tensor τ, so (2.10) can also be written as

$$(A_1 \xi_1) \cdot n = (A_2 \xi_2) \cdot n . \tag{2.11}$$

We claim that if $a \in R^d$ solves (2.8), then the associated ϕ solves (2.5). To see this, recall that the correspondence between (2.8) and (2.9) is

$$\xi_i = \xi + e(\phi) \quad \text{in material } i.$$

So if we define

$$\sigma(y) = \tilde{A}(y)[\xi + e(\phi)],$$

then $\sigma(y)$ is piecewise constant:

$$\sigma(y) = A_i \xi_i \quad \text{in material } i.$$

It is obvious that div $\sigma = 0$ within each layer, and (2.11) shows that $\sigma \cdot n$ is continuous across the interface. Therefore

$$\text{div}\,(\tilde{A}(y)[\xi + e(\phi)]) = 0$$

in the weak sense, globally. This is the optimality condition for (2.5). Hence ϕ solves (2.5), as asserted.

Lecture 3

Structural optimization for minimum compliance using optimal composites

Summary   This lecture explains how compliance optimization problems can be reduced to nonquadratic (but lower semicontinuous) variational problems, following the method of Kohn and Strang. It then describes the implementation of this approach for shape optimization in plane stress, following recent joint work with G. Allaire. In the limit as its weight tends to zero, the optimal design becomes a Michell truss.

We henceforth focus on shape optimization problems for two-dimensional structures in plane stress, with weight and compliance as the design criteria. We begin with a region Ω, loaded in some way on its boundary. The goal is to remove a certain amount of material from Ω, while increasing the compliance as little as possible. To avoid doing a constrained optimization, we shall actually solve

$$\min_{designs} \text{ compliance} + \lambda \cdot \text{weight}, \tag{3.1}$$

where λ is a fixed real number giving the relative importance of compliance vs. weight. As $\lambda \to 0$ weight is unimportant, and we remove almost no material. As $\lambda \to \infty$ weight is paramount, and we remove almost everything. (One easily verifies, e.g. using the Principle of Minimum Complementary Energy, that removing material increases the compliance.)

Lecture 1 presented a "relaxed formulation for numerical optimization", based on putting a homogeneous composite material in each finite element domain. Here we proceed slightly differently, introducing fine structure and optimizing the choice of composite *before* discretization. This leads to a nonlinear variational problem whose solution is the stress in the optimal design.

The mathematical formulation of the relaxed design problem is as follows. The domain Ω consists initially of a homogeneous elastic material with Hooke's law A_2. The region to be "removed" is treated as being replaced by a homogeneous elastic material A_1. (We shall pass to the limit $A_1 \to 0$ after doing the design optimization.) A *design* is described by a scalar function

$$\theta(x), \quad 0 \le \theta(x) \le 1,$$

representing the volume fraction of A_1, and a tensor-valued function

$$A(x) \in G_{\theta(x)},$$

representing the effective Hooke's law at $x \in \Omega$. Here we have set

$$G_\theta = \{\text{Hooke's laws of composites obtained by mixing} \atop A_1 \text{ and } A_2 \text{ in volume fractions } \theta, 1 - \theta, \text{ respectively}\}. \tag{3.2}$$

For any design,

$$\text{weight} = \int_\Omega [\rho_1 \theta(x) + \rho_2(1 - \theta(x))]dx,$$

where ρ_1 and ρ_2 are the densities of A_1 and A_2 respectively. Also, from the principle of Minimum Complementary Energy,

$$\text{compliance} = \inf_{\substack{\text{div } \tau=0 \\ \tau \cdot n=f}} \int_\Omega \langle A^{-1}(x)\tau, \tau \rangle dx.$$

Hence (3.1) can be expressed as

$$\min_{\substack{0 \le \theta(x) \le 1 \\ A(x) \in G_{\theta(x)}}} \min_{\substack{\text{div } \tau=0 \\ \tau \cdot n=f}} \int_\Omega \langle A^{-1}\tau, \tau \rangle + \lambda[\rho_1\theta + \rho_2(1 - \theta)]. \tag{3.3}$$

The next step is to change the order of minimization in (3.3). The constraints on $\theta(x)$ and $A(x)$ are local, i.e. they don't involve derivatives, so the minimization on θ and A can be done pointwise. We thus obtain

$$\min_{\substack{\text{div } \tau=0 \\ \tau \cdot n=f}} \min_{\substack{0 \le \theta(x) \le 1 \\ A(x) \in G_{\theta(x)}}} \int_\Omega \langle A^{-1}\tau, \tau \rangle + \lambda[\rho_1\theta + \rho_2(1 - \theta)]$$

$$= \min_{\substack{\text{div } \tau=0 \\ \tau \cdot n=f}} \int_\Omega F(\tau(x))dx, \tag{3.4}$$

with $F(\tau)$ defined by

$$F(\tau) = \min_{\substack{0 \le \theta \le 1 \\ A \in G_\theta}} \langle A^{-1}\tau, \tau \rangle + \lambda[\rho_1\theta + \rho_2(1 - \theta)] \tag{3.5}$$

for any (constant) symmetric second-order tensor τ.

It is clear from (3.5) what we need to know about composites in order to proceed. The crucial information is the "optimal lower bound on complementary energy"

$$\min_{A \in G_\theta} \langle A^{-1}\tau, \tau \rangle \tag{3.6}$$

as a function of θ and τ. Remarkably, it is possible to evaluate (3.6) explicitly. A formula for (3.5) then follows easily, by optimization over θ. The evaluation of (3.6) and (3.5) is a long story, which will be recounted in Lectures 4 and 5. Here let us simply give the result. We take A_2 (the given elastic material) to be isotropic, with bulk modulus κ and shear modulus μ; we set $A_1 = 0$, $\rho_1 = 0$ (the shape optimization limit of the material distribution problem); and we take $\rho_2 = 1$ (i.e. we absorb the density of material 2 into the Lagrange multiplier λ). Then the explicit formula for (3.5) is

$$F(\tau) = \begin{cases} \langle A_2^{-1}\tau, \tau \rangle + \lambda & \text{if } \rho(\tau) \ge 1 \\ \langle A_2^{-1}\tau, \tau \rangle + \lambda\rho(2 - \rho) & \text{if } \rho(\tau) \le 1, \end{cases} \tag{3.7}$$

with $\rho(\tau)$ defined by

$$\rho(\tau) = (\frac{\kappa + \mu}{4\kappa\mu})^{1/2}\lambda^{-1/2}(|\tau_1| + |\tau_2|), \tag{3.8}$$

where τ_1 and τ_2 are the eigenvalues of τ (the principal stresses).

The derivation of (3.7) includes a specification of the optimal composite $A(x)$, and a prescription for achieving it by a microstructure. The details will be given in Lecture 5, but we mention here that the optimal θ turns out to be

$$\theta = \begin{cases} 1 - \rho(\tau) & \text{if } \rho(\tau) \leq 1 \\ 0 & \text{if } \rho(\tau) \geq 1. \end{cases}$$

Thus the optimal design has no material where $\rho(\tau) = 0$, and it has the originally given material where $\rho(\tau) \geq 1$. Where $0 < \rho(\tau) < 1$ it consists of perforated composite materials whose microstructure, volume fraction of holes, and effective Hooke's laws vary from point to point.

In summary, finding the optimal design (for a given choice of λ) is equivalent to solving the variational problem (3.4). The solution τ is precisely the stress in the optimal design. The composition of the design can be read off from τ, using information gained in the course of evaluating (3.5).

The relaxed variational problem

Let us examine (3.4) more closely, to understand its character as a variational problem. We claim that it is *polyconvex*, i.e. that it has the form

$$\min_{\substack{\text{div } \tau = 0 \\ \tau \cdot n = f}} \int_\Omega \Phi(\tau(x), \det \tau(x)) dx$$

for a suitable *convex* function Φ. Indeed if we make the change of scale

$$\tilde{\tau} = (\frac{\kappa + \mu}{4\kappa\mu})^{1/2}\lambda^{-1/2}\tau$$

and set $\tilde{\rho}(\tilde{\tau}) = \rho(\tau) = |\tilde{\tau}_1| + |\tilde{\tau}_2|$, then elementary manipulation yields

$$F(\tau) = \lambda G(\tilde{\tau}) + 2\lambda\frac{\mu - \kappa}{\mu + \kappa}\det \tilde{\tau} \tag{3.9}$$

with

$$G(\tilde{\tau}) = \begin{cases} \tilde{\tau}_1^2 + \tilde{\tau}_2^2 + 1 & \text{if } \tilde{\rho}(\tilde{\tau}) \geq 1 \\ 2\tilde{\rho}(\tilde{\tau}) - 2|\det \tilde{\tau}| & \text{if } \tilde{\rho}(\tilde{\tau}) \leq 1. \end{cases} \tag{3.10}$$

The function G is known to be polyconvex (see Ref. [10]). The second term in (3.9) is just a constant times the determinant. So it follows that F is polyconvex.

The importance of polyconvexity is that it implies lower semicontinuity. In particular, *the minimum of* (3.4) *is achieved*. We have thus established the existence of a (relaxed) optimal design.

Numerical minimization

In recent joint work with G. Allaire, we used the finite element method to minimize (3.4) numerically. We found it convenient to represent τ by an Airy stress function:

$$\text{div } \tau = 0 \Leftrightarrow \tau = \begin{pmatrix} \psi_{22} & -\psi_{12} \\ -\psi_{12} & \psi_{11} \end{pmatrix},$$

in which $\psi_{ij} = \partial^2 \psi / \partial x_i \partial x_j$. The traction boundary condition $\tau \cdot n = f$ determines ψ and $\nabla \psi$ at $\partial \Omega$. Thus when expressed in terms of ψ, (3.4) looks something like a problem of plate theory. We used the Clough-Tocher finite elements for ψ, which are piecewise cubic and C^1.

The main difficulty in minimizing (3.4) arises from the fact that F is *not smooth*. As $\tau \to 0$ it has the singular behavior

$$F(\tau) = c(|\tau_1| + |\tau_2|) + \mathcal{O}(|\tau|^2).$$

(Recall that where τ is close to zero, θ is close to 1; these are regions where most of the material is being removed.) Moreover, F is also nonsmooth for $\tau \neq 0$, wherever $\rho(\tau) \leq 1$ and τ has rank one. (This singularity arises from a sudden change in the character of the optimal composite.) It would be interesting and worthwhile to apply techniques from nonsmooth optimization. We chose instead to do something more pedestrian. Noticing that

$$(|\tau_1| + |\tau_2|)^2 = ||\tau||^2 + 2|\det \tau|$$

with $||\tau||^2 = \tau_1^2 + \tau_2^2 = \text{tr}(\tau^2)$, we made the approximation

$$(|\tau_1| + |\tau_2|)^2 \approx ||\tau||^2 + 2(\epsilon^2 + (\det \tau)^2)^{1/2}. \tag{3.11}$$

Taking the square root of both sides gives a smooth approximation of $|\tau_1| + |\tau_2|$, which leads via (3.7)-(3.8) to a smooth approximation of $F(\tau)$.

We used the conjugate gradient method for the minimization. Because of the large number of degrees of freedom per element (due to ψ being piecewise cubic) it was difficult to take a mesh smaller that 16×16. Figures 3.1a,b show the density of holes in an optimally designed planar fillet, for two different choices of the Lagrange multiplier λ.

The Michell truss limit

Let us consider (formally) the limit $\lambda \to \infty$. This corresponds to insisting that the weight be very small. It seems reasonable to expect that $\tau = \tau_\lambda$ remains more or less bounded, since div $\tau = 0$ and $\tau \cdot n = f$ independent of λ. Recalling that

$$\rho(\tau) = \left(\frac{\kappa + \mu}{4\kappa\mu}\right)^{1/2} \lambda^{-1/2} (|\tau_1| + |\tau_2|),$$

we expect that $\rho(\tau) \ll 1$ when λ is large; hence

$$F(\tau) \sim 2\lambda\rho = (\frac{\kappa + \mu}{\kappa\mu})^{1/2}\lambda^{1/2}(|\tau_1| + |\tau_2|)$$

to principal order in λ. Thus in the limit $\lambda \to \infty$ we are effectively solving

$$\min_{\substack{\text{div } \tau = 0 \\ \tau \cdot n = f}} \int_\Omega |\tau_1| + |\tau_2|. \qquad (3.12)$$

This is none other than the well-known Michell truss problem. The usual interpretation is that (3.12) gives (a constant times) the weight of an optimal truss-like continuum. The truss members lie along the directions of principal stress, and their cross sections are proportional to the magnitudes of the principal stresses. The total volume occupied by the truss is small, so no special account need be taken of the regions where the truss members overlap.

It was always apparent that the theory of Michell trusses applied only to structures whose total density was small. One might have expected that the next step would be some correction, quadratic in τ, to account for overlapping of truss members. It is a big surprise, however, that the inclusion of quadratic corrections (more precisely, the use of (3.4) instead of (3.12)) yields the correct variational form of the optimal design problem at *any* value fraction!

Our derivation of (3.12) from (3.4) as $\lambda \to \infty$ is, of course, purely formal. It would be interesting to understand this in a mathematically rigorous way.

Lecture 4

An optimal lower bound on complementary energy

Summary   This lecture derives the optimal lower bound on $\langle A_*^{-1}\xi, \xi \rangle$, when A_* is the Hooke's law of a mixture of two materials A_1 and A_2 in fixed proportions $\theta_1, \theta_2 = 1 - \theta_1$. The proof of the bound uses the Hashin-Shtrikman variational principle; attainability is proved using sequential lamination.

We've seen that for compliance optimization, the essential information required about composites is the "optimal lower bound on complementary energy"

$$\min_{A_* \in G_\theta} \langle A_*^{-1}\xi, \xi \rangle. \qquad (4.1)$$

The difficulty is that we have no explicit characterization of G_θ, so as it stands (4.1) is uncomputable. Methods for evaluating such bounds have, however, become available in recent years, through the joint efforts of many people.

We shall proceed in two distinct steps. First, using the Hashin-Shtrikman variational principle we shall give a lower bound, i.e. a function $H(\theta, \xi, A_1, A_2)$ such that

$$\langle A_*^{-1}\xi, \xi \rangle \geq H(\theta, \xi, A_1, A_2) \tag{4.2}$$

for any composite made from A_1, A_2 in the given volume fractions. Then, using the method of sequential lamination, we shall show that H is the optimal lower bound. The bound H is given here as the extreme value of a finite dimensional concave maximization. An explicit formula will be derived in Lecture 5.

The lower bound

We assume that the component materials are "well-ordered", i.e. that $A_2 - A_1$ determines a positive quadratic form on strains. Note that the case of shape optimization corresponds to $A_1 \to 0$, so that $A_2 - A_1 \approx A_2$ is certainly positive. We do not require A_1 or A_2 to be isotropic at this stage. However, if they are anisotropic their orientations are fixed throughout the composite.

Our starting point is the dual variational principle for A_*:

$$\langle A_*^{-1}\xi, \xi \rangle = \min_{\substack{\text{div } \tau = 0 \\ \int \tau = 0}} \oint [\chi_1 \langle A_1^{-1}(\xi + \tau), \xi + \tau \rangle + \chi_2 \langle A_2^{-1}(\xi + \tau), \xi + \tau \rangle], \tag{4.3}$$

in which τ ranges over spatially periodic second-order tensor fields and $\oint$ denotes spatial averaging. Adding and subtracting the "reference energy" $\langle A_2^{-1}(\xi + \tau), \xi + \tau \rangle$, the right hand side can be rewritten as

$$\oint \chi_1 \langle (A_1^{-1} - A_2^{-1})(\xi + \tau), \xi + \tau \rangle + \oint \langle A_2^{-1}(\xi + \tau), \xi + \tau \rangle. \tag{4.4}$$

The second term of (4.4) expands as

$$\langle A_2^{-1}\xi, \xi \rangle + \oint \langle A_2^{-1}\tau, \tau \rangle.$$

The first term can be written as

$$\sup_{\eta(\nu)} \oint 2\chi_1 \langle \eta, (\xi + \tau) \rangle - \oint \chi_1 \langle (A_1^{-1} - A_2^{-1})^{-1}\eta, \eta \rangle.$$

Substitution into (4.3) leads to an expression of the form

$$\langle A_*^{-1}\xi, \xi \rangle = \inf_{\substack{\text{div } \tau = 0 \\ \int \tau = 0}} \sup_{\eta(\nu)} B(\tau, \eta)$$

with $B(\tau, \eta)$ concave in η and convex in τ. We can interchange the inf and the sup, by a general saddle point principle, leading to

$$\langle A_*^{-1}\xi, \xi \rangle = \sup_{\eta(\nu)} \inf_{\substack{\text{div } \tau = 0 \\ \int \tau = 0}} B(\tau, \eta).$$

Specialization to *constant* η then gives a lower bound. After some simplification the bound takes the form

$$\langle(A_*^{-1} - A_2^{-1})\xi, \xi\rangle \geq \sup_{\eta \text{ constant}} \{2\theta_1\langle\eta, \xi\rangle - \theta_1\langle(A_1^{-1} - A_2^{-1})^{-1}\eta, \eta\rangle$$

$$+ \min_{\substack{\text{div } \tau = 0 \\ \int \tau = 0}} \int 2\langle\eta, \tau\rangle\chi_1 + \langle A_2^{-1}\tau, \tau\rangle\}. \tag{4.5}$$

The next step is to evaluate the infimum over τ explicitly. This amounts to solving a problem of elastostatics with periodic boundary conditions in a uniform body with Hooke's law A_2. It is convenient to use Fourier analysis. The Fourier transform of τ at frequency k lies in

$$W(k) = \{\text{symmetric tensors } \Sigma \text{ such that } \Sigma \cdot k = 0\}. \tag{4.6}$$

Skipping the details, we find that the value of the infimum over τ in (4.5) is

$$-\sum_{k \neq 0} |\hat{\chi}_1(k)|^2 |\pi_{A_2^{-1/2}W(k)} A_2^{1/2}\eta|^2,$$

where k ranges over Z^n, and $\pi_V \xi$ denotes the orthogonal projection of ξ onto the subspace V. Thus (4.5) can be written as

$$\langle(A_*^{-1} - A_2^{-1})\xi, \xi\rangle \geq \sup_{\eta}\{2\theta_1\langle\eta, \xi\rangle - \theta_1\langle(A_1^{-1} - A_2^{-1})^{-1}\eta, \eta\rangle$$

$$-\sum_{k \neq 0} |\hat{\chi}_1(k)|^2\langle f(k)\eta, \eta\rangle\}, \tag{4.7}$$

where $f(k)$ is the (degenerate) Hooke's law defined by

$$\langle f(k)\eta, \eta\rangle = |\pi_{A_2^{-1/2}W(k)} A_2^{1/2}\eta|^2. \tag{4.8}$$

The final step is to derive a lower bound that is independent of the microstructure. Plancherel's formula gives

$$\sum_{k \neq 0} |\hat{\chi}_1(k)|^2 = \int(\chi_1 - \theta_1)^2 = \theta_1\theta_2.$$

If we set

$$g(\eta) = \sup_{|k|=1} |\pi_{A_2^{-1/2}W(k)} A_2^{1/2}\eta|^2, \tag{4.9}$$

then it is obvious that g is convex, and

$$\langle f(k)\eta, \eta\rangle \leq g(\eta).$$

Therefore

$$\sum_{k \neq 0} |\hat{\chi}_1(k)|^2\langle f(k)\eta, \eta\rangle \leq \theta_1\theta_2 g(\eta),$$

and (4.7) yields $\langle A_*^{-1}\xi, \xi\rangle \geq H$ with

$$H = \langle A_2^{-1}\xi, \xi\rangle + \theta_1 \sup_{\eta} \{2\langle\eta, \xi\rangle - \langle(A_1^{-1} - A_2^{-1})^{-1}\eta, \eta\rangle - \theta_2 g(\eta)\}. \tag{4.10}$$

This is the desired bound (4.2).

Sequential lamination

We shall show the optimality of this bound by constructing a sequentially laminated microstructure that achieves equality, for any given ξ. This is an inductive construction, which proceeds as follows (see Figure 4.1):

> A sequentially laminated composite of rank 1 is just an ordinary layered composite, obtained by mixing A_1 and A_2 in volume fractions ρ_1 and $1 - \rho_1$, using layers orthogonal to some unit vector n_1. Call the effective tensor $A_*^{(1)}$.   (4.11a)

> A sequentially laminated composite $A_*^{(r)}$ of rank r is obtained by mixing a sequentially laminated composite of rank $r - 1$, say $A_*^{(r-1)}$, with A_2, in volume fractions ρ_r and $1 - \rho_r$, using layers orthogonal to some unit vector n_r.   (4.11b)

A separation of scales is assumed, so that $A_*^{(r-1)}$ may be treated as a homogeneous material in the construction of $A_*^{(r)}$. We layer $A_*^{(r-1)}$ with A_2 and not A_1 at each stage, because A_2 was the "reference material" in (4.4). Notice that $A_*^{(r)}$ consists of "plate-like" inclusions of material 1 in a matrix of material 2.

We shall derive a convenient formula for the effective tensor of a sequentially laminated composite. The key is a formula for simple layering (of rank 1) which iterates easily. This can be obtained using the Hashin-Shtrikman variational principle, which turns out to be *exact* for simple layering.

Proposition *Let A_* be the effective Hooke's law of a simply layered composite obtained by mixing A_1 and A_2 in volume fractions $\rho, 1 - \rho$, with layers orthogonal to n. Then A_* is determined by*

$$\rho(A_*^{-1} - A_2^{-1})^{-1}\eta = (A_1^{-1} - A_2^{-1})^{-1}\eta + (1 - \rho)f(n)\eta \qquad (4.12)$$

for any second-order tensor η, with $f(n)$ defined by (4.8).

Proof Consider the calculation (4.3)-(4.10) as it applies to the simply layered case. We know from Lecture 2 that (for this case only) the microscopic strain and stress are actually constant. Therefore (4.5) becomes an equality rather than an inequality in the layered case, and (4.7) becomes

$$\langle(A_*^{-1} - A_2^{-1})\xi, \xi\rangle = \sup_{\eta} \{2\rho\langle\eta, \xi\rangle - \rho\langle(A_1^{-1} - A_2^{-1})^{-1}\eta, \eta\rangle$$
$$- \sum_{k \neq 0} |\hat{\chi}_1(k)|^2\langle f(k)\eta, \eta\rangle\}. \qquad (4.13)$$

Also, since the microstructure is layered,

$$\hat{\chi}_1(k) = 0 \quad \text{unless} \quad k\|n.$$

Hence (4.13) becomes

$$\langle (A_*^{-1} - A_2^{-1})\xi, \xi \rangle = \rho \sup_{\eta} \{ 2\langle \eta, \xi \rangle - \langle (A_1^{-1} - A_2^{-1})^{-1}\eta, \eta \rangle - (1 - \rho)\langle f(n)\eta, \eta \rangle \}.$$

Taking the Legendre transform of both sides yields

$$\rho \langle (A_*^{-1} - A_2^{-1})^{-1}\eta, \eta \rangle = \langle (A_1^{-1} - A_2^{-1})^{-1}\eta, \eta \rangle + (1 - \rho)\langle f(n)\eta, \eta \rangle$$

for every η. This is equivalent to (4.12). $\square$

Using (4.12), we easily deduce a formula for $A_*^{(r)}$ at any rank r. In the notation of (4.11a,b) we have

$$\rho_1 [(A_*^{(1)})^{-1} - A_2^{-1}]^{-1} = [A_1^{-1} - A_2^{-1}]^{-1} + (1 - \rho_1)f(n_1)$$

$$\rho_r [(A_*^{(r)})^{-1} - A_2^{-1}]^{-1} = [(A_*^{(r-1)})^{-1} - A_2^{-1}]^{-1} + (1 - \rho_r)f(n_r).$$

Some arithmetic gives

$$\beta_r [(A_*^{(r)})^{-1} - A_2^{-1}]^{-1} = [A_1^{-1} - A_2^{-1}]^{-1} + \sum_{i=1}^{r} (\beta_{i-1} - \beta_i)f(n_i)$$

with

$$\beta_0 = 1, \qquad \beta_r = \prod_{i=1}^{r} \alpha_i.$$

Notice that β_r is simply the volume fraction of A_1 in $A_*^{(r)}$. Since $1 = \beta_0 \geq \beta_1 \geq \ldots \geq \beta_r$ and

$$\sum_{i=1}^{r} (\beta_{i-1} - \beta_i) = 1 - \beta_r,$$

we conclude the following result.

Proposition *For any $\{m_i\}_{i=1}^{r}$, $0 \leq m_i \leq 1$, $\sum m_i = 1$; any $\theta_1, \theta_2 = 1 - \theta_1$; and any unit vectors $\{n_i\}_{i=1}^{r}$, there is a sequentially laminated microstructure made from A_1 and A_2 in volume fractions θ_1 and θ_2 whose effective tensor A_* is given by*

$$\theta_1 (A_*^{-1} - A_2^{-1})^{-1} = (A_1^{-1} - A_2^{-1})^{-1} + \theta_2 \sum_{i=1}^{r} m_i f(n_i). \qquad (4.14)$$

Optimality of the lower bound

Now let us prove that (4.2), with H given by (4.10), is the *optimal* lower bound on $\langle A_*^{-1}\xi, \xi \rangle$. We shall do this without actually evaluating the bound explicitly. Rather, the optimality conditions for the concave maximization (4.10) will be used to determine the parameters m_i, n_i for (4.14).

Consider the optimization in η that determines the bound:

$$\sup_{\eta}\{2\langle\eta,\xi\rangle - \langle(A_1^{-1} - A_2^{-1})^{-1}\eta,\eta\rangle - \theta_2 g(\eta)\}.$$

Since the expression in brackets is strictly concave, there is a unique extremal η^*. Since g is not smooth, the right tool for expressing the optimality condition is the subdifferential calculus (or, equivalently in this setting, the calculus of generalized gradients). It yields

$$2\xi - 2(A_1^{-1} - A_2^{-1})^{-1}\eta^* \in \theta_2\partial g(\eta^*), \tag{4.15}$$

where $\partial g(\eta^*)$ is the subdifferential of g at η^*. The function g is a maximum of quadratic forms:

$$g(\eta) = \sup_{|k|=1}\langle f(k)\eta,\eta\rangle \tag{4.16}$$

(see (4.8) and (4.9)). Its subdifferential at η^* is hence the convex hull of the tensors $2f(k)\eta$, as k ranges over extremals for (4.16). Therefore (4.15) is equivalent to

$$\xi - (A_1^{-1} - A_2^{-1})^{-1}\eta^* = \theta_2\sum_{i=1}^{r} m_i f(n_i)\eta^* \tag{4.17}$$

for some r, with $0 \le m_i \le 1$, $\sum m_i = 1$, and each n_i extremal for (4.16).

We assert that the sequentially laminated composite associated with this choice of $\{(m_i, n_i)\}$ achieves equality in the bound. Indeed, taking the inner product of (4.17) with η^* yields

$$\langle\xi,\eta^*\rangle - \langle(A_1^{-1} - A_2^{-1})^{-1}\eta^*,\eta^*\rangle = \theta_2 g(\eta^*).$$

Therefore the value of the bound (4.10) is actually

$$H = \langle A_2^{-1}\xi,\xi\rangle + \theta_1\langle\eta^*,\xi\rangle.$$

From (4.14) and (4.17), we see that the sequentially laminated composite has

$$\theta_1(A_*^{-1} - A_2^{-1})^{-1}\eta^* = \xi,$$

so

$$\theta_1\eta^* = (A_*^{-1} - A_2^{-1})\xi.$$

Thus

$$\langle A_*^{-1}\xi,\xi\rangle = \langle A_2^{-1}\xi,\xi\rangle + \theta_1\langle\eta^*,\xi\rangle = H.$$

In other words, A_* achieves equality in the bound (4.2).

Lecture 5

Explicit evaluation of the relaxed functional

Summary This lecture presents recent joint work with G. Allaire, deriving the explicit formula (3.7) for the relaxed functional F. We also discuss the character of the optimal microstructure.

We recall from Lecture 3 the form of the relaxed optimal design functional (3.5):

$$F(\tau) = \min_{\substack{0 \le \theta \le 1 \\ A \in G_\theta}} \langle A^{-1}\tau, \tau \rangle + \lambda[\rho_1\theta + \rho_2(1 - \theta)]. \tag{5.1}$$

Here

τ	$=$	a (constant) second-order tensor, representing the (macroscopic) stress at a given point in the structure;
A	$=$	a (constant) fourth-order tensor, representing the effective Hooke's law of the composite at that point;
G_θ	$=$	the set of all Hooke's laws achievable by mixtures of two given elastic materials A_1 and A_2 in volume fractions $\theta, 1 - \theta$ respectively;
ρ_j	$=$	the density of the jth material;
λ	$=$	a positive real number, giving the relative importance of compliance and weight.

The function $F(\tau)$ gives *compliance* $+\lambda \cdot$ *weight* as a function of the *stress* τ, for the optimal microstructure.

We focus on shape optimization in plane stress. Therefore we work entirely in space dimension 2, and we take the "weaker" material A_1 to be degenerate:

$$A_1 = 0, \quad \rho_1 = 0. \tag{5.2}$$

We also assume that before removing material the structure is composed of an isotropic elastic material with unit density:

$$A_2\xi = \kappa(\text{tr } \xi)I + 2\mu(\xi - \frac{1}{2}(\text{tr } \xi)I), \quad \rho_2 = 1. \tag{5.3}$$

With these choices of ρ_1 and ρ_2, (5.1) becomes

$$F(\tau) = \min_{0 \le \theta \le 1} \min_{A \in G_\theta} \langle A^{-1}\tau, \tau \rangle + \lambda(1 - \theta). \tag{5.4}$$

The crux of the matter is the minimization of $\langle A^{-1}\tau, \tau \rangle$ over $A \in G_\theta$. This was reduced in Lecture 4 to a finite dimensional concave maximization. When specialized to $A_1 = 0$, the optimal lower bound (4.10) becomes:

$$\min_{A \in G_\theta} \langle A^{-1}\tau, \tau \rangle = \langle A_2^{-1}\tau, \tau \rangle + \theta \sup_\eta [2\langle \eta, \tau \rangle - (1-\theta)g(\eta)], \tag{5.5}$$

with

$$g(\eta) = \sup_{|k|=1} |\pi_{A_2^{-1/2}W(k)} A_2^{1/2}\eta|^2. \tag{5.6}$$

Explicit calculation of $g(\eta)$

The first task is the evaluation of $|\pi_{A_2^{-1/2}W(k)} A_2^{1/2}\eta|^2$ as a function of the unit vector k. This is equivalent to finding the Fourier transform of the Green's function for isotropic elasticity. It is standard and elementary, but not simple. An efficient route to the answer is as follows. Recall from (4.6) the definition of $W(k)$:

$$W(k) = \{\Sigma : \Sigma \cdot k = 0\}.$$

Its orthogonal complement is

$$V(k) = \{\xi : \xi = v \otimes k + k \otimes v \text{ for some } v\}.$$

The orthogonal complement of $A_2^{-1/2}W(k)$ is therefore $A_2^{1/2}V(k)$, and we have

$$|\pi_{A_2^{-1/2}W(k)} A_2^{1/2}\eta|^2 = \langle A_2\eta, \eta \rangle - |\pi_{A_2^{1/2}V(k)} A_2^{1/2}\eta|^2. \tag{5.7}$$

The formula for $|\pi_{A_2^{1/2}V(k)} A_2^{1/2}\eta|^2$ is given in Ref. [49]. Specializing the result there to two dimensions gives

$$|\pi_{A_2^{1/2}V(k)} A_2^{1/2}\eta|^2 = 4\mu[|\eta k|^2 - \langle \eta k, k \rangle^2] \\ + \frac{1}{\kappa + \mu}[(\kappa - \mu) \operatorname{tr} \eta + 2\mu\langle \eta k, k \rangle]^2. \tag{5.8}$$

To calculate $g(\eta)$ we must maximize (5.7) over $|k| = 1$. This is equivalent to minimizing (5.8). It is convenient to express the right-hand side of (5.8) as

$$4\mu[|\eta k|^2 - \langle \eta k, k \rangle^2] + \frac{\mu}{\gamma + 2}[\gamma \operatorname{tr} \eta + 2\langle \eta k, k \rangle]^2$$

with $\gamma = (\kappa - \mu)/\mu$. By the method of Lagrange multipliers, any critical point k satisfies

$$\eta^2 k + [\frac{\gamma}{\gamma + 2} \operatorname{tr} \eta - \frac{2(\gamma + 1)}{\gamma + 2}\langle \eta k, k \rangle]\eta k = ck \tag{5.9}$$

for some real number c. We may work in a basis where η is diagonal with eigenvalues η_1, η_2. There are two cases: when k is an eigenvector of η, and when it is not.

Consider the latter case first. We claim that it gives the maximum, not the minimum, of (5.8). Indeed, if k is not an eigenvector of η then by (5.9) both η_1 and η_2 are roots of the polynomial

$$x^2 + [\frac{\gamma}{\gamma + 2} \operatorname{tr} \eta - \frac{2(\gamma + 1)}{\gamma + 2} \langle \eta k, k \rangle] x = c.$$

It follows that

$$-(\eta_1 + \eta_2) = \frac{\gamma}{\gamma + 2} \operatorname{tr} \eta - \frac{2(\gamma + 1)}{\gamma + 2} \langle \eta k, k \rangle,$$

whence

$$\operatorname{tr} \eta = \langle \eta k, k \rangle. \tag{5.10}$$

One verifies that (5.10) implies $\eta \in V(k)$, which yields

$$|\pi_{A_2^{1/2} V(k)} A_2^{1/2} \eta|^2 = |A_2^{1/2} \eta|^2.$$

This is clearly the maximum possible value of (5.8).

The remaining critical points are the eigenvectors of η. If k is an eigenvector with eigenvalue η_1 then (5.8) becomes

$$\frac{1}{\kappa + \mu} [(\kappa + \mu)\eta_1 + (\kappa - \mu)\eta_2]^2. \tag{5.11}$$

Similarly, when k is associated with η_2, (5.8) becomes

$$\frac{1}{\kappa + \mu} [(\kappa + \mu)\eta_2 + (\kappa - \mu)\eta_1]^2. \tag{5.12}$$

One verifies that if $|\eta_1| \leq |\eta_2|$ then

$$[(\kappa + \mu)\eta_1 + (\kappa - \mu)\eta_2]^2 \leq [(\kappa + \mu)\eta_2 + (\kappa - \mu)\eta_1]^2,$$

so (5.11) is preferred over (5.12). Substitution of (5.11) into (5.7) gives an explicit formula for $g(\eta)$:

$$g(\eta) = \langle A_2 \eta, \eta \rangle - \frac{1}{\kappa + \mu} [(\kappa + \mu)\eta_1 + (\kappa - \mu)\eta_2]^2.$$

Since $\langle A_2 \eta, \eta \rangle = (\kappa - \mu)(\eta_1 + \eta_2)^2 + 2\mu(\eta_1^2 + \eta_2^2)$, this formula simplifies to

$$g(\eta) = \frac{4\kappa\mu}{\kappa + \mu} \eta_2^2 . \tag{5.13}$$

We emphasize that (5.13) is asserted under the convention

$$|\eta_1| \leq |\eta_2|.$$

Explicit minimization over G_θ

To calculate (5.5) we must evaluate

$$\sup_\eta [2\langle \eta, \tau \rangle - (1 - \theta)g(\eta)] \tag{5.14}$$

with ξ and θ held fixed. It is easy to see that η should be simultaneously diagonal with τ. The optimum is achieved when $\eta_1 = \text{sgn}(\tau_1)t$, $\eta_2 = \text{sgn}(\tau_2)t$, with t achieving

$$\sup_t 2(|\tau_1| + |\tau_2|)t - \frac{4\kappa\mu}{\kappa + \mu}(1 - \theta)t^2.$$

The best t is

$$t = (|\tau_1| + |\tau_2|) \cdot \frac{\kappa + \mu}{4\kappa\mu(1 - \theta)}, \tag{5.15}$$

yielding

$$\frac{\kappa + \mu}{4\kappa\mu(1 - \theta)}(|\tau_1| + |\tau_2|)^2$$

for the maximum of (5.14). Substitution into (5.5) gives

$$\min_{A \in G_\theta} \langle A^{-1}\tau, \tau \rangle = \langle A_2^{-1}\tau, \tau \rangle + \frac{\theta}{1 - \theta}\frac{\kappa + \mu}{4\kappa\mu}(|\tau_1| + |\tau_2|)^2. \tag{5.16}$$

Optimization over θ

Combining (5.4) with (5.16), the relaxed functional is

$$F(\tau) = \langle A_2^{-1}\tau, \tau \rangle + \min_{0 \leq \theta \leq 1}[(-1 + \frac{1}{1 - \theta})\frac{\kappa + \mu}{4\kappa\mu}(|\tau_1| + |\tau_2|)^2 + \lambda(1 - \theta)]. \tag{5.17}$$

The optimal θ is determined by

$$1 - \theta = (\frac{\kappa + \mu}{4\kappa\mu})^{1/2}\lambda^{-1/2}(|\tau_1| + |\tau_2|) \tag{5.18}$$

if the quantity is ≤ 1, and $\theta = 0$ otherwise. If we set

$$\rho(\tau) = (\frac{\kappa + \mu}{4\kappa\mu})^{1/2}\lambda^{-1/2}(|\tau_1| + |\tau_2|)$$

then substitution into (5.16) gives

$$F(\tau) = \begin{cases} \langle A_2^{-1}\tau, \tau \rangle + \lambda & \text{if } \rho(\tau) \geq 1 \\ \langle A_2^{-1}\tau, \tau \rangle + \lambda\rho(2 - \rho) & \text{if } \rho(\tau) \leq 1 \end{cases}.$$

This is the same as (3.7).

Character of the optimal microstructure

If $\rho(\tau) \geq 1$ then there is no microstructure: at such points no material is removed. If $\rho(\tau) = 0$, which is the same as $\tau = 0$, then there is again no microstructure: at these points all the material is removed. Where $0 < \rho(\tau) < 1$, however, the optimal design has a perforated composite with volume fraction $\theta = 1 - \rho$ of holes.

According to Lecture 4, an optimal (sequentially laminated) microstructure can be read off from the optimality conditions for the Hashin-Shtrikman bound. In the present context the bound takes the form (5.5), and the optimality condition is

$$2\tau \in (1 - \theta)\partial g(\eta).$$

This is equivalent to

$$\tau = (1 - \theta) \sum_{i=1}^{p} m_i f(k_i)\eta, \tag{5.19}$$

where $0 \leq m_i \leq 1, \sum m_i = 1, f(k)$ is the Hooke's law associated to $|\pi_{A_2^{-1/2}W(k)} A_2^{1/2}\eta|^2$, and $\{k_i\}$ achieve the maximum in the definition of $g(\eta)$. We shall show that (5.19) always holds with $p = 2$, taking k_1 and k_2 to be the eigenvectors of τ with eigenvalues τ_1 and τ_2 respectively, and

$$m_1 = \frac{|\tau_2|}{|\tau_1| + |\tau_2|}, \qquad m_2 = \frac{|\tau_1|}{|\tau_1| + |\tau_2|}. \tag{5.20}$$

According to the construction of Lecture 4, (5.19)-(5.20) determine an optimal microstructure as follows. First we layer $A_1 = 0$ and A_2 in volume fractions $\rho_1 = 1 - (1 - \theta)m_1$ and $1 - \rho_1$ respectively, using layers orthogonal to k_1, to get a composite with Hooke's law C. Then we layer C with A_2 in volume fractions $\rho_2 = \theta/\rho_1$ and $1 - \rho_2$ respectively, using layers orthogonal to k_2, to get a composite A_*. This A_* achieves the optimal lower bound on $\langle A_*\tau, \tau \rangle$ at volume fraction θ. Notice that in the present context θ is itself determined by τ through (5.18). The microstructure is usually a rank-two laminate. However, it reduces to a rank-one laminate when $m_1 = 0$ or $m_2 = 0$, i.e. when τ has rank one, since in that case one of the two layerings is trivial. When it is a rank-two laminate, the optimal microstructure consists of perforation by a pattern of long, thin holes, the long direction being parallel to k_2 (see Figure 5.1). The microstructure is not unique: relabeling the eigenvalues and eigenvectors leads to a second, physically different microstructure with the same effective energy. Thus the slits can be aligned with either eigenvector of τ.

It remains to derive the explicit form of the optimality condition, (5.19)-(5.20). One verifies from (5.7)-(5.8) that

$$\begin{aligned} f(k)\eta = \quad & A_2\eta - 4\mu[(\eta k) \odot k - \langle \eta k, k \rangle k \odot k] \\ & -\frac{1}{\kappa + \mu}[(\kappa - \mu)\operatorname{tr}\eta + 2\mu\langle \eta k, k \rangle][(k - \mu)I + 2\mu k \odot k]. \end{aligned} \tag{5.21}$$

Consider first the case $\tau_1\tau_2 \leq 0$. Replacing τ by $-\tau$ if necessary, we may suppose that $\tau_1 \geq 0 \geq \tau_2$. Then the optimal η for (5.14) is

$$\eta = \begin{pmatrix} t & 0 \\ 0 & -t \end{pmatrix}, \quad t = \frac{\kappa + \mu}{4\kappa\mu(1-\theta)}(|\tau_1| + |\tau_2|) \tag{5.22}$$

(see (5.15)). In this case the only extremal k for the definition of $g(\eta)$ are the eigenvectors of τ. Both eigenvectors qualify, since $\eta_1^2 = \eta_2^2 = t^2$. For these η and k, (5.21) yields:

$$(1-\theta)f(k_1)\eta = -(|\tau_1| + |\tau_2|)k_2 \odot k_2$$
$$(1-\theta)f(k_2)\eta = (|\tau_1| + |\tau_2|)k_1 \odot k_1$$

So (5.19) reduces to

$$\begin{pmatrix} \tau_1 & 0 \\ 0 & \tau_2 \end{pmatrix} = (|\tau_1| + |\tau_2|)\begin{pmatrix} m_2 & 0 \\ 0 & -m_1 \end{pmatrix},$$

which yields (5.20).

The other case is $\tau_1\tau_2 \geq 0$. Replacing τ by $-\tau$ if necessary, we may suppose that $\tau_1 \geq 0$, $\tau_2 \geq 0$. Then the optimal η for (5.14) is

$$\eta = t \cdot I, \quad t = \frac{\kappa + \mu}{4\kappa\mu(1-\theta)}(|\tau_1| + |\tau_2|).$$

Since η is isotropic, every k is extremal in the definition of $g(\eta)$. We may nevertheless take k_1 and k_2 to be the eigenvectors of τ, as before. With this choice of η and k, (5.21) yields

$$(1-\theta)f(k_1)\eta = (|\tau_1| + |\tau_2|)k_2 \odot k_2$$
$$(1-\theta)f(k_2)\eta = (|\tau_1| + |\tau_2|)k_1 \odot k_1,$$

and (5.19) reduces to

$$\begin{pmatrix} \tau_1 & 0 \\ 0 & \tau_2 \end{pmatrix} = (|\tau_1| + |\tau_2|)\begin{pmatrix} m_2 & 0 \\ 0 & m_1 \end{pmatrix}.$$

The corresponding values of m_1 and m_2 are once again given by (5.20).

Historical and bibliographic notes

Lecture 1

There is by now a vast literature on what I call the "standard approach". Two recent surveys on the numerical side are

[1] Y. Ding, Shape optimisation of strucutres: a literature survey, *Comput. & Structures* **24** (1986), 985–1004.

[2] R. Haftka and R. Grandhi, Structural shape optimization – a survey, *Comput. Methods Appl. Mech. Engrg.* **57** (1986), 91–106.

One of the few implementations that permits changes of topology is described in

[3] A. Atrek, SHAPE: a program for shape optimization of continuum structures, in: *Computer Aided Design of Structures: Applications* (C. Brebbia and S. Hernandez), Springer-Verlag, 1989, 135–144.

Two recent surveys with a more theoretical viewpoint are

[4] N. Banichuk, *Problems and Methods of Optimal Structural Design*, Plenum Press, 1983.

[5] O. Pironneau, *Optimal Shape Design for Elliptic Systems*, Springer-Verlag, 1984.

The derivation of first-order optimality conditions has been studied by many mathematicians. The lectures of J.-P. Zolésio, M. Delfour, and several others at this NATO-ASI are concerned with problems of that type.

The idea of considering a relaxed formulation based on composite materials emerged in the 1970's, through the independent work of several groups, notably Cheng and Olhoff; Murat and Tartar; Cherkaev, Federov, and Lurie; Raitum; and Kohn and Strang. Representative articles from each group are:

[6] K.-T. Cheng and N. Olhoff, An investigation concerning optimal design of solid elastic plates, *Internat. J. Solids and Structures* **17** (1981), 305-323.

[7] F. Murat and L. Tartar, Calcul des variations et homogénéisation, in: *Les méthodes de l'homogénéisation: Théorie et applications en physique*, Collection de la Direction des études et recherche d'Électricité de France, Eyrolles, 1985, 316–369.

[8] K. Lurie, A. Cherkaev, and A. Fedorov, Regularization of optimal design problems for bars and plates I, II, *J. Optim. Theory Appl.* **37** (1982), 499–521 and 523–543.

[9] U. Raitum, On optimal control problems for linear elliptic equations, *Soviet Math. Dokl.* **20** (1979), 129–132.

[10] R. Kohn and G. Strang, Optimal design and relaxation of variational problems I–III, *Comm. Pure Appl. Math.* **39** (1986) 113–138, 139–182, 353–377.

Except for [6], this work was oriented more toward existence theorems than toward actual computation of optimal designs, and it focussed on scalar problems (torsional rigidity, conductivity, etc.) rather than elasticity. An analogous approach to a fully two-dimensional problem of plate theory was taken in

[11] A. Cherkaev and L. Gibiansky, Design of composite plates of extremal rigidity, Ioffe Physicotechnical Institute Report, 1984 (in Russian).

The idea of using a "partial relaxation" emerged from the engineering community. Early papers taking such an approach include

[12] G. Rozvany, N. Olhoff, K.-T. Chen, and J. Taylor, On the solid plate paradox in structural optimization, *J. Struct. Mech.* **10**, (1982), 1–32.

[13] G. Rozvany, T. Ong, W. Szeto, R. Sandler, N. Olhoff, and M. Bendsøe, Least-weight design of perforated plates I, II, *Internat. J. Solids and Structures* **23** (1987), 521–536 and 537–550.

Recent work with a more numerical orientation, still based on partial relaxation, includes

[14] M. Bendsøe and N. Kikuchi, Generating optimal topologies in structural design using a homogenization method, *Comput. Methods Appl. Mech. Engrg.* **71** (1988), 197–224.

[15] M. Bendsøe and N. Kikuchi, Optimal shape design as a material distribution problem, *Struct. Optim.* **1**, (1989), 193–202.

[16] M. Bendsøe and H. Rodrigues, Integrated topology and boundary shape optimization of 2-D solids, *Comput. Methods Appl. Mech. Engrg.*, to appear, 1992.

[17] K. Suzuki and N. Kikuchi, Shape and layout optimization using the homogenization method, *Comput. Methods Appl. Mech. Engrg.*, to appear, 1992.

The approach taken in these papers is essentially the one presented in Lecture 1, based on taking a different (uniform) composite in each finite element. The idea of using the relaxed problem to suggest a good "seed" for a standard code is explored in [16]. The optimal designs obtained in [17] seem to make relatively little use of composites, but a fundamental understanding of this phenomenon is lacking.

For theoretical questions such as the existence of optimal designs it makes little difference whether one is studying scalar problems (such as electrical or thermal conductivity) or vector ones (such as elasticity or plate theory). Hence the analysis of [7–10] applies even for the optimization of elastic structures. The characterization of the set G_θ (the set of all Hooke's laws at fixed volume fraction) is more difficult than its scalar analogue, however. This problem has been solved only in certain special cases, where it can be reduced to a fundamentally scalar question:

[18] K. Lurie and A. Cherkaev, *G*-closure of some particular sets of admissible material characteristics for the problem of bending of thin plates, *J. Optim. Theory Appl.* **42** (1984), 305–315.

[19] R. Lipton, On the effective elasticity of a two-dimensional homogenized incompressible elastic composite, *Proc. Roy. Soc. Edinburgh Sect. A* **110** (1988), 45–61.

[20] G. Francfort, Homogenization of a class of fourth order equations with application to incompressible elasticity, *Proc. Roy. Soc. Edinburgh Sect. A* **120** (1992), 25–46.

For optimization problems involving compliance alone full knowledge of G_θ is not required; rather, it is enough to know an optimal lower bound on complementary energy. See the notes to Lecture 3 for references on this.

The theory of G-convergence does not strictly speaking apply to the case of shape optimization. As pointed out in Lecture 1, one way around this is to consider instead the material distribution problem, optimizing first and then passing to the limit when one material becomes degenerate. There is also an alternative approach, at least for compliance optimization, based on the relaxation of variational problems. See [10] for a treatment of this type in a scalar setting.

Lecture 2

There are many standard texts on elastostatics and variational principles. They include

[21] J. Necas and I. Hlavacek, *Mathematical Theory of Elastic and Elasto-Plastic Bodies: an Introduction*, Elsevier, 1981.

[22] G. Duvaut and J.-L. Lions, *Inequalities in Mechanics and Physics*, Springer-Verlag, 1976

[23] K. Washizu, *Variational Methods in Elasticity and Plasticity*, Pergamon Press, 1968.

Saddle-point principles and convex duality are discussed in

[24] I. Ekeland and R. Temam, *Convex Analysis and Variational Problems*, North-Holland, 1976.

The literature on periodic homogenization is also extensive. Comprehensive treatments include

[25] A. Bensoussan, J.-L. Lions, and G. Papanicolaou, *Asymptotic Analysis for Periodic Structures*, North-Holland, 1978.

[26] E. Sanchez-Palencia, *Nonhomogeneous Media and Vibration Theory*, Lecture Notes in Physics **127**, Springer-Verlag, 1980.

The dual variational principles (2.5) and (2.6) for $\langle A_* \xi, \xi \rangle$ and $\langle A_*^{-1} \eta, \eta \rangle$ are discussed in

[27] P. Suquet, Une méthode duale en homogénéisation: application aux milieux élastiques, *J. Mech. Theor. Appl.*, special issue (1982), 79–98.

The more general approach based on G-convergence (also called H-convergence) was developed by DeGiorgi, Spagnolo, Murat, Tartar-Oleinik, and others. The literature is somewhat scattered; a comprehensive treatment is presented in

[28] V. Zhikov, S. Kozlov, O. Oleinik, and Ngoan, Averaging and G-convergence of differential operators, *Russian Math. Surveys* **34**:5 (1979), 69–147.

The assertion that "piecewise periodic composites are dense" will be proved in

[29] R. Kohn and G. DalMaso, The local character of G-closure, in preparation.

While homogenization is relatively new as an area of mathematical activity, the study of composite materials has a long history as an area of mechanics. One review is

[30] R. Christensen, *Mechanics of Composite Materials*, Wiley-Interscience, 1979.

The example of a layered elastic composite has long been understood. The main difficulty lies in finding a useful way to represent A_*. Viewpoints different from ours will be found in

[31] G. Backus, Long-wave elastic anisotropy produced by horizontal layering, *J. Geophys. Res.* **67** (1962), 4427.

[32] W. McConnell, On the approximation of elliptic operators with discontinuous coefficients, *Ann. Scuola Norm. Sup. Pisa Cl. Sci. (4)* **3** (1976), 121–137.

[33] L. Tartar, Remarks on homogenization, in: *Homogenization and Effective Moduli of Materials and Media* (J. L. Ericksen et al., eds.) Springer-Verlag, 1986, 228–246.

Lecture 3

The approach developed here, based on the principle of minimum complementary energy, is due to Kohn and Strang [10]. They also had another way of arriving at the "relaxed" functional F, based on polyconvexification and rank-one convexification. They executed it in [10] only for the special case $\kappa = \mu$ (which corresponds to Poisson's ratio being zero). In retrospect it can actually be used for any κ, μ.

The evaluation of F based on results from homogenization is from

[34] G. Allaire and R. Kohn, Optimal design for minimum weight and compliance in plane stress using extremal microstructures, preprint 1992.

That paper also discusses the numerical minimization of (3.4). A parallel development of this problem can also be found in

[35] A. Cherkaev, *Variational Methods of Optimization of Structures of Inhomogeneous Bodies*, Doctoral Dissertation, Leningrad University, 1989.

The same optimal design problem has been considered in [14–17], using methods that are philosophically similar to ours but very different in detail. The work of Cherkaev and Gibianski [11] on the material distribution problem in plate theory is closely analogous to the plane stress problem considered here.

The lower semicontinuity of F is no accident: our procedure for relaxation, based on the use of extremal composites, is guaranteed to yield a lower semicontinuous result, at least when $A_1 \neq 0$. This is discussed in [10]; a clearer proof (in a slightly different context) is given in

[36] R. Kohn and M. Vogelius, Relaxation of a variational method for impedance computed tomography, *Comm. Pure Appl. Math.* **40** (1987), 745–777.

The fact that F turns out to be *polyconvex* is, as far as we know, a convenient accident.

The literature on Michell trusses is extensive. The most rigorous treatment is

[37] J.-M. Lagache, Treillis de volume minimal dans une région donnée, *J. de Mécanique* **20** (1981), 415–448.

Other treatments include

[38] W. Hemp, *Optimum Structures*, Clarendon Press, 1973.

[39] W. Prager and G. Rozvany, Optimization of structural geometry, in: *Dynamical Systems* (A. Bednarek and L. Cesari, eds.), Academic Press, 1977, 265–294.

Lecture 4

The Hashin-Shtrikman variational principle was first presented for elasticity in

[40] Z. Hashin and S. Shtrikman, A variational approach to the theory of the elastic behavior of multiphase materials, *J. Mech. Phys. Solids* **11** (1963), 127–140.

A derivation closer to the one presented here was given in

[41] R. Hill, New derivations of some elastic extremum principles, in: *Progress in Applied Mechanics – The Prager Anniversary Volume*, Macmillan, 1963, 99–106.

The early applications of this principle focussed on composites with known symmetry, or random composites with known two-point correlation functions. The application to optimal bounds on $\langle A_*\xi, \xi \rangle$ or $\langle A_*^{-1}\xi, \xi \rangle$ was developed by Avellaneda, Kohn, Lipton, and Milton:

[42] M. Avellaneda, Optimal bounds and microgeometries for elastic two-phase composites, *SIAM J. Appl. Math.* **47** (1987), 1216–1228.

[43] R. Kohn and R. Lipton, Optimal bounds for the effective energy of a mixture of isotropic, incompressible, elastic materials, *Arch. Rational Mech. Anal.* **102** (1988), 331–350.

[44] G. Milton and R. Kohn, Variational bounds on the effective moduli of anisotropic composites, *J. Mech. Phys. Solids* **36** (1988), 597–629.

The use of sequentially laminated composites to saturate various bounds has a long history, going back at least to Bruggeman. A general discussion can be found in

[45] G. Milton, Modeling the properties of composites by laminates, in: *Homogenization and Effective Moduli of Materials and Media* (J. Ericksen et al. eds.), Springer-Verlag, 1986, 150–174.

The iterative procedure developed here was first presented for conductivity by Tartar:

[46] L. Tartar, Estimations fines des coefficients homogénéisés, in: *Ennio de Giorgi's Colloquium* (P. Kree, ed.), Pitman, 1985, 168–187.

A presentation for elasticity will be found in

[47] G. Francfort and F. Murat, Homogenization and optimal bounds in linear elasticity, *Arch. Rational Mech. Anal.* **94** (1986), 307–334.

A systematic treatment of the optimal energy bounds using the viewpoint of this lecture is given in

[48] G. Allaire and R. Kohn, Optimal bounds on the effective behavior of a mixture of two well-ordered elastic materials, *Quart. Appl. Math.* (1992), in press.

Lecture 5

The material in this lecture is drawn from [34]. For the derivation of (5.8), see

[49] R. Kohn, Relaxation of a double-well energy, *Contin. Mech. Thermodyn.* **3** (1991), 193–236.

A similar calculation, based on polyconvexification rather the Hashin-Shtrikman variational principle, is given in

[50] L. Gibiansky and A. Cherkaev, Microstructures of composites of extremal rigidity and exact estimates of the associated energy density, Ioffe Physicotechnical Institute Preprint **1145**, 1987.

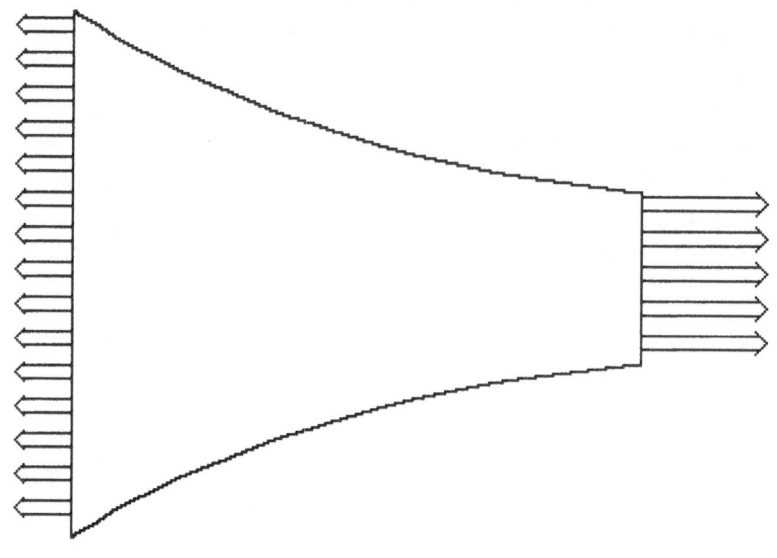

Figure 1.1: A planar fillet

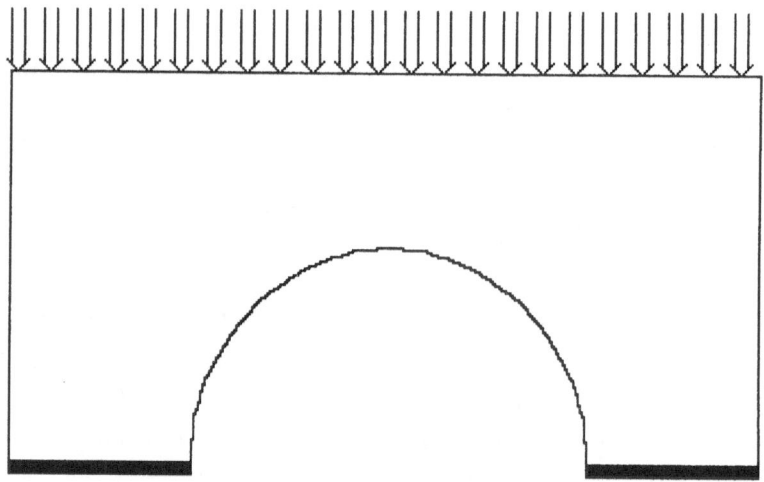

Figure 1.2: A planar bridge

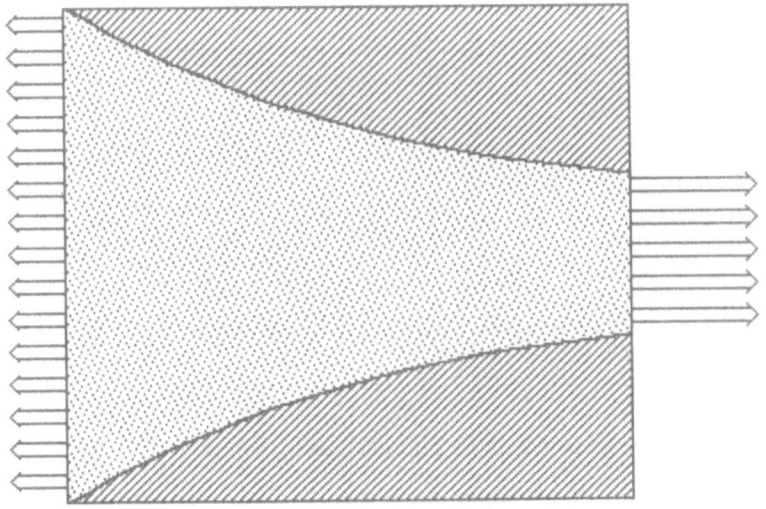

Figure 1.3: The planar fillet as a material distribution problem

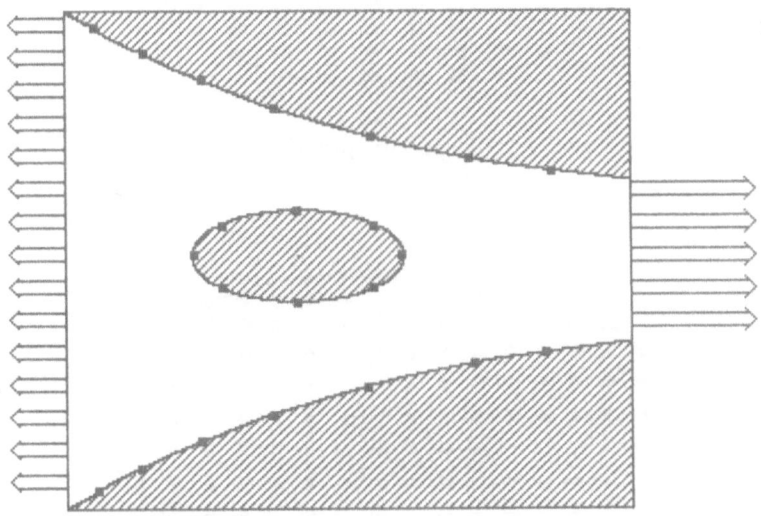

Figure 1.4: The standard approach

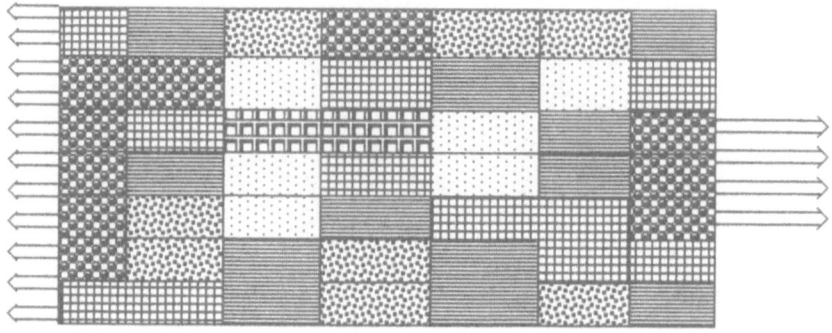

Figure 1.5: New approach using composites

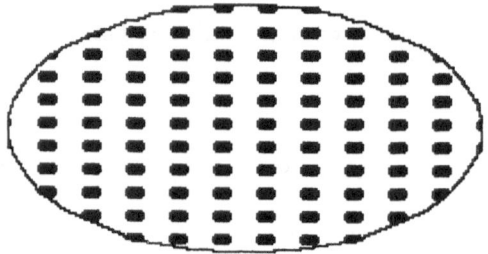

Figure 2.1: A spatially periodic composite

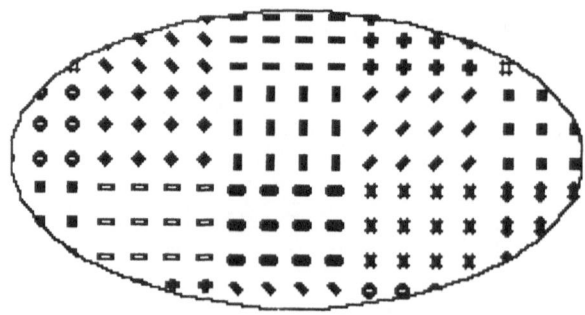

Figure 2.2: A composite with piecewise periodic structure

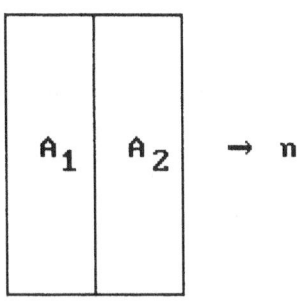

Figure 2.3: A unit cell of a layered composite

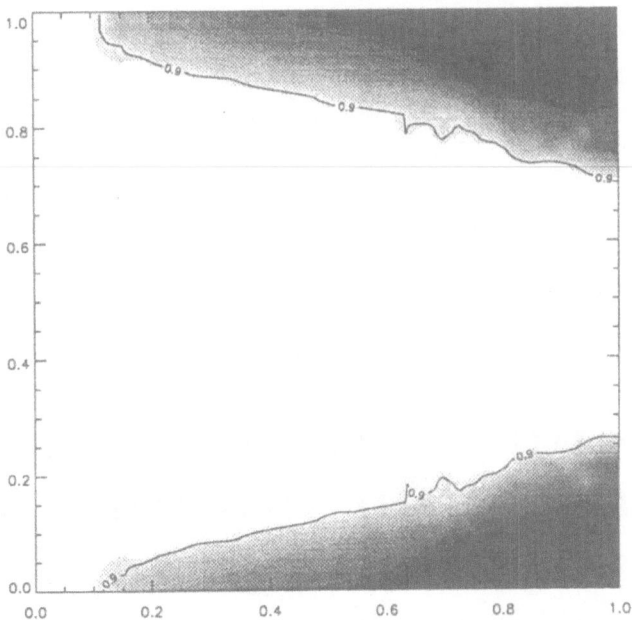

Figure 3.1a:

Optimal design of a planar fillet. Material is being removed from a rectangle, loaded as shown in Figure 1.3. The greyscale indicates the density of holes in the optimal design (white = no holes, black = no material). Here 17% of the material has been removed. The resulting design uses essentially no composites.

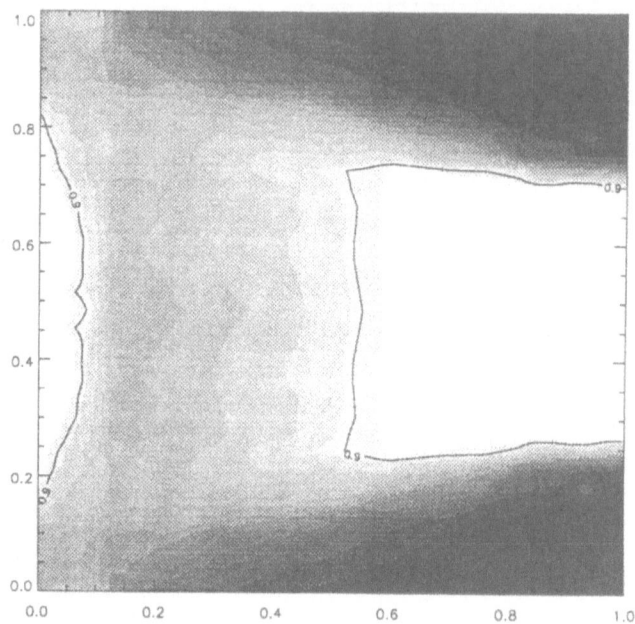

Figure 3.1b:

Same as 3.1a, but now 37% of the material has been removed.
This design makes extensive use of composite material.

Figure 4.1:

A sequentially laminated composite of rank two, with the
horizontal and vertical axes as layering directions.

Optimal Shape Design with Applications to Aerodynamics

Olivier PIRONNEAU

Université Pierre et Marie Curie (Paris VI)
4, pl. Jussieu
F-75252 Paris Cédex 05
et
I.N.R.I.A.
Domaine de Voluceau-Rocquencourt
B.P. 105
F-78153 Le Chesnais Cédex
France

Notes by A. VOSSINIS

Abstract

The purpose of these lectures is to show the use of numerical methods in shape optimization. We will also need to review some optimization methods and the finite element method (FEM). Specifically, we will consider: minimum drag problems for laminar flows; potential flows without lift; potential flows and Euler flows, riblets.

1 Laminar flows

1.1 Preliminaries

We consider a viscous, incompressible flow, which is governed by the Navier-Stokes equations:

$$\vec{u}_{,t} + \vec{u}\,\nabla\vec{u} + \nabla p - \nu\Delta\vec{u} = 0, \quad \forall x \in \Omega \tag{1}$$

$$\nabla \cdot \vec{u} = 0 \quad \forall t \in \Omega, \forall x \in \,]0, T[\tag{2}$$

where : Ω is an open bounded set, with smooth boundary $\Gamma = \partial\Omega, p(x,t)$ is the pressure in the flow, $\vec{u}(x,t)$ is the flow velocity. Equation (1) expresses the momentum balance in Ω and equation (2) expresses the mass conservation in Ω for an incompressible fluid. In order to have a well posed problem, we need initial and boundary conditions: Initial condition:

$$\vec{u}(x,0) = \vec{u}_o(x), \, \forall x \text{ in } \Omega \tag{3}$$

211

M. C. Delfour and G. Sabidussi (eds.), Shape Optimization and Free Boundaries, 211–251.
© 1992 *Kluwer Academic Publishers.*

Boundary condition:

$$\vec{u}(x,t) = \vec{u}_\Gamma(x,t), \quad \forall x,t \text{ in } \Gamma \times]0,T[\quad \text{(Dirichlet boundary condition)} \qquad (4)$$

We give two problems of optimal design, where we can use equations (1)–(4) as state equations.

a) *Wing optimization:* Define $\mathcal{O}$ a class of admissible wings. Find a wing profile $s \in \mathcal{O}$ such that the drag is minimal (see Figure 1).

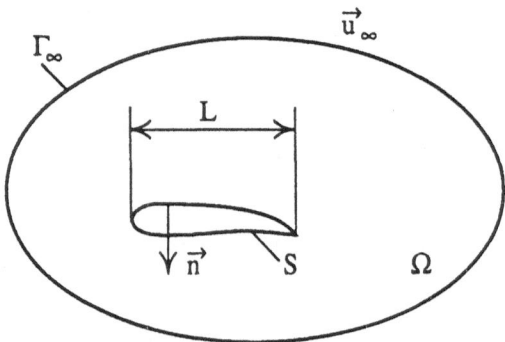

Figure 1

The wing moves with a certain velocity $\vec{u}_\infty$ in an unbounded domain, approximated by a bounded domain Ω with boundary Γ_∞. The velocity at Γ_∞ is $\vec{u} \simeq 0$.

The drag is $\vec{D} = \int_S \underline{\sigma} \cdot \vec{n} d\gamma$, where

$$
\begin{aligned}
\underline{\sigma} &= \underline{\sigma}(x,t) = \frac{\nu}{2}(\nabla \vec{u} + \nabla \vec{u}^T) - pI \text{ (stress tensor)} \\
\vec{n} &= \text{normal external to } \Omega \\
\gamma &= \text{arc length.}
\end{aligned}
$$

We wish to find S, such that the work due to $\vec{D}$, $\vec{D} \cdot \vec{u}_\infty$, is minimal.

b) *Problem for a wind tunnel:* We want to solve the same problem as that of a wing, but now Ω is the domain defined by the wind tunnel.

Figure 2 describes the geometry and the boundary conditions:

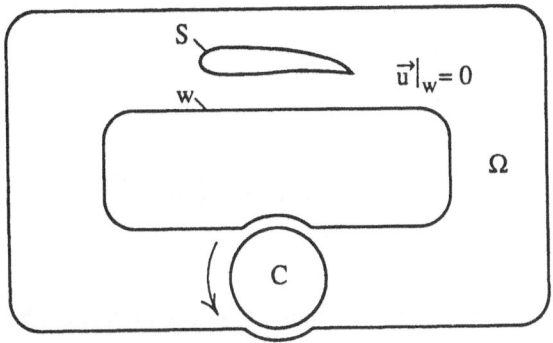

$$\text{pump: } \vec{u}|_C = \vec{\omega} \times \vec{r}$$

Figure 2

We want to minimize the work of the drag in both cases. More precisely, we shall minimize $\vec{D} \cdot \vec{u}_\infty$, as the following proposition gives a formula for it.

Proposition 1 *For stationary flows in the geometry of Figure 1 governed by equations (1)–(4) we have the formula:*

$$\vec{D} \cdot \vec{u}_\infty = \nu \int_\Omega |\nabla \vec{u}|^2 dx.$$

Proof We multiply equation (1) by $\vec{u}$ and integrate on Ω:

$$\int_\Omega \vec{u}_{,t} \vec{u} dx + \int_\Omega \vec{u} \nabla \vec{u} \cdot \vec{u} dx + \int_\Omega \vec{u} \cdot \nabla p dx - \nu \int_\Omega \nabla \vec{u} \cdot \vec{u} dx = 0 \Rightarrow$$

$$\frac{1}{2} \int_\Omega (\vec{u}^2)_{,t} dx + \frac{1}{2} \int_\Omega \vec{u} \cdot \nabla (\vec{u}^2) dx + \int_\Omega \vec{u} \cdot \nabla p dx + \nu \int_\Omega |\nabla \vec{u}|^2 dx - \nu \int_\Gamma \frac{\partial \vec{u}}{\partial n} \cdot \vec{u} d\gamma = 0,$$

where we used Green's formula on the last integral.

Since the flow is stationary, $\frac{1}{2} \int_\Omega (\vec{u}^2)_{,t} dx = 0$.

By Green's formula and (2), we have:

$$\frac{1}{2} \int_\Omega \vec{u} \cdot \nabla (\vec{u}^2) dx = \int_\Gamma \vec{u} \cdot \vec{n} \frac{\vec{u}^2}{2} d\gamma \quad \text{and} \quad \int_\Omega \vec{u} \cdot \nabla p dx = \int_\Gamma \vec{u} \cdot \vec{n} p d\gamma \quad \text{with} \quad \Gamma = \Gamma_\infty \cup S.$$

Using the boundary conditions and some calculations, we find that:

$$\int_S \vec{u}_\infty \cdot \vec{n} \frac{\vec{u}_\infty^2}{2} dx + \int_S \vec{u}_\infty \cdot \vec{n} p d\gamma + \nu \int_\Omega |\nabla \vec{u}|^2 dx = \nu \int_\Gamma \frac{\partial \vec{u}}{\partial \vec{n}} \cdot \vec{u} d\gamma.$$

Since $\vec{u}_\infty = constant$ and $\int_S \vec{n} d\gamma = 0$, we find $\int_S \vec{u}_\infty \cdot \vec{n} \frac{u_\infty^2}{2} d\gamma = 0$ and therefore after a few manipulations, we see that

$$2 \int_\Gamma \frac{\partial \vec{u}}{\partial \vec{n}} d\gamma = \int_\Gamma (\nabla \vec{u} + \nabla \vec{u}^T) \cdot \vec{n} d\gamma$$

and we find

$$\vec{D} \cdot \vec{u}_\infty = - \int_S \vec{u}_\infty \cdot \vec{n} p d\gamma + \nu \int_S \vec{u}_\infty \cdot \frac{\partial \vec{u}_\infty}{\partial \vec{n}} d\gamma.$$

$\square$

So according to this proposition, the minimum drag problem is also:

Find $S \in \mathcal{O}$ which minimizes

$$E(\Omega) = \nu \int_\Omega |\nabla \vec{u}|^2 dx$$

subject to: $\vec{u}$ solution of

$$\begin{cases} \vec{u} \nabla \vec{u} + \nabla p - \nu \Delta \vec{u} = 0 \ \text{ in } \ \Omega \\ \nabla \cdot \vec{u} = 0 \ \text{ in } \ \Omega \\ \vec{u}|_\Gamma = \vec{u}_\Gamma \end{cases} \tag{5}$$

The above problem is an optimal shape problem for laminar flow with minimum drag or energy.

When the Reynolds number $|\vec{u}_\infty| \cdot L/\nu \ll 1$, we can simplify equations (5) and take the optimal shape design problem in Stokes flow, using the following

Proposition 2 *If $|\vec{u}_\infty| \cdot L/\nu \ll 1$ where $L = $ diameter of S (see Figure 1) then (5) is well approximated by the Stokes problem:*

$$\begin{cases} -\nu \Delta \vec{u} + \nabla p = 0 \ \text{ in } \ \Omega \\ \nabla \cdot \vec{u} = 0 \ \text{ in } \ \Omega \\ \vec{u}|_\Gamma = \vec{u}|_\Gamma \end{cases} \tag{6}$$

$\square$

For a proof of this proposition see Landau-Lifshitz [11] or Bachelor [12].

So, let us consider the following optimal shape design problem:

Find $\min_{S \in \mathcal{O}} \nu \int_\Omega |\nabla \vec{u}|^2 dx$, subject to: $\vec{u}$ solution of (6).

Applications of this formulation are found merely in biology, where the corresponding problem is the optimal swimming of microorganisms for which $L \ll 1$.

1.2 Optimal shape design problem in Stokes flow

Find

$$\min_{\Omega \in \mathcal{O}} E(\Omega) = \min_{\Omega \in \mathcal{O}} \int_{\Omega} |\nabla \vec{u}|^2 dx$$

subject to:

$$\begin{cases} -\nu \Delta \vec{u} + \nabla p = 0 & \text{in } \Omega \\ \nabla \cdot \vec{u} = 0 & \text{in } \Omega \\ \vec{u}|_S = 0 \\ \vec{u}|_{\Gamma_\infty} = \vec{u}_\infty \end{cases} \tag{7}$$

An example of $\mathcal{O}$ is:

$$\mathcal{O} = \{\Omega : \text{ volume } \tilde{S} = 1, \partial\Omega = S \cup \Gamma_\infty\},$$

where $\tilde{S} = $ the interior defined by the closed boundary S.

On the existence of solutions of this problem, see [7], [8]. We remark that the uniqueness of the solution is an open problem.

For deriving the optimality conditions, we need to do some calculus of variations on $E(\Omega)$. So, we consider that Ω is the optimum solution and $\Omega' \in \mathcal{O}$ is a domain "near" Ω (see Figure 3) defined by its boundary $\Gamma' = \partial\Omega'$ with

$$\Gamma' = \{x + \alpha(x)\vec{n}(x) : \alpha = \text{ regular, small, } \forall x \in \Gamma\}$$

$$\Gamma = \partial\Omega.$$

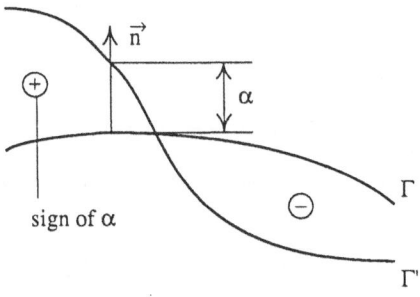

Figure 3

For every admissible α, we have:

$$E(\Omega') \geq E(\Omega). \tag{8}$$

Define

$$\delta\Omega = \Omega \cup \Omega' - \Omega \cap \Omega',$$

and associate a convention to it, that

$$\int_{\delta\Omega} f dx \equiv \int_{\Omega'-\Omega\cap\Omega'} f dx - \int_{\Omega-\Omega\cap\Omega'} f dx. \tag{9}$$

Define also

$$\delta\vec{u} = \vec{u}(\Omega') - \vec{u}(\Omega) \tag{10}$$

extending $\vec{u}$ smoothly in $\tilde{S}$. Then

$$\delta E = E(\Omega') - E(\Omega) = \nu\delta\left(\int_{\Omega} |\nabla\vec{u}|^2 dx\right) =$$
$$\nu\int_{\delta\Omega} |\nabla\vec{u}|^2 dx + 2\nu\int_{\Omega} \nabla\delta\vec{u} \cdot \nabla\vec{u} dx + o(\delta\Omega, \delta\vec{u}). \tag{11}$$

When $\delta\Omega$ is smooth and $\nabla\vec{u}$ is continuous, then

$$\nu\int_{\delta\Omega} |\nabla\vec{u}|^2 dx = \nu\int_{\Gamma} \alpha|\nabla\vec{u}|^2 d\gamma + o(||\alpha||_{c^2}) = \nu\int_{\Gamma} \alpha\left|\frac{\partial\vec{u}}{\partial\vec{n}}\right|^2 d\gamma + o((||\alpha||_{c^2}) \tag{12}$$

(cf. [9]).

In order to complete the computation of δE, we need the following lemma:

Lemma 1 *Equations* (7) *imply*

$$-\nu\Delta\delta\vec{u} + \nabla\delta p = 0 \quad in \quad \Omega \tag{13}$$

$$\nabla \cdot \delta\vec{u} = 0 \quad in \quad \Omega \tag{14}$$

$$\delta\vec{u}|_{\Gamma_\infty} = 0 \tag{15}$$

$$\delta\vec{u}|_S = -\alpha\frac{\partial\vec{u}}{\partial\vec{n}} \tag{16}$$

Proof We have the relations:

$$\delta(-\nu\Delta\vec{u} + \nabla p) = 0 \Rightarrow -\nu\Delta\delta\vec{u} + \nabla\delta p = 0 \tag{17}$$

$$\delta(\nabla \cdot \vec{u}) = 0 \Rightarrow \nabla \cdot \delta\vec{u} = 0 \quad in \quad \Omega\cap\Omega' \tag{18}$$

$$\delta\vec{u}|_{\Gamma_\infty} = 0. \tag{19}$$

if $\vec{u}_\infty$ is constant and Γ_∞ the same in Ω and Ω'. Finally

$$\delta\vec{u}|_S \simeq \delta\vec{u}|_{\partial(\Omega\cap\Omega')} = \begin{cases} +\vec{u}(\Omega')|_S, & \text{if } \alpha > 0 \\ -\vec{u}(\Omega)|_{S'}, & \text{if } \alpha < 0. \end{cases} \tag{20}$$

Using Taylor's expansion, we obtain:

$$\vec{u}(\Omega')|_S = \vec{u}(\Omega')(x' - \alpha\vec{n}) = \vec{u}(\Omega')(x') - \alpha\frac{\partial\vec{u}}{\partial\vec{n}}(\Omega')|_{S'} + o(|\alpha|)$$
$$= -\alpha\frac{\partial\vec{u}}{\partial\vec{n}}(\Omega')|_{S'} + o(|\alpha|), \text{ since } \vec{u}(\Omega')(x') = 0 \tag{21}$$

and

$$\vec{u}(\Omega)|_S = 0. \tag{22}$$

So,

$$\delta\vec{u}|_S = -\alpha \left.\frac{\partial\vec{u}}{\partial\vec{n}}\right|_S. \tag{23}$$

□

Finally, if all equalities are up to higher order terms, we have:

$$\nu \int_{\delta\Omega} |\nabla\vec{u}|^2 dx = \nu \int_S \alpha \left|\frac{\partial\vec{u}}{\partial\vec{n}}\right|^2 d\gamma. \tag{24}$$

Remark A lot of regularity is needed to perform the previous calculation, but it can be justified by other methods also, such as in Simon [14], Murat-Simon [15], and

$$\nu \int_\Omega \nabla\delta\vec{u}\,\nabla\vec{u}dx = \nu \int_\Omega (-\Delta\vec{u})\delta\vec{u}dx + \nu \int_\Gamma \frac{\partial\vec{u}}{\partial\vec{n}}\delta\vec{u}d\gamma =$$

$$\int_\Omega p\nabla\cdot\delta\vec{u}dx - \int_\Gamma p\delta\vec{u}\cdot\vec{n}d\gamma + \nu\int_\Gamma\frac{\partial\vec{u}}{\partial\vec{n}}\delta\vec{u}d\gamma = \tag{25}$$

$$\int_\Gamma \left(\nu\frac{\partial\vec{u}}{\partial\vec{n}} - p\vec{n}\right)\cdot\delta\vec{u}d\gamma = -\int_S \nu\alpha\left|\frac{\partial\vec{u}}{\partial\vec{n}}\right|^2 d\gamma,$$

because, if $\vec{s}$ denotes the tangent component,

$$\delta\vec{u}|_{\Gamma_\infty} = 0, \quad \delta\vec{u}|_S = -\alpha\frac{\partial\vec{u}}{\partial\vec{n}} \quad \text{and} \quad \vec{n}\cdot\frac{\partial\vec{u}}{\partial\vec{n}} = -\vec{s}\cdot\frac{\partial\vec{u}}{\partial\vec{s}} = 0 \quad \text{on} \quad \Gamma. \tag{26}$$

Hence,

$$\nu \int_\Omega \nabla\delta\vec{u}\cdot\nabla\vec{u}dx = -\nu\int_S \alpha\left|\frac{\partial\vec{u}}{\partial\vec{n}}\right|^2 d\gamma$$

and

$$\delta E \simeq -\nu\int_S \alpha\left|\frac{\partial\vec{u}}{\partial\vec{n}}\right|^2 d\gamma. \tag{27}$$

We have proved the

Proposition 3 *The variation of E with respect to* Ω *is:*

$$\delta E = -\nu\int_S \alpha\left|\frac{\partial\vec{u}}{\partial\vec{n}}\right|^2 d\gamma + o(\alpha). \tag{28}$$

□

Consequences

1) Supposing that $\mathcal{O} = \{S : S \supset C\}$, since $\left|\frac{\partial\vec{u}}{\partial\vec{n}}\right|^2 > 0, C$ is the solution. Indeed any $\alpha < 0$ will give $\delta E > 0$. In other words, any fairing around C will increase the drag in Stokes flow.

2) If $O = \{S : Vol\ \tilde{S} = 1\}$, then $\delta E \geq 0$ for every α with

$$\int_\Gamma \alpha d\gamma = o(|\alpha|). \tag{29}$$

Hence, if $\vec{u}$ is smooth, (27) and (29) imply

$$\left|\frac{\partial \vec{u}}{\partial \vec{n}}\right|_{|_\Gamma} = \text{constant}. \tag{30}$$

Lighthill (cf. [10]) showed that near the leading and the trailing edge the only possible axisymmetric flow which can give $\left|\dfrac{\partial \vec{u}}{\partial \vec{n}}\right| = \text{constant on } S$, is an S conical of half angle equal to 60°.

To compute an axisymmetric surface S which satisfies (7) and (30) we could try one iteration of a gradient method starting from the ellipsoid with minimum drag to which is added a conical front and rear and near the leading and trailing stagnation point.

The result is shown in Figure 4. A decrease of drag of 5% was found with respect to the optimum ellipsoid.

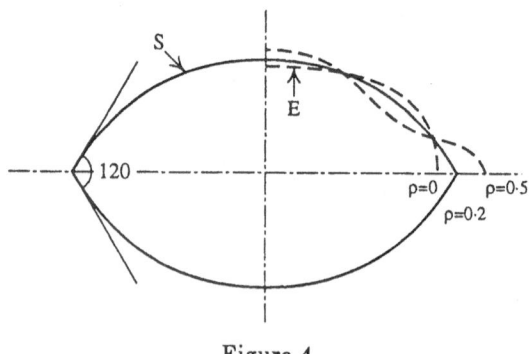

Figure 4

1.3 Optimal shape design in laminar flow

We consider the minimum drag energy problem where the state equation is the Navier-Stokes equations.

The mathematical formulation of such a problem is:

Find

$$\min_{S \in O} E(\Omega) = \nu \int_\Omega |\nabla \vec{u}|^2 dx$$

with $\vec{u}$ subject to

$$- \nu \Delta \vec{u} + \nabla p + \vec{u} \nabla \vec{u} = 0, \quad \text{in} \quad \Omega \tag{31}$$

$$\nabla \cdot \vec{u} = 0, \quad \text{in} \quad \Omega \tag{32}$$

$$\vec{u}|_S = 0 \tag{33}$$

$$\vec{u}|_{\Gamma_\infty} = \vec{u}_\infty = \text{constant} \tag{34}$$

and $\mathcal{O} = \{S : vol\ \tilde{S} = 1\}, \Gamma = \partial \Omega = S_\infty \cup \Gamma$ smooth.

We shall derive the optimality condition for this problem. In order to do this we must express the variation of $E(\Omega)$ in terms of the variation of Ω.

We consider that Ω is the optimal solution and that Ω' is a domain obtained by a small perturbation of Ω defined as before: Ω' has the boundary

$$\partial \Omega' = S' \cup \Gamma_\infty \quad \text{where} \quad S' = \{x + \alpha \vec{n} : x \in S\}.$$

We call $\vec{u}'$ the solution of (31)–(34) on Ω' and we define:

$$\delta E = E(\Omega') - E(\Omega) = \nu \int_{\delta \Omega} |\nabla \vec{u}|^2 dx + 2\nu \int_\Omega \nabla \vec{u} \nabla \delta \vec{u} dx + o(\delta \vec{u}, \alpha). \tag{35}$$

We can prove the following in the same way we proved Lemma 1:

Lemma 2 *Equations* (31)–(34) *imply that*

$$- \nu \Delta \delta \vec{u} + \nabla \delta p + \vec{u} \nabla \delta \vec{u} + \delta \vec{u} \nabla \vec{u} = 0 \tag{36}$$

$$\nabla \cdot \delta \vec{u} = 0 \tag{37}$$

$$\delta \vec{u}|_{\Gamma_\infty} = 0 \tag{38}$$

$$\delta \vec{u}|_S = -\alpha \frac{\partial \vec{u}}{\partial \vec{n}} \tag{39}$$

$\square$

Lemma 2 is not sufficient to get rid of the second term in the right side of (35). So, we introduce the adjoint equation:

Let $(\vec{P}, q)$ be the solution of:

$$- \nu \Delta \vec{P} + \nabla q - \vec{u} \nabla \vec{P} - (\nabla \vec{P}) \vec{u} = -2\nu \Delta \vec{u} \quad \text{in} \quad \Omega \tag{40}$$

$$\nabla \cdot \vec{P} = 0 \quad \text{in} \quad \Omega \tag{41}$$

$$\vec{P}|_\Gamma = 0 \tag{42}$$

In order to compute δE we use Lemma 2 and equations (40)–(42).

Multiplying (40) by $\delta\vec{u}$ and integrating on Ω, we obtain:

$$
\begin{aligned}
-2\nu \int_\Omega \Delta\vec{u}\cdot\delta\vec{u}dx = \; & -\nu\int_\Omega \Delta\vec{P}\cdot\delta\vec{u}dx - \int_\Omega \nabla\vec{P}\vec{u}\cdot\delta\vec{u}dx \\
& -\int_\Omega \vec{u}\nabla\vec{P}\cdot\delta\vec{u}dx + \int_\Omega \nabla q\cdot\delta\vec{u}dx.
\end{aligned}
\tag{43}
$$

We use Green's formula and integrations by parts in (43):

$$
-\int_\Omega \nabla\vec{P}\vec{u}\cdot\delta\vec{u}dx = \int_\Omega \vec{u}\nabla\delta\vec{u}\cdot\vec{P}dx - \int_\Gamma (\vec{P}\cdot\delta\vec{u})(\vec{u}\cdot\vec{n})d\gamma = \int_\Omega \vec{u}\nabla\delta\vec{u}\cdot\vec{P}dx \tag{44}
$$

because $\nabla\cdot\vec{u} = 0$ in Ω and $\vec{P}|_\Gamma = 0$. Now

$$
\begin{aligned}
-\int_\Omega \vec{u}\nabla\vec{P}\cdot\delta\vec{u}dx &= \int_\Omega \delta\vec{u}\nabla\vec{u}\cdot\vec{P}dx + \int_\Omega \vec{P}\cdot\vec{u}\nabla\cdot\delta\vec{u}dx - \int_\Gamma (\vec{P}\cdot\vec{u})(\delta\vec{u}\cdot\vec{n})d\gamma \\
&= \int_\Omega \delta\vec{u}\nabla\vec{u}\cdot\vec{P}dx.
\end{aligned}
\tag{45}
$$

as $\nabla\cdot\delta\vec{u} = 0$ in Ω and $\vec{P}|_\Gamma = 0$.

The last integral in (43) is zero, because:

$$
\int_\Omega \nabla q\cdot\delta\vec{u}dx = -\int_\Omega q\nabla\cdot\delta\vec{u}dx + \int_\Gamma q\delta\vec{u}\cdot\vec{n}d\gamma = 0
$$

as $\nabla\cdot\delta\vec{u} = 0$ in Ω and

$$
\delta\vec{u}\cdot\vec{n} = \begin{cases} 0 & \text{on } \Gamma_\infty \\ -\alpha\dfrac{\partial\vec{u}}{\partial n}\cdot\vec{n} = 0 & \text{on } S. \end{cases}
$$

Using these results in (43), we obtain:

$$
\begin{aligned}
-2\nu\int_\Omega \Delta\vec{u}\cdot\delta\vec{u}dx = & \\
= \nu\int_\Omega \nabla\vec{P}\cdot\nabla\delta\vec{u}dx &- \nu\int_\Gamma \frac{\partial\vec{P}}{\partial n}\cdot\delta\vec{u}d\gamma + \int_\Omega \vec{u}\nabla\delta\vec{u}\cdot\vec{P}dx + \int_\Omega \delta\vec{u}\nabla\vec{u}\cdot\vec{P}dx \\
= -\nu\int_\Omega \Delta\delta\vec{u}\cdot\vec{P}dx &+ \nu\int_\Gamma \vec{P}\cdot\frac{\partial(\delta\vec{u})}{\partial n}d\gamma - \nu\int_\Gamma \frac{\partial\vec{P}}{\partial n}\cdot\delta\vec{u}d\gamma \\
+ \int_\Omega \vec{u}\nabla\delta\vec{u}\cdot\vec{P}dx &+ \int_\Omega \delta\vec{u}\cdot\nabla\vec{u}\cdot\vec{P}dx.
\end{aligned}
\tag{46}
$$

If we multiply (36) by $\vec{P}$, integrate on Ω and use Green's formula, we obtain:

$$
\begin{aligned}
-\nu\int_\Omega \Delta\delta\vec{u}\cdot\vec{P}dx &+ \int_\Omega \vec{u}\nabla\delta\vec{u}\cdot\vec{P}dx + \int_\Omega \delta\vec{u}\nabla\vec{u}\cdot\vec{P}dx \\
&+ \int_\Omega \nabla\delta p\cdot\vec{P}dx = -\int_\Gamma \delta p\vec{P}\cdot\vec{n}d\gamma = 0,
\end{aligned}
\tag{47}
$$

since $\nabla \cdot \vec{P} = 0$ in Ω and $\vec{P}|_\Gamma = 0$. In addition, $\nu \int_\Gamma \vec{P} \cdot \dfrac{\partial(\delta \vec{u})}{\partial n} d\gamma = 0$, so (46) gives

$$- 2\nu \int_\Omega \Delta \vec{u} \cdot \delta \vec{u} dx = -\nu \int_\Gamma \frac{\partial \vec{P}}{\partial n} \cdot \delta \vec{u} d\gamma = \nu \int_S \alpha \frac{\partial \vec{P}}{\partial \vec{n}} \cdot \frac{\partial \vec{u}}{\partial n} d\gamma \qquad (48)$$

because of (38) and (39).

Using Green's formula in the left side of (48), we find:

$$- 2\nu \int_\Omega \Delta \vec{u} \cdot \delta \vec{u} dx = 2\nu \int_\Omega \nabla \vec{u} \cdot \nabla \delta \vec{u} dx - 2\nu \int_\Gamma \frac{\partial \vec{u}}{\partial n} \delta \vec{u} d\gamma. \qquad (49)$$

The equality of (48) and (49) gives:

$$2\nu \int_\Omega \nabla \vec{u} \cdot \nabla \delta \vec{u} dx = \nu \int_S \alpha \left(\frac{\partial \vec{P}}{\partial n} \cdot \frac{\partial \vec{u}}{\partial n} - 2 \left| \frac{\partial \vec{u}}{\partial n} \right|^2 \right) d\gamma. \qquad (50)$$

From equation (24) we know that:

$$\nu \int_{\delta\Omega} |\nabla \vec{u}|^2 dx = \nu \int_S \alpha \left| \frac{\partial \vec{u}}{\partial n} \right|^2 d\gamma.$$

Using (24) and (50) in (35), we find that:

$$\delta E = \nu \int_S \alpha \left(\frac{\partial \vec{P}}{\partial n} - \frac{\partial \vec{u}}{\partial n} \right) \cdot \frac{\partial \vec{u}}{\partial n} d\gamma. \qquad (51)$$

We have proved the

Proposition 4 *The variation of E with respect to Ω is:*

$$\delta E = \nu \int_S \alpha \left(\frac{\partial \vec{P}}{\partial n} - \frac{\partial \vec{u}}{\partial n} \right) \cdot \frac{\partial \vec{u}}{\partial n} d\gamma + o(\alpha).$$

$\square$

For the chosen admissible set $\mathcal{O}$ we have that $\delta E \geq 0$ for every α with $\int_\Gamma \alpha d\gamma = 0$. So, the optimality condition for this problem is: $\dfrac{\partial \vec{u}}{\partial n} \cdot \left(\dfrac{\partial \vec{P}}{\partial n} - \dfrac{\partial \vec{u}}{\partial n} \right) = \text{constant on } S$.

We remark about the equations (40)–(42) and (31)–(34) that when $\nu \to \infty$, the solution $\vec{P} \to 2u$, so the previous result on the optimality condition for the Stokes problem is recovered. This is how far we can go without using a computer.

If we wish to use a computer in order to solve this problem, we must use a *gradient method*.

A gradient method is based on the following observation:

Suppose we have to find $\min_{x \in \mathbb{R}^N} J(x)$. Taylor's expansion of this function gives:

$$J(z + \delta z) = J(z) + grad_z J \cdot \delta z + o(|\delta z|),\tag{52}$$

so taking $\delta z = -\lambda\, gradJ(z), \lambda > 0$, we find:

$$J(z + \delta z) - J(z) = -\lambda|gradJ(z)|^2 + o(|\delta z|),\tag{53}$$

that is, if $|gradJ(z)| \neq 0$ and $\lambda \ll 1$, we have

$$\lambda|gradJ(z)|^2 \gg o(\lambda|gradJ(z)|)$$

and we obtain:

$$J(z - \lambda gradJ(z)) < J(z).$$

In other words, the sequence defined by:

$$z^o \text{ given}$$

$$z^{n+1} = z^n - \lambda gradJ(z^n), n = 0, 1, 2, \ldots\tag{54}$$

in such that $J(z^n)$ converges to a local minimum of $J(z)$.

An inprovement of the method is, with every computation of z^n, to compute

$$\lambda^n = \text{ solution of } \min_{\lambda \in \mathbb{R}} J(z^n - \lambda gradJ(z^n))\tag{55}$$

and to use λ^n instead of λ to compute z^{n+1} according to (54).

We have to remark, however, that minimizing a one parameter function is not simple and is usually done by trial-and-error methods.

An *application* of a gradient method to the optimum design problem in laminar flow would be:

Start with $S^o, n = 0$
loop

1. Solve (31)–(34) for $(\vec{u}, p)$ around S^n
2. Solve (40)–(42) for $(\vec{P}, q)$ around S^n

3. Choose $\alpha = -\dfrac{\partial \vec{u}}{\partial n} \cdot \left. \left(\dfrac{\partial \vec{P}}{\partial n} - \dfrac{\partial \vec{u}}{\partial n} \right) \right|_{S^n}$

and take

$\alpha' = \alpha - \dfrac{1}{|S^n|} \displaystyle\int_{S^n} \alpha\, d\gamma$

4. Set $S^{n+1} = S^n + \alpha' \vec{n}$ except if α is too small

It is left as an exercise to show that

$$\int_{S^n} \alpha' \frac{\partial \vec{u}}{\partial n} \cdot \left(\frac{\partial \vec{P}}{\partial n} - \frac{\partial \vec{u}}{\partial n} \right) d\gamma < 0.\tag{56}$$

This method also has certain difficulties.

First of all, we have to solve the Navier-Stokes equations and this is not simple because of the presence of the boundary layer where $|\vec{u}|$ has a large gradient (see Figure 5):

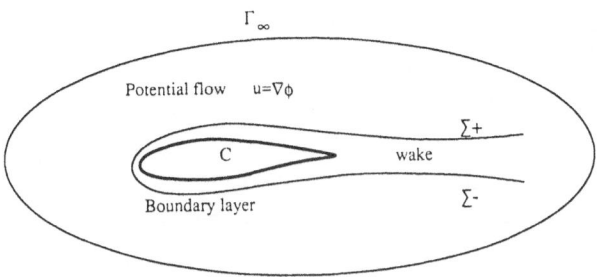

Figure 5

A boundary layer Σ develops around C when the Reynolds number is large. In order to solve (31)–(34) we observe that:

$$-\Delta \vec{u} \;=\; \nabla \times \nabla \times \vec{u} - \nabla\left(\nabla \cdot \vec{u}\right) \tag{57}$$

$$\vec{u}\nabla\vec{u} \;=\; -\vec{u} \times \nabla \times \vec{u} + \nabla\left(\frac{\vec{u}^2}{2}\right) \tag{58}$$

so taking:

$$p = p_\infty - \frac{|\vec{u}|^2}{2}$$
$$\nabla \times \vec{u} = 0$$
$$\nabla \cdot \vec{u} = 0$$

will satisfy (31)–(32) but not the boundary conditions (33)–(34).

We can use however this observation to derive the following *numerical method*:

a) In $C = \Omega - \Sigma$ find ϕ such that:
$$\Delta\phi = 0 \tag{59}$$

$$\left.\frac{\partial\phi}{\partial n}\right|_{\Gamma_\infty} = \vec{u}_\infty \cdot \vec{n} \tag{60}$$

$$\left.\frac{\partial\phi}{\partial n}\right|_{\partial\Sigma} = 0 \tag{61}$$

b) Solve Navier-Stokes equations in Σ only with Γ_∞ replaced by $\partial\Sigma$ and $\vec{u}|_{\partial\Sigma} = \nabla\phi|_{\partial\Sigma}$. It can be shown that this step can be carried out by one relaxation sweep in the direction of the flow only.

Another disadvantage is that the convergence of the method has the form of Figure 6:

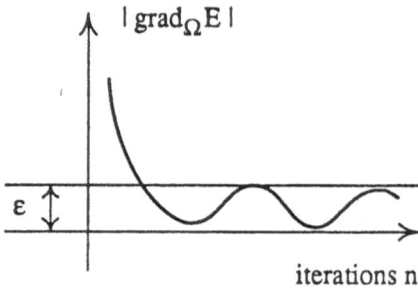

Figure 6

The functional $E(\Omega^n)$ oscillates near the minimum. This is due to numerical noise, ε, that is the error on the computation of $\left.\frac{\partial \vec{u}}{\partial \vec{n}}\right|_S$ and $\left.\frac{\partial \vec{P}}{\partial \vec{n}}\right|_S$. These errors depend merely on the discretization errors of the numerical methods used to solve Navier-Stokes and the adjoint state equations.

These errors can prevent the computation of the optimal solution below a certain precision, sometimes not so low (Figure 6).

Since we use a discretization to solve equations (31)–(34) and (40)–(42), we should change the problem and try to compute the optimality condition for the discrete problem.

More precisely, if we discretize Ω by the points $\{q^i\}, i = 1, \ldots, N$, we have that $E(\Omega)$ is approximated by $E_h(\Omega) = E_h(q^i)$ and we have to compute the derivatives

$$\frac{\partial E_h}{\partial q_j^i}, i = 1, \ldots, N, j = 1, 2 \tag{62}$$

in order to compute the variation of E_h with respect to Ω.

An *alternative* to this computation is to use Newton's method in order to find S such that

$(\vec{u}, p)$ is solution of Navier-Stokes equations;

$(\vec{P}, q)$ is solution of adjoint state equations;

$\frac{\partial \vec{u}}{\partial n} \cdot \left(\frac{\partial \vec{P}}{\partial n} - \frac{\partial \vec{u}}{\partial n} \right)$ is constant;

$$and \ \frac{\partial^2 E}{\partial \alpha^2} \geq 0 \ (\text{because this algorithm could find a maximum instead of a minimum})$$

In order to introduce the discretized optimization methods, we consider a simpler problem.

1.4 Potential flow optimization problem

If we look for the optimal wing, why not find the optimal Σ first (see Figure 5)? So, we state the *inverse problem in potential flow*:

$$\min_{\Sigma} \int_{\partial \Sigma} |\nabla \phi - \vec{u}|^2 dx$$

subject to

$$\Delta \phi = 0 \ \text{ in } \ \Omega - \Sigma \tag{63}$$

$$\frac{\partial \phi}{\partial n}\Big|_{\Gamma} = g, \tag{64}$$

where $g|_{\partial \Sigma} = 0$, $g|_{\Gamma_\infty} = \vec{u}_\infty \cdot \vec{n}$, $\Gamma = \partial \Sigma \cup \Gamma_\infty$, and where $\vec{u}$ is a guess for the optimal velocity on $\partial \Sigma$.

A similar problem to this one, is found by optimizing a nozzle. We have chosen it to illustrate the method, because it is simpler.

1.5 Nozzle optimization problem

Find S such that

$$\min_{S \in \mathcal{O}} \int_{D} |\nabla \phi - \vec{u}|^2 dx$$

subject to:

$$- \Delta \phi = 0 \ \text{ in } \ \Omega \tag{65}$$

$$\frac{\partial \phi}{\partial n}\Big|_{\Gamma} = g \tag{66}$$

where g is a function with $\int_{\Gamma} g d\gamma = 0$, $g|_{S} = 0$ and $\Gamma = \partial \Omega$.

The geometry and the function g are shown in Figure 7:

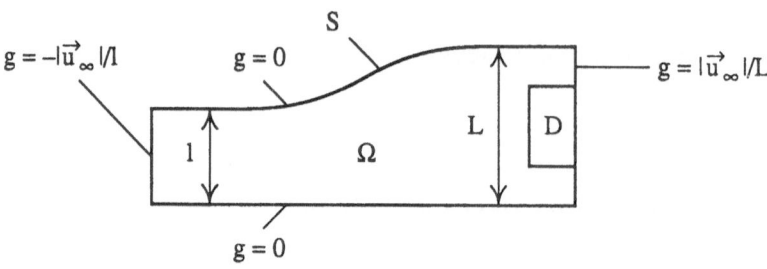

Figure 7

Figure 7 shows half of the nozzle, where the horizontal boundary is a symmetry line.

We shall derive the optimality condition using the adjoint state equation for the continuous problem first and we will give an idea of what happens in the discrete case.

Consider the variational formulation of (65):

$$\text{Find } \phi \in H^1(\Omega)/\mathbb{R} \text{ such that}$$

$$\int_\Omega \nabla\phi \cdot \nabla w \, dx = \int_\Gamma g w \, d\gamma, \forall w \in H^1(\Omega). \tag{67}$$

We have also:

$$\delta E = 2 \int_D (\nabla\phi - \vec{u}) \cdot \nabla\delta\phi \, dx. \tag{68}$$

We differentiate (67) and we obtain:

$$\int_\Omega \nabla\delta\phi \cdot \nabla w \, dx + \int_{\delta\Omega} \nabla\phi \cdot \nabla w \, dx = 0, \tag{69}$$

because $g|_S = 0$ and the rest of the boundary is supposed fixed.

When the perturbation of S to S' is "small", so we can write:

$$S' = \{x + \alpha\vec{n} : x \in S\}$$

and we obtain:

$$\int_{\delta\Omega} \nabla\phi \cdot \nabla w = \int_S \alpha\nabla\phi \cdot \nabla w \, d\gamma. \tag{70}$$

We introduce the adjoint state equation:

Find $p \in H^1(\Omega)$ such that:

$$\int_\Omega \nabla p \cdot \nabla w \, dx = 2 \int_D (\nabla\phi - \vec{u}) \cdot \nabla w \, dx, \forall w \in H^1(\Omega). \tag{71}$$

Taking $w = \delta\phi$ in (71) we find (see (68)):

$$\delta E = \int_\Omega \nabla p \cdot \nabla \delta\phi dx. \tag{72}$$

Taking $w = p$ in (69) we find (using (70) too):

$$\int_\Omega \nabla p \cdot \nabla \delta\phi dx = -\int_S \alpha \nabla\phi \cdot \nabla p d\gamma \tag{73}$$

and therefore:

$$\delta E = -\int_S \alpha \nabla\phi \cdot \nabla p d\gamma + o(\alpha). \tag{74}$$

We have to find an expression similar to (74) for the discrete case.

For discretization we use the *Finite Element Method* of degree 1 on triangles.

More precisely, we divide Ω into triangles and approximate Ω by $\Omega_h = \cup_k T_k$, where the T_k are triangles.

The division of Ω is such that the vertices of $\partial\Omega_h$ belong to $\partial\Omega$ and for the intersection of 2 triangles we allow only one of the three possibilities:

$$T_i \cap T_j = \begin{cases} \emptyset \\ \text{or 1 edge,} \qquad i \neq j \\ \text{or 1 vertex} \end{cases}$$

An admissible triangulation would be one like in Figure 8.

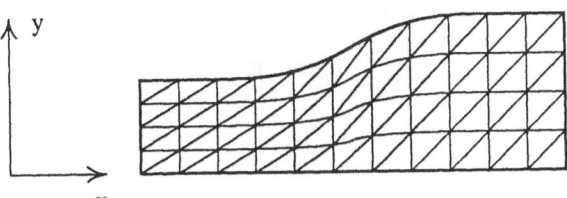

Figure 8

We consider the function space:

$$H_h = \{w_h \in C^o(\bar{\Omega}_h) : w_h|_{T_k} \in P^1\}$$

where P^1 is the set of polynomials of degree ≤ 1. We see that in one dimension these functions are as in Figure 9:

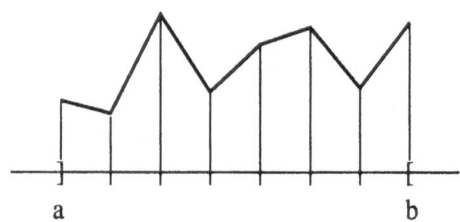

$$a \qquad\qquad\qquad\qquad b$$

Figure 9

So, the functions of H_h are determined by their values on the vertices of discretization. Moreover, we have the following:

Proposition 5 *The dimension of H_h equals the number of vertices q^i of the discretization, and every function ϕ_h belonging to H_h is completely determined by $\phi_h(q^i)$.*   □

We define the *hat functions*:

$$w^i \in H_h$$
$$w^i(q^j) = \delta_{ij}.$$

In one dimension these functions are as shown in Figure 10:

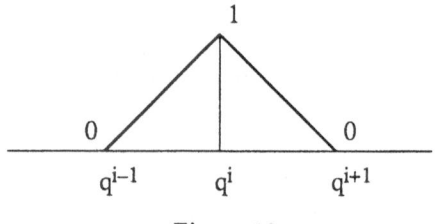

Figure 10

We state the

Proposition 6 $H_h = span\{w^i\}$.   □

We define the *discrete form* of the variational formulation (67):

Find $\phi_h \in H_h$ such that:

$$\int_\Omega \nabla\phi_h \cdot \nabla w^j \, dx = \int_\Gamma g w^j \, d\gamma, \forall j \tag{75}$$

So, the *discretized optimization problem* is:

Find

$$\min_{S_h} \int_D |\nabla\phi_h - \vec{u}|^2 dx = \min_{S_h} E(\Omega_h)$$

with ϕ_h subject to (75), S_h discrete approximation of S.

Since we want to find S_h, the optimization parameters are the vertices $q^j \in S_h$. But, if we change an internal vertex $q^i \notin S_h$, then $E(\Omega_h)$ changes.

Thus, in fact, $E(\Omega_h)$ is a function of all vertices $q^1, \ldots, q^N$.

If we denote $E(\Omega_h) = E(q^1, \ldots, q^N)$, we have to find the derivatives

$$\frac{\partial E}{\partial q_j^i}, i = 1, \ldots, N, j = 1, 2. \tag{76}$$

Let us show how the calculus of variation of $E(\Omega_h)$ works in the discrete case.

Calculus of variations of $E(\Omega_h)$

We suppose that q^k is an internal or boundary node of the triangulation and that $q^k \notin D$ (the case $q^k \in D$ is uselessly complicated but contains no major additional difficulties).

We move q^k to the position $q^k + \delta q^k$.

We write formally:

$$\delta E = 2 \int_D (\nabla\phi_h - \vec{u}) \cdot \nabla \delta\phi_h dx \tag{77}$$

and from (75), we obtain by differentiation:

$$\int_{\Omega_h} \nabla \delta\phi_h \cdot \nabla w^j dx + \int_{\delta\Omega_h} \nabla\phi_h \cdot \nabla w^j dx + \int_{\Omega_h} \nabla\phi_h \cdot \nabla \delta w^j dx = 0 \tag{78}$$

because $g|_S = 0$ and w^j varies if we change the triangulation (this variation gives the last integral in (78)).

We observe also that $\delta\phi_h \notin H_h$, because the two solutions which give the variation $\delta\phi_h$ are not obtained in the same triangulation.

Since by definition $\phi_h \in H_h$, we have:

$$\phi_h(x) = \sum_{i=1}^{N} \phi_i w^i(x) \tag{79}$$

so

$$\delta\phi_h(x) = \sum_{i=1}^{N} \delta\phi_i w^i(x) + \sum_{i=1}^{N} \phi_i \delta w^i(x). \tag{80}$$

We call

$$\delta\tilde{\phi}_h = \sum_{i=1}^{N} \delta\phi_i w^i(x) \in H_h.$$

But the second sum in (80) does not belong to H_h.

In order to use the equation (78), we have to state the following lemmas:

Lemma 2   *When q^k moves to $q^k + \delta q^k$, then the function $w^i(x)$ changes to $w^i(x) + \delta w^i(x)$ with*

$$\delta w^i(x) = -w^k(x)\nabla w^i \cdot \delta q^k + o(\delta q^k). \tag{81}$$

$\square$

We can explain Lemma 2 better by referring to Figure 11:

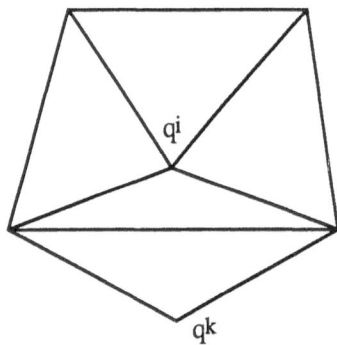 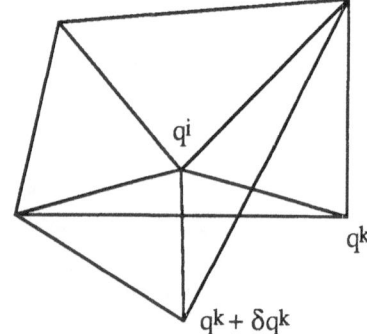

Figure 11.a   Figure 11.b

If q^k moves, then w^i associated with q^i In this case we change the supp w^i if we
does not change, since q^i does not belong   move q^k, so w^i changes by the δw^i given
to supp w^i.   by Lemma 2

Lemma 3   *If $f|_{T_k} \in C^1$, then*

$$\int_{\delta\Omega_h} f\,dx = \int_{\Omega_h} \delta q^k \cdot \nabla(fw^k)dx + o(\delta q^k), \tag{82}$$

where $\int_{\Omega_h} f dx = \sum_k \int_{T_k} f\,dx$ by definition. □

For the proof of these two lemmas, see O. Pironneau [1].

The preceding lemmas allow to compute all the integrals in (78) and to obtain an expression for $\delta E(\Omega_h)$, if we define the (discrete) adjoint state equation:

Find $p_h \in H_h$ such that

$$\int_\Omega \nabla p_h \cdot \nabla w^j dx = 2\int_D (\nabla \phi_h - \vec{u}) \cdot \nabla w^j dx, \quad \forall j. \tag{83}$$

We can prove that the discrete variation of $E(\Omega_h)$ is given by the following

Theorem 1 *When q^k moves to $q^k + \delta q^k$, $E(\Omega_h)$ varies by*

$$\begin{aligned}
\delta E(\Omega_h) \;=\; & \int_\Omega (\nabla \phi_h \cdot \nabla w^k)(\nabla p_h \cdot \delta q^k) dx \\
& + \int_{\Omega_h} (\nabla w^k \cdot \nabla p_h)(\nabla \phi_h \cdot \delta q^k) dx \\
& - \int_{\Omega_h} (\nabla \phi_h \cdot \nabla p_h)(\nabla w^k \cdot \delta q^k) dx + o(\delta q^k).
\end{aligned} \tag{84}$$

□

According to Theorem 1, we can write

$$\delta E = \chi^k \cdot \delta q^k \tag{85}$$

and therefore a *gradient method* is:

Change q^k to $q^k - \lambda\chi^k$, with $\lambda \simeq$ optimum step size. Stop when $\chi^k \simeq 0, \forall k.$

Independence from E

Note that the adjoint state p depends on the criterion t. On the other hand, if the software is to be provided as a black box to the industry it must be such that it is easy to:

– change the design criterion;

– add geometrical contraints.

Suppose that we minimize a functional of the general form:

$$E(\phi, \Omega) = \int_D f(\phi) dx, \quad \phi = \{\phi^j\}, j = 1, \ldots, r. \tag{86}$$

Since the second member of the adjoint state equation (83) is δE, we must be able to compute $\frac{\partial E}{\partial \phi_j}$ independently of $E(\phi, \Omega)$.

This computation can be done by finite differences because:

$$\frac{\partial E}{\partial \phi_j} \simeq \frac{E(\phi_h + \delta\phi_h u^j, \Omega_h) - E(\phi_h, \Omega_h)}{\delta\phi_j} \tag{87}$$

This computation is not expensive. The number of elementary computations is of order N. Indeed, if N is the number of the mesh nodes, the calculation cost is of the order N, which is the same cost as the solution of a laplacian (cf. Arumugan [4]).

Add geometrical constraints

To add geometrical constaints is easy if we give a parametrized description of the domain and its triangulation.

If the boundary to optimize is described by r parameters α_j, we can define it by a curve (ex. spline) defined by α_j, and then generate the triangulation with vertices $\{q^i\}, i = 1, \ldots, N$, on the curve.

Since in this case only the parameters α_j move independently, we must compute the variation of E with respect to α_j. But

$$\frac{\partial E}{\partial \alpha_j} = \sum_{k,i} \frac{\partial E}{\partial q_i^k} \cdot \frac{\partial q_i^k}{\partial \alpha_j}, i = 1, \ldots, N, k = 1, 2. \tag{88}$$

Therefore, we must be able to compute $\frac{\partial q_i^k}{\partial \alpha_j}$ and this is done also by finite differences:

$$\frac{\partial q_i^k}{\partial \alpha_j} \simeq \frac{q_i^k(\alpha_j + \delta\alpha_j) - q_i^k(\alpha_j)}{\delta\alpha_j} \tag{89}$$

which is not computationally expensive.

Remark One could think that we can compute everything by finite differences, even

$$\frac{\partial E}{\partial q_i^k} \simeq \frac{E(q_i^k + \delta q_i^k) - E(q_i^k)}{\delta q_i^k} \tag{90}$$

but this is far too expensive, since we have to solve the state equation every time we compute $E(q_i^k)$. So the computational cost of (90) is $2N * O(N) \simeq O(N^2)$ which is the cost of solution of N partial differential equations.

Another trouble we face with the discretized problem is *how to move the internal mesh nodes* in order to follow the motion of the boundary nodes.

In the nozzle optimization problem, this is quite easy, since all the nodes may move in the y direction (see Figure 8) and the mesh is constructed by dilating the mesh of a rectangle according to the dilation of one rectangle edge to the moving boundary.

It is easily seen that the situation is not so simple for the mesh around a complex multi-bodied wing and even more for a three dimensional mesh. There is always the problem of how we can make the internal nodes follow the motion of boundary nodes.

We would like to end the lecture on the nozzle optimization problem by showing some results obtained by F. Angrand and A. Vossinis.

The results of Angrand show the optimization of a nozzle when

$$E = \int_S (|\nabla\phi|^2 - p_o)^2 dx$$

is minimum, i.e. we try to recover a certain pressure distribution p_o on the profile, using a gradient method; the results can by found in [1], [2].

The results of Vossinis are on the optimization of a nozzle when

$$E = \int_S (|\nabla\phi - \vec{u}|^2 dx$$

i.e. we try to recover a certain velocity distribution on the profile using a gradient method and methods with no gradient computation, namely GMRES, Powell's method and Least Squares Method (subroutine V A02AD of Harwell). We remark that these last methods can be used as black boxes.

Results are shown for two types of problem:

- $\vec{u}$ is generated by a known profile S and we try to find again this profile. In that case min $E = 0$. These types of problems will be called inverse problems.

- $\vec{u}$ is an arbitrary velocity distribution and we seek the profile which realizes it. Now we cannot assert that min $E = 0$.

We see that GMRES worked very well on inverse problems as a black box and it was almost as good as a gradient method when min $E \neq 0$. Still this conclusion may change if the number of unknown parameters defining the boundary is large. Results of Angrand are shown in Figures A.1 and A.2 and the ones of Vossinis in Figures A.3 and A.4 of Appendix A.

2 Compressible, potential flows

Let us describe now a wing optimization problem when the flow is potential, but the fluid is compressible.

2.1 Optimization of a wing in potential and compressible flow

The equations governing such a flow are:

$$\vec{u} = \nabla\phi \tag{91}$$

$$\nabla \cdot (\rho\vec{u}) = 0 \qquad \text{in } \Omega \tag{92}$$

$$\rho = (1 - |\vec{u}|^2)^{\frac{1}{\gamma-1}}, \gamma = 1.4 \tag{93}$$

where: $\vec{u}$ = flow velocity, ρ = mass density, p = pressure, γ = fraction of specific heats.

We recall that equation (92) expresses the conservation of mass, and that

$$p = \rho^\gamma. \tag{94}$$

Combining equations (91)–(93) we obtain the following non-linear partial differential equation (PDE):

$$\nabla \cdot [(1 - |\nabla\phi|^2)^{\frac{1}{\gamma-1}} \nabla\phi] = 0 \quad \text{in } \Omega. \tag{95}$$

This equation is elliptic if

$$|\nabla\phi| < \frac{\gamma - 1}{2(\gamma + 1)} \tag{96}$$

and hyperbolic otherwise. The geometry and boundary conditions we consider are shown in Figure 12:

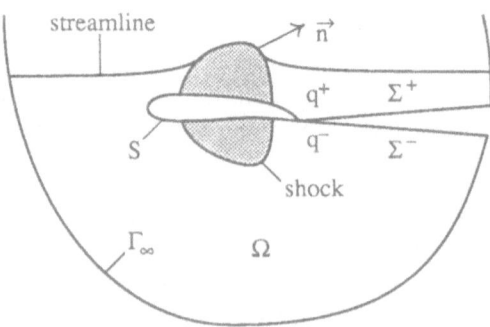

Figure 12

Behind the trailing edge q, ϕ is discontinuous. Let Σ be the streamline which starts from q. In gray the supersonic zone.

This streamline is divided in two lines Σ^+ and Σ^- and we have the conditions:

$$\phi_{\Sigma^+} - \phi_{\Sigma^-} = \alpha \text{ constant,} \tag{97}$$

and

$$|\nabla\phi|_{q^+} = |\nabla\phi|_{q^-} \tag{98}$$

and (95) is integrated on $\Omega - \Sigma$.

So, the boundary conditions are of Neumann type on Γ_∞ and S, and of Dirichlet periodic type on Σ. We note that the constant α is unknown but we have an additional equation (98).

The PDE (95) is hyperbolic in the shaded areas (Figure 12) and elliptic outside. Shocks may occur at the interface. The shocks are such that $|\nabla\phi|$ is discontinuous, but $\rho\nabla\phi \cdot \vec{n}$ is continuous.

In order to handle the shocks we have to add an artificial viscosity ε and to replace (95) by

$$\nabla \cdot [(1 - |\nabla\phi|^2)^{\frac{1}{\gamma-1}} \nabla\phi] + \varepsilon\Delta\rho = 0 \tag{99}$$

in $\Omega - \Sigma$.

Now the drag has two terms: one term due to viscosity (which is not modelled by (99)) and another one due to the discontinuity of the pressure through the shocks (pressure drag).

The pressure drag is equal to:

$$D_p = \int_S (p - p_\infty) d\gamma. \tag{100}$$

So the optimization problem is:

$$\min_S D_p \tag{101}$$

where p is obtained by (94), where $\vec{u} = \nabla\phi$ and ϕ is solutions of (99).

A first difficulty we confront in the solution of this problem is the solution of (99). It is difficult to prove the existence of solutions of (99).

However, (99) can be solved numerically using the Finite Element Method (FEM). We refer to Angrand [2] for the solution of the optimization problem (101), with a different criterion, where (99) is solved by the FEM, and a gradient method is used to solve the optimization problem. Angrand faced severe difficulties to handle the geometrical constraints.

We recall that geometrical constraints can be included into the formulation if we use a spline representation of the wing profile. We give an example of such constraints in Figure 13. The results obtained by Angrand are shown in Figures A.5, A.6 and A.7 in Appendix A.

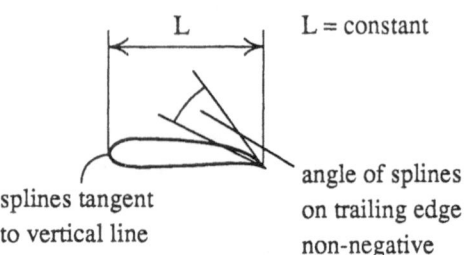

Figure 13

Jameson has solved the same problem using a transformation of coordinates (cf. [5]).

According to Jameson, we transform the domain outside the airfoil described by carte-sian coordinates $z = (x, y)$ to the domain interior to the unit circle described by the polar coordinates (ρ, θ) in the σ–plane as in Figure 14. Conformal transformations are used.

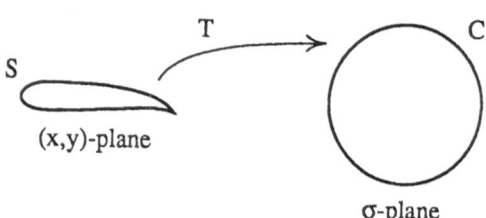

Figure 14

Then, we can either minimize the pressure drag (100) or, supposing that we want a certain velocity distribution on C, we introduce the cost function

$$E = \frac{1}{2} \int_C (|\vec{u}| - |\vec{u}_d|)^2 d\theta \qquad (102)$$

which will be minimized as a function of

$$h = \left| \frac{dz}{d\sigma} \right|. \qquad (103)$$

This problem is an optimal control problem where the control is the mapping modulus h.

The main problem of this method is that the inverse transformation from the σ–plane to the (x, y)–plane may give a non-admissible profile, for example an open airfoil at the trailing edge.

It is proved that the profile is closed under certain conditions. Suppose that we have a power expansion of $\log\left(\frac{dz}{d\sigma}\right)$, that is:

$$\log\left(\frac{dz}{d\sigma}\right) = \sum_{n=0}^{\infty} C_n \sigma^{-n} \tag{104}$$

where $z = x + iy$ and only negative powers are taken (otherwise $\frac{dz}{d\sigma}$ becomes unbounded for large σ).

Then, the wing profile is closed if and only if $C_o = C_1 = 0$.

Another problem is that if the flow is transsonic, the function $\vec{u} - \vec{u}_d$ is not differentiable, so we have to choose another cost function of the form

$$E = \frac{1}{2}\int_C\left(\lambda_1 Q^2 + \lambda_2\left(\frac{dQ}{d\theta}\right)^2\right)d\theta \tag{104a}$$

with Q the smoothed velocity deviation given by the equation

$$\lambda_1 Q - \lambda_2 \frac{d^2 Q}{d\theta^2} = |\vec{u}| - |\vec{u}_d| \tag{104b}$$

(Q periodic function), and λ_1, λ_2 parameters.

Numerical experiments however showed that one has to include the pressure drag in the cost function when the Mach number is greater than 0.8. So, the function to minimize becomes

$$E = \frac{1}{2}\int_C\left(\lambda_1 Q^2 + \lambda_2\left(\frac{dQ}{d\theta}\right)^2\right)d\theta + \lambda_3 \cdot C \int_C (p - p_\infty)d\gamma \tag{104c}$$

where λ_3 is a parameter used as a weight term for the pressure drag.

Excellent results have been obtained by Jameson using a gradient method.

We have to remark that a continuous computation of the gradient $\frac{\partial E}{\partial h}$ was made and used in a discrete problem. As we saw in 1.3 this causes a numerical noise, but it worked very well in the method proposed by Jameson.

3 Optimum design of riblets

We consider a flat plate in a fluid flow parallel to it (Figure 16).

Experiments show that if little groves are digged along the surface of the plate, the drag is reduced. We would like to find the shape of these groves, called riblets, so that the drag is minimized. Figure 16 shows a section of a plate with riblets perpendicular to the flow u_∞.

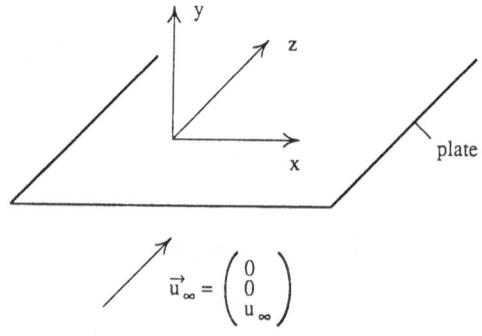

Figure 15

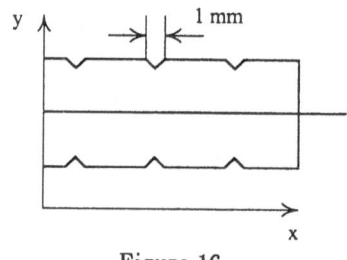

Figure 16

We consider again an incompressible flow. So, the problem statement is:

Find:

$$\min_{\Sigma} \vec{u}_\infty \cdot \int_\Sigma \left[\frac{\nu}{2}(\nabla\vec{u} + \nabla\vec{u}^T) - p \right] \vec{n}\,d\gamma \tag{105}$$

with $(\vec{u}, p)$ solution of

$$\frac{\partial\vec{u}}{\partial t} + \vec{u}\nabla\vec{u} + \nabla p - \nu\Delta\vec{u} = 0 \quad \text{in} \quad \Omega \tag{106}$$

$$\nabla\cdot\vec{u} = 0 \quad \text{in} \quad \Omega \tag{107}$$

$$\vec{u}|_\Sigma = 0 \tag{108}$$

$$\vec{u}|_{\Gamma_\infty} = \vec{u}_\infty \tag{109}$$

Experiments show that riblets reduce the drag by 5%.

We shall consider that the solution of (106)–(109) is of the form:

$$\vec{u} = \begin{pmatrix} 0 \\ 0 \\ u(x, y) \end{pmatrix} \quad \text{and} \quad p = p(z). \tag{110}$$

Equations (110) imply that $\vec{u}\nabla\vec{u} = 0$ and, supposing stationary flow, equation (106) becomes

$$\nabla p - \nu\Delta\vec{u} = \begin{pmatrix} \frac{\partial p}{\partial x} \\ \frac{\partial p}{\partial y} \\ \frac{\partial p}{\partial z} - \nu\Delta_{x,y}\vec{u} \end{pmatrix} = 0. \tag{111}$$

Equations (111) can be satisfied for $p = kz$ ($k = $ constant), so we obtain from (111):

$$-\nu\Delta u + k = 0. \tag{112}$$

So, the state equation is simplified and the boundary conditions are those of Figure 17.

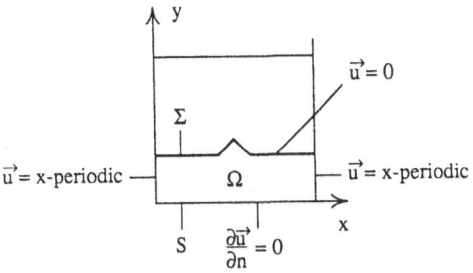

Figure 17

We consider in Figure 17 only one cell, i.e. a part of the plate containing one riblet; suppose also that S is the surface of the boundary layer. To express the fact that the flow is uniform near S, we use an homogeneous Neumann condition on it.

The simplified problem is:

Find

$$\min_\Sigma \left(-\nu \int_\Sigma \frac{\partial\vec{u}}{\partial\vec{n}} d\gamma \right) \tag{113}$$

with $(\vec{u}, k)$ solution of:

$$-\nu\Delta\vec{u} + k = 0 \quad \text{in} \quad \Omega \tag{114}$$

$$\nabla \cdot \vec{u} = 0 \quad \text{in} \quad \Omega \tag{115}$$

$$\vec{u} = 0 \quad \text{on} \quad \Sigma \tag{116}$$

$$\frac{\partial \vec{u}}{\partial \vec{n}} = 0 \quad \text{on} \quad S \tag{117}$$

$$\vec{u} = x - \text{periodic} \quad \text{on} \quad \partial\Omega - \Sigma - S. \tag{118}$$

It is important to note that the pressure gradient k is determined by the condition:

$$\int_\Omega \vec{u} dx = d, \tag{119}$$

where d is the given total flux.

Without loss of generality, we can assume that $\nu = 1$. Then, we have to find Σ admissible, such that:

$$\min_\Sigma \left(-\int_\Sigma \frac{\partial \vec{u}}{\partial \vec{n}} d\gamma \right) \tag{120}$$

subject to:

$$-\Delta\vec{u} = k \tag{121}$$

and (115)–(118).

The problem given above has two difficulties:

a) We have to minimize a boundary integral;

b) k is determined by (119).

We shall transform the problem in order to solve these difficulties.

Because of the boundary conditions (116)–(118), the integral to minimize becomes:

$$-\int_\Sigma \frac{\partial \vec{u}}{\partial \vec{n}} d\gamma = -\int_{\partial\Omega} \frac{\partial \vec{u}}{\partial \vec{n}} d\gamma. \tag{122}$$

Further, using Green's formula, (122) gives:

$$\int_\Sigma \frac{\partial \vec{u}}{\partial \vec{n}} d\gamma = -\int_{\partial\Omega} \frac{\partial \vec{u}}{\partial \vec{n}} d\gamma = \int_\Omega \nabla\vec{u} \cdot \nabla 1 dx + \int_\Omega (\Delta u) \cdot 1 dx = -k|\Omega|, \tag{123}$$

because of (121) and we denote $|\Omega| = \int_\Omega dx$. Equation (119) gives:

$$kd = \int_\Omega \vec{u}k dx = -\int_\Omega \vec{u} \cdot \Delta \vec{u} dx = \int_\Omega |\nabla \vec{u}|^2 dx - \int_{\partial\Omega} \vec{u} \cdot \frac{\partial \vec{u}}{\partial \vec{n}} d\gamma = \int_\Omega |\nabla \vec{u}|^2 dx, \qquad (124)$$

where we used (121), Green's formula and the boundary conditions (116)–(118), so

$$\int_{\partial\Omega} \vec{u} \cdot \frac{\partial \vec{u}}{\partial n} d\gamma = 0.$$

So, the integral to minimize becomes:

$$-\int_\Sigma \frac{\partial \vec{u}}{\partial n} d\gamma = k|\Omega| = \frac{|\Omega|}{d} \int_\Omega |\nabla \vec{u}|^2 dx. \qquad (125)$$

This handles the first difficulty.

For the second one, let us consider:

$$\vec{v} = \frac{\vec{u}}{k}. \qquad (126)$$

Having $\vec{v}$ as unknown speed, (121) and (115)–(118) give:

$$-\Delta \vec{v} = 1 \quad \text{in} \quad \Omega \qquad (127)$$

$$\nabla \cdot \vec{v} = 0 \quad \text{in} \quad \Omega \qquad (128)$$

$$\vec{v}|_\Sigma = 0 \qquad (129)$$

$$\frac{\partial \vec{v}}{\partial n}\bigg|_S = 0 \qquad (130)$$

$$\vec{v} = x - \text{ periodic} \quad \text{on } \partial\Omega - \Sigma - S. \qquad (131)$$

Using Green's formula and (127)–(131), we obtain:

$$\int_\Sigma \frac{\partial \vec{v}}{\partial n} d\gamma = \int_{\partial\Omega} \frac{\partial \vec{v}}{\partial n} d\gamma = \int_\Omega \nabla \vec{v} \cdot \nabla 1 dx + \int_\Omega \nabla \vec{v} \cdot 1 dx = -|\Omega| \qquad (132)$$

and

$$\int_\Omega \vec{v} dx = \int_\Omega (-\Delta \vec{v}) \cdot \vec{v} dx = \int_\Omega |\nabla \vec{v}|^2 dx - \int_{\partial\Omega} \vec{v} \cdot \frac{\partial \vec{v}}{\partial n} d\gamma = \int_\Omega |\nabla \vec{v}|^2 dx. \qquad (133)$$

Therefore, according to (125), the integral to minimize is:

$$\frac{|\Omega|}{d} \int_\Omega |\nabla \vec{u}|^2 dx = \frac{|\Omega|}{d} k^2 \int_\Omega |\nabla \vec{v}|^2 dx =$$

$$= |\Omega| k \int_\Omega |\nabla \vec{v}|^2 dx \frac{k}{d} = k \int_\Omega |\nabla \vec{v}|^2 dx \cdot \frac{1}{\int_\Omega \vec{v} dx} = \qquad (134)$$

$$= k = \frac{d|\Omega|}{\int_\Omega \vec{v} dx} = \frac{d|\Omega|}{\int_\Omega |\nabla \vec{v}|^2 dx},$$

where we used equations (132) and (133) and the fact that

$$\int_\Omega \vec{u}dx = d \Rightarrow \int_\Omega k\vec{v}dx = d \Rightarrow \tag{135}$$

$$\int_\Omega \vec{v}dx = \frac{d}{k} \ . \tag{136}$$

We conclude that instead of minimizing (125), we can maximize $\frac{1}{d|\Omega|}\int_\Omega |\nabla\vec{v}|^2 dx$.

The optimization problem is now:

$$\max_{\Sigma\in\mathcal{O}} \frac{1}{d|\Omega|} \int_\Omega |\nabla\vec{v}|^2 dx, \tag{137}$$

where $\vec{v}$ is the solution of (127)–(131).

In order to solve this problem, we have to define a class of admissible solutions Σ. If there are no constraints on Σ, then the solution $\Sigma \rightarrow -\infty$ in y gives on (113)–(118) a criterion tending to 0 if k is fixed. It is not difficult to see that the same thing happens if d is fixed instead of k.

Therefore, we must include a constraint of the type

$$|\Omega| = f(\Sigma) \tag{138}$$

and the simplest one is to impose a constant average thickness of the boundary layer:

$$\int_\Sigma yd\gamma = \text{ constant.} \tag{139}$$

Equations (138) and (139) imply that we have to define the set of admissible Σ:

$$\mathcal{O} = \{\Sigma : |\Omega| = \beta = \text{ constant}\}.$$

We continue by deriving the optimality condition for (137).

We consider that Σ is a solution and Ω the corresponding domain. We obtain Σ' and the corresponding Ω' by a small perturbation of Σ, namely we take:

$$\Sigma' = \{x + \lambda\alpha(x)\vec{n}(x) : x \in \Sigma\},$$

where $\vec{n}$ is the normal to Σ, $\alpha(x) = $ regular and $\lambda = $ small scalar. The solutions of (127)–(131) in Ω' and Ω, respectively, are $\vec{v}(\Omega')$ and $\vec{v}(\Omega)$ and we define $\delta\vec{v} = \vec{v}(\Omega') - \vec{v}(\Omega)$.

We have the relations:

$$-\Delta\delta\vec{v} = 0 \ \ \text{in} \ \ \Omega' \cap \Omega \tag{140}$$

$$\left.\frac{\partial\delta\vec{v}}{\partial\vec{n}}\right|_{S'\cap S} = 0, \tag{141}$$

where $S' = \partial\Omega' - \Sigma'$.

As before it is easy to show that

$$\delta\vec{v}(x) = -\lambda\alpha\frac{\partial\vec{v}(\Omega')}{\partial\vec{n}} + o(\lambda), \forall x \in \Sigma$$

and

(142)

$$\vec{v}'_\alpha = \lim_{\lambda \to 0}\frac{\vec{v}(\Omega') - \vec{v}(\Omega)}{\lambda}$$

with

$$-\Delta\vec{v}'_\alpha = 0 \quad \text{in} \quad \Omega \tag{143}$$

$$\vec{v}'_\alpha|_\Sigma = -\alpha\frac{\partial\vec{v}}{\partial\vec{n}} \tag{144}$$

$$\frac{\partial\vec{v}'_\alpha}{\partial\vec{n}}\bigg|_{\partial\Omega-\Sigma} = 0. \tag{145}$$

The cost function is $(d = 1)$,

$$J(\Sigma) = \frac{1}{|\Omega|}\int_\Omega|\nabla\vec{v}|^2 dx, \tag{146}$$

and therefore:

$$J(\Sigma') - J(\Sigma) = -\frac{\lambda}{|\Omega|^2}\int_\Sigma\alpha d\gamma\int_\Omega|\nabla\vec{v}|^2 dx + \frac{\lambda}{|\Omega|}\int_\Sigma\alpha|\nabla\vec{v}|^2 d\gamma + o(\alpha) \leq 0 \tag{147}$$

which implies (using eq. (127)–(131)):

$$\int_\Sigma\alpha\left|\frac{\partial\vec{v}}{\partial\vec{n}}\right|^2 d\gamma \leq \frac{1}{|\Omega|}\int_\Sigma\alpha d\gamma\int_\Omega|\nabla\vec{v}|^2 dx + o(\alpha). \tag{148}$$

The set $\mathcal{O}$ of admissible Σ we considered implies that the admissible set of α is:

$$\{\alpha : \int_\Sigma\alpha d\gamma = 0\}.$$

So, equation (153) gives the following:

$$\left|\frac{\partial\vec{v}}{\partial\vec{n}}\right| = \text{constant.}$$

Proposition 7 *If Σ is a regular solution of (137) and $\vec{v} \in (H^2(\Omega))^2$ a solution of (127)–(131), then we have:*

$$\left|\frac{\partial\vec{v}}{\partial\vec{n}}\right| = \text{constant on } \Sigma \tag{149}$$

$\square$

We can prove in the same way that, for the minimization of

$$E(\Sigma) = \frac{|\Omega|}{d} \int_{\Omega} |\nabla \vec{u}|^2 dx \tag{150}$$

with $\vec{u}$ subject to (121) and (115)–(118), we have the following

Proposition 8 *If Σ is a regular solution which minimizes* (150) *and $\vec{u}$ solves* (121) *and* (115)–(118), *then*

$$\left|\frac{\partial \vec{u}}{\partial \vec{n}}\right| = \text{ constant on } \Sigma \tag{151}$$

$\square$

In both cases, we must find Σ such that:

$$-\Delta \vec{u} = k \quad \text{in} \quad \Omega \tag{152}$$

$$\vec{u}|_{\Sigma} = 0 \tag{153}$$

$$\frac{\partial \vec{u}}{\partial \vec{n}}\bigg|_{\Sigma} = c \text{ (constant)} \tag{154}$$

$$\frac{\partial \vec{u}}{\partial \vec{n}}\bigg|_{S} = 0 \tag{155}$$

$$\vec{u} = x - \text{ periodic.} \tag{156}$$

One evident solution is the flat plate. In this case, we know the analytical solution. We refer to the geometry of Figure 18, in order to describe the solution of (152)–(156):

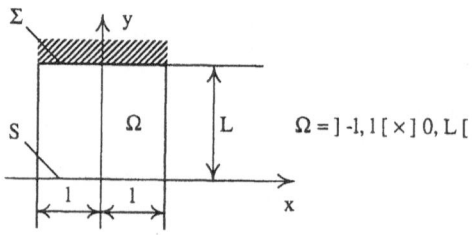

Figure 18

$$\vec{u} = -\frac{3d}{4lL^3}(y^2 - L^2)\vec{e}_z, \tag{157}$$

with $\vec{e}_z$ the unit vector on the z-axis, and

$$\frac{\partial u}{\partial n}\Big|_\Sigma = -\frac{3d}{2lL^2}. \tag{158}$$

Another trivial solution of (152)–(156) is the semi-cylinder. If we call l the cylinder radius, the solution $\vec{u}$ in polar coordinates is:

$$u(r,\theta) = -\frac{4d}{\pi l^4}(r^2 - l^2) \tag{159}$$

and

$$\frac{\partial \vec{u}}{\partial \vec{n}}\Big|_\Sigma = -\frac{8d}{\pi l^3}. \tag{160}$$

The drag for the flat plate is found to be

$$D_p = \frac{3dv}{2L^2} = \frac{3dv}{2l^2}\left(\frac{1}{L}\right)^2 \tag{161}$$

and for the semi-cylinder:

$$D_c = \frac{4dv}{l^2}. \tag{162}$$

Since $|\Omega|$ must be the same in both cases, we have:

$$\frac{l}{L} = \frac{4}{\pi}, \tag{163}$$

so:

$$\frac{D_p}{D_c} = \frac{6}{\pi^2}. \tag{164}$$

We conclude that the flat plate is a better solution than the semi-cylinder ($D_p \simeq 0.4 D_c$).

We ask now if we can do better. Firstly, let us find out if the flat plate is not a minimum. We have to compute $\frac{d^2 J}{d\lambda^2}\Big|_{\lambda=0}$, that is, the second derivative of J with respect to α (for the expression of $J(\Sigma)$ see equation (146)).

Observing that Taylor's expansion of $\vec{v}(\Omega')$ gives:

$$\vec{v}(\Omega') = \vec{v}(\Omega) + \lambda \vec{v}'_\alpha + \frac{\lambda^2}{2} v''_{\alpha\alpha}, \tag{165}$$

we compute:

$$\begin{aligned}
J''_{\lambda\lambda} &= \frac{d^2}{d\lambda^2}\int_{\Omega(\lambda)}|\nabla\vec{v}(\lambda)|^2 dx \\
&= 2\int_\Omega |\nabla\vec{v}'_\alpha|^2 dx + 2\int_\Omega \nabla\vec{v}\cdot\nabla\vec{v}''_{\alpha\alpha}dx \\
&\quad + \frac{d^2}{d\lambda^2}\int_{\Omega(\lambda)}|\nabla\vec{v}|^2 dx + 4\int_\Sigma \alpha\nabla\vec{v}'_\alpha\cdot\nabla\vec{v}d\gamma.
\end{aligned} \tag{166}$$

It can been shown that:

$$\frac{d^2}{d\lambda^2}\int_{\Omega(\lambda)}|\nabla\vec{v}|^2 dx = \int_\Sigma \alpha^2 \frac{\partial|\nabla\vec{v}|^2}{\partial\vec{n}}d\gamma \tag{167}$$

(cf. [3]).

From equations (127)–(131), we obtain:

$$\Delta\vec{v}''_{\alpha\alpha} = 0 \tag{168}$$

$$\left.\frac{\partial\vec{v}''_{\alpha\alpha}}{\partial\vec{n}}\right|_\Sigma = 0 \tag{169}$$

which, with (129), imply that

$$\int_\Omega \nabla\vec{v}\cdot\nabla\vec{v}''_{\alpha\alpha}dx = 0. \tag{170}$$

Therefore, (166) gives (with use of (167) and (170)):

$$\begin{aligned}
J''_{\lambda\lambda} &= 2\int_\Omega |\nabla\vec{v}'_\alpha|^2 dx + 4\int_\Sigma \alpha\frac{\partial\vec{v}'_\alpha}{\partial\vec{n}}\cdot\frac{\partial\vec{v}}{\partial\vec{n}}d\gamma \\
&\quad + \int_\Sigma \alpha^2\frac{\partial|\nabla\vec{v}|^2}{\partial\vec{n}}d\gamma.
\end{aligned} \tag{171}$$

We use (144)–(145) to simplify the second integral:

$$\begin{aligned}
\int_\Sigma \alpha\frac{\partial\vec{v}}{\partial\vec{n}}\cdot\frac{\partial\vec{v}'_\alpha}{\partial\vec{n}}d\gamma &= -\int_\Sigma \vec{v}'_\alpha\cdot\frac{\partial\vec{v}'_\alpha}{\partial\vec{n}}d\gamma \\
&= -\int_{\partial\Omega} \vec{v}'_\alpha\cdot\frac{\partial\vec{v}'_\alpha}{\partial\vec{n}}d\gamma = -\int_\Omega |\nabla\vec{v}'_\alpha|^2 d\gamma.
\end{aligned} \tag{172}$$

Finally, we find that:

$$J''_{\lambda\lambda} = -2\int_\Omega |\nabla\vec{v}'_\alpha|^2 dx + \int_\Sigma \alpha^2\frac{\partial|\nabla\vec{v}|^2}{\partial\vec{n}}d\gamma. \tag{173}$$

This computation proves the

Proposition 9 *The second derivative of*

$$J(\Sigma) = \int_{\Omega(\lambda)} |\nabla\vec{v}(\lambda)|^2 dx$$

is

$$\left.\frac{d^2 J}{d\lambda^2}\right|_{\lambda=0} = -2\int_\Omega |\nabla\vec{v}'_\alpha|^2 dx + \int_\Sigma \alpha^2\frac{\partial|\nabla\vec{v}|^2}{\partial\vec{n}}d\gamma.$$

□

In order for the flat plate to be a local maximum of $J(\Sigma)$, we must have $J''_{\lambda\lambda} < 0$.

Since for the flat plate

$$v = (L^2 - y^2)/2 \tag{174}$$

it is difficult to find the sign of $J''_{\lambda\lambda}$.

We can develop $\alpha(x)$ in a Fourier series, because Σ must be periodic. As it is proved in [3], the angle of contact of S and Σ must be $90°$, so:

$$\frac{d\alpha}{dx}(\pm l) = 0. \tag{175}$$

Thus, the Fourier series for $\alpha(x)$ is:

$$\alpha(x) = \sum_{k=1}^{\infty} \alpha_k \cos\left(k\pi \frac{x}{l}\right), \tag{176}$$

where in (175) and (176), x denotes the abscissa.

From equations (143)–(145), we find:

$$v'_\alpha = L \sum_k \alpha_k \cos\left(k\pi \frac{x}{l}\right) \cosh\left(k\pi \frac{y}{l}\right) \cosh\left(k\pi \frac{L}{l}\right) \tag{177}$$

and therefore:

$$
\begin{aligned}
J''_{\lambda\lambda} &= \int_{-l}^{l} (-2\alpha) L^2 \frac{\pi}{l} \sum_k k\alpha_k \cos\left(k\pi \frac{x}{l}\right) \tanh\left(k\pi \frac{L}{l}\right) + 2\int_{-l}^{l} \alpha^2 L dx \\
&= 2Ll \sum_k \alpha_k^2 (1 - z \cdot \tanh z)\big|_{z=k\pi \frac{l}{L}}.
\end{aligned} \tag{178}
$$

From equation (178) we see that $J''_{\lambda\lambda}$ changes sign depending on whether the quantity $-\pi(L/l)k \tan h(k\pi L/l) + 1$ is positive or negative.

So, we have the

Theorem 2 *If $\pi L/l > \rho$, where ρ the root of $1 - x \tan hx$, then the flat plate gives a local maximum to $J(\Sigma)$ for $|\Omega| = $ constant.* □

Till now we considered only analytical solutions. A numerical solution can be obtained using a Finite Element discretization and a Gradient Method.

For a detailed analysis of the discrete problem, see [3] and [4].

Results are shown in Figure A.8 of Appendix A.

We note that the solution depends on the ratio L/l. But l is an output of a full 3D Navier-Stokes calculation; thus the analysis should be completed with a study of the dependence of l on Σ.

References

[1] O. Pironneau, *Optimal Shape Design for Elliptic Systems*, Springer-Verlag, 1984.

[2] F. Angrand, Méthodes numériques pour des problèmes de conception optimale en aérodynamique, Thesis, Université Paris VI, 1980.

[3] G. Arumugam, O. Pironneau, On the problems of riblets as a drag reduction device, *Optimal Control Appl. Methods* 10 (1989).

[4] G. Arumugam, Optimisation de forme pour quelques problèmes de Mécanique des Fluides, Thesis, Université Paris VI, 1987.

[5] A. Jameson, *Automatic Design of Transonic Airfoils to Reduce the Shock Induced Pressure Drag*, Princeton University, Princeton, N.J., USA.

[6] R. Mäkinen, Finite element design sensitivity analysis for nonlinear potential problems, University of Jyväskylä, Finland.

[7] D. Chenais, On the existence of solution in a domain identification problem, *J. Math. Anal. Appl.* 52 (2) (1975), 189–219.

[8] O. Pironneau, On optimum design in fluid mechanics, *J. Fluid Mech.* 64 (1974), 97–110.

[9] O. Ladyzhenskaya, *The Mathematical Theory of Viscous Incompressible Flow*, Gordon & Breach, New York, 1963.

[10] O. Pironneau, On optimum profiles in Stokes flow, *J. Fluid Mech.* 59 (1973), 117–128.

[11] L.D. Landau, E.M. Lifshitz, *Fluid Mechanics*, Pergamon Press, 1987.

[12] G.K. Batchelor, *An Introduction to Fluid Dynamics*, Cambridge University Press, 1967.

[13] A. Vossinis, Optimization algorithms for optimum shape design problems, to appear.

[14] J.-C. Simon, Contrôle optimal par rapport au domaine, Thesis, Université Paris VI, 1977.

[15] J.-C. Simon, Differentiation with respect to the domain in boundary value problems, *Numer. Funct. Anal. Optim.* 2 (1980), 649–687.

Appendix A

Figure A.1

Intermediate nozzles (from Angrand [2])

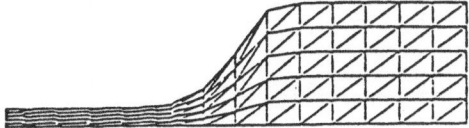

Figure A.2

The computed optimum profile after 30 iterations (from Angrand [2])

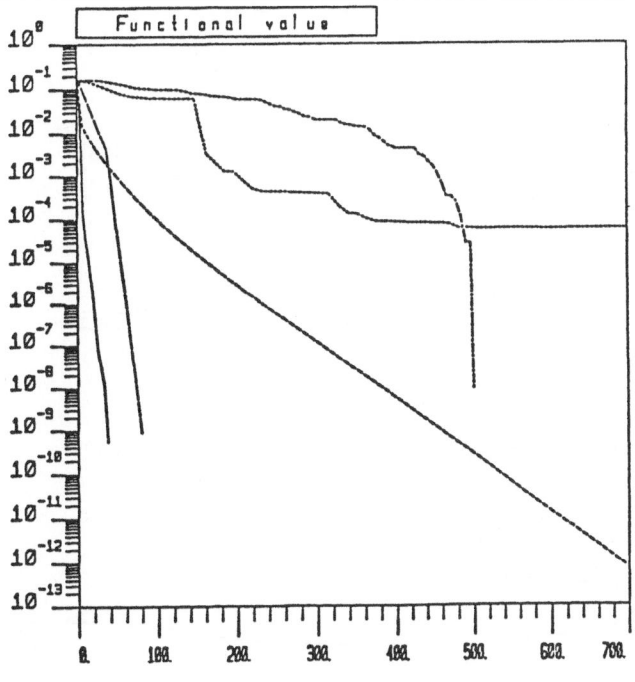

Figure A.3

Convergence curves for an inverse problem (from Vossinis [13])

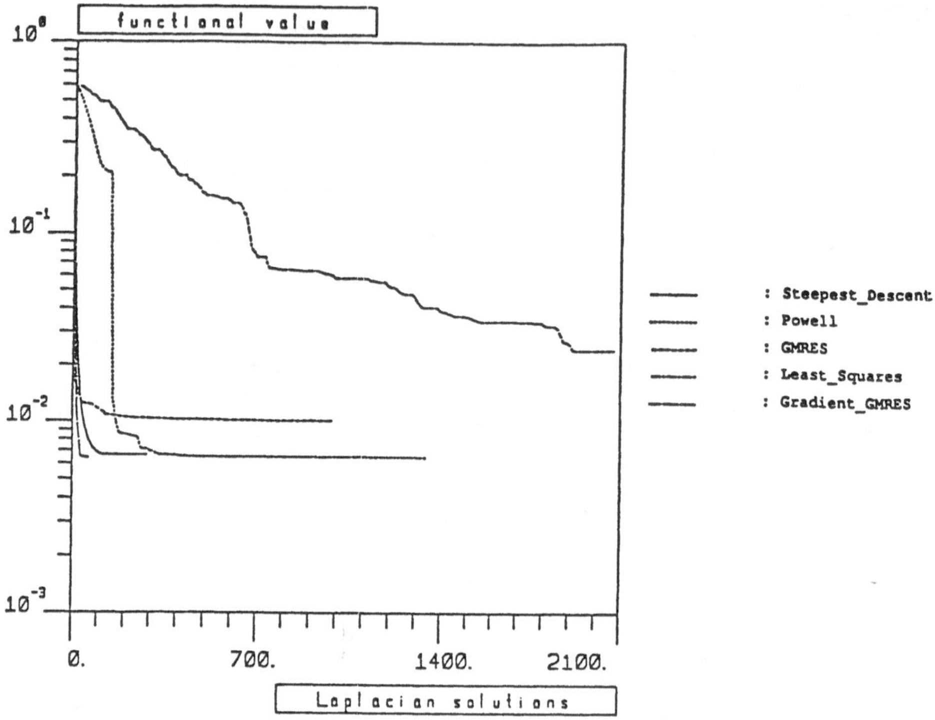

functional value

10^0

10^{-1}

10^{-2}

10^{-3}

0.   700.   1400.   2100.

Laplacian solutions

—————— : Steepest_Descent
—————— : Powell
—————— : GMRES
—————— : Least_Squares
—————— : Gradient_GMRES

Figure A.4

Convergence curves for a real (from Vossinis [13])

Figure A.5

Initial and final airfoil S^o, S^{30} (from Angrand [2]).

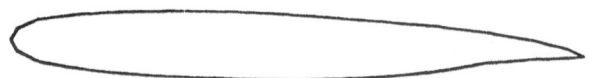

Figure A.6

Final airfoil S^{30} (from Angrand [2])

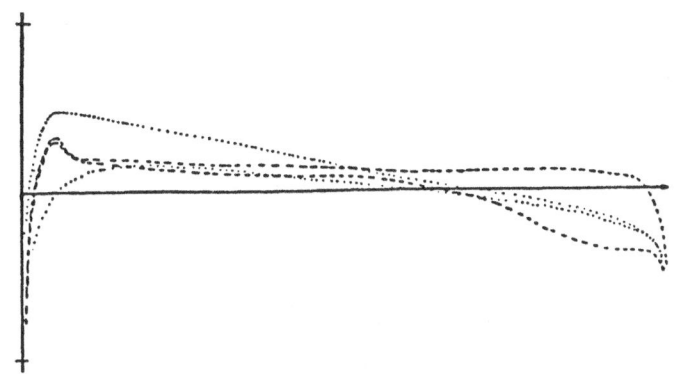

Figure A.7

Pressure distribution on the skin $S^o(\ldots)$ and S^{30} (__) (from Angrand [2])

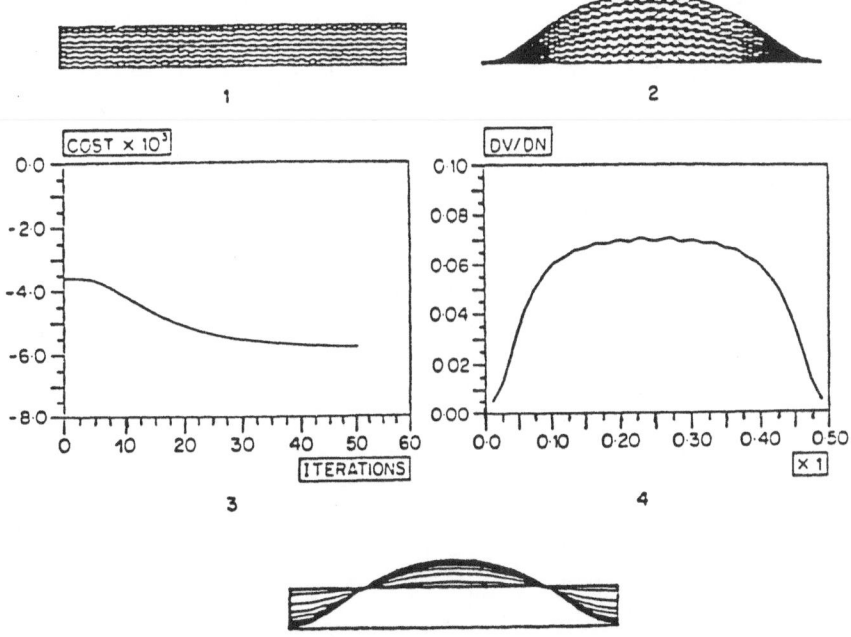

Results after 60 iterations starting from the domain shown in $1(L = 0.06,\ l = 1/2)$. The cost function $\int_\Omega |\nabla v|^2 dx$ has initial value $= 0.36 \cdot 10^{-4}$ and final value $= 0.58 \cdot 10^{-4}$. Figure 2 shows the shape of iteration 60; Figure 3 the values of the cost function at each iteration; Figure 4 the value of $\partial v/\partial n$; and Figure 5 shows some intermediate shapes.

Figure A.8

Optimization of riblets (from Arumugan [4])

Approximation and Localization of Attractors

Keith PROMISLOW[1]
Institute for Applied Mathematics and Scientific Computing
Indiana University–Bloomington
618 East Third Street
Bloomington, IN 47401
USA

Roger TEMAM
Institute for Applied Mathematics and Scientific Computing
Indiana University–Bloomington
618 East Third Street
Bloomington, IN 47401
USA

and

Laboratoire d'Analyse Numérique
Université de Pairs-Sud
F-91405 Orsay Cédex
France

Abstract

On overview of recent work on approximation of attractors as a model of turbulence for a class of dissipative evolution equations, including Navier-Stokes, Thermohydraulics, Combustion, the Bénard problem, Kuramoto-Sivashinsky, Cahn-Hilliard, and Ginzburg-Landau, is presented. This overview is comprised of motivational background on turbulent phenomena, existence results on absorbing sets and attractors, estimations of the dimensions of the attractors, sufficient conditions for the existence of Inertial Manifolds, the development of the theory of Approximate Inertial Manifolds, and the results of applications of this theory to various numerical schemes. We observe a connection with the theory of Slow Manifolds in Meteorology.

---

[1]This text was prepared by the first author based upon the lectures of the second author given at the Séminaire de mathématiques supérieures/NATO ASI at the Université de Montréal.

M. C. Delfour and G. Sabidussi (eds.), Shape Optimization and Free Boundaries, 253–285.
© 1992 *Kluwer Academic Publishers.*

Introduction

The focus of these lectures is the relationship between the attractors of a certain class of dissipative partial differential equations and turbulent behaviour as it has been observed in the physical systems which these equations model. Specifically we will deal with the Navier-Stokes, Thermohydraulics, Combustion, the Bénard problem, Kuramoto-Sivashinsky, Cahn-Hilliard, and Ginzburg-Landau equations; but our primary emphasis will be on the Navier-Stokes equation, and the turbulent fluid flows which it models.

The presentation is divided into two sections. In the first we investigate the idea of an absorbing set, typically a ball, in the phase space into which the orbits enter in a finite time and remain inside thereafter. This concept leads directly, through compactness arguments, to the attractor, which may be thought of as encompassing the long-term, permanent states of the system. In particular, the states which are thought of as turbulent are contained here. Thus to understand the structure of the attractor is to gain information about the possible nature of turbulence. This section concludes with estimates on the Hausdorff and Fractal dimensions of the attractor; in a sense the most basic localization information possible.

The second section presents an overview of the recent work on finer localizations and modelings of the attractor in which the authors have been actively involved. Specifically, the motivation for and the basic steps in the development of Inertial Manifolds and Approximate Inertial Manifolds is presented. The Inertial Manifold (IM) results from the embedding of the fractal attractor into a larger set: a smooth, invariant manifold of finite dimension which attracts the orbits at an exponential rate. Thus we exchange the attractor with its precise locating of the turbulent regimes, for a larger, yet still finite dimensional set which has better behaviour with respect to the flow of solutions. An Approximate Inertial Manifold (AIM), of order η, is a yet smoother and simpler manifold which attracts the orbits not onto itself but into a neighborhood of thickness η about itself; the localization is no longer finite dimensional, but the simplicity of the equation of the manifold makes it suitable for explicit calculations, including numerical schemes. The paper concludes with computational and theoretical results from the applications of AIMs to numerical analysis.

1   Turbulence and attractors

In this section we describe the nature of the physical aspects of turbulence and give some theoretical results of the approach of dynamical systems to the problem of turbulence. For a visual presentation of the onset of turbulence see Figure 1.1.

The material presented in this lecture applies in particular to the following mathematical models:

(i) *Equations of Fluid Dynamics: Incompressible Navier-Stokes in 2 or 3D:*

$$U = (U_1, U_2, U_3) \quad U = U(x, t), \quad p = p(x, t):$$

$$\partial U/\partial t + (U \circ \nabla)U - \nu \Delta U + \nabla p = f, \tag{1.1}$$

$$\text{div } U = 0.$$

(ii) *Thermohydraulics:*

U as above, $T = T(x, t)$: Equation (1.1) plus

$$\partial T/\partial t + (U \circ \nabla)T - \kappa \Delta T = f. \tag{1.2}$$

(iii) *Chemical Kinematics and Combustion:*

Navier-Stokes (1.1);

Equations for the concentration of the constituant chemicals $\Psi = \Psi(x, t)$:

$$\partial \Psi/\partial t + (U \circ \nabla)\Psi - \lambda \Delta \Psi = -\omega(T, \Psi); \tag{1.3}$$

Heat equation with source:

$$\partial T/\partial t + (U \circ \nabla)T - \kappa \Delta T = \omega(T, \Psi).$$

(iv) *Kuramoto-Sivashinsky Equations:*

$$\partial u/\partial t + \nu \partial^4 u/\partial x^4 + \partial^2 u/\partial x^2 + 1/2(\partial u/\partial x)^2 = 0. \tag{1.4a}$$

or, for $v = \partial u/\partial x$,

$$\partial v/\partial t + \nu \partial^4 v/\partial x^4 + \partial^2 v/\partial x^2 + v(\partial v/\partial x) = 0. \tag{1.4b}$$

(v) *Cahn-Hilliard:*

$$\partial u/\partial t + \Delta(\Delta u - g(u)) = 0 \tag{1.5}$$

with

$$g(u) = \alpha u^3 - \beta u \quad \alpha, \beta > 0.$$

(vi) *Ginzburg-Landau:*

$$\partial u/\partial t - (\lambda + i\alpha)\Delta u + (\kappa + i\beta)|u|^2 u - \gamma u = 0 \tag{1.6}$$

$$\lambda, \alpha, \beta, \gamma, \kappa \in \mathbb{R} \quad \text{with} \quad \lambda, \kappa > 0$$

$$u = u(x,t) \in \mathbb{C}, \quad i = \sqrt{-1}.$$

(vii) *Reaction-Diffusion:*

$$u' - d\Delta u + f(u) = 0 \tag{1.7}$$

with $d > 0$ and f a polynomial of odd degree with a positive leading coefficient.

We proceed now with the mathematical formulation of the general setting. In general we consider an evolution equation in an infinite dimensional Hilbert space H. That is, we have a function u and a time $T > 0$ satisfying,

$$u : [0,T] \to H$$

$$du/dt = F(u(t)), \quad 0 < t < T \tag{1.8}$$

$$u(0) = u_0$$

where F maps some part D of H into H; $F : D \subset H \to H$.

Let us consider examples from the previously given models:

(i) *Navier-Stokes:*

$H = \{u \in L^2(\Omega)^3 |\ \text{div}\ u = 0, u \ \text{satisfies certain boundary conditions}\}$

$u(t) = U(t) = \{x \in \Omega \to U(x,t)\}.$

(ii) *Thermohydraulics:*

$$H \subset L^2(\Omega)^4$$

$$u(t) = (U(t), T(t)).$$

(iii) *Combustion:*

$$H \subset L^2(\Omega)^{4+m}$$

where m is the number of constituant reactants,

$$u(t) = (U(t), T(t), \Psi(t)).$$

Now let us take a more specific but still sufficiently general mathematical formulation.

$$du/dt + Au + R(u) = 0, \tag{1.9}$$

$$u(0) = u_0.$$

Here $A : D(A) \subset H \to H$ is an unbounded, self-adjoint, symmetric, positive-definite linear operator. $R(u)$ is the non-linear operator, in the case of Navier-Stokes $R(u) = B(u, u) - f$ where B is a bilinear from and $f \in L^2(\Omega)$.

The first and most natural question to ask is that of the existence and uniqueness for the solutions of (1.9). In short, there is no general theorem but rather many particular theorems for the many different equations showing the respective existence, uniqueness, and regularity. Moreover, there are notable cases of nonresults: the existence of strong solutions of 3D Navier-Stokes has been an open problem for over 50 years.

The route by which we will examine turbulence in dynamical systems is through the study of solutions as the time t goes to infinity. An example to keep in mind is the flow of a fluid past a spherical object at different Reynolds numbers.

To study this type of behaviour in a more general setting let us rewrite the basic equation as

$$u'(t) = F_\mu(u(t)), \quad t > 0 \tag{1.10}$$

$$u(0) = u_0$$

where μ is a bifurcation parameter; in the case above, μ represents the fluid velocity at infinity or, equivalently, the corresponding Reynolds number. The basic cases, which may not all appear for a particular equation or which may appear in a more complicated sequential order, are:

(i) *Trivial dynamics* $0 < \mu < \mu_1$: There exists a unique stationary solution u_s, depending on μ, such that

$$F(u_s) = 0,$$

and

$$u(t) \to u_s \quad \text{as} \quad t \to \infty$$

for any choice of u_0.

(ii) *Bifurcation of stationary solution* $\mu_1 < \mu < \mu_2$: There exist several stationary solutions $u^1, u^2, \ldots, u^n$, such that

$$F(u^i) = 0 \quad \text{for} \quad i = 1, \ldots, n,$$

and

$$u(t) \to u^i \quad \text{as} \quad t \to \infty,$$

with i depending on u_0.

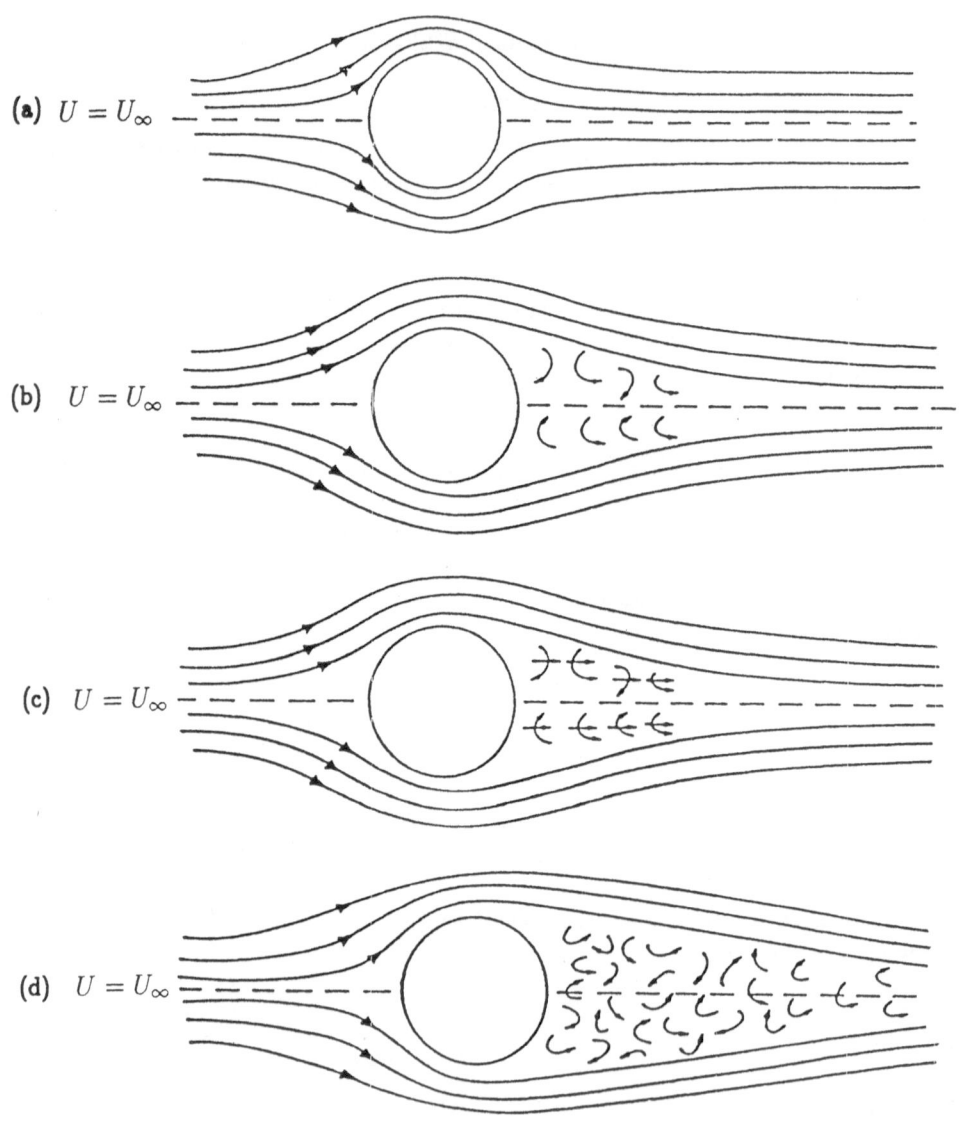

Figure 1.1

(a) Small fluid velocity and laminar flow – a stationary solution. (b) Larger fluid velocity, the flow experiences a bifurcation of the stationary solution to steady, laminar flows with a complicated region behind the sphere. (c) Yet larger velocity, Hopf bifurcations occur and time-dependent solutions appear as the vortices are continually moving to the right and disappearing. (d) Finally, at sufficiently high Reynolds number, there is fully turbulent wake.

(iii) *Hopf bifurcation* $\mu_2 < \mu < \mu_3$: There exists a function $\varphi : \mathbb{R} \to H$ and a time $T > 0$ such that

$$\varphi(t + T) = \varphi(t),$$

that is, φ is T-periodic,

$$\varphi'(t) = F_\mu(\varphi(t)),$$

and

$$|u(t) - \varphi(t)| \to 0 \quad \text{as} \quad t \to \infty.$$

(iv) *Quasi-periodic solutions* $\mu_3 < \mu < \mu_4$: This case is as in the periodic case except that

$$\varphi(t) = \Phi(t/T_1, \ldots, t/T_n),$$

where Φ is 2π-periodic with respect to each variable and the $T_i > 0$ are rationally independent. That is, the Fourier transform of φ has n rationally independent peaks.

(v) *General case* $\mu_4 < \mu$: There is a set $\mathcal{A} \subset H$, such that $\text{dist}(u(t), \mathcal{A}) \to 0$ as $t \to \infty$, see Figure 1.2. A more precise definition of $\mathcal{A}$, the attractor, is given below; note that the attractor is often a very complicated, fractal subset of H.

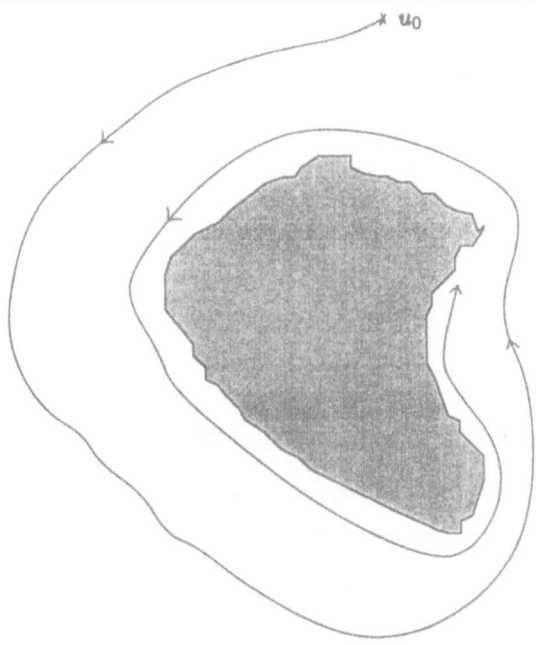

Figure 1.2

The orbit, starting at u_0, converges slowly towards the attractor.

The first tool which we will need for the study of the attractor is the semigroup $\{S(t)\}_{t \in \mathbb{R}_+}$ of the solution. This is the family of maps

$$S(t) : H \to H$$

$$S(t)u_0 = u(t) \quad t > 0,$$

which enjoy the property

$$S(t + s) = S(t)S(s) \quad t, s > 0$$

$$S(0) = I.$$

In general, $S(t)$ is injective, which implies backward uniqueness of the solutions; however $S(t)$ is not generally surjective, since $S(t)$ is a smoothing operator, $S(t)H$ is a "small" subset of H. The semigroup represents the effects of the flow on the points in the state space, similar to tracing out an orbit in a vector field diagram.

The semigroup leads us to the idea of the invariant set.

Definition $X \subset H$ is an *invariant set* for the semigroup $S(t)$ if $S(t)X = X$ for all $t > 0$.

Examples of invariant sets

- Stationary solutions, $X = \{u_s\}$ or $X = \{u^1, u^2, \ldots, u^n\}$.

- The orbit of a periodic solution φ,

$$X = \{\varphi(t) : t \in [0, T]\}.$$

- A complete orbit of a solution u,

$$X = \{u(t) : t \in \mathbb{R}\}.$$

- The stable ($+$) and unstable ($-$) manifolds of a stationary point

$$\mathcal{V}_+(u_s) = \{u_* \text{ that belong to a complete orbit } |S(t)u_* \to u_s \text{ as } t \to \infty\}$$

$$\mathcal{V}_-(u_s) = \{u_* \text{ that belong to a complete orbit } |S(t)u_* \to u_s \text{ as } t \to -\infty\}.$$

- A homoclinic or heteroclinic curve

  - a homoclinic curve is the coincidence of an unstable manifold of a stationary point with a stable maniflod of the same point;

  - a heteroclinic curve is the coincidence of an unstable manifold of a stationary point with a stable manifold of a different stationary point.

An important concept in the study of dissipative evolution equations is the absorbing set. An *absorbing set* is a bounded set $\mathcal{B}$ such that for any bounded set $B \subset H$, there exists a time $t_0 = t_0(B)$ such that

$$S(t)B \subset \mathcal{B} \quad \text{for all } t \geq t_0.$$

The existence of an absorbing set is related to the dissipativity of the dynamical system; indeed one says that the system dissipates the bounded sets. The importance of an absorbing set lies in its localization of the long-term dynamics of the system, especially if the set $\mathcal{B}$ is a compact subset of H.

Next we define the ω-limit set of the semigroup $S(t)$. Given $u_0 \in H$,

$$\omega(u_0) \equiv \bigcap_{t \geq 0} \overline{\bigcup_{\tau > t} S(\tau)u_0}$$

More generally, if $X \subset H$,

$$\omega(X) \equiv \bigcap_{t \geq 0} \overline{\bigcup_{\tau > t} S(\tau)X}$$

Note that the ω-limit set of X, which could be empty a priori, represents the image of the set X under the flow as $t \to \infty$; the points which $S(t)X$ comes "close to" infinitely often. From its definition, $\omega(X)$ is invariant for the semigroup $S(t)$.

An *attractor* is a set $\mathcal{A} \subset H$ such that

(i) $\mathcal{A}$ is an invariant set, i.e. $S(t)\mathcal{A} = \mathcal{A}$ for all $t > 0$;

(ii) $\mathcal{A}$ possesses an open neighborhood U such that, for every $u_0 \in U, S(t)u_0 \to \mathcal{A}$ as $t \to \infty$: dist$(S(t)u_0, \mathcal{A}) \to 0$ as $t \to \infty$.

The largest open set $U \supseteq \mathcal{A}$ which satisfies (ii) is called the *basin of attraction* of $\mathcal{A}$. We say that $\mathcal{A}$ *uniformly attracts* a set $B \subset U$ if dist$(S(t)B, \mathcal{A}) \to 0$ as $t \to \infty$. Finally, the *global* or *maximal attractor* is a compact attractor whose basin of attraction is H and which uniformly attracts the bounded sets of H. Note that the global attractor is unique if it exists.

We now give an illustrative example of an attractor. Consider the Minea system, introduced in Minea (1976), a two or three dimensional ODE,

$$H = \mathbb{R}^{2 \text{ or } 3} \quad u = u(x, y, z)$$

$$
\begin{aligned}
x' + x + \delta(y^2 + (z^2)) &= 1 \\
y' + y - \delta xy &= 0 \\
(z' + z - \delta xz &= 0).
\end{aligned}
$$

In this system $\delta > 0$, plays the role of the bifurcation parameter or Reynolds number. The fixed points of the two dimensional version of this system are, for $\delta > 1$,

$$\{x = 1, y = 0\} \quad \text{and} \quad \{x = 1/\delta, y^2 = (\delta - 1)/\delta^2\},$$

and the attractor in this case is described in the legend of Fig. 1.3.

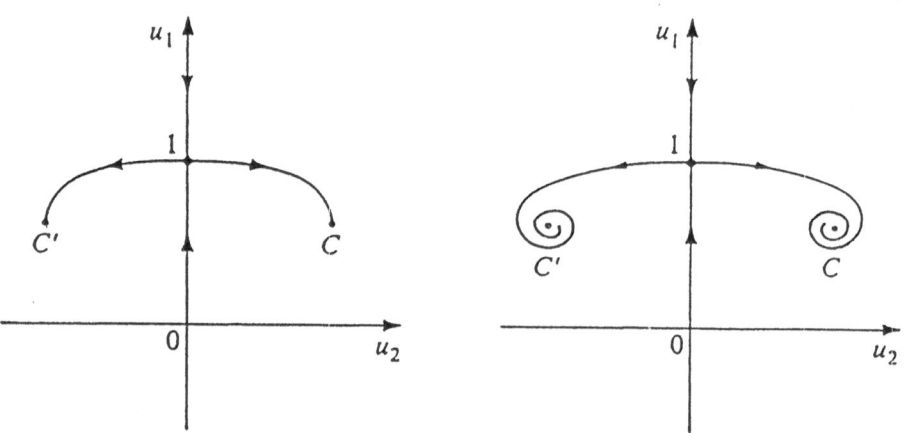

Figure 1.3

(a) $1 < \delta < 9/8$, the global attractor is composed of the two heteroclinic curves which connect the stationary points 1 to C, and 1 to C'. (b) $9/8 < \delta$, the attractor has become more complicated, the curves are bounded away from C and C'; but instead approach ellipses encompasing C and C', points inside these ellipses spiral away from C and C' and towards the ellipses. The size of the ellipses increases with d. The global attractor consists of the two ellipses and their interiors as well as the unstable manifolds of the point 1 which approach the ellipses. (c) $0 < \delta < 1$, not shown here, the only fixed point is $\{x = 1, y = 0\}$ which is a stationary solution and the global attractor.

A remark about computation: recalling the previous discussion of possible long time dynamics of a dissipative system, as well as the example above, if one is in a trivial case where there are only one or several stationary solutions which attract all the orbits, then computation is relatively simple. However, with higher values of bifurcation parameters it it important to realize the fundamental qualitative changes taking place in the dynamics:

- permanent regimes may not correspond to stationary solutions;

- delicate dependence on initial data;

- many basins of attraction.

For example, in the Minea system in the case (a), if the initial data is chosen with a small positive y value and a large positive x value, then the orbit will move down along the x-axis until it reaches the vicinity of the point $(1,0)$, where it will remain for an arbitrarily long period of time before moving along the curve towards C'. Thus one could be very easily mislead into thinking that the orbit converges to the stationary solution $(1,0)$. All of these effects require greater computer precision, hence increased computer time.

We now pass to the problem of the existence of the global attractor. We will provide sufficient conditions for the existence of a global attractor for a class of dissipative evolution equations and show that these conditions are satisfied by the Navier-Stokes equation.

Theorem 1 *Assuming that*

(1) *there exists an absorbing set B for the semigroup;*

(2) *the $S(t)$ are uniformly compact for t large, that is, for all bounded B there exists a time $t_0(B)$ such that*

$$\bigcup_{t \geq t_0} S(t)B$$

is relatively compact.

Then the global attractor exists for $S(t)$.

Theorem 2 *Assume that the semigroup has the form*

$$S(t) = S_1(t) + S_2(t),$$

where $S_1(t)$ satifies (1) and (2) of Theorem 1 and $S_2(t)$ converges to zero exponentially as $t \to \infty$. Then the global attractor exists for $S(t)$.

We will now show that Theorem 1 holds for the following class of Navier-Stokes type equations:

$$u' + Au + Ru = 0 \tag{1.11}$$
$$Ru = B(u, u) - f.$$

Hypotheses on A

A is an unbounded, seft-adjoint, positive definite, linear operator on H, with domain $D(A) \subset H$ and has a compact inverse. Thus A has a complete set of eigenvectors $\{w_j\}$ such that

$$Aw_j = \lambda_j \quad 0 < \lambda_1 \leq \lambda_2 \leq \ldots \quad \text{and} \quad \lambda_j \to \infty \text{ as } j \to \infty.$$

We define the powers of A, A^s with domain $D(A^s)$ and inner product $(A^{s/2}\circ, A^{s/2}\circ)$. The space V is defined as $D(A^{1/2})$ and its inner product will be denoted by $((\circ, \circ))$.

Hypotheses on B

The bilinear form B satisfies the following relations:

$$(B(u,v),v) = 0 \qquad \text{for all} \quad u,v \in V \tag{1.12}$$

$$|(B(u,v),w)| \le c_1 |u|^{1/2}||u||^{1/2}||v|| \, |w|^{1/2}||w||^{1/2} \quad \text{for all} \ \ u,v,w \in V$$

$$|(B(u,v)|_{L^2} \le c_2 |u|^{1/2}||u||^{1/2}||v||^{1/2}|Av|^{1/2} \qquad \text{for all} \ \ u \in V, v \in D(A) \tag{1.13}$$

$$|(B(u,v)|_{L^2} \le c_3 |u|^{1/2}|Au|^{1/2}||v|| \qquad \text{for all} \ \ u \in D(A), v \in V$$

We now show the existence of an absorbing set in H, thus verifing assumption (1) of Theorem 1. Take the inner product of equation (1.11) with u, yielding,

$$\frac{1}{2}(d/dt)|u|^2 + \nu||u||^2 = (f,u) \le |f||u|$$

$$\le |f| \, ||u||/\lambda_1^{1/2}$$

$$\le |f|^2/2\nu\lambda_1 + \nu/2||u||^2$$

$$(d/dt)|u|^2 + \nu||u||^2 \le |f|^2/\nu\lambda_1$$

$$(d/dt)|u|^2 + \nu\lambda_1|u|^2 \le |f|^2/\nu\lambda_1$$

$$|u(t)|^2 \le |u(0)|^2 \exp(-\nu\lambda_1 t) + (|f|/\nu\lambda_1)^2(1 - \exp(-\nu\lambda_1 t)).$$

Taking $\rho_0 = (|f|/\nu\lambda_1)^2$ and fixing $\rho \ge \rho_0$, we have

$$|u(t)|^2 \le |u_0|^2 \exp(-\nu\lambda_1 t) + (\rho_0)^2).$$

If $u_0 \in B \subset B_H(0,R)$, then for $t \ge t_0(B) \equiv 1/\nu\lambda_1 \log(R^2/(\rho^2 - \rho_0^2))$ we have that $u(t) \in B_H(0,\rho)$. That is, $B_H(0,\rho)$ is an absorbing set in H.

Let us continue by showing the existence of an absorbing set in V. To this end, take the V inner product of (1.11) with u to obtain

$$\frac{1}{2}(d/dt)||u||^2 + \nu|Au|^2 = (f,Au) + B(u,u,Au)$$

$$(d/dt)||u||^2 + \nu|Au|^2 \le (2/\nu)|f|^2 + 2c/\nu^3|u|^2||u||^4.$$

With the uniform Gronwall lemma we obtain

$$||u(t)||^2 \le (a_3/r + a_2)\exp(a_1) \quad \text{for} \ \ t \ge t_0(B) + r, \quad r \ \text{any positive number}.$$

Thus we have an absorbing set in V, and by the compact injection of V into H we have that the $S(t)$ are uniformly compact for t larger than $t_0(B) + r$; hence Theorem 1 applies. Note that for the 3D Navier-Stokes the existence of an absorbing ball in V is still open. It was shown in Constantin-Foias-Temam (1985) that this is equivalent to the well-posedness of the 3D Navier-Stokes equations.

A natural and important question, having shown the existence of a global attractor, is "Where is it?" or, less specifically, "How big is it?" An answer to the first, and more difficult, question is the topic of the second part of this lecture. We obtain a partial answer to the second question by calculating upper and lower bounds on the Hausdorff dimension of the attractor.

In the calculation of upper bounds on the Hausdorff dimension of a global attractor, the method of Lyapunov exponents, also presented by Constantin-Foias-Temam (1985), is used. Here the idea is to examine the evolution of an m-dimensional volume element under the semigroup $S(t)$. Then, using Lieb-Thirring inequalities, uniform bounds are obtained on the limiting volume of any m dimensional element. Typically an estimate m_0 is obtained such that for $m > m_0$ any m-dimensional volume decays exponentially to zero under the flow $S(t)$ as $t \to \infty$. This m_0 is an upper bound on the dimension of the attractor.

Let us take as an example the 2D Navier-Stokes equation. Denote by G the nondimensional Grashof number

$$G = \frac{|f|_{L^2}}{\nu^2 \lambda_1},$$

then the Reynolds number Re has the form

$$Re = \sqrt{G} = \frac{|f|_{L^2}^{1/2}}{\nu \lambda_1^{1/2}}$$

and it has been shown, Constantin-Foias-Temam (1988), that in the case of periodic boundary conditions, the Hausdorff dimension of the global attractor $\mathcal{A}$ is bounded like

$$\dim_H \mathcal{A} \le c G^{2/3}.$$

Similar results have been obtained for a wide variety of equations, such as those mentioned at the beginning of this lecture; moreover the results obtained by this method are physically relevant and in some cases are optimal.

One way to obtain lower bounds on the dimension of the attractor is to look at the dimension of the unstable manifolds about the stationary solutions. The attractor must contain these manifolds, hence they provide a lower bound on its dimension. As an illustrative example we will again take the Navier-Stokes equation in two space dimensions, but this time in the Babin-Vishik formulation. Here one considers periodic boundary conditions on a 2D rectangle of dimensions $2\pi L$ and $2\pi L/\alpha$, where $0 < \alpha \le 1$. It has been shown, Ghidaglia-Temam (1990), that this problem has a stationary solution with approximately c/α unstable eigenvalues. That is,

$$\dim \mathcal{V}_-(u_s) \geq c/\alpha,$$

implying that

$$\dim \mathcal{A} \geq c(\nu, f)/\alpha.$$

In addition to the above, Kolmogorov's Law, see Landau-Lifschitz (1953), about the nature of the Fourier spectrum for solutions of the 2 or 3 dimensional Navier-Stokes equations provides an independent calculation for the dimension of the attractor for this equation. We take the physical space to be $(0, 2\pi)^3$ and write the solution u in terms of its Fourier expansion,

$$u(x,t) = \sum_{k \in \mathbb{Z}^3} \hat{u}(k,t)e^{ikx} \quad \hat{u}(-k,t) = \overline{\hat{u}}(k,t).$$

The energy contained in the kth frequency is

$$E(k,t) = c \sum_{|k|=k} |\hat{u}(k,t)|^2 k^2$$

which, by Kolmogorov's law, decays like $k^{-5/3}$ for k large. Moreover, we define the inertial range to be the wave numbers between k_0, the lowest wave number, and k_d, the cutoff wave number defined below. The spectrum beyond k_d is called the dissipative range since the energy here is quickly dissipated by viscosity effects.

From this information we may conclude that

(i) $\hat{u}(k,t)$ is negligible for $|k| = k > k_d$ where the cutoff wave number is given by

$$k_d = (\nu^3/\varepsilon)^{-1/4}, \quad \text{where } \varepsilon = \nu \langle \langle |\text{grad } u(x,t)|^2 \rangle \rangle;$$

(ii) $E(k,t) = ck^{-5/3}$ for $k_0 < k < k_d$, the inertial range.

Since the dimension of the attractor can be no larger than the number of modes with significant energy, we can use these facts to estimate its dimension. This number of modes is the same as the number of 3-tuples $k = (k_1, k_2, k_3)$ such that $|k|$ is in the inertial range. This number is approximately

$$(k_d/k_0)^n,$$

where $n = 2$ or 3, depending on the space dimension.

It has been shown in Constantin-Foias-Manley-Temam (1985) that this is *exactly* the dimension of the attractor, and that it agrees with the previously given estimates. Further improvements in space dimension $n = 2$, using Kraichman's theory of two dimensional turbulence, Kraichman (1967), appear in Constantin-Foais-Temam (1988) and Ghidaglia-Temam (1990).

2 Inertial manifolds and their approximation

2.1 Inertial manifolds

Let us reconsider the problem of the flow of a fluid around a spherical obstacle, but from a phenomenological aspect. The flow can be considered as a superposition of large structures or eddies and small structures. The important point in this view of turbulence is that the large structures are seen as carrying the smaller ones. As the large structures break down or decompose, there is a net flow of energy into the smaller structures, thus determining their form. From a mathematical point of view the large structures are viewed as stable and reproducible while the small structures are viewed as unstable and unreproducible.

The ideas above, as well as the finite dimensionality of the attractor, have lead to the posing and partial resolution of the following questions:

- Can one embed the attractor in a finite dimensional manifold?

- If not, can one approximate the attractor by smooth, finite-dimensional manifolds?

- In this last case, can the dynamics on the attractor be approximated uniformly in time by the dynamics on the finite dimensional, approximate manifold?

Definition An *inertial manifold* (IM) for the semigroup $\{S(t)\}_{t>0}$ is a Lipschitz manifold $\mathcal{M}$, positively invariant for the flow, which attracts all the orbits at an exponential rate. That is,

(i) $S(t)\mathcal{M} \subset \mathcal{M}$, for all $t > 0$,

(ii) $\mathrm{dist}(S(t)u_0, \mathcal{M}) \leq Ke^{-ct}$, where c and K depend boundedly on u_0.

Remarks (i) $\mathcal{M}$ is not unique, indeed it can be modified arbitrarily outside the absorbing set.

(ii) If $\mathcal{A}$ and $\mathcal{M}$ exist, then $\mathcal{A} \subset \mathcal{M}$.

(iii) $\mathcal{M}$ is a global analogue of the central manifold.

Here is a comparison of the attractor and the inertial manifold

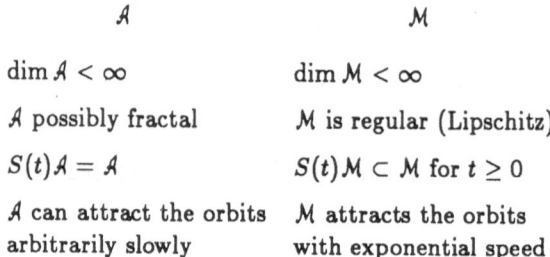

$\mathcal{A}$	$\mathcal{M}$
dim $\mathcal{A} < \infty$	dim $\mathcal{M} < \infty$
$\mathcal{A}$ possibly fractal	$\mathcal{M}$ is regular (Lipschitz)
$S(t)\mathcal{A} = \mathcal{A}$	$S(t)\mathcal{M} \subset \mathcal{M}$ for $t \geq 0$
$\mathcal{A}$ can attract the orbits arbitrarily slowly	$\mathcal{M}$ attracts the orbits with exponential speed

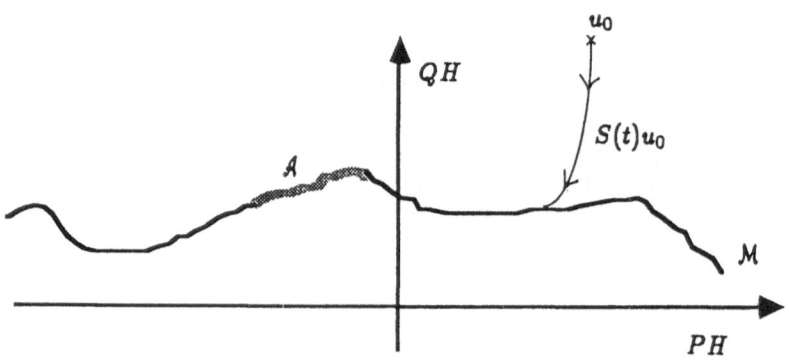

Figure 2.1

The orbit $S(t)u_0$, converges exponentially towards the inertial manifold and then towards the attractor by a path close to the inertial manifold (and included on the manifold for all practical purposes).

An overview of construction of an inertial manifold

- We have already introduced the eigenvalues of A:

$$Aw_j = w_j, \quad 0 < \lambda_1 \leq \lambda_2 \leq \ldots, \quad \lambda_j \to \infty \text{ as } j \to \infty.$$

The w_j's are an orthonormal basis of H, and $u \in H$ implies that

$$u = \sum_{i=1}^{\infty} a_i w_i.$$

- Define $P_m : H \to P_m H = \operatorname{span} \{w_1, \dots, w_m\}$

$$P_m u = \sum_{i=1}^{\infty} a_i w_i \quad Q_m = I - P_m$$

$$u = y_m + z_m, \quad y_m = P_m u, \quad z_m = Q_m.$$

- By projection of our general equation (1.9) we obtain

$$y_m' + A y_m + P_m R(y_m + z_m) = 0$$
$$z_m' + A z_m + Q_m R(y_m + z_m) = 0. \tag{2.1}$$

- We look for the manifold $\mathcal{M}$ as the graph of a function

$$z = \Phi(y)$$

$$\Phi : y \in P_m H \to z \in Q_m H.$$

The function Φ produces a model of turbulence, a law of enslavement of the small wave-lengths to the large, see Haken (1983):

$$\hat{u}(k, t) = \Phi_k(\hat{u}(1, t), \hat{u}(2, t), \dots, \hat{u}(m, t)) \quad k = m + 1, \dots .$$

Overview of existence theorems for inertial manifolds

There are several different approaches by different authors to show the existence of an inertial manifold. The one which we will highlight is that by Foias-Sell-Temam (1985). This demonstration uses a fixed point method, constructing a function $\mathcal{T}$ which acts on maps from PH into QH and has as its fixed point the function Φ whose graph is the inertial manifold. The difficulties of the proof lie in showing that $\mathcal{T}$ is a strict contraction. For this, several hypotheses are required; these or similar hypotheses are common to all the existence proofs yet developed.

Hypotheses for existence:

(i) Several technical hypotheses on the nonlinear term, regularity of the solutions, existence of absorbing ball and global attractor.

(ii) A requirement that m, the dimension of the space $P_m H$, be large enough.

(iii) The spectral gap condition, the most restrictive hypothesis, requiring the existence of gaps in the spectrum of the linear operator A, namely

$$\lambda_{m+1} - \lambda_m \geq K(\lambda_{m+1}^{1/2} + \lambda_m^{1/2}).$$

A skeletal list of references for this theory contains the following works: Debussche (1989), Constantin-Foias-Nicolaenko-Temam (1989), Demengel-Ghidaglia (1990), Foias-Sell-Temam (1985 and 1988), Mallet-Paret and Sell (1988), Mañé (1984), Marion (1990), and Temam (1988).

Examples

(i) *Kuramoto-Sivashinsky* (1.4a *or* b). One of the most difficult problems for this equation is to show the dissipativity, i.e., the existence of an absorbing set. This was established for odd solutions (u odd or v even) by Nicolaenko-Scheurer-Temam (1985 and 1986). However, Il'yashenko (1990) has recently proved this result for general solutions of the Kuramoto-Sivashinsky equation. The following discussion applies to odd solutions.

For this equation the spectral gap condition is satisfied; the eigenvalues of the linear operator $A = -\partial^4/\partial x^4$ in the periodic case behave like

$$\lambda_m \sim cm^4, \tag{2.2}$$

so that, loosely speaking, we have

$$\lambda_{m+1} - \lambda_m \sim m^3,$$

and

$$c(\lambda_{m+1}^{1/2} + \lambda_m^{1/2}) \sim m^2.$$

Hence for m large enough,

$$\lambda_{m+1} - \lambda_m \geq c(\lambda_{m+1}^{1/2} + \lambda_m^{1/2}).$$

Hence one can show, using (2.2), that there exist arbitrarily large m's such that the spectral gap condition and (ii) are satisfied. Thus, there exists an inertial manifold for the Kuramoto-Sivashinsky equation (1.4).

(ii) *Reaction-diffusion equations* (1.7). In space dimension n, the eigenvalues of the linear operator $A = -\Delta$ with periodic boundary conditions have the form,

$$\nu(m_1^2 + \cdots + m_n^2) \quad \text{for } m_1, \ldots, m_n \in \mathbb{Z}_+.$$

However, because of the nature of the non-linear term, the spectral gap condition reduces to

$$\lambda_{m+1} - \lambda_m \geq K,$$

where K is a positive constant.

So we must find sufficiently large gaps in the spectrum. In space dimension one this is obvious since the eigenvalues are simply the squares of the integers; in space dimension two this result follows from number theory, see for example Hardy and Wright (1962).

On the contrary, for $n \geq 4$, any number can be written as the sum of n squares; thus the result does not hold in general. In summary, for space dimensions 1 and 2 there exists an inertial manifold for the Reaction-Diffusion equations (1.7).

(iii) *Navier-Stokes equation in* 2D (1.1). Here the spectrum of the linear operator $A = -\nu\Delta$ behaves like

$$\lambda_m = cm,$$

and the spectral gap condition cannot be verified with the methods used above. The existence of an inertial manifold for the Navier-Stokes equation has *not* yet been demonstrated, nor has it been ruled out.

Slow manifolds

In meteorology, an idea related to inertial manifolds, the *slow manifold*, has developed independently (and previously). In the framework of short term weather forecasting, one takes the equation

$$u' + Au + R(u) = 0$$
$$u = \{U = \text{ velocity of air}, \ T = \text{ temperature of air}, \ldots\}$$

and assumes that the long term dynamics are "somewhat" stationary; that is, they are close to a stationary solution u_s such that $Au_s + R(u_s) = 0$. Now, given such a function u_s, the equation is rewritten in terms of $v = u - u_s$,

$$v' + Av + R(v + u_s) - R(u_s) = 0$$

or

$$v' + Av + R'(u_s)v + \{R(v + u_s) - R(u_s) - R'(u_s)v\} = 0. \tag{2.3}$$

The physical assumption in that a permanent regime has been reached so that the observed orbits of (2.3) lie on its attractor and in particular on its slow manifold S, a sort of inertial manifold for (2.3), see Tribbia (1979) and Phillips (1981). A precise study of the connection between slow manifolds and inertial manifolds will appear elsewhere, see Debussche-Temam (1991).

To take advantage of this manifold in the short term, algorithms have been developed to take a measured initial condition u_0, usually containing considerable error, and project it onto the slow manifold S, obtaining a new initial value $Pu_0 = \hat{u}_0$. Numerical schemes which employ this method have given better short time weather prediction and are commonly used in weather prediction world-wide.

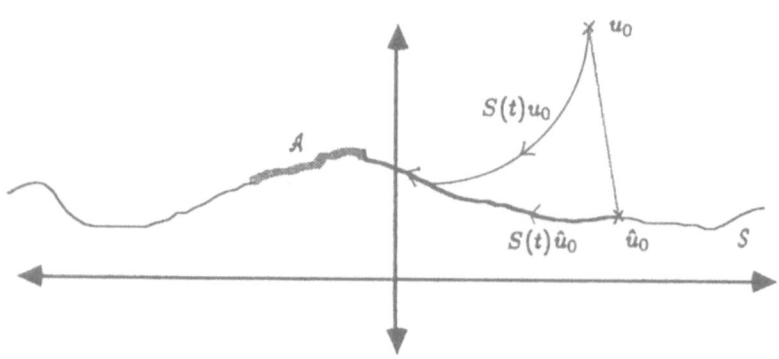

Figure 2.2

The initial value u_0 is projected onto the slow manifold, resulting in better short-time prediction.

2.2   Approximation of inertial manifolds

In the numerical calculation of turbulent flows, there are several inherent difficulties.

Difficulty 1   The size of the computation.

The number of modes necessary for a complete description of a 3D turbulent flow is equal to the dimension of its attractor, i.e.,

$$(k_d/k_0)^3 \approx (Re)^{3/4} \approx (10^6)^{9/4} \approx 10^{13.5}.$$

For a Cray II super computer, central memory is of the order 2.5×10^6. In 2D one has

$$(k_d/k_0)^3 \approx Re \approx 10^6;$$

thus one can now begin to compute full 2D turbulent flows.

Difficulty 2   Large time computations.

Given the equation

$$u' = F(u(t))$$
$$u(0) = u_0.$$

For $0 < t < 1/\varepsilon$, let $\tau = \varepsilon t$. Thus for $0 < \tau < 1$, $\bar{u}(\tau/\varepsilon)$ satisfies

$$d\bar{u}/dt(\tau) = 1/\varepsilon F(u(\tau))$$
$$\bar{u}(0) = u_0.$$

This is a stiff equation; the large time problem has difficulties similar to that for a stiff equation. Moreover, if we make the following classical discretization in time,

$$u(t) \in H \qquad \rightarrow \quad u_n(t) \in \mathbb{R}^n$$

$$u'(t) = F(u(t)) \quad \rightarrow \quad (u^{n+1} - u^n)/\Delta t = F_n(u^{n+1})$$

$$u(0) = u_0 \qquad \rightarrow \quad u^0 = u_0$$

then we obtain error estimates like

$$||u(t) - u_n(t)|| \leq ch^p \quad \text{for } 0 \leq t \leq T$$

but

$$c = c(T) \sim c_1 \exp(c_2 T).$$

Hence error bounds explode for large t–either the algorithm is not efficient or the numerical analysis is not tight.

As an attempt to avoid these difficulties, the concept of *approximate inertial manifold* (AIM) was introduced. An AIM of order η is a smooth, finite dimensional manifold which attracts all the orbits into a neighborhood of thickness η about the manifold, in finite time and with exponential speed. AIMs are derived from approximate interaction laws between the small and large structures. Whereas the exact interaction law Φ for the Inertial Manifold, when it exits, is obtained by nonconstructive techniques, these approximate interaction laws are simpler, explicit, and are computationally viable. However, since they are approximate, their manifolds attract the orbits not onto themselves but into a thin neighborhood.

Construction of AIMs

As in the construction of Inertial Manifolds, the solution u is partitioned into large and small length scales

$$u(x,t) = y(x,t) + z(x,t)$$

$$y \in PH \quad \text{and} \quad z \in QH.$$

The general system is then projected to obtain (2.1). However there are two important differences in the AIM case; first, there are a variety of choices for the projection P.

(i) *Fourier/Spectral method.* This is the decomposition used in the Inertial Manifold case. Given a complete, orthonormal family of eigenvectors $\{w_j\}$ of the linear operator

A, the solution u is written as an infinite sum and the projection, P, is onto the span of the first m eigenvectors.

(ii) *Finite elements/Hierarchial bases.*   The finite elements decomposition is based upon triangulations T of the physical space $\Omega \subset \mathbb{R}^n, n = 2$ or 3 typically. Given a triangulation T_1 of Ω, a subspace W_1 of $H = H_0^1(\Omega)$ is defined to be functions of prescribed regularity which are piecewise polynomial, of degree at most m, on the triangular components of T_1 and are zero on ∂T_1. Given a finer triangulation $T_2 \supseteq T_1$, we define a subspace W_2 of $W_1^{\perp}$ as the functions of prescribed regularity which are piecewise polynomial, of degree at most m, on the triangular components of T_2 and are zero on the vertices of the triangles of T_1. Thus the projection P has range W_1 and its orthogonal complement, Q, is approximated by a projection onto W_2.

There are also decompositions based upon the theories of:

(iii) *Finite differences and incremental unknowns.*

(iv) *Wavelets.*

The second difference between the methods is that the manifolds are not restricted to being graphs above PH,

$$z = \Phi(y),$$

but can also be analytic or algebraic sets

$$\Phi(y, z) = 0.$$

Hence the attractor belongs respectively to the set,

$$|z - \Phi(y)| < \eta$$

or

$$|\Phi(y, z)| < \eta.$$

The essential idea in the construction of AIMs is that after a short transition time, independent of the initial data,

$$z \ll y.$$

Then, by asymptotic expansion, using $|z|/|y|$ as the small parameter, we construct the manifolds by a series of successively finer approximations. Note that there is another small parameter, $1/Re$, which is not the proper one for the expansion.

First order approximation

We consider the Navier-Stokes equation (1.10), with the bilinear term B. Since we assume $z \ll y$, we have that

$$|B(z, z)| \ll |B(z, y)| \ll |B(y, y)|.$$

Also, for m large,

$$z' \approx z \ll Az \sim \lambda_{m+1} z.$$

Hence, from the equation

$$z' + Az + QB(y + z, y + z) = Qf,$$

we may, as a first approximation, drop the z' term and the B terms involving z, to find

$$Az + QB(y, y) = Qf$$

$$z = \Phi(y) = A^{-1}(Qf - QB(y, y)).$$

It has been shown in Foias-Manley-Temam (1988) that this Φ is an AIM of the 2D Navier-Stokes of the order η,

$$\eta = (\lambda_1/\lambda_{m+1})^{3/2} \log(\lambda_{m+1}/\lambda_1).$$

From here, in Temam (1989), this idea was extended to obtain an infinite sequence of AIMs, $\mathcal{M}_j$, of increasingly finer order, η_j, where

$$\eta_j \leq \delta^{j/2+1} L^{j/2+1/2}$$

$$\delta = (\lambda_1/\lambda_{m+1}), \quad L = 1 - \log(\delta).$$

So far we have defined the following AIMs:

$$z_m^0 = \Phi_{m,0}(y) \equiv 0$$

$$z_m^1 = \Phi_{m,1}(y) \equiv A^{-1}(Q_m f - Q_m B(y, y)).$$

We seek to form AIMs of order η_2 and η_3. In this case we again neglect the time derivative, but not the first order terms in z,

$$Az_m^2 + Q_m\{B(y, y) + B(y, z_m^1) + B(z_m^1, y)\} = Q_m f.$$

Thus,

$$z_m^2 = \Phi_{m,2}(y) \equiv A^{-1} Q_m\{f - B(y, y) - B(y, z_m^1) - B(z_m^1, y)\}.$$

For the third approximation we arrive at the need to treat the time derivative in a nontrivial manner. In this case we use the equation for the Φ_1 manifold,

$$Az_m^1 + Q_m B(y, y) = Q_m f$$

and take time derivatives,

$$d/dt(z_m^1) = A^{-1} Q_m\{B(y, y') + B(y', y)\}$$

and use

$$y' = -Ay - P_m B(y_m + z_m^1) + P_m f.$$

Thus we obtain an approximation for z' (call it $\{z'\}$) and the equation which defines z_m^3 and Φ_3

$$\{z_m'\} + Az_m^3 + Q\{B(y,y) + B(y,z_m^2) + B(z_m^2,y) + B(z_m^2,z_m^2)\} = Q_m f.$$

Let us now verify our first assumption, that $z \ll y$ for λ_m large. We write the projected equations, omitting the subscript m to simplify the notation,

$$y' + \nu Ay + PB(y + z, y + z) = Pf$$

$$z' + \nu Az + QB(y + z, y + z) = Qf$$

and take the inner product of the second equation with z to obtain

$$\frac{1}{2}(d/dt)|z|^2 + \nu||z||^2 = (f,z) - (B(y + z, y), z) - (B(y + z, z), z).$$

The last term in this equation, $(B(y + z, z), z)$, is zero by the orthogonality property (1.12). Writing $y + z = u$, we obtain

$$\frac{1}{2}(d/dt)|z|^2 + \nu||z||^2 \le |f||z| + |B((u,y),z)|.$$

Now we use the inequality (1.13) and fact that if

$$z = \sum_{j=m+1}^{\infty} a_j w_j(x) \quad \text{then} \quad ||z||^2 = \sum_{j=m+1}^{\infty} \lambda_j |a_j|^2$$

so

$$||z||^2 \ge \lambda_{m+1} \sum_{j=m+1}^{\infty} |a_j|^2 = \lambda_{m+1}|z|^2,$$

so that, writing Λ for λ_{m+1}, we obtain

$$\frac{1}{2}(d/dt)|z|^2 + \nu||z||^2 \le |f|\Lambda^{-1/2}||z|| + c|u| \, ||u|| \, ||z||$$
$$\le (|f|\Lambda^{-1/2} + c|u| \, ||u||)||z||$$
$$\le \nu/2||z||^2 + 1/2\nu(|f|\Lambda^{-1/2} + c|u| \, ||u||)^2$$

But we have shown that $|u|$ and $||u||$ are bounded for t large enough. Thus

$$d/dt|z|^2 + \nu||z||^2 \le 1/\nu(|f|\Lambda^{-1/2} + c|u| \, ||u||)^2 \equiv K_1$$

$$d/dt|z|^2 + \nu\Lambda|z|^2 \le K_1.$$

By the Gronwall inequality we obtain

$$|z(t)|^2 \le |z(0)|^2 \exp(-\nu\Lambda t) + K_1/\nu\Lambda.$$

Assuming $|z_0| \le R$ and
$$t \ge 1/\nu\Lambda \log[R^2\nu\Lambda/K_1],$$
then
$$|z(t)| \le c(\lambda_1/\Lambda)^{1/2} \equiv c\delta^{1/2}.$$
In fact, the bound can be improved to
$$|z(t)| \le c\delta L,$$
where $L = 1 - \log(\delta)$ is a logarithmic correction. This behaviour is depicted below.

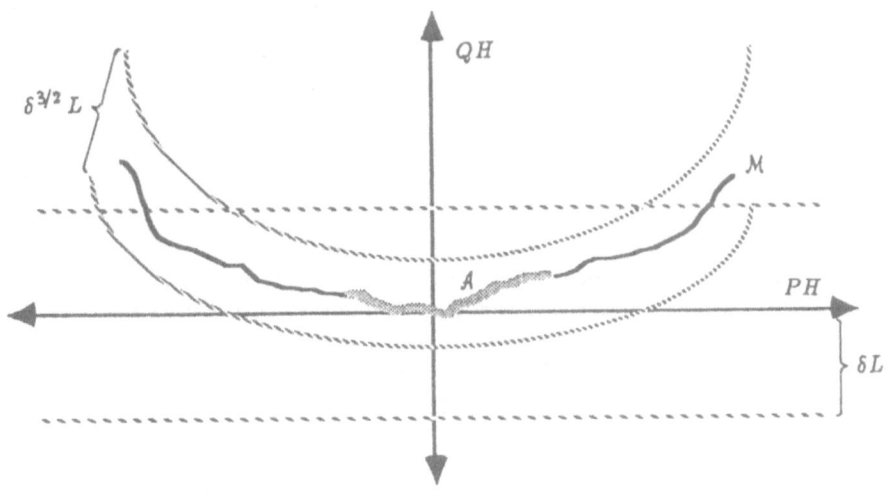

Figure 2.3

The attractor lies on the inertial manifold and is contained in thin neighborhoods of the approximate inertial manifolds.

A necessarily incomplete list of recent works on AIMs includes: Debussche (1989), Foias-Manley-Temam (1985), Foias-Temam (1990), Jolly-Kevrekidis-Titi (1989), Marion-Temam (1989), Promislow (1989), Promislow-Temam (1990), Temam (1989).

2.3 Numerical results involving AIMs: nonlinear Galerkin methods

In order to obtain efficiency in the numerical computation it is necessary to

- Treat the small and large length scales differently.

- Construct approximate solutions close to the attractor.

In the traditional Galerkin method, one takes as many modes as possible, sets the rest to zero, and projects the equation onto the space spanned by these modes

$$P_m H = \text{span}\{w_1, \ldots, w_m\}$$

$$u_m \in P_m H$$

$$du_m/dt + \nu A u_m + P_m B(u_m) = P_m f.$$

This corresponds to the first AIM, $z = \Phi_0(y) = 0$. For the nonlinear Galerkin methods, the projection is taken onto the inertial manifold $\mathcal{M}_1$.

$$y_m = \sum_{i=1}^{m} a_i w_i, \qquad z_m = \sum_{i=m+1}^{\infty} a_i w_i,$$

$$y' + Ay + P_m(B(y,y) + B(y,z) + B(z,y)) = P_m f$$
$$Az + (P_{2m} - P_m)B(y,y) = (P_{2m} - P_m)f.$$

The term $B(z,z)$ is dropped in the y equation, as otherwise convergence cannot be shown. Where the Galerkin method would use $2m$ modes for the y term and set the rest to zero, nonlinear Galerkin would take m modes for y and m modes for z, setting the rest to zero. The advantage of the nonlinear Galerkin method is that, by simplifying the equation for the z terms, speed is increased without sacrificing accuracy.

Numerical evidence shows, see for example Jauberteau-Rosier-Temam (1989), that in a case of $\dim u = 128^2$ modes with y containing 40% and z containing 60% of the modes, we have

$$|z|_{L^2}/|y|_{L^2} \sim 10^{-6}.$$

The primary motivations/justifications for using NLG are:

1. The attractor is closer to $\mathcal{M}_1$ than to $\mathcal{M}_0$.

2. The algorithm has been proven to converge, Marion-Temam (1989).

3. The stability conditions after time discretization are improved.

4. Significant gain in computing time.

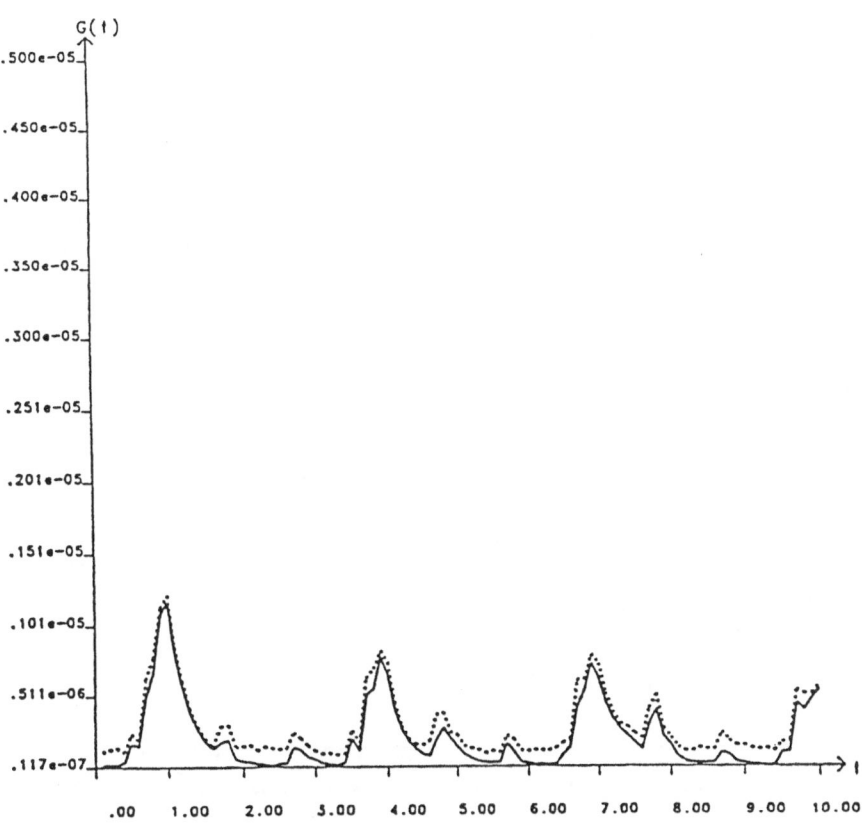

Figure 2.4

Comparison of the errors obtained by the two methods

$$G(t) = \log_{10}\left(\frac{\|u - u_{ex}\|_{L_\infty}}{\|u_{ex}\|_{L_\infty}}\right)$$

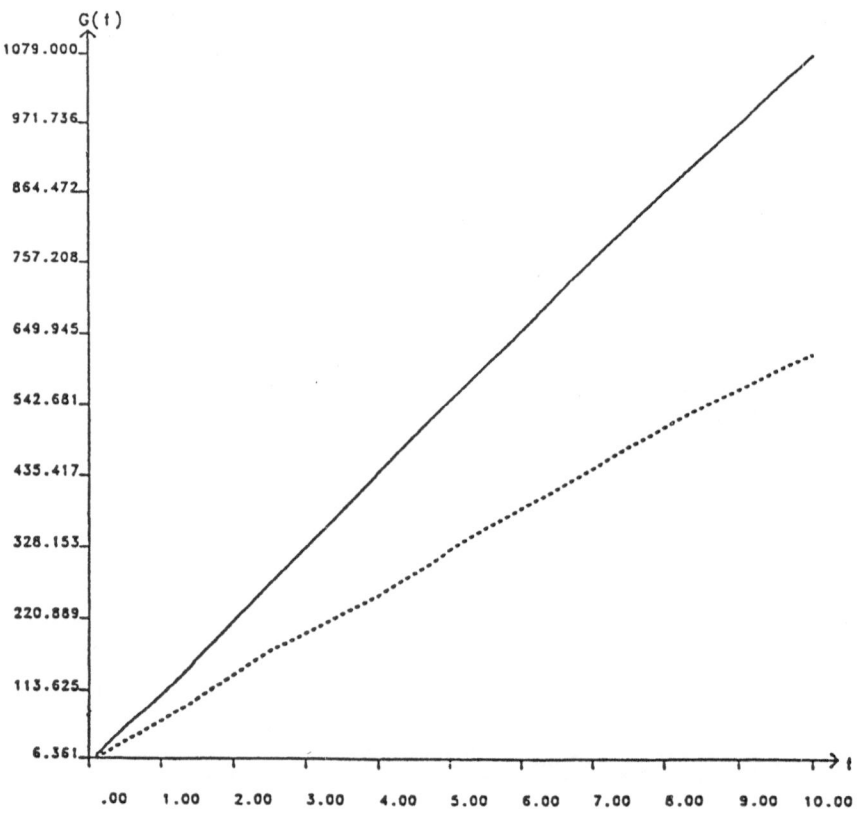

Figure 2.5

Comparison of CPU times (in seconds) needed by the two methods

(- - - Nonlinear Galerkin; —— Usual Galerkin).

Note that the improved stability condition results from the fact that the small eddies are treated implicitly, hence they do not depend on a refined resolution. The large eddies are treated explicitly and the stability condition is that for the large eddies,

$$\Delta t/(\Delta x)^2 \le 2/\nu$$

instead of

$$\Delta t/(\Delta x)^2 \le 1/2\nu.$$

Hence the time step can be taken to be 4 times larger.

The theoretical results quoted before give a sufficient condition on m, the number of modes, for the inertial manifold to exist. But this condition often requires m to be very large, placing k_m, the wave number, well inside the dissipative range. On the other hand, numerical results like that given above show that $|z_m| \ll |y_m|$ for very moderate values of m. Thus a strategy has been developed in Dubois-Jauberteau-Temam (1990) to dynamically allocate the respective size of the y and z components. Essentially this works as follows:

- Total size of $y + z$ chosen according to physical considerations and available computing power and memory space.

- The relative sizes of y and z are chosen according to repeated numerical tests based upon

  - comparison of z to discretization or round off error with respect to time;
  - comparison of $|z|/|y|$ to a prescribed tolerance value.

Finally, we outline a proof of the convergence of the *NLG* method for the Navier-Stokes equation as $m \to \infty$, as given in Marion-Temam (1989). We recall the representations of y and z and the equations which they solve,

$$y = y_m = \sum_{i=1}^{m} a_j w_j$$

$$z = z_m = \sum_{j=m+1}^{m} a_j w_j$$

$$y' + \nu Ay + P\{B(y,y) + B(y,z) + B(z,y)\} = Pf,$$

$$\nu Az + QB(y,y) = Qf.$$

Take the inner product of the y equation with y and that of the z equation with z, adding the two together:

$$\frac{1}{2}(d/dt)|y|^2 + \nu(||y||^2 + ||z||^2) +$$
$$(B(y,y),y) + (B(y,z),y) + (B(z,y),y) + B(y,y),z) = (f,y+z).$$

But $(B(y,y),y) = (B(z,y),y) = 0$ by the orthogonality property, and $(B(y,z),y) = -(B(y,y),z)$ by the skew-symmetry property. So we observe that

$$\frac{1}{2}(d/dt)|y|^2 + \nu(||y||^2 + ||z||^2) \le |f|/2\nu\lambda_1 + \nu/2(||y||^2 + ||z||^2)$$

$$(d/dt)|y|^2 + \nu(||y||^2 + ||z||^2) \le |f|/\nu\lambda_1$$

$$|y_m(t)|^2 \le |y_m(0)|^2 \exp(-\nu\lambda_1 t) + 2(|f|/\nu\lambda_1)^2$$

$$|y_m(t)|^2 \le |u_0|^2 \exp(-\nu\lambda_1 t) + 2(|f|/\nu\lambda_1)^2.$$

From these we may conclude that

$$\int_t^{t+r} ||z_m(s)||^2 ds \le K_2.$$

This implies that

$$\int_t^{t+r} |z_m(s)|^2 ds \le K_2/\Lambda \to 0 \quad \text{as} \quad m \to \infty.$$

After a passage to the limit in this equation as $m \to \infty$, we obtain,

Theorem *As $m \to \infty$,*

$$y_m \to u$$

in $L^2(0,T;V), L^p(0,T;H)$ for all $p \ge 1$, for all $T \le \infty$, strongly, in $L^\infty(\mathbb{R}_+;H)$ weak;*
and

$$z_m \to 0$$

in $L^2(0,T;V)$ strongly, in $L^\infty(\mathbb{R}_+;H)$ weak.*

References

A. Babin and M. Vishik [1983], Attractors of partial differential evolution equations and their dimension, *Russian Math. Surveys* **38**, 151–213.

P. Constantin, C. Foias, O. Manley and R. Temam [1985], Determining modes and fractal dimension of turbulent flows, *J. Fluid Mech.* **150**, 427–440.

P. Constantin, C. Foias, B. Nicolaenko and R. Temam [1988], *Integral Manifolds and Inertial Manifolds for Dissipative Partial Differential Equations*, Springer-Verlag, New York, Applied Math. Sci. Series, Vol. **70**.

P. Constantin, C. Foias and R. Temam [1985], *Attractors Representing Turbulent Flows*, Memoirs of AMS **53**, No 314, 67 + vii pages.

P. Constantin, C. Foias and R. Temam [1988], On the dimension of the attractors in two-dimensional turbulence, *Physica D* **30**, 284–296.

A. Debussche [1989], Quelques problèmes concernant le comportement pour les grands temps des équations dissipatives, Thèse de 3e cycle, Université de Paris-Sud, Orsay.

A. Debussche and R. Temam [1991], Inertial manifolds and the slow manifolds in meteorology, *Differential Integral Equations* **4**, 897–931.

F. Demengel and J.M. Ghidaglia [1989], Some remarks on the smoothness of inertial manifolds, Institute for Applied Mathematics and Scientific Computing, Preprint No. **8908**.

T. Dubois, F. Jauberteau and R. Temam [1990], Solutions of the incompressible Navier-Stokes equations by the Nonlinear Galerkin Method, submitted to *J. Comput. Phys.*

C. Foias, O. Manley and R. Temam [1988], Modelling of the interaction of small and large eddies in two-dimensional turbulent flows, *RAIRO Modél. Math. Anal. Numér.* (M^2AN) **22**, 93–114.

C. Foias, G. Sell and R. Temam [1985], Variétés inertielles pour l'équation de Kuramoto-Sivashinski, *C.R. Acad. Sci. Paris Série I* **301**, 285–288.

C. Foias, G. Sell and R. Temam [1988], Inertial manifolds for nonlinear evolutionary equations, *J. Differential Equations* **73**, 309–353.

J.M. Ghidaglia and R. Temam [1991], Lower bounds on the dimension of the attractor for the Navier-Stokes equations in space dimension 3 in: *Mechanics, Analysis, and Geometry: 200 Years After Lagrange* (M. Francaviglia and D. Holmes, eds.), Elsevier, Amsterdam.

H. Haken [1983], *Advanced Synergetics*, Springer-Verlag, New York.

G. Hardy and E. Wright [1962], *An Introduction to the Theory of Numbers*, Oxford University Press, Oxford.

Yu. S. Il'yashenko [1990], Global analysis of the phase portrait for the Kuramoto-Sivashinsky equation, *IMA Preprint Series*, Minneapolis, No. **655**.

F. Jauberteau, C. Rosier and R. Temam [1989/90], The nonlinear Galerkin method in computational fluid dynamics, *Appl. Numer. Math.* **6**, 361–370.

F. Jauberteau, C. Rosier and R. Temam [1990], A nonlinear Galerkin method for the Navier-Stokes equations, in Proceedings Conference on "Spectral and High Order methods for partial differential equations", ICOSAHOM'89, Como, Italy, 89 *Comput. Methods Appl. Mech. Engrg.* **80**, 245–260.

M. Jolly, I. Kevrekidis and E. Titi [1989], Approximate inertial manifolds for the Kuramoto-Sivashinsky equation: analysis and computations, *MSI Preprint* No. **89-52**, Cornell University, Ithaca, NY.

R. H. Kraichman [1967], Inertial ranges in two-dimensional turbulence, *Phys. Fluids* **10**, 1417–1423.

L. Landau and E. Lifschitz [1966], *Fluid Mechanics*, Pergamon Press, Oxford, 120–123.

J. Mallet-Paret and G. Sell [1988], Inertial Manifolds for Reaction-Diffusion equations in higher space dimensions, *IMA Preprint Series*, Minneapolis, No. **331**.

R. Mañé [1981], On the dimension of the compact invariant sets of certain nonlinear maps, *Dynamical Systems and Turbulence, Warwick 1980*, Lecture Notes in Mathematics **898**, Springer-Verlag, New York, 230–242.

R. Mañé [1977], Reduction of semilinear parabolic equations to finite dimensional C^1 flows, *Geometry and Topology*, Lecture Notes in Mathematics **597**, Springer-Verlag, New York, 361–378.

M. Marion [1989], Inertial manifolds associated with partly dissipative reaction-diffusion systems, *J. Math. Anal. Appl.* **143**, 295-326.

M. Marion and R. Temam [1989], Nonlinear Galerkin methods, *SIAM J. Numer. Anal.* **26**, 1139–1157.

M. Marion and R. Temam [1990], Nonlinear Galerkin methods: The finite elements case, *Numer. Math.* **57**, 205–226.

G. Minea [1976], Remarques sur l'unicité de la solution stationnaire d'une équation de type Navier-Stokes, *Rev. Roumaine Math. Pures Appl.* **21**, 1071–1075.

B. Nicolaenko, B. Scheurer, R. Temam [1985], Some global dynamical properties of the Kuramoto-Sivashinsky equations: Non-linear stability and attractors, *Physica D* **16**, 155–183.

B. Nicolaenko, B. Scheurer, R. Temam [1986], Attractors for the Kuramoto-Sivashinsky equations, in *Nonlinear Systems of Partial Differential Equations in Applied Mathematics*, Lectures in Applied Math. **23**, Part 2, 169–170, AMS, Providence.

N. Phillips [1981], Variational analysis and the slow manifold, *Monthly Weather Review* **109**, No. 12, 2415–2426.

K. Promislow [1990], Induced trajectories and approximate inertial manifolds for the Ginzburg-Landau partial differential equation, *Physica D* **41**, 232–252.

K. Promislow and R. Temam [1991], Localization and approximation of attractors for the Ginzburg-Landau equation, *J. Dynamics Differential Equations* **3**, 491–514.

R. Temam [1988], *Infinite Dimensional Dynamical Systems in Mechanics and Physics*, Springer-Verlag.

R. Temam [1989], Attractors for the Navier-Stokes equations, localization and approximation, *J. Fac. Sci. Univ. Tokyo Sect. I.A Math.* **36**, 629–647.

R. Temam [1989], Induced trajectories and approximate inertial manifolds, *RAIRO Modél. Math. Anal. Numér.* (M^2AN) **23**, 541–561.

E. Titi [1990], On approximate inertial manifolds for the Navier-Stokes equations, *J. Math. Anal. Appl.* **149**, 540–557.

J. Tribbia [1979], Nonlinear initialization on an equatorial Beta-plane, *Monthly Weather Review* **107**, No. 6, 704–713.

Shape Sensitivity Analysis of Variational Inequalities

Jan SOKOLOWSKI[1]

Systems Research Institute

Polish Academy of Sciences

ul. Newelska 6

PL 01-447 Warszawa

Poland

Abstract

Results on the differential stability of the solutions of the variational inequalities are presented. The concept of a polyhedric convex set is used. The material derivative method is applied in order to derive the results on the shape sensitivity analysis.

A detailed description of the material derivative method in the shape optimization of variational inequalities can be found in J. Sokolowski and J.-P. Zolésio, *Introduction to Shape Optimization. Shape Sensitivity Analysis*, Springer Series in Computational Mathematics, vol. 16, Springer-Verlag 1992.

1 Introduction

In the present paper the differential stability of solutions of the variational inequalities with respect to the perturbations of the domain of integration in considered. First let us consider an example of the variational inequality in L^2 space.

Let $\Omega \subset R^n$ be a given domain and let $w = w(\Omega)$ denote the solution of the following elliptic equation

$$-\Delta w = f, \quad \text{in} \ \ \Omega$$
$$\frac{\partial w}{\partial n} = u, \quad \text{on} \ \ \Gamma.$$

Denote by $\mathcal{U}$ the set of admissible controls and by $J(u)$ the cost functional and consider the following optimal control problem

$$\min_{u \in \mathcal{U}} J(u)$$

where

$$\mathcal{U} = \{u \in L^2(\Gamma) | u_{\min} \le u(x) \le u_{\max}, \quad \text{a.e. on} \ \ \Gamma\}$$

---

[1]Partially supported by grant 21207 9101 of the State Committee for Scientific Research of the Republic of Poland.

M. C. Delfour and G. Sabidussi (eds.), Shape Optimization and Free Boundaries, 287–319.

$$J(u) = \frac{1}{2}\int_\Omega (w - z_d)^2 dx + \frac{1}{2}\int_\Gamma (u)^2 d\Gamma;$$

here $u_{min}, u_{max} \in R, z_d(.) \in L^2(\Omega)$ are given.

The optimality conditions for the optimal control problem lead to a variational inequality over the convex set $\mathcal{U}$. To derive the variational inequality we introduce the Lagrangian

$$\mathcal{L}(w, u, p) = \frac{1}{2}\int_\Omega (w - z_d)^2 dx + \frac{1}{2}\int_\Gamma (u)^2 d\Gamma$$
$$+ \int_\Omega [\nabla w.\nabla p - fp]dx + \int_\Gamma updГ.$$

From the stationarity condition $\frac{\partial\mathcal{L}}{\partial w} = 0$ we obtain the adjoint state equation

$$-\Delta p = -(w - z_d), \quad \text{in} \quad \Omega$$

$$\frac{\partial p}{\partial n} = 0, \quad \text{on} \quad \Gamma.$$

The necessary and sufficient optimality condition

$$u \in \mathcal{U} : \left(\frac{\partial\mathcal{L}}{\partial u}, v - u\right) \geq 0, \quad \forall v \in \mathcal{U}$$

takes the form of the following variational inequality

$$u \in \mathcal{U} : \int_\Gamma (u + p)(\varphi - u)d\Gamma \geq 0, \quad \forall\varphi \in \mathcal{U}.$$

The variational inequality is equivalent to the following equation

$$u = P_{\mathcal{U}}(-p),$$

i.e., the unique optimal control is the metric projection in $L^2(\Gamma)$ onto $\mathcal{U}$ of the trace of the adjoint state $-p_{|\Gamma}$. The metric projection $P_{\mathcal{U}}$ is directionally differentiable [74]. It can be shown [74] that for the control problem under consideration the shape differentiability of an optimal control is equivalent to the differentiability of the metric projection $P_{\mathcal{U}}$ onto the set of admissible controls $\mathcal{U}$. In fact we have the same conclusion in the general case of variational inequalities in the Sobolev spaces, e.g. for the unilateral problems [83]. Therefore a part of the paper concerns the directional differentiability of the metric projection onto the convex sets in the Sobolev spaces.

We refer the reader to [74], [75] for the related results on the shape sensitivity analysis of optimal control problems for the distributed parameter systems.

Only some specific variational inequalities are considered in the present paper. We refer the reader to [59]–[64] for the results on the fourth order elliptic variational inequalities. Sensitivity analysis of the variational inequalities describing the unilateral problems in elasticity, including variational inequalities of the second kind, frictionless contact problem, and contact problem with given friction, is performed in [77]–[79]. The related results on the shape sensitivity analysis of the contact problem are presented in [90].

The paper is organized as follows. In section 2 we recall the definition and some properties of the capacity. In section 3 an obstacle problem is considered. The domain derivative of the solution of the obstacle problem is derived, as well as the shape derivative of some shape functionals. Section 4 concerns the differential stability of the metric projection in Hilbert space onto a convex, closed subset. The class of the so-called polyhedric convex sets is characterized. The examples presented include among others the projection onto a ball and the variational inequality describing a shallow shell with the obstacle. In section 5 the shape sensitivity analysis is performed for the variational inequalities describing an elastic-plastic torsion problem and an obstacle problem for the Kirchhoff plate.

Finally a list of references related to the topic of the paper is included.

2 Preliminaries

Capacity

We recall some properties of the Sobolev spaces and the notion of capacity [19], [97]. The Sobolev spaces $H_0^1(\Omega)$ and $H_0^2(\Omega)$ are the closures of $C_0^\infty(\Omega)$ with norms

$$||\varphi||^2_{H_0^1(\Omega)} = \int_\Omega |\nabla \varphi|^2 dx$$

$$||\varphi||^2_{H_0^2(\Omega)} = \int_\Omega |\Delta \varphi|^2 dx$$

respectively. If $\varphi \in H_0^2(\Omega)$, from the definition $D^\alpha \varphi \in H_0^1(\Omega)$ for each α with $|\alpha| = 1$. Functions in $H_0^1(\Omega)$ are defined quasi everywhere and are quasi-continuous. These notions are made precise below.

The C_1-capacity of a compact set F is defined as

$$C_1(F) = \int \{ \int |\nabla \varphi|^2 dx : \varphi \geq 1 \quad \text{on} \quad F, 0 \leq \varphi \in C_0^\infty(\Omega^d) \},$$

similarly C_2-capacity

$$C_2(F) = \inf \{ \int |\Delta \varphi|^2 dx : \varphi \geq 1 \quad \text{on} \quad F, 0 \leq \varphi \in C_0^\infty(\Omega^d) \}.$$

The capacity of a Borel set is then defined as the supremum of the capacities of its compact subsets. A statement holds C_i-q.e., $i = 1, 2$, if it holds except for a set of C_i-capacity zero. With this definition we have the following results:

1. Let $\varphi \in H_0^1(\Omega)$, and $\{\varphi_n\} \subset C_0^\infty(\Omega)$ converge to φ in $H_0^1(\Omega)$. Then a subsequence of $\{\varphi_n\}$ converges C_1-q.e. and this is a representative of φ.

2. Let $\varphi \in H_0^1(\Omega)$. Then φ has a quasi-continuous representative: There is a representative $\bar{\varphi}$ such that given $\varepsilon > 0$, there is an open set $U(\varepsilon)$ of C_1-capacity less than ε such that the restriction of $\bar{\varphi}$ to the complement of $U(\varepsilon)$ is continuous.

3. Any two quasi-continuous representatives of $\varphi \in H_0^1(\Omega)$ agree C_1-q.e.

4. Every set of positive Lebesque measure has positive C_1-capacity.

We shall write "q.e." instead of "C_1-q.e.".

Finally we recall the notion of fine continuity [19].

A set $A \subset R^n$ is said to be thin at x_0 relative to the capacity C_1 if

$$\int_0^1 \left(\frac{C_1[A \cap B(x_0, r)]}{C_1[B(x_0, r)]} \right)^{\frac{1}{2}} \frac{dr}{r} < \infty$$

where $B(x_0, r) = \{x \in R^n | \ |x - x_0| \leq r\}$.

A function $u(.)$ is finely continuous at x_0 if there exists a set A that is thin at x_0 and

$$\lim_{A \not\ni x \to x_0} u(x) = u(x_0).$$

If $u \in H^1(R^n)$ then $u(.)$ is finely continuous at all points except for a set of C_1-capacity zero.

We use standard notation throughout the paper [19], [97]. In particular for a given family of domains $\{\Omega_t\}, t \geq 0$, see section 5 for details, and a given family of elements $y(\Omega_t) \in H^2(\Omega_t)$, the domain derivative in the direction of a vector field $V(.,.)$ is denoted by $y' = y'(\Omega; V) \in H^1(\Omega)$. In the same way the shape derivative of a shape functional $J(\Omega)$ is denoted by $dJ(\Omega; V)$. We refer the reader to [87] for the related properties of the domain derivatives as well as of the shape derivatives of functionals.

3   Obstacle problem

In this section we present the results obtained for the shape differentiability of the solutions of the obstacle problem.

Let y denote the deflection of an elastic membrane of the reference configuration $\Omega \subset R^2$ subjected to the pressure f. We assume that there is an obstacle ψ for the membrane and therefore we have the unilateral condition

$$y \geq \psi. \tag{3.1}$$

The deflection y is determined by minimizing the energy

$$I(\phi) = \frac{1}{2} \int_\Omega |\nabla \phi|^2 dx - \int_\Omega f\phi dx \tag{3.2}$$

subject to (3.1). In order to have a unique solution we assume that $y = 0$ on $\partial\Omega$ and therefore y minimizes the functional (3.2) over the convex closed set

$$K = \{\phi \in H_0^1(\Omega) | \phi(x) \geq \psi(x) \quad \text{in } \Omega\}$$

which is nonempty provided $\psi \leq 0$ on $\partial\Omega$. The necessary and sufficient optimality condition for the minimum problem under study takes the form of a variational inequality:

Find and element $y = y(\Omega) \in K = K(\Omega)$ such that

$$\int_\Omega \nabla y . \nabla(\phi - y) dx \geq \int_\Omega f(\phi - y) dx, \quad \forall \phi \in K$$

(here we assume that $f \in L^2(\Omega)$), or equivalently by the solution of the following complementarity problem:

$$y - \psi \geq 0, \quad \text{in } \Omega$$
$$-\Delta - f \geq 0, \quad \text{in } \Omega$$
$$(y - \psi)(\Delta y + f) = 0, \quad \text{in } \Omega.$$

It follows from a regularity result of Brézis and Stampacchia [11] (see also [40]) that

$$y \in H^2(\Omega) \cap H_0^1(\Omega).$$

We denote by $Z \subset \Omega$ the so-called coincidence set, $\psi \in H^1(\Omega)$,

$$Z = \{x \in \Omega | y(x) = \psi(x)\}.$$

Let us note that $\phi + y \in K$ for any $\phi \in H_0^1(\Omega)$, $\phi \geq 0$, therefore

$$\int_\Omega \nabla y \nabla \phi dx - \int_\Omega f\phi dx \geq 0, \quad \forall \phi \geq 0;$$

hence there exists the nonnegative Radon measure μ given by

$$\int \phi d\mu \equiv \int_\Omega (-\Delta y - f)\phi dx, \quad \forall \phi \in H_0^1(\Omega)$$
$$= \int_\Omega (\nabla y . \nabla \phi - f\phi) dx$$

such that

$$\mu(\Omega \backslash Z) = 0.$$

Observe that in general the set Z is not closed.

We denote by $\Pi : H^{-1}(\Omega) \ni f \to y \in H_0^1(\Omega)$ the nonlinear mapping associated with the unilateral problem under consideration.

It can be shown [48], [29] that the mapping Π is directionally differentiable and the differential of Π in the direction $h \in H^{-1}(\Omega)$, denoted by $\Pi'(h)$, minimizes the quadratic functional

$$J(\phi) = \frac{1}{2} \int_\Omega |\nabla\phi|^2 dx - \int_\Omega h\phi dx$$

over the convex cone

$$S = \{\phi \in H_0^1(\Omega) | \phi \geq 0 \text{ q.e. on } Z, \int_\Omega \phi d\mu = 0\}.$$

Using this result and the material derivative method it can be shown (Sokolowski and Zolésio [85], [87]) that the domain derivative y' of the solution y of the unilateral problem is the unique minimizer of the functional

$$j(\phi) = \frac{1}{2} \int_\Omega |\nabla\phi|^2 dx$$

over the cone

$$S_v(\Omega) = \{\phi \in H^1(\Omega) | \phi = -v\frac{\partial y}{\partial n} \text{ on } \Gamma, \phi \geq 0 \text{ q.e. on } Z, \int_\Omega \phi d\mu = 0\}$$

provided the obstacle ψ is sufficiently smooth, e.g. $\psi \in H^2(\Omega)$, here $v = V.n$ is the normal component of the vector field $V(0,.)$ on the boundary $\Gamma = \partial\Omega$.

On the other hand the solution y of the unilateral problem can be characterized as the metric projection in Sobolev space $H_0^1(\Omega)$ of an element $F \in H_0^1(\Omega)$ onto the convex set $K \subset H_0^1(\Omega)$

$$y = P_K F$$

$$\int_\Omega |\nabla(y - F)|^2 dx = \min\{\int_\Omega |\nabla(u - F)|^2 dx | u \in K\},$$

where $F = \Delta^{-1}f$, i.e.,

$$F \in H_0^1(\Omega)$$

$$\int_\Omega \nabla F.\nabla\phi dx = \int_\Omega f\phi dx, \quad \forall\phi \in H_0^1(\Omega).$$

Here we use the same symbol for the duality pairing between $H^{-1}(\Omega)$ and $H_0^1(\Omega)$ and for the scalar product in $L^2(\Omega)$.

Finally we derive the shape derivatives [87] of the following shape functionals:

energy:

$$E_K(\Omega) = \min_{\phi \in K}\{\frac{1}{2} \int_\Omega |\nabla\phi|^2 dx - \int_\Omega f\phi dx\}; \qquad (3.3)$$

· and the quadratic cost functional:

$$J(\Omega) = \frac{1}{2} \int_\Omega (y(\Omega) - z_d)^2 dx, \qquad (3.4)$$

where $y(\Omega)$ denotes the solution of the obstacle problem and $z_d \in H^1(R^2)$ is given. Let us note that $E_K(\Omega) = \{\frac{1}{2}\int_\Omega |\nabla y(\Omega)|^2 dx - \int_\Omega fy(\Omega)dx\}$.

In order to derive the form of the shape derivative $dE(\Omega;V)$ of the functional (3.3) we do not use the domain derivative y' since

$$dE(\Omega;V) = -\int_\Gamma [\frac{1}{2}\left|\frac{\partial y(\Omega)}{\partial n}\right|^2 + f]V.n d\Gamma.$$

On the other hand, the shape derivative of the cost functional (3.4) takes the following form

$$dJ(\Omega;V) = \int_\Omega (y(\Omega) - z_d)y'(\Omega;V)dx + \frac{1}{2}\int_\Gamma (z_d)^2 V.n d\Gamma. \tag{3.5}$$

The mapping

$$V \to y'(\Omega;V)$$

is linear provided

$$S = \{\phi \in H_0^1(\Omega)|\phi = 0 \text{ q.e. on } Z\},$$

i.e., S is a linear subspace; in such a case it follows that

$$S_v(\Omega) = \{\phi \in H^1(\Omega)|\phi = -v\frac{\partial y}{\partial n} \text{ on } \Gamma, \phi = 0 \text{ q.e. on } Z\}$$

and the adjoint state can be introduced

$$p \in H_0^1(\Omega \backslash Z): \quad \int_\Omega \nabla p . \nabla \varphi dx = \int_\Omega (y(\Omega) - z_d)\varphi dx, \quad \forall \varphi \in H_0^1(\Omega \backslash Z)$$

provided the set $Z \subset \Omega$ is compact. Using the adjoint state the equation (3.5) can be rewritten in the following form

$$dJ(\Omega;V) = \int_\Gamma [\frac{1}{2}(z_d)^2 + 2\frac{\partial y}{\partial n}\frac{\partial p}{\partial n}]vd\Gamma.$$

4 Differential stability of metric projection in Hilbert space

We recall briefly the main results on the differential stability of metric projection in Hilbert space onto a closed and convex subset which are used in the shape sensitivity analysis of variational inequalities [59]–[64], [80], [82]–[91]. We start with the following examples.

Example 1 Let us consider an elementary example of the projection mapping in R onto the set $K = [0,1]$.

We have

$$\forall x \in R: \quad P_K(x) = \begin{cases} 1, & x > 1 \\ x, & 0 \le x \le 1 \\ 0, & x < 0. \end{cases}$$

It is easy to see that the mapping $x \to P_K(x)$ is differentiable everywhere except at $x = 0$ and $x = 1$. At the point $y = 0$ we have for $h = \pm 1$ and for small enough $\varepsilon > 0$,

$$P_K(y + \varepsilon h) = P_K y + \varepsilon h^+,$$

where

$$h^+ = \begin{cases} h, & h \geq 0 \\ 0, & h < 0. \end{cases}$$

Therefore for sufficiently small $\varepsilon > 0$,

$$[P_K(y + \varepsilon h) - P_K(y)]/\varepsilon = h^+ = \lim_{\varepsilon \to 0}[P_K(y + \varepsilon h) - P_K(y)]/\varepsilon.$$

Hence at $y = 0$ we have

$$P_K(y + \varepsilon h) = P_K(y) + \varepsilon Q(h) + o(\varepsilon)$$

where the mapping $Q(.) : R \to R$ is defined by $Q(h) = h^+, \forall h \in R$. In the notation of this chapter the mapping $Q(.)$ is called the conical differential of the projection $P_K(.)$ at $y = 0$. Let us recall how the mapping $P_K(.)$ is defined by a variational inequality. Since for a given $x \in R$ we have

$$(P_K(x) - x)^2 \leq (v - x)^2, \forall v \in K$$

then by a standard argument it follows that the element $P_K(x)$ is given by the unique solution of the following variational inequality:

$$K \ni P_K(x) : (P_K(x) - x)(v - P_K(x)) \geq 0, \quad \forall v \in K.$$

Example 2 Let $K \subset R^N$ be a compact, convex set with nonempty interior and with smooth boundary ∂K of class C^2. We assume that there is given a convex function $\psi \in C^2(R^N)$ such that $\psi(\bar{x}) \leq 0$ for some $\bar{x} \in R^N$ and

$$\partial K = \{x \in R^N | \psi(x) = 0\}.$$

It can be verified using known results, see e.g. Malanowski [47], that the projection in R^N onto the set K is directionally differentiable, i.e., for given elements $f, h \in R^N$ and sufficiently small $\varepsilon > 0$,

$$P_K(f + \varepsilon h) = P_K(f) + \varepsilon Q(h) + o(\varepsilon)$$

where the element $Q(h) \in R^N$ is given by the unique solution of the following variational inequality:

$$Q \in S_K(f)$$
$$(AQ, x - Q)_{R^N} \geq (h, x - Q)_{R^N}, \forall x \in S_K(f).$$

Here we denote

$$A = I + \lambda D^2 \psi(u)$$

$$u = P_K(f)$$

$$\lambda = \begin{cases} \|f - u\|_{R^N} / \|D\psi(u)\|_{R^N}, & f \ni K; \\ 0, & \text{otherwise.} \end{cases}$$

$$S_K(f) = \begin{cases} x \in R^N : D\psi(u).x \leq 0, \lambda D\psi(u).x = 0, & \text{if } f \notin \text{int } K \\ R^N, & \text{if } f \in \text{int } K. \end{cases}$$

In this case $Q(.) : R^N \to R^N$ is the metric projection onto $S_K(f)$ with respect to the scalar product in R^N which depends on the multiplier λ. $\qquad \square$

We recall the abstract results on the directional differentiability of projection in Hilbert space onto a convex, closed subset. Let H be a separable Hilbert space, $K \subset H$ a convex and closed subset. Let there be given a bilinear form

$$a(.,.) : H \times H \to R$$

which is coercive and continuous, i.e.,

$$a(v, v) \geq \alpha \|v\|_H^2, \alpha > 0, \forall v \in H$$

$$|a(v, z)| \leq M \|v\|_H \|z\|_H, \forall v, z \in H.$$

We assume for simplicity that the bilinear form is symmetric

$$a(v, z) = a(z, v) \quad \forall v, z \in H.$$

We denote by $P_K(f)$ the a-projection in H of an element $f \in H$ onto the convex set K. The element $y = P_K(f)$ minimizes the quadratic functional

$$I(v) = \frac{1}{2}a(v - f, v - f)$$

over the set K. Therefore y satisfies the following variational inequality:

Find $y \in K$ such that
$$a(y - f, v - f) \geq 0, \quad \forall v \in K. \tag{4.1}$$

It can be shown that the mapping $P_K(.) : H \to K \subset H$ is Lipschitz continuous

$$\|P_K(f_1) - P_K(f_2)\|_H \leq \frac{M}{\alpha} \|f_1 - f_2\|_H, \forall f_1, f_2 \in H. \tag{4.2}$$

Therefore by a generalization of the Rademacher theorem (Mignot [48]) it follows that there exists a dense subset $\Xi \subset H$ such that for any $f \in \Xi$ we can find a linear continuous mapping $P_K'(.) = P_K'(F, .) : H \to H$ such that

$$\forall h \in H : P_K(f + \varepsilon h) = P_K(f) + \varepsilon P_K'(h) + o(\varepsilon)$$

where $\|o(\varepsilon)\|_H \varepsilon \to 0$. In the sequel we will use the concept of the so-called conical differential of the projection $P_K(.)$.

Definition 1 The mapping $P_K(.)$ is *conically differentiable* at $f \in H$ if there exists a continuous mapping

$$Q(.) : H \to H,$$

$$Q(\alpha h) = \alpha Q(h), \forall \alpha > 0, \forall h \in H,$$

such that for all sufficiently small $\varepsilon > 0$,

$$\forall h \in H : P_K(f + \varepsilon h) = P_K(f) + \varepsilon Q(h) + o(\varepsilon) \tag{4.3}$$

where $\|o(\varepsilon)\|_H \varepsilon \to 0$ with $\varepsilon \to 0$, uniformly on compact subsets of H.

In order to derive the form of the mapping $Q(.)$ defined by (4.3) for a class of sets K we need the following notation. For a given element $y \in K$ we denote by $C_K(y)$ the radial cone

$$C_K(y) = \{\phi \in H | \exists \varepsilon > 0 \text{ such that } y + \varepsilon \phi \in K\}.$$

In general the convex cone $C_K(y)$ in not closed, we denote by $T_K(y) = \overline{C_K(y)}$ its closure in H. $T_K(y)$ is a tangent cone. $N_K(y)$ denotes the normal cone to K at $y \in K$ of the form

$$N_K(y) = \{\phi \in H | a(\phi, z - y) \le 0, \forall z \in K\}.$$

The normal cone is convex and closed. Finally we denote by $S_K(f) \subset H$ the convex, closed cone of the form

$$S_K(f) = \{v \in T_K(y) | a(f - y, v) = 0\}, \tag{4.4}$$

where $y = P_K(f), f \in H$ is a given element. Let us assume that there is given a continuous mapping

$$f(.) : [0, \delta) \to H$$

which is right differentiable at 0, i.e., there exists an element $f'(0) \in H$ such that

$$\lim_{\tau \to 0} \|(f(\tau) - f(0))\tau - f'(0)\|_H = 0.$$

Denote

$$y(\tau) = P_K(f(\tau)), \tau \in [0, \delta)$$

$$\gamma(\tau) = (y(\tau) - y(0))/\tau,$$

and observe that in view of (4.2) we have

$$\|\gamma(\tau))\|_H \le M/\alpha, \quad \forall \tau \in (0, \delta).$$

It can be shown, we refer the reader to [29] for the proof, that every weak limit point γ of $\gamma(\tau)$ for $\tau \to 0$ satisfies

$$\gamma \in S_K(f(0)).$$

We denote

$$[f - y]^\perp = \{\phi \in H | a(y - f, \phi) = 0\}$$

the orthogonal subspace in H to the element $f - y \in H$.

Proposition 1 *Let $K \subset H$ be a closed, convex subset of the Hilbert space H. Then for any $f \in H$, any element $w \in C_K(u) \cap [f - u]^\perp, u = P_K(f)$, we have*

$$P_K(f + tw) = P_K(f) + tw;$$

therefore

$$P_K(f + tw) = P_K(f) + tw + o(t), \forall w \in \overline{C_K(u) \cap [f - u]^\perp}.$$

Proof is given in [29].

Theorem 1 *Let $f \in H$ be a given element, denote $u = P_K f$. Let us suppose that for any element $w \in S_K(u)$ it follows that*

$$P_K(f + tw) = P_K(f) + tw + o(t).$$

Then for $t > 0, t$ small enough,

$$\forall h \in H : P_K(f + th) = P_K f + t P_S(h) + o(t),$$

where $P_S(.)$ is the projection in H onto $S_K(u)$.

Proof is given in [29].

The explicit form of the tangent cone $T_K(u) = \overline{C_K(u)}$ for some convex sets is derived in the following examples.

Example 3 Denote

$$H = L^2(\Omega), \Omega \subset R^N \text{ is a given domain,}$$
$$K = \{\phi \in L^2(\Omega) | \phi(x) \geq 0 \text{ for a.e. } x \in \Omega\}.$$

Let

$$f \in L^2(\Omega);$$

then

$$u = P_K f = f^+ = \max\{f, 0\}$$
$$\Xi = \{x \in \Omega | f(x) \leq 0\}.$$

We have

$$\overline{C_K(u)} = \{\phi \in L^2(\Omega) | \phi(x) \geq 0 \text{ for a.e. } x \in \Xi\}.$$

Example 4 (Mignot [48]) Let

$$H = H_0^1(\Omega)$$
$$K = \{\phi \in H_0^1(\Omega) | \phi(x) \geq 0 \text{ for a.e. } x \in \Omega\}.$$

For any element $f \in H_0^1(\Omega)$

$$u = P_K f \in K : \int_\Omega \nabla(u - f)\nabla(\phi - u)dx \geq 0, \forall \phi \in K.$$

Then

$$\overline{C_K(u)} = \{\phi \in H_0^1(\Omega)|\phi(x) \geq 0 \text{ for q.e. } x \in \Xi\},$$

where $\Xi = \{x \in \Omega|u(x) = 0\}$.

Here q.e. means C_1-quasi-everywhere, i.e., everywhere except possibly on a set of C_1-capacity zero. The definition of capacity is given in section 2.

Example 5 (Rao and Sokolowski [60]) Let us consider the space $H = H^2(\Omega) \cap H_0^1(\Omega)$ with the scalar product

$$(y, z)_H = \int_\Omega \Delta y \Delta z dx, \quad \forall y, z \in H.$$

Denote by K the following convex cone

$$K = \{\varphi \in H|\varphi \geq \psi \text{ in } \Omega\}.$$

Let $u \in K$ be a given element, denote $\Xi = \{x \in \Omega|u(x) = \psi(x)\}$.

Theorem 2 *Assume that $\psi \in H$ and Ξ is compact. Then*

$$T_K(u) = \{\varphi \in H|\varphi \geq 0 \quad C_2\text{-q.e. on } \Xi\}.$$

Proof of Theorem 2 is given in [60]. □

For a family of convex sets in Hilbert space the metric projection is conically differentiable.

Definition 2 A convex, closed set $K \subset H$ is called *polyhedric* if the following condition is satisfied for all $f \in H$:

$$S_K(f) = \text{cl}\{v \in C_K(y)|a(f - y, v) = 0\} \tag{4.5}$$

where cl $A = \overline{A}$ denotes closure of a set $A, y = P_K f$, and the cone $S_K(f)$ is defined by (4.4).

Let us observe that the inclusion

$$\text{cl}\{v \in C_K(y)|a(f - y, v) = 0\} \subset S_K(f)$$

holds for any element $f \in H$.

The sets K in the examples 1, 3, 4, are polyhedric; however, in general the sets K in the examples 2, 5 fail to be polyhedric.

For any polyhedric set $K \subset H$ the form of the conical differential of the metric projection onto K is derived by Mignot [48] and Haraux [29].

Corollary 1 *Let $f(.) : [0, \delta) \to H$ be right-differentiable at $\tau = 0$ in H, and suppose that for the set $K \subset H$ the following condition is satisfied:*

$$T_K(f) \cap [f - g]^{\perp} = \mathrm{cl}(C_K(f) \cap [f - g]^{\perp}) \tag{4.6}$$

for $f = f(0), g = P_K(f(0))$.

Then for sufficiently small $\tau > 0$,

$$P_K(f(\tau)) = P_K(f(0)) + \tau P_{S_K(f(0))}(f'(0)) + o(\tau)$$

where $\|o(\tau)\|_{H^{\tau}} \to 0$ with $\tau \to 0$. □

In particular the condition (4.6) implies that the projection $P_K(.)$ is conically differentiable at $f = f(0) \in H$ and we have

$$Q(h) = P_{S_K(f)}(h), \quad \forall h \in H.$$

Let us observe that in general

$$Q(h) \neq -Q(-h)$$

for the mapping $Q(.)$.

Since we consider the case of a symmetric bilinear form $a(.,.) : H \times H \to R$, we can use the notion of projection $P_K(.)$ in H onto K, as in Theorem 1.

The conclusion of Theorem 1 remains valid for a nonsymmetric, coercive bilinear from $a(.,.)$ provided the Hilbert space H is the so-called Dirichlet space [19]. Let us recall here that from the Stampacchia Theorem [40] it follows that in the nonsymmetric case there exists a unique solution of the variational inequality (4.1).

We present an example of the set $K \subset L^2(\Omega, R^N)$ that is not polyhedric, nevertheless we provide the form of the conical differential of the metric projection onto K [73].

Example 6 Let us consider the metric projection in the space $H = L^2(\Omega, R^N)$ onto the set

$$K = \{v \in L^2(\Omega, R^N) | \frac{1}{2} \sum_{i=1}^{N} a_i v_i^2(\xi) \leq 1 \text{ for a.e. } \xi \in \Omega\}, \tag{4.7}$$

where $a_i > 0, i = 1, \dots N$, are given constants. We denote

$$\psi(x) = \frac{1}{2} \sum_{i=1}^{N} a_i x_i^2 - 1, \quad x \in R^N.$$

Let

$$U = \{x \in R^N | \psi(x) \leq 0\}.$$

Denote by $P_U(.) : R^N \to R^N$ the metric projection in R^N onto U. Let $f(.) \in L^\infty(\Omega, R^N)$ be a given element, consider $\xi \in \Omega$ and denote by $u(\xi)$ the projection of $f(\xi) \in R^N$ onto U:

$$u(\xi) = P_U(f(\xi)).$$

Let $\lambda(\xi)$ be the associated Lagrange multiplier

$$\lambda(\xi) = \begin{cases} ||f(\xi) - u(\xi)||_{R^N} / ||D\psi(u(\xi))||_{R^N}, & \text{if } f(\xi) \notin U \\ 0, & \text{if } f(\xi) \in U \end{cases}$$

and introduce the symmetric matrix

$$A(\xi) = [(1 + \lambda(\xi)a_i)\delta_{ij}]_{N \times N},$$

where $\delta_{ij} = 1$ for $i = j, \delta_{ij} = 0$ for $i \neq j$.

In the particular case of the set $U \subset R^N$, the cone (4.5) takes the form

$$S_U(f(\xi)) = \{x \in R^N | D\psi(u(\xi))x \leq 0, \lambda(\xi)D\psi(u(\xi))x = 0\}.$$

It can be easily verified that the condition (4.6) is not satisfied in that case. It is clear that the projection $P_U(.)$ is differentiable provided the associated Lagrange multiplier $\lambda \neq 0$. The right-derivative $q = q(\xi) \in R^N$ at $f(\xi) \in R^N$ in the direction $h \in R^N$ is given by the unique solution of the following variational inequality

$$q \in S_U(f(\xi))$$

$$\langle A(\xi)q - h, v - q \rangle_{R^N} \geq 0, \forall v \in S_U(f(\xi)).$$

Therefore the projection $P_K(.)$ in $L^2(\Omega, R^N)$ onto the set (4.7) is right differentiable at f in any direction $h \in L^2(\Omega, R^N)$. The right-derivative $q(.) \in L^2(\Omega, R^N)$ is given by

$$q(.) \in S_K(f) = \{v(.) \in L^2(\Omega, R^N) | D\psi(u(\xi))v(\xi) \leq 0, \lambda(\xi)D\psi(u(\xi))v(\xi) = 0, \text{ a.e. in } \Omega\}$$

$$\int_\Omega \langle A(\xi)q(\xi) - h(\xi), v(\xi) - q(\xi) \rangle_{R^N} d\xi \geq 0, \forall v(.) \in S_K(f).$$

In this example the set K is not polyhedric and it follows that $q \neq P_{S_K(f)}(h)$. It shows that Theorem 1 cannot be extended to the convex sets which do not satisfy the condition (4.6). □

Example 7 (Projection onto a ball) Let us recall [81], [96] that the following result on the directional differentiability holds for the metric projection onto the unit ball in the Hilbert space $H, ||f||$ denotes the norm of $f \in H$.

Proposition 2 *The metric projection P_K in Hilbert space H onto the convex set*

$$K = \{v \in H| ||v|| \leq 1\}$$

has the following form:

$$P_K f = \begin{cases} f/\|f\|, & \text{if } \|f\| > 1; \\ f, & \text{otherwise.} \end{cases}$$

If $\|f\| = 1$ *then* [96] *for small enough* $\varepsilon > 0$,

$$\forall v \in H : P_K(f + \varepsilon v) = P_K f + \varepsilon P_S v + o(\varepsilon),$$

where

$$S = T_K(f) = \{v \in H | (f, v) \le 0\}.$$

For $f \notin K$ *the metric projection is differentiable, i.e., for* ε *small enough*

$$\forall v \in H : P_K(f + \varepsilon v) = P_K f + \varepsilon R v + o(\varepsilon),$$

where $R : H \to H$ *is a linear mapping of the following form*

$$\forall v \in H : R v = \frac{1}{\|f\|} P_S v,$$

and S *is the following linear subspace*

$$S = T_K(P_K(f)) \cap [f - P_K(f)]^\perp = \{v \in H | (f, v) = 0\}.$$

$\square$

Example 8 (Shallow shell with obstacle) Following [59] we provide an example of a polyhedric convex set.

Let us consider the convex set

$$K = \{\varphi = (\varphi_1, \varphi_2, \varphi_3) \in H | \mathcal{R}\varphi \ge \psi \text{ in } \Omega\},$$

where $H = H_0^2(\Omega) \times H_0^1(\Omega) \times H_0^1(\Omega), \psi(x), x \in \Omega$ denotes the obstacle, and $\mathcal{R}$ is the following linear mapping

$$\mathcal{R}\varphi = \varphi_1 - a_2\varphi_2 - a_3\varphi_3;$$

here $a_2 = \frac{\partial \psi}{\partial x_1}, a_3 = \frac{\partial \psi}{\partial x_2}$.

We assume for the sake of simplicity that the obstacle is sufficiently smooth, hence

$$\mathcal{R}\varphi \in H_0^1(\Omega), \forall \varphi \in H.$$

Denote by

$$H_{\mathcal{R}}^1(\Omega) = \{\phi \in H_0^1(\Omega) | \phi = \mathcal{R}\varphi, \text{ for some } \varphi \in H\}$$

the image in $H_0^1(\Omega)$ of the mapping $\mathcal{R}$. Let us consider the metric projection P_K in H onto the convex set K with respect to the scalar product

$$(\varphi, \phi)_H = \int_\Omega (\Delta\varphi_1\Delta\phi_1 + \nabla\varphi_2.\nabla\phi_2 + \nabla\varphi_3.\nabla\phi_3)dx.$$

Let $T_K(\phi)$ denote the tangent cone to K at $\phi \in K$. First we derive the form of the tangent cone $T_K(\phi)$ for any $\phi \in K$.

Theorem 3 *The tangent cone $T_K(\phi)$ takes the following form:*

$$T_K(\phi) = \{\varphi \in H | \mathcal{R}\varphi \geq 0 \text{ q.e. on } \Xi\},$$

where

$$\Xi = \{x \in \Omega | \mathcal{R}\phi = \psi\}$$

is the coincidence set.

Here q.e. means "quasi everywhere" with respect to the capacity of the space $H^1_{\mathcal{R}}(\Omega)$ equipped with the smallest norm for which the mapping $\mathcal{R} : H \to H^1_{\mathcal{R}}(\Omega)$ is continuous.

Proof [59].   We assume that the coincidence set $\Xi \subset \Omega$ is compact. Denote by M the following closed convex cone

$$M = \{\varphi \in H | \mathcal{R}\varphi \geq 0 \text{ q.e. on } \Xi\}.$$

Denote

$$C_K(\phi) = \{\varphi \in H | \exists t > 0 \text{ such that } \phi + t\varphi \in K\}.$$

It is clear that $C_K(\phi) \subset M$ hence $T_K(\phi) \subset M$. Let $\Upsilon \in M$ be a given element and let Υ_0 denote the orthogonal projection of Υ onto the convex cone $T_K(\phi)$. Then

$$(\Upsilon_0 - \Upsilon, \varphi)_H \geq 0, \forall \varphi \in T_K(\phi) \tag{4.8}$$

$$(\Upsilon_0 - \Upsilon, \Upsilon_0)_H = 0. \tag{4.9}$$

We claim that

$$(\Upsilon_0 - \Upsilon, \varphi)_H = 0 \text{ if } \mathcal{R}\varphi = 0. \tag{4.10}$$

Indeed, if $\mathcal{R}\varphi = 0$, then $\pm\varphi \in C_K(\phi)$ so that (4.10) follows from (4.8). Define on $H^1_{\mathcal{R}}(\Omega)$ the positive linear functional, well defined in view of (4.10),

$$Lv = (\Upsilon_0 - \Upsilon, \varphi)_H, v = \mathcal{R}\varphi \in H^1_{\mathcal{R}}(\Omega).$$

Then there is a non-negative Radon measure λ on Ω such that

$$Lv = \int \mathcal{R}\varphi d\lambda = \int v d\lambda.$$

We claim that λ is concentrated on Ξ. Indeed, if $\varphi_0 \in C^\infty_0(\Omega \backslash \Xi)$ then clearly $\pm\varphi = \pm(\varphi_0, 0, 0)$ belongs to $T_K(\phi)$ so that from (4.8) it follows that

$$\int \varphi_0 d\lambda = L\varphi_0 = (\Upsilon_0 - \Upsilon, \varphi)_H = 0.$$

Finally in view of (4.9)

$$0 \leq (\boldsymbol{\varUpsilon}_0 - \boldsymbol{\varUpsilon}, \boldsymbol{\varUpsilon}_0 - \boldsymbol{\varUpsilon})_{\boldsymbol{H}} = -(\boldsymbol{\varUpsilon}_0 - \boldsymbol{\varUpsilon}, \boldsymbol{\varUpsilon})_{\boldsymbol{H}} = -\int \mathcal{R}\boldsymbol{\varUpsilon} d\lambda.$$

Now λ is concentrated on Ξ, $\mathcal{R}\boldsymbol{\varUpsilon} \geq 0$ on Ξ so that the last quantity is non-positive. Hence we must have

$$\boldsymbol{\varUpsilon}_0 = \boldsymbol{\varUpsilon}.$$

Since the element $\boldsymbol{\varUpsilon} \in \boldsymbol{T}_{\boldsymbol{K}}(\phi)$ is arbitrary , it follows that

$$\boldsymbol{T}_{\boldsymbol{K}}(\phi) = \mathcal{M}.$$

Let $\boldsymbol{g} \in \boldsymbol{H}$ be a given element, denote $\phi = P_{\boldsymbol{K}}\boldsymbol{g}$ and let μ be the non-negative Radon measure defined as follows

$$\int \mathcal{R}\varphi d\mu = (\phi - \boldsymbol{g}, \varphi)_{\boldsymbol{H}}, \forall \varphi \in \boldsymbol{H}.$$

$\square$

Theorem 4 *The convex cone K is polyhedric.*

The proof of Theorem 4 is given in [59].

5   Shape sensitivity analysis

Let there be given a vector field

$$V(.,.) \in C^1(0, \delta; C^2(R^2; R^2)),$$

where $\delta > 0$ is a given constant. We define a family $\{\Omega_t\} \subset R^2, t \in [0, \delta)$, of domains as follows [87]:

$$\begin{aligned} \Omega_t &= T_t(V)(\Omega) \\ &= \{x \in R^2 | \exists X \in \Omega \text{ such that } x(0) = X, x = x(t)\}, \end{aligned}$$

where $x(t) \in R^n, t \in [0, \delta)$, is given by the unique solution of the following system

$$\begin{cases} \frac{dx}{dt} = V(t, x(t)), & t \in (0, \delta) \\ x(0) = X \end{cases}$$

We refer to the set of lectures of Michel Delfour [14] and Jean-Paul Zolésio [102] in this volume for an introduction to the material derivative method.

In order to obtain the material (domain) derivatives of solutions to the variational inequalities the results of section 4 are combined with the following lemma [70].

Lemma 1 *Let the bilinear form $a_\tau(.,.) : H \times H \to R$ be coercive and continuous uniformly with respect to $\tau \in [0, \delta)$. Denote by $A_\tau \in \mathcal{L}(H; H')$ the linear operator defined as follows:*

$$a_\tau(\phi, \varphi) = \langle A_\tau \phi, \varphi \rangle, \forall \phi, \varphi \in H$$

and suppose that

$$A_\tau = A_0 + \tau A' + o(\tau) \quad in \quad \mathcal{L}(H; H').$$

Let $f_\tau \in H'$ satisfy

$$f_\tau = f_0 + \tau f' + o(\tau) \quad in \quad H'$$

for sufficiently small $\tau > 0$. Furthermore assume that $K \subset H$ is a convex closed set and that for the solutions of the variational inequality

$$\Pi f \in K : a_0(\Pi f, \varphi - \Pi f) \geq \langle f, \varphi - \Pi f \rangle, \quad \forall \varphi \in K$$

the following differential stability result holds for sufficiently small $\varepsilon > 0$,

$$\forall h \in H : \Pi(f_0 + \varepsilon h) = \Pi f_0 + \varepsilon \Pi' h + o(\varepsilon) \quad in \quad H,$$

where the mapping $\Pi' : H' \to H$ is continuous and positively homogeneous.

Then the solutions of the following variational inequality

$$y_\tau \in K : a_\tau(y_\tau, \varphi - y_\tau) \geq \langle f_\tau, \varphi - y_\tau \rangle, \quad \forall \varphi \in K$$

are differentiable with respect to τ, i.e., for $\tau > 0, \tau$ small enough,

$$y_\tau = y_0 + \tau y' + o(\tau), \quad in \quad H,$$

where

$$y' = \Pi'(f' - A' y_0).$$

□

The results on the shape sensitivity analysis of variational inequalities lead to the numerical methods of optimization for the related shape optimization problems, see e.g. [53] for the numerical results. Further numerical results for the second order elliptic unilateral problems for the particular geometry of the domain of integration in the form of a graph in R^2 can be found in [30].

Elastic-plastic torsion problem

Let $\Omega \subset R^2$ be a given bounded domain with smooth boundary $\Gamma = \partial\Omega$. Let us consider the following variational inequality:

$$\text{Find} \quad u \in K(\Omega) \quad \text{such that}$$

$$\int_\Omega \nabla u(x) \nabla(\phi(x) - u(x))dx \geq \mu \int_\Omega (\phi(x) - u(x))dx, \forall \phi \in K(\Omega), \tag{5.1}$$

where $\mu > 0$ is a given constant and $K(\Omega)$ is a closed and convex subset of Sobolev space $H_0^1(\Omega)$ of the form

$$K(\Omega) = \{\phi \in H_0^1(\Omega)| \quad |\nabla\phi(x)| \leq k \quad \text{for a.e. } x \in \Omega\}; \tag{5.2}$$

here $k > 0$ is a given constant. It can be shown [25] that there exists a unique solution to (5.1). Denote by $P \subset \Omega$ the so-called plastic region:

$$P = \{x \in \Omega| \quad |\nabla u(x)| = k\}.$$

Then $E = \Omega \backslash P$ is the so-called elastic region and we have

$$-\Delta u(x) = \mu, \quad \text{in } E.$$

The elastic region E and the plastic region P are not known a priori and should be determined, therefore (5.1) is a free boundary problem. Let us recall that the solution of the variational inequality (5.1) satisfies the following regularity condition [11], [25]

$$u \in H^2(\Omega) \cap H_0^1(\Omega).$$

On the other hand it can be shown [11] that the solution of variational inequality (5.1) is simultaneousely a unique solution of the following variational inequality

$$u \in \tilde{K}(\Omega)$$

$$\int_\Omega \nabla u(x) \nabla(\phi(x) - u(x))dx \geq \mu \int_\Omega (\phi(x) - u(x))dx, \quad \forall \phi \in \tilde{K}(\Omega);$$

here

$$\tilde{K} = \{\phi \in H_0^1(\Omega)| \quad \phi(x) \leq \rho(x) \quad \text{for a.e. } x \in \Omega\}$$

$$\rho(x) = \min_{\xi \in \partial\Omega} ||x - \xi||_{R^2}, \quad x \in \overline{\Omega}$$

and we assume that $k = 1$ in (5.2).

Material derivative $\dot{u}(\Omega)$

We denote by $\rho_t(.)$ the distance function:

$$\rho_t(x) = \inf_{\xi \in \partial\Omega_t} ||x - \xi||_{R^2}, \quad x \in \Omega_t$$

where $\Omega_t = T_t(V)(\Omega)$, $V(.,.)$ being a given vector field.

We assume that the following condition is satisfied:

$$\rho t(.) \in H_0^1(\Omega_t) \text{ for sufficiently small } t > 0,$$

and there exists an element $\dot{\rho}(.) \in H_0^1(\Omega)$ such that

$$\lim_{t \to 0} ||(\rho_t \circ T_t - \rho_0)/t - \dot{\rho}||_{H_0^1(\Omega)} = 0.$$

Here $\dot{\rho}$ denotes the material derivative of the distance function ρ in the direction of the vector field $V(.,.)$.

Let us consider the following variational inequality defined in the domain Ω_t

$$u_t \in K(\Omega_t):$$
$$\int_{\Omega_t} \nabla u_t(x) \nabla(\phi(x) - u_t(x)) dx \geq \mu \int_{\Omega_t} (\phi(x) - u_t(x)) dx, \quad \forall \phi \in K(\Omega_t).$$

Theorem 5 *For sufficiently small* $t > 0$,

$$u_t \circ T_t = u_0 + t\dot{u} + o(t),$$

where $||o(t)||_{H_0^1(\Omega)}/t \to 0$ *with* $t \to 0$.

The material derivative $\dot{u} \in H_0^1(\Omega)$ is given by the unique solution of the following variational inequality

$$\dot{u} \in S(\Omega):$$
$$\int_\Omega \nabla \dot{u}(x) \nabla(\phi(x) - \dot{u}(x)) dx \geq \int_\Omega F'(x)(\phi(x) - \dot{u}(x)) dx$$
$$\int_\Omega -\langle A'(x) \nabla(\phi(x) - \dot{u}(x)) \rangle_{R^2} dx, \quad \forall \phi \in S(\Omega)$$

where

$$S(\Omega) = \{\phi \in H_0^1(\Omega) | \phi(x) \leq \dot{\rho}(x) \text{ q.e. on } P,$$
$$\int_P \nabla u_0 \nabla \phi(x) dx = \mu \int_P \phi(x) dx\}$$
$$F'(x) = \mu \text{ div} V(0, x), x \in \Omega$$
$$A'(x) = -DV(0, x) - {}^*DV(0, x) + \text{div} V(0, x) I.$$

$\square$

Remark 1 In order to use the material derivative $\dot{u}$ in shape optimization, the form of the material derivative $\dot{\rho}(x), x \in P$, should be derived. It follows that

$$\dot{\rho}(x) = -\langle n(z(x)), V(0, x) - V(0, z(x)) \rangle_{R^2}, x \in P,$$

where

$$z(x) = \arg\min\{||\xi - x||_{R^2} | \xi \in \partial\Omega\}, x \in P,$$

and $n(\xi), \xi \in \partial\Omega$, denotes the outward unit normal vector on $\partial\Omega$.

Domain derivative u'

Let $\tilde{u}_t(x), x \in \Omega_t, t \in [0,\delta)$, denote an extension of the element $u_t \in H_0^1(\Omega_t)$ to R^2 defined by

$$\tilde{u}_t(x) = \begin{cases} u_t(x), & x \in \Omega_t, t \in [0,\delta); \\ 0, & x \in R^2 \backslash \Omega_t, t \in [0,\delta). \end{cases}$$

It can be shown [87] that for sufficiently small $t > 0$,

$$\tilde{u}_{t|\Omega} = u_0 + tu' + o(t), \quad \text{in } H^1(\Omega),$$

where $\|o(t)\|_{H_0^1(\Omega)}/t \to 0$ with $t \to 0$.

We derive the form of the domain derivative u'.

Theorem 6 *The domain derivative $u' = u'(\Omega, V)$ of the solution $u = u(\Omega)$ of variational inequality (5.1) in the direction of a vector field $V(.,.)$ is given by the unique solution of the following variational inequality:*

$$u' \in S_v(\Omega)$$

$$\int_\Omega \nabla u'(x) \nabla(\phi(x) - u'(x)) dx \geq 0, \quad \forall \phi \in S_v(\Omega),$$

where

$$S_v(\Omega) = \{\phi \in H^1(\Omega) | \phi(x) = v(x) \text{ on } \partial\Omega, \phi(x) \geq \rho'(x) \text{q.e. on } P,$$

$$\int_P (\Delta\rho(x) + \mu)(\phi(x) - \rho'(x)) dx = 0\}$$

$$v(x) = \langle V(0,x), n(x) \rangle_{R^2}, \quad x \in \partial\Omega$$

$$\rho'(x) = v(z(x)), \quad x \in P$$

$$\Delta\rho(x) = 1/(\rho(x) - R(z(x))), \quad x \in P.$$

$R(.)$ is the radius of curvature of $\partial\Omega$ provided the set

$$\{S_v(\Omega) - S_v(\Omega)\} \cap H^2(\Omega)$$

is dense in the set $\{S_v(\Omega) - S_v(\Omega)\} \subset H^1(\Omega)$. □

Kirchhoff plate

Let $\Omega \in R^2$ be a given domain with smooth boundary $\Gamma = \partial\Omega$. We derive the form of the domain derivative of the solution of an obstacle problem in the Sobolev space $H_0^2(\Omega)$ in the direction of a vector field $V(.,.)$.

Let us assume that $a(.) \in H^2(R^n)$ is a given element such that $a(x) \le 0$ in some open neighborhood in R^2 of the manifold $\partial\Omega \subset R^2$. Thus for $t > 0, t$ small enough, the set

$$K(\Omega_t) = \{\varphi \in H_0^2(\Omega) | \varphi(x) \ge a(x) \text{ in } \Omega\}$$

is a non-empty, closed convex subset of the Sobolev space $H_0^2(\Omega)$. We use the same symbol $a(.)$ to denote the restriction of $a(.) \in H^2(R^2)$ to the domain $\Omega_t, t \in [0, \delta)$. Let there be given an element $f \in H^2(R^2)$. We denote by $w_t \in H^2(\Omega_t), t \in [0, \delta)$, the unique solution of the following variational inequality

$$\begin{cases} w_t \in K(\Omega_t) \\ \int_{\Omega_t} \Delta w_t \Delta(\varphi - w_t) dx \ge \int_{\Omega_t} f(\varphi - w_t) dx, \quad \forall \varphi \in K(\Omega_t). \end{cases}$$

For $t = 0$, we denote $w(x) = w_0(x), x \in \Omega$. Furthermore we denote by $\tilde{w}_t(x), x \in R^n$, an extension of the element $w_t(x), x \in \Omega, t \in [0, \delta)$, defined as follows

$$\tilde{w}_t(x) = \begin{cases} w_t(x), & x \in \Omega_t, t \in [0, \delta) \\ 0, & x \in R^2 \backslash \Omega_t, t \in [0, \delta). \end{cases}$$

It can be shown [60] that the element $\tilde{w}_{t|\Omega} \in H^2(\Omega)$ is right-differentiable with respect to t, at $t = 0$. We denote $V(0) = V = (0, .)$.

First we introduce:

Definition 3 A compact set F is *admissible* if for any element $\varphi \in H_0^2(\Omega), \varphi = 0$ on F implies $\varphi \in H_0^2(\Omega \backslash F)$

Theorem 7 *Assume that sptµ is admissible in the sense of Definition 3. Let $\nabla w. V(0) \in H^2(\Omega)$. Then for $t > 0, t$ small enough,*

$$\tilde{w}_{t|\Omega} = w + tw' + o(t), \quad in \ H^2(\Omega),$$

where $\|o(t)\|_{H^2(\Omega)}/t \to 0$ *with* $t \downarrow 0$.

The shape derivative $w' \in H^2(\Omega)$ *is given by the unique solution of the following variational inequality:*

$$w' \in S_v = \{\varphi \in H^2(\Omega) \cap H_0^1(\Omega) | \frac{\partial\varphi}{\partial n} = -v\frac{\partial^2 w}{\partial n^2} \text{ on } \partial\Omega,$$

$$\varphi(x) \ge 0 \text{ q.e. on } \Xi^+, \varphi(x) = 0 \text{ q.e. on } \Xi^0\}$$

$$\int_\Omega \Delta w' \Delta(\varphi - w') dx \ge 0, \quad \forall \varphi \in S_v(\Omega).$$

Here we denote $v(x) = \langle V(O, x), n(x) \rangle_{R^d}; x \in \partial\Omega, n(x), x \in \partial\Omega$, *is the unit outward normal vector on* $\partial\Omega$ *and* $\Xi^0 = $ *fine sptµ for an arbitrary space dimension* $d > 3$ [19], [57], $\Xi^0 = $ sptµ *for* $d = 1, 2, 3$, *where the Radon measure* $\mu \ge 0$ *is defined by*

$$\int \varphi d\mu = \int_\Omega \{\Delta w \Delta\varphi - \varphi f\} dx,$$

$$\Xi^+ = \Xi \backslash \Xi^0$$

and

$$\Xi = \{x \subset \Omega | w(x) = a(x)\}.$$

Proof of Theorem 7 is given in [60].

6 Sensitivity analysis of shape estimation problems

We present here the results [80] on the shape sensitivity analysis of the elliptic state equation and of the associated shape functional in the variable domain setting, i.e., using the shape derivative of the solution of the elliptic equation. Let $Q \subset R^{n-1}, n \geq 2$, be a given domain with smooth boundary ∂Q. We denote by $\Omega = \Omega_f \subset R^n$ the domain of the following form

$$\Omega = \{(x', x) | 0 < x_n < f(x'), x' \in Q\},$$

where $x' = (x_1, \ldots, x_{n-1})$, and $f(.)$ is a given function in the following set

$$K = \{f \in H_0^s(Q) | 0 < \psi_1(x') \leq f(x') \leq \psi_2(x'), \forall x' \in Q\},$$

here $\psi_i(.) \in H_0^s(Q), i = 1, 2$, are given elements such that the set K is nonempty, $s > n-1$; we select e.g. $s = 3$.

Let us consider the following elliptic equation

$$-\Delta y = F, \quad \text{in } \Omega_f$$

$$y = 0, \quad \text{on } \partial\Omega_f.$$

Denote

$$\Omega_t = \{(x', x) | 0 < x < f(x') + th(x'), x' \in Q\},$$

where $h(.) \in H_0^s(Q)$ is given, t is small; furthermore we denote by

$$h_\nu(x') = \frac{h(x')}{(1 + |\nabla f(x')|^2)^{\frac{1}{2}}}$$

the normal component of the vector field $(0, \ldots, 0, h(x')) \in R^n$ on $\Gamma(f)$. Let

$$-\Delta y_t = F, \quad \text{in } \Omega_t$$

$$y_t = 0, \quad \text{on } \partial\Omega_t,$$

We have the domain derivative $y' = y'(h_\nu)$

$$y' = \frac{\partial Y}{\partial t}(x, t)|_{t=0},$$

where

$$Y(x,t) = \begin{cases} y_t(x) & \text{for } x \in \Omega_t, t \geq 0 \\ 0 & \text{for } x \notin \Omega_t, t \geq 0. \end{cases}$$

The domain derivative is given by the unique solution of the following elliptic equation

$$\begin{aligned} -\Delta y' &= 0, & \text{in } \Omega_f \\ y' &= -h_\nu \frac{\partial y}{\partial n}, & \text{on } \Gamma(f) \\ y' &= 0, & \text{on } \partial\Omega\backslash\Gamma(f). \end{aligned}$$

For the shape functional

$$J_\alpha(f) = \frac{1}{2}\int_{\Omega_f} (y(f;.) - z(.))^2 dx + \frac{\alpha}{2}\|f\|^2_{H^s_0(Q)}$$

the directional derivative in a direction h takes the form

$$dJ_\alpha(f;h) = (\mathcal{G}_0(f)h)_{H^s_0(Q)} + \alpha(f,h)_{H^s_0(Q)}$$

here we assume $z = 0$ on $\Gamma(f)$

$$= \int_{\Omega_f} (y(f;.) - z(.))y'(h_\nu)dx + \alpha(f,h)_{H^s_0(Q)}.$$

In the standard way we introduce the adjoint state

$$-\Delta p = y - z \quad \text{in } \Omega_f$$

$$p = 0, \quad \text{on } \partial\Omega_f.$$

Then

$$dJ_\alpha(f;h) = \int_{\Gamma(f)} h_\nu \frac{\partial p}{\partial n}\frac{\partial y}{\partial n} d\Gamma + \alpha(f,h)_{H^s_0(Q)}.$$

Finally we evaluate the second derivative

$$\begin{aligned} d^2 J_\alpha(f;h,v) &= \lim_{t\downarrow 0}\frac{1}{t}(dJ_\alpha(f+tv;h) - dJ_\alpha(f;h)) \\ &= (\mathcal{H}_0(f)h,v)_{H^s_0(Q)} + \alpha(h,v)_{H^s_0(Q)} \end{aligned}$$

$$= -\int_{\Gamma(f)}[(h_\nu \frac{(\nabla f(x'),\nabla v(x'))}{(1+|\nabla f(x')|^2)} + v_\nu \frac{(\nabla f(x'),\nabla h(x'))}{(1+|\nabla f(x')|^2)} + 2\kappa_m v_\nu h_\nu)\frac{\partial p}{\partial n}\frac{\partial y}{\partial n}]d\Gamma$$

$$+ \int_{\Gamma(f)}[h_\nu v_\nu \frac{\partial}{\partial n}(\frac{\partial y}{\partial n}\frac{\partial p}{\partial n})]d\Gamma + \int_{\Gamma(f)} h_\nu[\frac{\partial p'}{\partial n}\frac{\partial y}{\partial n} + \frac{\partial p}{\partial n}\frac{\partial y'}{\partial n}]d\Gamma + \alpha(h,v)_{H^s_0(Q)}$$

where κ_m is the mean curvature on $\Gamma(f)$, and the domain derivatives $y'(v_\nu), p'(v_\nu)$ satisfy the following elliptic equation

$$\begin{aligned} -\Delta y' &= 0, & \text{in } \Omega_f \\ y' &= -v_\nu \frac{\partial y}{\partial n}, & \text{on } \Gamma(f) \\ y' &= 0, & \text{on } \partial\Omega\backslash\Gamma(f) \end{aligned}$$

and

$$-\Delta p' = y' - z', \quad \text{in } \Omega_f$$
$$p' = -v_\nu \frac{\partial y}{\partial n}, \quad \text{on } \Gamma(f)$$
$$p' = 0, \quad \text{on } \partial\Omega\backslash\Gamma(f).$$

Here z' is the domain derivative of the observation $z(.)$; we can assume $z' = 0$.

Remark 2 Since on $\Gamma(f)$ we have $y'(h_\nu) = -h_\nu \frac{\partial y}{\partial n}, p'(h_\nu) = -h_\nu \frac{\partial p}{\partial n}$, it follows that

$$\int_{\Gamma(f)} h_\nu [\frac{\partial p'}{\partial n}\frac{\partial y}{\partial n} + \frac{\partial p}{\partial n}\frac{\partial y'}{\partial n}]d\Gamma = -\int_{\Gamma(f)} [y'(h_\nu)\frac{\partial p'}{\partial n}(v_\nu) + p'(h_\nu)\frac{\partial y'}{\partial n}(v_\nu)]d\Gamma.$$

We suppose

$$z_\varepsilon = z + \varepsilon\vartheta.$$

Hence the derivative η of the adjoint state p with respect to ε is the unique solution of the following elliptic equation

$$-\Delta\eta = -\vartheta, \quad \text{in } \Omega_f$$
$$\eta = 0, \quad \text{on } \partial\Omega_f.$$

For any $\alpha > 0$ there exists an element f_ε^* which minimizes the perturbed cost functional

$$J_{\alpha,\varepsilon}(f) = \frac{1}{2}\int_{\Omega_f} (y(f;\cdot) - z_\varepsilon(\cdot))^2 dx + \frac{\alpha}{2}\|f\|_{H_0^s(Q)}^2$$

over the set $K \subset H_0^s(Q), s > n - 1$; we assume that $z = 0$ on $\Gamma(f_0^*)$.

Theorem 8 *Assume that there exists $\beta > 0$ such that*

$$d^2 J_\alpha(f_0^*; v, v) \geq \beta\|v\|_{H_0^s(Q)}, \forall v \in \{S - S\},$$

and suppose that the following condition [80] is satisfied:

$$T_K(f) \cap [f - g]^\perp = \text{cl}(C_K(f) \cap [f - g]^\perp)$$

for

$$f = f_0^* = P_K(-\frac{1}{\alpha}\mathcal{G}_0(f_0^*)), \quad g = -\frac{1}{\alpha}\mathcal{G}_0(f_0^*).$$

Then for sufficiently small $\varepsilon > 0$,

$$f_\varepsilon^* = f_0^* + \varepsilon q + o(\varepsilon), \quad \text{in } H_0^s(Q),$$

where $\|o(\varepsilon)\|_{H_0^s(Q)}/\varepsilon \downarrow 0$ with $\varepsilon \downarrow 0$ and the element q is given by the unique solution of the following optimality system, where we denote $f = f_0^*$:

Find (y', p', q) such that the following system is satisfied.

State equation:

$$-\Delta y' = 0, \qquad in \ \ \Omega_f$$

$$y' = -q_\nu \frac{\partial y}{\partial n}, \quad on \ \ \Gamma(f)$$

$$y' = 0, \qquad on \ \ \partial\Omega\backslash\Gamma(f).$$

Adjoint state equation:

$$-\Delta p' = y', \quad in \ \ \Omega_f$$

$$p' = -q_\nu \frac{\partial p}{\partial n}, \quad on \ \ \Gamma(f), p' = 0, \quad on \ \ \partial\Omega\backslash\Gamma(f).$$

Optimality conditions:

$$q \in S = T_K(f_0^*) \cap [f_0^* + \frac{1}{\alpha}\mathcal{G}_0(f_0^*)]^\perp, f_0^* = f,$$

$$-\int_{\Gamma(f)} [((h-q)_\nu \frac{(\nabla f(x'), \nabla q(x'))}{(1+|\nabla f(x')|^2)} + q_\nu \frac{(\nabla f(x'), \nabla(h-q)(x'))}{(1+|\nabla f(x')|^2)} + 2\kappa_m q_\nu (h-q)_\nu) \frac{\partial p}{\partial n} \frac{\partial y}{\partial n}] d\Gamma$$

$$+ \int_{\Gamma(f)} [(h-q)_\nu q_\nu \frac{\partial}{\partial n}(\frac{\partial y}{\partial n} \frac{\partial p}{\partial n})] d\Gamma + \int_{\Gamma(f)} (h-q)_\nu [\frac{\partial p'}{\partial n} \frac{\partial y}{\partial n} + \frac{\partial p}{\partial n} \frac{\partial y'}{\partial n}] d\Gamma$$

$$+ \alpha((h-q), q)_{H_0^s(Q)} + \int_{\Gamma(f)} [(h-q)_\nu \frac{\partial \eta}{\partial n} \frac{\partial y}{\partial n}] d\Gamma \geq 0, \forall h \in S.$$

Here $T_K(v)$ denotes the tangent cone to K at $v \in K, [f-v]^\perp$ is the hyperplane orthogonal in $H_0^s(Q)$ to $f - v$. The cone S takes the following form:

$$S = \{\varphi \in H_0^s(Q) | \varphi \geq 0 \ \ on \ \ \Xi_1, \varphi \leq 0 \ \ on \ \ \Xi_2, (f_0^* + \frac{1}{\alpha}\mathcal{G}_0(f_0^*), \varphi)_{H_0^s(Q)} = 0\},$$

where

$$\Xi_i = \{x \in Q | f_0^*(x) = \psi_i(x)\}, i = 1, 2.$$

Proof of Theorem 8 is given in [80].

References

[1] Banks, H.T. and Kojima, F., Boundary shape identification problems in two-dimensional domains related to thermal testing of materials, *Quart. Appl. Math.* **17** (1989), 273–294.

[2] Barbu, V., *Optimal Control of Variational Inequalities*, Lecture Notes in Mathematics **100**, Pitman, London (1984).

[3] Bendsoe, M.P., Olhoff, N. and Sokolowski, J., Sensitivity analysis of problems of elasticity with unilateral constraints, *J. Struct. Mech.* **13** (1985), 201–222.

[4] Bendsoe, M.P. and Sokolowski, J., Sensitivity analysis and optimization of elastic-plastic structures, *J. Engrg. Optimization*, ASI Special Issue (1987), 31–38.

[5] Bendsoe, M.P. and Sokolowski, J., Design sensitivity analysis of elastic-plastic analysis problem, *Mech. Structures Mach.* 1 (16) (1988), 81–102..

[6] Bendsoe, M.P. and Sokolowski, J., Sensitivity analysis and optimal design of elastic plates with unilateral point supports, *Mech. Structures Mach.* 15 (1987), 383–393.

[7] Benedict, R., Sokolowski, J. and Zolésio, J.P., Shape optimization for contact problems, in: *System Modelling and Optimization* (P. Thoft-Christensen, ed.), Lecture Notes in Control and Information Sciences 59, Springer-Verlag (1984), 790–799.

[8] Benedict, R. and Taylor, J.E., Optimal design for elastic bodies in contact, in: *Optimization of Distributed Parameter Structures* (J. Cea and E.J Haug, eds.), Sijthoff & Noordhoff (1981), 1553–1599.

[9] Bielski, W.R. and Telega, J.J., A contribution to contact problems for a class of solids and structures, *Arch. Mech.* 37 (1985), 303–320.

[10] Brézis, H. and Browder, F., Some properties of higher order Sobolev spaces, *J. Math. Pures Appl.* 61 (1982), 245–259.

[11] Brézis, H. and Stampacchia, G., Sur la régularité de la solution d'inéquations elliptiques, *Bull. Soc. Math. France* 96 (1968), 153–180.

[12] Coffman, C.V. and Grover, C.L., Obtuse cones in Hilbert spaces and applications to partial differential equations, *J. Funct. Anal.* 35 (1980), 369–396.

[13] Da Carmo, M.P., *Differential Geometry of Curves and Surfaces*, Prentice-Hall, Englewood Cliffs, NJ (1976).

[14] Delfour, M., Shape derivatives and differentiability of Min Max, *this volume*.

[15] Delfour, M. and Zolésio, J.-P., Shape sensitivity analysis via min max differentiability, *SIAM J. Control Optim.* 26 (1988), 834–862.

[16] Delfour, M.C. and Zolésio, J.-P., Shape sensitivity analysis via a penalization method, *Ann. Mat. Pura Appl. (4)* 151 (1988), 179–212.

[17] Dems, K. and Mroz, Z., Variational approach by means of adjoint systems to structural optimization and sensitivity analysis. II. Structure shape variation. *Internat. J. Solids and Structures* 20 (1984), 527–552.

[18] Dervieux, A., Résolution de problèmes à frontière libre, Thèse d'État, Université de Paris VI, 1981.

[19] Doob, J.L., *Classical Potential Theory and its Probabilistic Counterpart*, Springer-Verlag, New York, 1985.

[20] Duvaut, G. and Lions, J.L., *Inequalities in Mechanics and Physics*, Grundlehren der mathematischen Wissenschaften **219**, Springer-Verlag, Berlin (1976).

[21] Ekeland, I. and Temam, R., *Convex Analysis and Variational Problems*, North-Holland, Amsterdam (1976).

[22] Fiacco, A.V., *Introduction to Sensitivity and Stability Analysis in Nonlinear Programming*, Academic Press, New York (1983).

[23] Fichera, G., *Existence Theorems in Elasticity* (Handbuch der Physik Vol. VIa), Springer-Verlag (1972), 374–389.

[24] Fitzpatrick, S. and Phelps, R.R., Differentiability of the metric projection in Hilbert space, *Trans. Amer. Math. Soc.* **207** (1982), 483–501.

[25] Friedman, A., *Variational Principles and Free-Boundary Problems*, John Wiley and Sons (1982).

[26] Goto, Y., Fujii, N., and Muramatsu, Y., Second order necessary optimality conditions for domain optimization problem with a Neumann problem, Lecture Notes in Control and Information Sciences **113**, Springer-Verlag (1988), 259–268.

[27] Grisvard, R.V., *Elliptic Problems in Non-Smooth Domains*, Pitman, London (1985).

[28] Haftka, R.T. and Grandhi, R.V., Structural shape optimization – a survey, *Comput. Methods Appl. Mech. Engrg.* **57** (1986), 91–106.

[29] Harvaux, A., How to differentiate the projection on a convex set in Hilbert space. Some applications to variational inequalities, *J. Math. Soc. Japan* **29** (1977), 615–631.

[30] Haslinger, J. and Neittaanmaki, P., *Finite Element Approximation for Optimal Shape Design*, John Wiley and Sons, Chichester (1989).

[31] Haug, E.J. and Cea, J. (eds.), *Optimization of Distributed Parameter Structures*, Sijthoff and Noordhoff (1981).

[32] Haug, E.J., Choi, K.K., and Komkov, V., *Design Sensitivity Analysis of Structural Systems*, Academic Press, New York (1986).

[33] Hedberg, L.I., Spectral synthesis in Sobolev spaces, and uniqueness of solutions of the Dirichlet problem, *Acta Math.* **147** (1981), 237–264.

[34] Hlavacek, I., Haslinger, J., Necas, J. and Lovisek, J., *Numerical Solution of Variational Inequalities*, Springer Series in Comput. Physics, Springer-Verlag (1982).

[35] Hlavacek, I. and Necas, J., Optimization of the domain in elliptic unilateral boundary value problems by finite element method, *RAIRO Modél. Math. Anal. Numér.* **16** (1982), 351–373.

[36] Hoffmann, K.-H. and Sprekels, J., On the identification of parameters in general variational inequalities by asymptotic regularization, *SIAM J. Math. Anal.* 5 (**17**) (1986), 1198–1217.

[37] Kappel, F., Kunisch, K. and Schappacher, W. (eds.), *Distributed Parameter Systems*, Lecture Notes in Control and Information Sciences **75**, Springer-Verlag (1985).

[38] Kappel, F., Kunisch, K. and Schappacher, W. (eds.), *Distributed Parameter Systems*, Lecture Notes in Control and Information Sciences **102**, Springer-Verlag (1987).

[39] Kikuchi, N. and Oden, J.T., *Contact Problems in Elasticity*, SIAM, Philadelphia (1987).

[40] Kinderlehrer, D. and Stampacchia, G., *An Introduction to Variational Inequalities and Their Applications*, Academic Press (1980).

[41] Khludnev, A.M., Existence and regularity of solutions of unilateral boundary value problems in linear theory of shallow shells (in Russian), *Differentsial'nye Uravneniya* **11** (1984), 1968–1975.

[42] Kosinski, W., *Introduction to Field Singularities and Wave Analysis*, PWN – Ellis Horwood (1986).

[43] Lasiecka, I. and Sokolowski, J., Sensitivity analysis of control constrained optimal control problem for wave equation, *SIAM J. Control Optim.* **29** (1991), 1128–1149.

[44] Lasiecka, I. and Triggiani, R. (eds.), *Control Problems for Systems Described by Partial Differential Equations and Applications* (Proceedings of the IFIP-WG7.2 Working Conference, Gainesville, Florida, February 3–6, 1986) Lecture Notes in Control and Information Sciences **97**, Springer-Verlag (1987).

[45] Lions, J.L., *Optimal Control of Systems Governed by Partial Differential Equations*, Springer-Verlag, New York (1971).

[46] Lions, J.L., *Perturbations singulières dans les problèmes aux limites et en contrôle optimal*, Lecture Notes in Mathematics **323**, Springer-Verlag, Berlin (1973).

[47] Malanowski, K., *Stability of Solutions to Convex Problems of Optimization*, Lecture Notes in Control and Information Sciences **93**, Springer-Verlag (1987).

[48] Mignot, F., Contrôle dans les inéquations variationnelles elliptiques, *J. Funct. Anal.* **22** (1976), 25–39.

[49] Mignot, F. and Puel, J.-P., Optimal control of some variational inequalities, *SIAM J. Control Optim.* **22** (1984), 466–478.

[50] Moreau, J.J., Quadratic programming in mechanics: dynamics of one-sided constraints, *SIAM J. Control* **4** (1966), 153–158.

[51] Murat, F. and Simon, J., *Sur le contrôle par un domaine géométrique*, Publications du Laboratoire d'Analyse Numérique, Université de Paris VI (1976).

[52] Necas, J. and Hlavacek, I., *Mathematical Theory of Elastic and Elasto-Plastic Bodies*, Elsevier, Amsterdam (1981).

[53] Neittaanmaki, P., Sokolowski, J. and Zolésio, J.-P., Optimization of the domain in elliptic variational inequalities, *Appl. Math. Optim.* **18** (1988), 85–98.

[54] Olhoff, N. and Taylor, J.E., On structural optimization, *J. Appl. Mech.* **50** (1983), 1139–1151.

[55] Panagiotopoulos, P.D., *Inequality Problems in Mechanics and Applications*, Birkhäuser, Boston (1985).

[56] Pironneau, O., *Optimal Shape Design for Elliptic Systems*, Springer Series in Computational Physics, Springer-Verlag, New York (1984)

[57] Rabier, P.J. and Oden, J.T., Solution to Signorini-like contact problems through interface models – I. Preliminaries and formulation of a variational equality, *Nonlinear Anal.* **11** (1987), 1325–1350.

[58] Rabier, P.J. and Oden, J.T., Solution to Signorini-like contact problems through interface models – II. Existence and uniqueness theorems, *Nonlinear Anal.* **12** (1988), 1–17.

[59] Rao, M. and Sokolowski, J., Polyhedricity of convex sets in Sobolev space $H_0^2(\Omega)$, to appear.

[60] Rao, M. and Sololowski, J., Sensitivity analysis of unilateral problems in $H_0^2(\Omega)$ and applications, to appear in: *Proc. Conf. on Free Boundaries, Montréal 1990*, Longman-Pitman.

[61] Rao, M. and Sokolowski, J., Shape sensitivity analysis of state constrained optimal control problems for distributed parameter systems, Lecture Notes in Control and Information Sciences **114**, Springer-Verlag (1989), 236–245.

[62] Rao, M. and Sokolowski, J., Differential stability of solutions to parametric optimization problems, *Math. Methods Appl. Sci.* **14** (1991), 281–294.

[63] Rao, M. and Sokolowski, J., Sensitivity analysis of shallow shell with an obstacle, in: *Modelling and Inverse Problems of Control for Distributed Parameter Systems* (A. Kurzhanski and I. Lasiecka, eds.), Lecture Notes in Control and Information Sciences **154** (1991), 135–144.

[64] Rao, M. and Sokolowski, J., Sensitivity analysis of Kirchhoff plate with obstacle, INRIA, Rapport de recherche (1987).

[65] Rousselet, B., *Quelques résultats en optimisation de domaines*, Thèse d'État, Université de Nice (1982).

[66] Saguez, C., *Contrôle optimal de systèmes à frontière libre*, Thèse d'État, Compiegne (1980).

[67] Simon, J., Differentiability with respect to the domain in boundary value problems, *Numer. Funct. Anal. Optim.* **2** (1980), 649–687.

[68] Sokolowski, J., Optimal control in coefficients for weak variational problems in Hilbert space, *Appl. Math. Optim.* **7** (1981), 283–293.

[69] Sokolowski, J., Control in coefficients for PDE, in: *Theory of Nonlinear Operators* (Abh. der Akad. d. Wiss. d. DDR), Akademie Verlag, Berlin (1981), 287–295.

[70] Sokolowski, J., Sensitivity analysis for a class of variational inequalities, in: *Optimization of Distributed Parameter Structures*, Vol. 2 (E.J. Haug and J. Cea, eds.), Sijthoff & Noordhoff, Rockville MD (1981), 1600–1609.

[71] Sokolowski, J., Optimal control in coefficients of boundary value problems with unilateral constraints, *Bull. Polish Acad. Tech. Sci.* **31** (1983), 71–81.

[72] Sokolowski, J., Sensitivity analysis of Signorini variational inequality, in: *Banach Center Publications* (B. Bojarski, ed.), Polish Scientific Publisher, Warsaw (1988).

[73] Sokolowski, J., Differential stability of solutions to constrained optimization problems, *Appl. Math. Optim.* **13** (1985), 97–115.

[74] Sokolowski, J., Sensitivity analysis of control constrained optimal control problems for distributed parameter systems, *SIAM J. Control Optim.* **25** (1987), 1542–1556.

[75] Sokolowski, J., Shape sensitivity analysis of boundary optimal control problems for parabolic systems, *SIAM J. Control Optim.* **26** (1988), 763–787.

[76] Sokolowski, J., *Sensitivity Analysis and Parametric Optimization of Optimal Control Problems for Distributed Parameter Systems*, Habilitation Thesis, Warsaw Technical University Publications, Prace Naukowe, Elektronika, z. 73 (Polish) 1985.

[77] Sokolowski, J., Shape sensitivity analysis of nonsmooth variational problems, Lecture Notes in Control and Information Sciences **100** (J.-P. Zolésio ed.), Springer-Verlag, Berlin (1988), 265–285.

[78] Sokoloswki, J., Sensitivity analysis of contact problems with prescribed friction, *Appl. Math. Optim.* **18** (1988), 99–117.

[79] Sokolowski, J., Sensitivity analysis of contact problem with friction, in: *Free Boundary Problems: Theory and Applications* (K.-H. Hoffmann and J. Sprekels, eds.), Pitman Research Notes in Mathematics **185** (1990), 329–334.

[80] Sokolowski, J., Differential stability of solutions to shape optimization problems, to appear in *Mech. Structures Mach.* (special volume dedicated to Jean Céa).

[81] Sokolowski, J. and Sprekels, J., Dynamical shape control and stabilization of non-linear rod, to appear in *Math. Methods Appl. Sci.*

[82] Sokolowski, J. and Zolésio, J.-P., Shape sensitivity analysis for variational inequalities, Lecture Notes in Control and Information Sciences **38** (R.F. Drenick and F. Kozin, eds.), Springer-Verlag (1982), 401–407.

[83] Sokolowski, J. and Zolésio, J.-P., Dérivation par rapport au domaine dans les problèmes unilatéraux, INRIA, Rapport de recherche No 132 (1982).

[84] Sokolowski, J. and Zolésio, J.-P., *Shape sensitivity analysis of elastic structures*, The Danish Center of Applied Mathematics and Mechanics, Lyngby, Report No. 289 (1984).

[85] Sokolowski, J. and Zolésio, J.-P., Dérivée par rapport au domaine de la solution d'un problème unilatéral, *C.R. Acad. Sci. Paris Série I* **301** (1985), 103–106.

[86] Sokolowski, J. and Zolésio, J.-P., Shape sensitivity analysis of unilateral problems, *SIAM J. Math. Anal.* **18** (1987), 1416–1437.

[87] Sokolowski, J. and Zolésio, J.-P., *Introduction to Shape Optimization. Shape Sensitivity Analysis*, Springer Series in Computational Mathematics vol. 16, Springer-Verlag, 1992.

[88] Sokolowski, J. and Zolésio, J.-P., Shape sensitivity analysis of an elastic-plastic torsion problem, *Bull. Polish Acad. Sci. Tech. Sci.* **33** (1985), 579–586.

[89] Sokolowski, J. and Zolésio, J.-P., Sensitivity analysis of elastic torsion problem, Lecture Notes in Control and Information Sciences **84** (A. Prekopa et al., eds.), Springer-Verlag (1986), 845–853.

[90] Sokolowski, J. and Zolésio, J.-P., Shape sensitivity analysis of contact problems with prescribed friction, *Nonlinear Analysis* **12** (1988), 1399–1411.

[91] Sokolowski, J. and Zolésio, J.-P., Shape design sensitivity analysis of plates and plane elastic solids under unilateral constraints, *J. Optim. Theory Appl.* **54** (1987), 361–382.

[92] Telega, J.J., Limit analysis theorems in the case of Signorini's boundary conditions and friction, *Arch. Mech.* **37** (1985), 549–562.

[93] Temam, R., *Mathematical Problems in Plasticity*, Gauthier-Villars, Paris (1985).

[94] Temam, R., *Infinite Dimensional Dynamical Systems in Mechanics and Physics*, Springer-Verlag (1988).

[95] Troicki, V.A. and Pietukhov, L.V., *Shape Optimization in Elacticity* (in Russian), Nauka, Moscow (1982).

[96] Zarantonello, F.H., Projections on convex sets in Hilbert space and spectral theory, in: *Contributions to Nonlinear Functional Analysis*, Publ. No. 27, Math. Res. Center Univ. Wisconsin, Academic Press, New York (1971), 237–424.

[97] Ziemer, P.W., *Weakly Differentiable Functions*, Springer-Verlag, New York, 1989.

[98] Zolésio, J.-P., *Identification de domaines par déformations*, Thèse d'État, Université de Nice (1979).

[99] Zolésio, J.-P., The material derivative (or speed) method for shape optimization, in: *Optimization of Distributed Parameter Structures* (E. J. Haug and J. Cea, eds.), Sijthoff & Noordhoff (1981), 1089–1151.

[100] Zolésio, J.-P., Shape controlability of free boundaries, Lecture Notes in Control and Information Sciences **59** (P. Thoft-Christensen ed.), Springer-Verlag (1985), 354–361.

[101] Zolésio, J.-P., *Boundary Control and Boundary Variations*, Lecture Notes in Control and Information Sciences **100**, Springer-Verlag, Berlin (1988).

[102] Zolésio, J.-P., Introduction to shape optimization problems and free boundary problems, *this volume*.

Diffusion with Strong Absorption

Ivar STAKGOLD

Department of Mathematical Sciences
501 Ewing Hall
University of Delaware
Newark, DE 19716
USA

Abstract

After a review of the qualitative properties of the heat equation without absorption or with linear absorption, consideration is given to problems with nonlinear absorption and, in particular, to those with a fractional power absorption rate (strong absorption). In the latter case, there can appear new phenomena such as extinction in finite time, localization of support, thermal fronts and dead cores. These phenomena are analyzed in various geometries and conditions for their existence are established. The relation between the steady and evolutionary problems is investigated. The final section deals with gas-solid reactions and the range of validity of the pseudo-steady-state approximation.

Chapter 1 – Introduction to the heat equation

1 The equation of heat conduction

The equation of heat conduction is a mathematical statement of the physical law of conservation of energy in a special setting. Let Ω be a domain in which heat conduction is taking place and let D be an *arbitrary bounded* subdomain of Ω. We now take a heat balance per unit time for the domain D. The heat G_D generated within D (say, by an exothermic chemical reaction) must be balanced by the sum of the heat F_D flowing out of D through ∂D and the rate of change of the total enthalpy H_D (amount of heat) in D. The basic conservation law then takes the form

$$G_D = F_D + \frac{dH_D}{dt}. \tag{1.1}$$

We shall assume that there are volume densities $g(x,t)$ and $h(x,t)$ defined for x in Ω in terms of which we can write for any subdomain D:

$$G_D = \int_D g(x,t)dx, \quad H_D = \int_D h(x,t)dx.$$

M. C. Delfour and G. Sabidussi (eds.), Shape Optimization and Free Boundaries, 321–345.
© 1992 *Kluwer Academic Publishers. Printed in the Netherlands.*

The flow of heat across a surface element $\vec{n}\,ds$ located at (x,t) will be written in terms of a heat flux vector $\vec{f}(x,t)$ as $\vec{f}\cdot\vec{n}\,ds$, so that

$$F_D = \int_{\partial D} \vec{f}\cdot\vec{n}\,ds,$$

for *all* subdomains D of Ω. We apply the divergence theorem to F_D to transform it into a volume integral. After combining the three volume integrals in (1.1) and using the postulated continuity of the integrand, we obtain

$$h_t + \operatorname{div}\,\vec{f} = g \quad \text{for all } x \in \Omega. \tag{1.2}$$

Here $h = Cu$ where $u(x,t)$ is the temperature and C is the specific heat per unit volume. The flux $\vec{f}$ is related to the temperature through Fourier's law

$$\vec{f} = -k\,\operatorname{grad}\,u$$

where k is the thermal conductivity. Then (1.2) becomes the partial differential equation

$$(Cu)_t - \operatorname{div}\,(k\,\operatorname{grad}\,u) = g. \tag{1.3}$$

In our work we shall take C and k as constants, but we point out that the case of nonlinear diffusion where $k = k(u)$ will be treated in the lectures by J.L. Vazquez; another set of problems that will not be discussed here involve a phase transition such as the melting of ice, where h has a jump discontinuity (equal to the latent heat) at the melting temperature.

Equation (1.3) applies equally well to diffusion of a substance of concentration u, where now $C = 1$ and k is the diffusion constant. We shall feel free to appeal to either physical intepretation of (1.3) as the occasion suggests.

Confining ourselves to the case where C and k are constants, we obtain

$$u_t - \Delta u = g, \tag{1.4}$$

where we have divided by k, rescaled the time and redefined g. The sign of g determines whether we have heat *sources* (wherever $g > 0$) or *sinks* ($g < 0$). It will be convenient to denote $u_t - \Delta u$ by Lu. Obviously L is a linear operator.

If g is a given as a function of x and t, (1.4) becomes a linear parabolic equation. For general information on the linear heat equation, see [4], [8] and [20]. In many problems, however, g is prescribed as a function of u; for problems of absorption, the case $g(u) = -\lambda^2 u^p, p > 0, \lambda^2 > 0$, is particularly important, and (1.4) becomes

$$Lu = -\lambda^2 u^p, \tag{1.5}$$

which is a nonlinear parabolic equation unless $p = 1$ (so-called heat equation with linear absorption).

To have a well-posed problem (that is a problem with one and only one solution depending continuously on the data) we must impose suitable initial and boundary conditions on (1.4) or (1.5). Typically, we associate with (1.4) a prescribed initial value of the temperature in Ω and a given boundary temperature on $\partial\Omega$ for all times $t > 0$; if g is given as a function of x, t, we then have a linear initial-boundary value problem:

$$
\begin{cases}
Lu = g(x,t), & x \in \Omega, \quad t > 0; \\
u(x,0) = u_0(x), & x \in \Omega; \\
u(x,t) = b(x,t), & x \in \partial\Omega, \quad t > 0.
\end{cases}
\tag{1.6}
$$

Our physical intuition tells us that the solution of (1.6) at any time cannot depend on subsequent values of g and b. If we are only interested in the time interval up to T, we could as well consider the problem

$$
\begin{cases}
Lu = g, & x \in \Omega, \quad 0 < t < T; \\
u(x,0) = u_0(x), & x \in \Omega; \\
u(x,t) = b(x,t), & x \in \partial\Omega, \quad 0 < t < T.
\end{cases}
\tag{1.7}
$$

Thus in (1.6) or in (1.7), the differential equation holds in a cylinder in space-time

$$
Q_T = \Omega \times (0,T),
$$

where T could be finite or infinite and with u given on the so-called *parabolic boundary* Σ_T of the cylinder (lower base plus lateral surface). Note that we do not specify u on the upper base ($x \in \Omega, t = T$).

2 The maximum principle

Theorem 1 *Let Ω be a bounded domain in R^n and let $u(x,t)$ be continuous in $\overline{Q}_T$ and satisfy the differential inequality*

$$
Lu \leq 0 \quad in \quad Q_T.
\tag{1.8}
$$

If $u \leq M$ on $\overline{\Sigma}_T$, then $u(x,t) \leq M$ in $\overline{Q}_T$.

Remark The physical interpretation is simple. Inequality (1.8) tells us that no heat is added in the interior. It therefore follows that the interior temperature cannot exceed the maximum of the combined initial and boundary temperatures. To prove the theorem, we start with the easier case of a strict inequality in (1.8).

Lemma *Let $v(x,t)$ be continuous in $\overline{Q}_T$ and satisfy*

$$
Lv < 0 \quad in \quad Q_T,
\tag{1.9}
$$

then v cannot have a (local) maximum in Q_T nor on the upper base $(x \in \Omega, t = T)$.

Proof of Lemma If v had a (local) maximum at an interior point (x,t), then $v_t = 0$ and $\Delta v \leq 0$, which contradicts (1.9). If v had a maximum at $x \in \Omega, t = T$, then $\Delta v \leq 0$ and $v_t \geq 0$ which also violates (1.9).

Proof of Theorem 1 Since Ω is bounded, it can be enclosed in an n-ball of finite radius a with center at the origin. Let $u(x,t)$ satisfy the hypotheses in Theorem 1, and define $v(x,t) = u(x,t) + \epsilon|x|^2$, where $\epsilon > 0$. Then

$$Lv \leq -2n\epsilon < 0,$$

so that v satisfies the inequality in the Lemma. Thus,

$$u(x,t) \leq v(x,t) \leq \max_{(x,t)\in\overline{\Sigma}_T} v \leq M + \epsilon a^2,$$

and, by letting $\epsilon \to 0$, we find $u \leq M$.

Corollary 1 (Minimum principle) *Let $u(x,t)$ be continuous on $\overline{Q}_T$ and satisfy*

$$Lu \geq 0 \quad in \quad Q_T.$$

If $u \geq m$ on $\overline{\Sigma}_T$, then $u(x,t) \geq m$ in $\overline{Q}_T$.

Proof Let $w = -u$ and apply Theorem 1.

Corollary 2 *If $u(x,t)$ satisfies the homogeneous heat equation $u_t - \Delta u = 0$ in Q_T, then $m \leq u(x,t) \leq M$, where m and M are the respective minimum and maximum of u on $\overline{\Sigma}_T$.*

Corollary 3 (Uniqueness) *Let u and v be two solutions of (1.7). Then $u \equiv v$ in Q_T.*

Proof Let $w = u - v$, then $w = 0$ on Σ_T and satisfies $Lw = 0$ in Q_T. By Corollary 2, $w \equiv 0$ in $\overline{Q}_T$, so that $u = v$.

Remark Since T is arbitrary, it is clear that any two solutions of (1.6) must coincide for all x and t.

Remark Obviously a constant satisfies the homogeneous heat equation in Q_T so that it is possible for the max (and min) of u to be taken on in the interior. Is this the only case where the maximum, say, is taken on in the interior? The answer is no (unlike the elliptic case of the Laplace equation). If, for instance $u_0(x) \equiv 1$ in Ω and $b(x,t) = 1$ for x on $\partial\Omega$ and $t \leq T_0$ and $b(x,t) = 0$ for x on $\partial\Omega$ and $T_0 < t < T$, then $u(x,t) \equiv 1$ for $t \leq T_0$ so that max u is taken on also at any point in Q_T with $t \leq T_0$. But, fortunately, this is the only situation in which an interior maximum can occur. We have (proof omitted, see Protter and Weinberger [16]):

Theorem 2 *If $Lu = 0$ in Q_T and the maximum (or minimum) of u on $\overline{Q}_T$ occurs at a point (x_0, t_0) with $x_0 \in \Omega$ (that is, an interior point), then $u \equiv$ constant for all $t \leq t_0$.*

Let Ω be a bounded domain and consider the problem

$$Lu = g(x,t) \text{ in } Q_T, \quad u \text{ given on } \Sigma_T, \tag{1.10}$$

for two different sets of data. We have the following

Comparison Theorem *If $g_1 \geq g_2$ in Q_T, and $u_1|_{\Sigma_T} \geq u_2|_{\Sigma_T}$, then $u_1(x,t) \geq u_2(x,t)$ in $\overline{Q}_T$.*

Proof We see that $w = u_2 - u_1$ satisfies $Lw \leq 0$ in Q_T and $w \leq 0$ on Σ_T. The maximum principle gives $w \leq 0$ in Q_T, the desired result.

An interesting consequence of Theorem 2 is the loss of compactness of initial support. Consider Problem (1.6) with $g = b = 0$ and $u_0(x) \geq 0, u_0 \not\equiv 0$ and supp $u_0 = D \subset \Omega$. Think of u_0 being positive in some small part of Ω and 0 elsewhere. Then $u(x,t) > 0$ for all points $x \in \Omega, t > 0$. Thus, instantaneously, the support of $u(x,t)$ becomes $\overline{\Omega}$. This also applies to the case where $\Omega = R^n$. If $u_0(x) \geq 0$ and $u_0(x) \not\equiv 0$, the solution of $Lu = 0, u(x,0) = u_0(x)$ is positive for $t > 0$ and *all* x. (Instantaneous loss of compactness of support; infinite speed of propagation.)

3 Positivity results – infinite speed of propagation

We consider the homogeneous heat equation with initial temperature $u_0(x) \geq 0$ and $u_0(x) \not\equiv 0$.

a) Let us first look at a bounded domain Ω with vanishing boundary temperature $(u|_{\partial\Omega} = 0)$. Then $u(x,t) > 0$ for every $x \in \Omega$ and $t > 0$. This follows from the maximum principle. Note that the result holds even if $u_0(x)$ has very small support.

b) For $\Omega = R^1$, the solution of the initial value problem is given explicitly using Green's function:

$$u(x,t) = \frac{1}{\sqrt{4\pi t}} \int_{-\infty}^{\infty} e^{-(x-y)^2/4t} u_0(y) dy. \tag{1.11}$$

It is clear that $u > 0$ for all $x \in R^1$ and $t > 0$, despite the fact that the initial temperature $u_0(x)$ may have small support. Thus, *compact support is lost instantaneously* and the temperature is positive on the whole line for $t > 0$ (*infinite speed of propagation*). Note that (1.11) provides the solution to the initial value problem if

$$|u_0(x)| \leq C \exp(Ax^B), \quad B < 2. \tag{1.12}$$

If $u_0(x) = Ce^{Ax^2}$, then (1.11) blows up in the finite time $T = \frac{1}{4A}$. We shall confine ourselves to cases where $u_0(x)$ satisfies (1.12).

Even if we have linear absorption there is instantaneous loss of compact support. Indeed, consider the problem

$$u_t - u_{xx} = -\lambda^2 u, \quad -\infty < x < \infty, \ t > 0; \quad u(x,0) = u_0(x) \geq 0. \tag{1.13}$$

Setting $u = v e^{-\lambda^2 t}$, we see that v satisfies the heat equation *without* absorption and with initial value $u_0(x)$. Thus, $v > 0$ for all $t > 0$ and all x in R, and so is $u(x,t)$.

Similar properties hold in R^n.

Let us now list some of the properties of the initial value problem

$$\begin{aligned}
u_t - \Delta u &= 0, \quad x \in \Omega, \quad t > 0; \\
u(\partial \Omega, t) &= 0, \quad t > 0; \\
u(x,0) &= u_0(x) \geq 0.
\end{aligned} \tag{1.14}$$

I. $u(x,t) > 0$ for every $x \in \Omega, t > 0$.

II. $\max\limits_{x \in \bar{\Omega}} u(x,t)$ is monotonically decreasing in t.

III. If $\Delta u_0 \leq 0$, then $u(x,t)$ is monotonically decreasing in time. [First, observe that $u(x,t) \leq u_0(x)$ by the Comparison Theorem. Then let $v(x,t) = u(x, t + \Delta t)$ and, again by comparison, $v(x,t) \leq u(x,t)$]. Note that if $\Delta u_0 > 0$ at x_0, then $u(x_0, t)$ is initially increasing (since $u_t(x_0, t) > 0$ from the differential equation). This may even be true at points where $\Delta u_0 = 0$; suppose, for instance, that $u_0 = 0$ in some subdomain of Ω. Then, $u(x,t)$ is initially increasing in that subdomain since $u(x,t) > 0$ for $t > 0$.

IV. $\lim\limits_{t \to \infty} u(x,t) = 0$ (see section 4).

4 "Explicit" solutions and asymptotic behavior

Let Ω be a bounded domain and consider the problem

$$Lv = g(x), \quad v(x,0) = v_0(x), \quad v(\partial \Omega, t) = b(x). \tag{1.15}$$

For large t, we expect the solution to be essentially independent of time and to approach the steady-state $U(x)$, the solution of the elliptic problem

$$-\Delta U = g(x), \quad x \in \Omega; \quad U(\partial \Omega) = b(x). \tag{1.16}$$

Writing $v(x,t) = U(x) + u(x,t)$, we see that $u(x,t)$ satisfies

$$Lu = 0, \quad u(x,0) = u_0(x), \quad u(\partial \Omega, t) = 0, \tag{1.17}$$

where $u_0(x) = v_0(x) - U(x)$. Regarding $U(x)$ as known, (1.15) is reduced to solving the homogeneous heat equation (1.17) with vanishing boundary data. This is easily done by an expansion in eigenfunctions of the negative Laplacian. Let (λ_k, ϕ_k) be the eigenpairs satisfying

$$-\Delta \phi_k = \lambda_k \phi_k, \quad x \in \Omega; \quad \phi_k(\partial \Omega) = 0, \tag{1.18}$$

where the λ_k's are the eigenvalues indexed, with due regard to multiplicity, so that

$$\lambda_1 < \lambda_2 \leq \lambda_3 \leq \lambda_4 \leq \cdots$$

and the eigenfunctions satisfy

$$\int_\Omega \phi_k^2(x)dx = 1, \quad \phi_1(x) > 0.$$

$$\langle \phi_i, \phi_j \rangle = \int_\Omega \phi_i(x)\phi_j(x)dx = 0, \quad i \neq j.$$

The solution of (1.17) is then given by

$$u(x,t) = \sum_{k=1}^\infty \langle u_0, \phi_k \rangle e^{-\lambda_k t} \phi_k(x). \tag{1.19}$$

Note that $\lim_{t\to\infty} u(x,t) = 0$, as expected, so that the solution $v(x,t)$ of (1.15) tends to the steady state $U(x)$ as $t \to \infty$. It is an easy exercise to also write $U(x)$ in terms of the eigenpairs (λ_k, ϕ_k). The leading term in (1.19) for large t is $\langle u_0, \phi_1 \rangle e^{-\lambda_1 t}\phi_1(x)$ which is positive if $\langle u_0, \phi_1 \rangle > 0$. Thus, even if $u_0(x)$ changes sign (as long as $\langle u_0, \phi_1 \rangle \neq 0$), $u(x,t)$ is of one sign for large t. This should be constrasted with the case $u_0 = \phi_2$, when $u(x,t) = e^{-\lambda_2 t}\phi_2(x)$ (and hence, changes sign on Ω). Note, that in any event, $u(x,t) \to 0$ as $t \to \infty$.

Additional asymptotic information on $u(x,t)$ can be obtained by introducing

$$w(x,t) = \int_0^t u(x,\tau)d\tau,$$

which is seen to satisfy

$$Lw = u_0(x); \quad w(x,0) = 0, \quad w(\partial\Omega,t) = 0.$$

Hence $w(x,t)$ tends as $t \to \infty$ to the solution of the Poisson problem

$$-\Delta V = u_0(x), \quad V(\partial\Omega) = 0, \tag{1.20}$$

and

$$\int_0^\infty u(x,\tau)d\tau = V(x),$$

which gives us some global information on $u(x,t)$.

Chapter 2 – The heat equation with nonlinear absorption

1 Formulation and preliminary results

We shall consider the equation

$$Lu \doteq u_t - \Delta u = -\lambda^2 u_+^p, \quad p > 0, \tag{2.1}$$

where $u(x,t)$ can be regarded either as a temperature in a heat conduction problem or as the concentration of a diffusing gas. The term $-\lambda^2 u_+^p$ characterizes the absorption; here $u_+ = \max(u,0)$. Thus, there is a threshold (chosen as $u = 0$) below which no absorption takes place. Since both the initial and boundary conditions on u will be nonnegative, the solution $u(x,t)$ can also be shown to be nonnegative; against that background, we may choose to omit the $+$ subscript in (2.1) without confusion.

If $p = 1$, we have linear absorption and then, as we have seen, the solutions behave qualitatively like those of the heat equation without absorption. The situation is similar if $p > 1$ but much different if $0 < p < 1$ (*strong absorption*). When $0 < p < 1$ the absorption is larger for small values of u than in the case $p = 1$. This leads to new phenomena such as extinction in finite time ($u \equiv 0$ after some time T) and compact support of the solution when the initial value on R^n has compact support.

We first investigate some simple special problems associated with (2.1):

a) *Spatially homogeneous solutions.* Considering (2.1) on R^n with $u(x,0) = $ constant, we expect the solution to be independent of x. We seek $u(t)$ that satisfies

$$u_t = -\lambda^2 u_+^p, \quad t > 0; \quad u(0) = u_0 > 0. \tag{2.2}$$

By simple scaling we see that

$$u(t) = u_0 B_p(\lambda^2 u_0^{p-1} t),$$

where $B_p(t)$ is the solution of (2.2) with $\lambda = u_0 = 1$.

An easy calculation gives

$$B_p(t) = [1 - (1-p)t]_+^{\frac{1}{1-p}}, \quad p \neq 1$$
$$B_1(t) = e^{-t}. \tag{2.3}$$

For $p > 1$ the $+$ subscript is unnecessary as the bracket is positive for all $t \geq 0$. Clearly $B_p(t)$ is decreasing and tends to zero as $t \to \infty$; the same is true for $B_1(t)$.

When $p < 1$, then (2.3) gives

$$B_p(t) = \begin{cases} [1 - (1-p)t]^{\frac{1}{1-p}}, & t < \frac{1}{1-p} \\ 0, & t \geq \frac{1}{1-p}, \end{cases} \tag{2.4}$$

so that B_p is monotone decreasing and identically zero for $t \geq \frac{1}{1-p}$. We have *extinction in finite time*. Note that $B_p(t)$ as given by (2.4) is a classical solution of the differential equation (2.2), the two pieces in the definition merging smoothly at the extinction time $1/(1-p)$. The nonlinearity u^p is non-lipschitzian at the origin when $0 < p < 1$; while the forward problem (2.2) has a unique solution, this is not true of the backward problem

$$u_t = u_+^p, \quad t > 0, \quad u(0) = 0,$$

which has the solutions $u \equiv 0$ and

$$u = \begin{cases} 0, & t \le a \\ [(1-p)(t-a)]^{1/1-p} & , t > a, \end{cases}$$

where a is an arbitrary positive number.

b) *Stationary solutions in R^n.*

We seek nonnegative solutions on R^n of

$$\Delta u = u_+^p, \quad p > 0 \tag{2.5}$$

which depend only the distance from the origin. As a first attempt we look for power-law solutions of the form $C|x|^\alpha$. Since the Laplacian reduces the power by two, we will need $\alpha - 2 = \alpha p$ or $\alpha = 2/(1-p)$. Recalling the radial Laplacian, we find $C^{p-1} = \alpha(n + \alpha - 2)$ so that

$$C = C_{p,n} = \left[\frac{(1-p)^2}{2[n(1-p) + 2p]} \right]^{1/1-p}. \tag{2.6}$$

Note that the solution $C_{p,n}|x|^{2/1-p}$ is regular at the origin if $0 < p < 1$, but not if $p > 1$. It is easy to verify that the solution for $0 < p < 1$ is, in fact, a classical solution. The same is true for $C_{p,n}|x - a|^{2/1-p}$.

For the equation

$$\Delta u = \lambda^2 u_+^p, \tag{2.7}$$

we find, by rescaling the x coordinate, that

$$u = C_{p,n}|\lambda(x - a)|^{2/1-p}. \tag{2.8}$$

Of course (2.5) and (2.7) have many other solutions some of which are strictly positive on R^n (See Herrero-Velázquez [14], for instance).

The one-dimensional case (when $0 < p < 1$) leads easily to some interesting additional solutions constructed from (2.8). We could, for instance, consider the functions

$$u = C_{p,1}[\lambda(x - a)]_+^{2/1-p}, \quad u = C_{p,1}[\lambda(a - x)]_+^{2/1-p} \tag{2.9}$$

which vanish for $x \le a$ and $x \ge a$, respectively, yet are classical solutions of (2.7). We can also construct a solution which contains a dead core (a bounded interval in which u vanishes); it suffices to take, with $a < x < b$,

$$u(x) = C_{p,1}[\lambda(a - x)_+]^{2/1-p} + C_{p,1}[\lambda(x - b)_+]^{2/1-p}. \tag{2.10}$$

The function (2.10) is sketched below; it is symmetric about $x = \dfrac{a+b}{2}$.

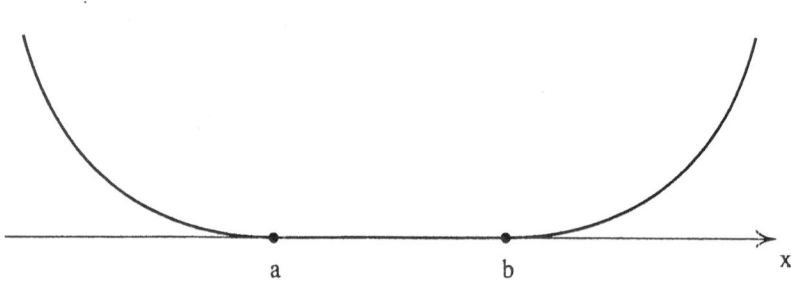

2 Extinction and localization

Consider the initial value problem on R^n:

$$Lu = -u_+^p; \quad u(x,0) = u_0(x) \geq 0,$$

where $u_0(x)$ is *bounded*. We then have, by comparison, $u(x,t) \leq v$, where v is the solution corresponding to the constant initial value $||u_0||$ and $||\ \ ||$ stands for the supremum of $|u_0(x)|$. But v is spatially homogeneous and is given by $||u_0||B_p(||u_0||^{p-1}t)$ which, when $0 < p < 1$, is identically zero for $t \geq \frac{||u_0||^{1-p}}{1-p}$. Therefore $u(x,t) \equiv 0$ for $t \geq \frac{||u_0||^{1-p}}{1-p}$, when $0 < p < 1$; *extinction in finite time!*

Herrero and Velázquez [14] show that, if $u_0(x)$ is unbounded but of smaller order than (2.8) at infinity, then for each x, there exists $T(x)$ such that $u(x,t) = 0$ for $t \geq T(x)$. On the other hand, if $u_0(x) \geq C|x-a|^{2/1-p}$ with $C > C_{p,n}$, then $u(x,t) \to \infty$ as $t \to \infty$. The paper also contains more delicate asymptotic results when $u_0(x)$ has precisely the critical behavior $C_{p,n}|x-a|^{2/1-p}$ at infinity.

If $0 < p < 1$ and u_0 has compact support, so does $u(x,t)$ for any t. We already know there is extinction in finite time, but prior to that time $u(x,t)$ has compact support (*localization*). Evans and Knerr [7] showed that even if $u_0(x) > 0$ for all $x \in R^n$ (but vanishes sufficiently fast at $|x| = \infty$), then $u(x,t)$ has compact support for all $t > 0$. Of course if $u_0(x) \equiv$ constant, then $u(x,t)$ is also spatially homogeneous and positive for all $t < u_0^{1-p}/(1-p)$.

3 Thermal waves

Let us consider (2.1) with $0 < p < 1$ on an interval in R^1 with zero Dirichlet boundary data:

$$u_t - u_{xx} = -\lambda^2 u^p, \quad 0 < x < l, \quad t > 0; \quad u(x,0) = u_0(x) \geq 0, \quad u(0,t) = 0, u(l,t) = 0.$$

Let us assume that $u_0(x)$ is concave with $u_0(0) = u_0(l) = 0$, and that u_0 is symmetric about $x = l/2$. A simple argument shows again that we have extinction in finite time. Let T be the time of extinction, that is,

$$T = \inf\{t : u(x,t) \equiv 0\}.$$

As $t \to T-$, $u(x,t) \to 0$ uniformly in x, but what about the support of $u(x,t)$ for $t = T - \epsilon$? Friedman and Herrero [9] showed that at $t = T - \epsilon$, u vanishes except for a small interval about $x = l/2$. There are cooling fronts emanating from $x = 0$ and $x = l$ (with $u = 0$ behind the fronts) and meeting at $x = l/2, t = T$. The support of $u(x,t)$ is an interval $\frac{l}{2} - \xi(t) \leq x \leq \frac{l}{2} + \xi(t)$ with $\lim_{t \to T-} \xi(t) = 0$.

Next, we would like to point to the existence of such fronts in the one-dimensional problem:

$$u_t = u_{xx} = -\lambda^2 u^p, \quad -\infty < x < \infty, \quad u(x,0) = u_0(x) \geq 0. \tag{2.11}$$

The function $u = C_{p,1}[-\lambda x]_+^{2/1-p}$ has been shown to satisfy the steady-state problem (see (2.9)), so that it also satisfies the evolution problem (2.11) with the initial value $C_{p,1}[-\lambda x]_+^{2/1-p}$ which is positive for $x < 0$ and identically zero for $x > 0$. Suppose we consider (2.11) for $u_0(x)$ positive for $x < 0$ and zero for $x > 0$. Then, we expect $u(x,t)$ to vanish for $x > \xi(t)$ and be positive for $x < \xi(t)$. Thus, we have a front traveling to the right if $\xi'(t) > 0$ and to the left if $\xi'(t) < 0$. In a recent paper, Grundy and Peletier [12] show that if $u_0(x)$ is flatter that $|x|^{2/1-p}$ as $x \to 0-$, the front initially travels to the left; if $u_0(x)$ is not as flat as $|x|^{2/1-p}$, the front travels to the right.

Chapter 3 – The steady-state dead core

1 Existence and uniqueness

We consider the steady-state problem

$$-\Delta u = -\lambda^2 u^p, \quad x \in \Omega; \quad u(\partial\Omega) = b(x), \tag{3.1}$$

where $p > 0$, $\lambda \geq 0$, $b(x) \geq 0$ ($\|b\| = 1$) are given, and Ω is a bounded domain in R^n. We have written the equation (3.1) with two minus signs to feature the positive operator $-\Delta$ and therefore to recognize the right side as an absorption term. This will simplify keeping inequalities straight when we deal later with supersolutions and subsolutions.

It follows from the maximum principle that every solution of (3.1) is nonnegative in Ω. If $p \geq 1$, $u > 0$ in Ω; if, however, $0 < p < 1$ and λ^2 is sufficiently large the solution may vanish identically in a region D known as a dead core. For all $p > 0$, it appears from (3.1) by dividing by λ^2 that u will be small except near the boundary. Our goal is different: we want to show u is identically zero in some region D and this can happen only if $0 < p < 1$ and λ^2 large enough.

Existence of a solution is a consequence of Amann's extension of the method of super- and subsolutions to handle nonlinearities that are not smooth (see [1]). We recall that a *supersolution* $\bar{u}(x)$ of (3.1) is defined through the inequalities

$$-\Delta \bar{u} \geq -\lambda^2 \bar{u}^p, \quad x \in \Omega; \quad \bar{u}(\partial\Omega) \geq b(x), \tag{3.2}$$

whereas a *subsolution* $\underline{u}$ of (3.1) satisfies (3.2) with both inequalities reversed. Amann's theorem states that if we can find a supersolution $\bar{u}$ and a subsolution $\underline{u}$ satisfying $\underline{u}(x) \leq \bar{u}(x)$, then there exists at least one solution $u(x)$ of (3.1) with

$$\underline{u}(x) \leq u(x) \leq \bar{u}(x) \quad \text{on} \quad \bar{\Omega}. \tag{3.3}$$

It is easy to see that $\underline{u} \equiv 0$ is a subsolution and $\bar{u} \equiv ||b|| = 1$ is a supersolution of (3.1). Therefore (3.1) has at least one solution $u(x)$ with $0 \leq u(x) \leq 1$. Of course, these bounds also hold for any nonnegative solution of (3.1) by a straightforward application of the maximum principle.

The solution of (3.1) is easily seen to be unique. Suppose u_1, u_2 are solutions of (3.1) with $u_2 > u_1$ on a part M of Ω. In view of the fact that $u_1 = u_2$ on $\partial\Omega$, M can be chosen so that $u_1 = u_2$ on ∂M (which may, of course, include part or all of $\partial\Omega$). Then by the monotonicity of u^p, we have

$$\Delta(u_2 - u_1) = \lambda^2(u_2^p - u_1^p) > 0 \text{ in } M; \quad u_2 - u_1 = 0 \text{ on } \partial M.$$

The maximum principle for the Laplacian then states that $u_2 - u_1 < 0$ in M, which is a contradiction.

Because of uniqueness, we may be able to improve our bounds for $u(x)$. Suppose $\underline{u}$ is a subsolution satisfying $0 \leq \underline{u} \leq 1$ and $\bar{u}$ is a supersolution satisfying $0 \leq \bar{u} \leq 1$; then, necessarily $\underline{u}(x) \leq u(x) \leq \bar{u}(x)$.

2 Comparison theorems

With Ω, p and $b(x)$ fixed, consider (3.1) for two values of λ with $\lambda_1 < \lambda_2$. Denoting the corresponding solutions by $u_1(x)$ and $u_2(x)$, then

$$u_1(x) \geq u_2(x).$$

Physically, this result is obvious: if we increase absorption we decrease the solution. The proof follows from the observation that $u_2(x)$ is a lower solution of (3.1) with $\lambda = \lambda_1$. As a corollary, we see that the corresponding dead cores, if any, satisfy $D_2 \supset D_1$. Friedman and Phillips [10], have shown the stronger result

$$\text{dist}\,(\partial D_2, D_1) > 0. \tag{3.4}$$

A second comparison theorem is also easy to prove: suppose $b(x) \equiv \text{constant}$ (say, b), λ and p are fixed and $\Omega_2 \supset \Omega_1$. Then the respective solutions $u_2(x), u_1(x)$ of (3.1) satisfy $u_1(x) \geq u_2(x)$ on Ω_1. Again the proof is easy, since $u_2(x)$ is a lower solution to (3.1) on Ω_1 and the result follows. The respective dead cores, if any, satisfy $D_2 \supset D_1$.

3 Explicit solutions

We now study (3.1) for the case $b \equiv 1$; note that any other constant boundary value can be rescaled by an appropriate change in λ. The problem under consideration is

$$\Delta u = \lambda^2 u^p, \quad x \in \Omega; \quad u(\partial \Omega) = 1. \tag{3.5}$$

If Ω is a ball of radius R with center at the origin, the solution can be found explicitly from (2.8) for special values of λ, p and R. We see that if $0 < p < 1$ and

$$\lambda R = \left(\frac{1}{C_{p,n}}\right)^{\frac{1-p}{2}} \doteq P_{p,n} = \frac{[2n(1-p)+4p]^{1/2}}{1-p}, \tag{3.6}$$

then the solution is $u = C_{p,n}|\lambda x|^{2/1-p}$. In this case we have a one-point dead core. It follows from (3.4) that there is a dead core of positive measure if $\lambda R > P_{p,n}$ and no dead core if $\lambda R < P_{p,n}$.

In R^1, for an interval of length $2R$, formula (3.6) gives a one-point dead core if

$$\lambda R = P_{p,1} = \frac{\sqrt{2}\sqrt{p+1}}{1-p}.$$

If $\lambda R > P_{p,1}$, then we can get an explicit expression for u (see (2.9)),

$$u(x) = C_{p,1}\lambda^{2/1-p}[(x-l)_+^{2/1-p} + (-x-l)_+^{2/1-p}], \tag{3.7}$$

with l defined from $\lambda(R-l) = P_{p,1}$. Note that (3.7) vanishes on $-l \leq x \leq l$ and satisfies $u(\pm R) = 1$. If $\lambda R < P_{p,1}$, the solution is positive on $-l \leq x \leq l$, but there is no simple explicit expression for it.

The slab $-R < x_1 < R$ in R^n has, of course, precisely the same solution we have just found (identifying x with the coordinate x_1). The semi-infinite domain $x > 0$ can also be easily treated. Assuming $u(x)$ bounded, the solution of

$$u_{xx} = \lambda^2 u^p, \quad x > 0; \quad u(0) = 1$$

is given by

$$C_{p,1}[\lambda(-x+d)_+]^{2/1-p}, \quad 0 < p < 1, \tag{3.8}$$

where $d = P_{p,1}/\lambda$. Thus the support of the solution is $0 \leq x \leq P_{p,1}/\lambda$. If $p = 1$, the solution is $e^{-\lambda x}$ and if $p > 1$,

$$u(x) = \left[1 + \frac{\lambda x(p-1)}{\sqrt{2(p+1)}}\right]^{-\frac{2}{p-1}}.$$

Note that if $p \geq 1, u(x) > 0$ for all x but tends monotonically to zero as $x \to \infty$. It follows, by comparison, that (3.1) for a bounded domain cannot have a dead core if $p \geq 1$ no matter how big λ is.

4 Estimates for the dead core

Throughout this section, we take $0 < p < 1$. Consider (3.5) on a bounded convex domain Ω. At a point Q on the boundary construct a supporting hyperplane (unique if $\partial\Omega$ is smooth) and let F_Q be the corresponding half-space containing Ω. The solution of (3.5) for the half-space F_Q is, by the second comparison theorem, smaller than the solution $u(x)$ of (3.5) for Ω. Therefore, by (3.8), $u(x) > 0$ for all points of Ω at a distance smaller than $P_{p,1}/\lambda$ from Q. Of course, we can do this for all points Q on $\partial\Omega$, so that $u(x) > 0$ for all points which are at a distance smaller than $P_{p,1}/\lambda$ from the boundary. If D_Ω denotes the dead core and

$$S(d) = \{x \in \Omega : \text{dist}(x, \partial\Omega) \geq d\}.$$

then, if Ω is convex,

$$D_\Omega \subset S(P_{p,1}/\lambda). \tag{3.9}$$

For nonconvex domains, this argument fails but the gradient bounds derived in [19] and in Section 5 can be used to obtain a similar $S(B/\lambda)$ bound but B is now smaller (independent of λ, but dependent on the maximum of $-K$, where K is the average curvature). To see how the results of Section 5 are used, let x_0 be a point on ∂D_Ω and x_1 on $\partial\Omega$; join the two points by a straight line with r a coordinate increasing from x_0 to x_1. Then, if $K_0 = 0$, (3.26) gives

$$\frac{du}{dr} \leq |\text{grad } u| \leq \frac{\sqrt{2\lambda}}{\sqrt{p+1}} u^{\frac{p+1}{2}},$$

and, integrating from x_0 to x_1, we obtain

$$|x_1 - x_0| \geq \frac{2\sqrt{p+1}}{\lambda\sqrt{2}(1-p)}, \tag{3.10}$$

which is the same as (3.9), but is valid for a larger class of domains, those with nonnegative mean curvature.

If, however $|K_0| > 0$, then we can use (3.26), replacing u^{p+1} by u, to obtain

$$|x_1 - x_0| \geq \frac{2}{\sqrt{\frac{2\lambda^2}{p+1} + A(\lambda, |K_0|)}}. \tag{3.11}$$

Since the right side is $O(\frac{1}{\lambda})$ for large λ, we find a formula similar to (3.10) but with a smaller coefficient.

Our result (3.9) does not guarantee the existence of a dead core, only its absence if $S(P_{p,1}/\lambda)$ is empty. We can, however, find a bound in the opposite direction for D_Ω. Let Ω be an arbitrary, bounded domain, not necessarily convex; let x_0 be a point of Ω and let d_0 be its distance from $\partial\Omega$. The ball $B_0(d_0)$ with center at x_0 and radius d_0 is then contained in Ω.

Now x_0 will belong to the dead core of $B_0(d_0)$ if $\lambda d_0 \geq P_{p,n}$, see (3.6), and, *a fortiori*, in the dead core of Ω. Therefore

$$D_\Omega \supset S(P_{p,n}/\lambda), \quad \text{for any } \Omega, \tag{3.12}$$

and, when combined with (3.9),

$$S(P_{p,n}/\lambda) \subset D_\Omega \subset S(P_{p,1}/\lambda), \quad \Omega \text{ convex.} \tag{3.13}$$

Thus the dead core is at a distance $O(1/\lambda)$ from the boundary and ultimately as $\lambda \to \infty$, engulfs all of Ω. A similar result holds even if Ω is not convex. For further results on the steady-state dead core, see [2], [5], [10].

5 Gradient bounds

Consider the Dirichlet problem

$$\Delta u = f(u), \quad x \in \Omega; \quad u(\partial\Omega) = A > 0, \tag{3.14}$$

where $f(u) \geq 0$ for $u \geq 0$. We fix our attention on a specific nonnegative solution $u(x)$ of (3.14). Then of course $u(x) \leq A$ and u has a minimum value u_m. Our goal is to find an estimate of $|\text{grad } u|$. For the *one-dimensional case* we can multiply (3.14) by u_x and immediately find that $u_x^2 - 2F(u) = \text{constant} = -2F(u_m)$ where

$$F(u) \doteq \int_0^u f(z)dz. \tag{3.15}$$

In more than one dimension, we are unable to find an explicit expression for grad u, but it is nevertheless useful to study the behavior of $|\text{grad } u|^2 - 2F(u)$. For technical convenience, we actually consider the function

$$\phi^\epsilon(x) = |\text{grad } u|^2 - 2F(u) - \epsilon u = \sum_{j=1}^n u_j^2 - 2F(u) - \epsilon u, \tag{3.16}$$

where $u_j = \partial u / \partial x_j$ and $\epsilon > 0$ is a parameter to be adjusted. We shall show that ϕ^ϵ must attain its *maximum at an interior point where* $|\text{grad } u| = 0$. It will then follow from (3.16) that

$$\begin{aligned} |\text{grad } u|^2 &\leq 2[F(u) - F(u_m)] + \epsilon(u - u_m) \\ &\leq 2F(u) + \epsilon u. \end{aligned} \tag{3.17}$$

To prove (3.15), let us differentiate (3.14):

$$\begin{aligned} \phi_i^\epsilon &= \sum_j 2u_j u_{ji} - 2f(u)u_i - \epsilon u_i, \\ \phi_{ii}^\epsilon &= \sum_j 2u_{ji} + \sum_j 2u_j(u_{ii})_j - 2f'u_i^2 - (\epsilon + 2f)u_{ii}, \end{aligned} \tag{3.18}$$

$$\Delta\phi^\epsilon = \sum_{i,j} 2u_{ji}^2 - f(2f + \epsilon). \tag{3.19}$$

Now suppose ϕ^ϵ has a maximum at an interior point ς where grad $u \neq 0$. Choose the coordinate x_1 in the direction of grad u; thus, $u_1 > 0$ and $u_j = 0, j > 1$. Then from (3.18) we find $u_{11}(\varsigma) = f + \epsilon/2$ and

$$\Delta\phi^\epsilon(\varsigma) \geq 2[f^2 + f\epsilon + \frac{\epsilon^2}{4}] - 2f^2 - \epsilon f = f\epsilon + \frac{\epsilon^2}{2} > 0,$$

which contradicts the assumption. Thus ϕ^ϵ can have an interior maximum only where grad $u = 0$.

Next we rule out a boundary maximum. Since u is constant on $\partial\Omega$,

$$\phi_\nu^\epsilon = 2u_\nu u_{\nu\nu} - [2f(1) + \epsilon]u_\nu = -u_\nu[-2u_{\nu\nu} + 2f(1) + \epsilon]. \tag{3.20}$$

From the differential equation on the boundary, we have, since the tangential derivatives vanish,

$$f(1) = (\Delta u)_{\partial\Omega} = u_{\nu\nu} + K(n-1)u_\nu,$$

where K is the *mean curvature* of the boundary. Substitution in (3.20) gives

$$\phi_\nu^\epsilon = -u_\nu[\epsilon + 2K(n-1)u_\nu]. \tag{3.21}$$

Obviously $u_\nu \geq 0$; if $u_\nu = 0$, we would have a local maximum on the boundary, thereby contradicting the differential equation; therefore $u_\nu > 0$ and, if $K \geq 0$, then $\phi_\nu^\epsilon < 0$ on $\partial\Omega$ showing that ϕ^ϵ cannot have a maximum there. We therefore have the predicted result (3.17),

$$|\text{grad } u|^2 \leq 2F(u) + \epsilon u, \quad \text{for every } \epsilon > 0; \tag{3.22}$$

hence

$$|\text{grad } u|^2 \leq 2F(u) \quad (\text{if } K \geq 0). \tag{3.23}$$

Note that if Ω is convex then certainly $K \geq 0$. Of course, we can have $K \geq 0$ $(n \geq 3)$ without having Ω convex.

If all we know is that the mean curvature is bounded below $(K \geq -|K_0|)$, then, by choosing $\epsilon > 2|K_0|(n-1)M$, where M is the maximum of u_ν on the boundary, we again find that ϕ^ϵ must have its maximum where grad u vanishes. Hence, we have, from (3.22),

$$|\text{grad } u|^2 \leq 2F(u) + \epsilon u, \quad \text{for } \epsilon > 2|K_0|(n-1)M,$$

or

$$|\text{grad } u|^2 \leq 2F(u) + 2|K_0|(n-1)Mu. \tag{3.24}$$

Inequality (3.24) applied on the boundary gives

$$M^2 \leq 2F(1) + 2|K_0|(n-1)M$$

or

$$M \leq 2|K_0|(n-1) + \sqrt{2F(1)}$$

and

$$|\text{grad } u|^2 \leq 2F(u) + 2|K_0|(n-1)\left[\sqrt{2F(1)} + 2|K_0|(n-1)\right]u. \tag{3.25}$$

For the problem considered in the previous sections

$$F(u) = \lambda^2 u^{p+1}/p + 1$$

and (3.25) becomes

$$|\text{grad } u|^2 \leq \frac{2\lambda^2 u^{p+1}}{p+1} + A(\lambda, |K_0|)u \tag{3.26}$$

where A is $O(\lambda)$ as $\lambda \to \infty$. If $K \geq 0$, we can drop the term Au in (3.26).

Chapter 4 – The parabolic dead core

1 Formulation and general properties

We shall consider the following problem (see (2.1)),

$$\begin{cases} Lu = -\lambda^2 u^p, & x \in \Omega, \ t > 0, \\ u(x,0) = u_0(x), & x \in \Omega, \ \text{where } 0 \leq u_0(x) \leq 1; \\ u(x,t) = 1, & x \in \partial\Omega, \ t > 0. \end{cases} \tag{4.1}$$

We expect $u(x,t)$ to tend to the steady-state solution (now called $U(x)$) of (3.5). When $u_0(x) \geq U(x)$, the solution of (4.1) will satisfy $u(x,t) \geq U(x)$ for all t. If $u_0(x) > U(x) > 0$ in Ω, then $u(x,t) > U(x)$ for all t and $x \in \Omega$. If $U(x)$ contains a dead core $D(0 < p < 1$ and λ^2 large enough), then it will turn out that $u(x,t)$ itself has a dead core D_t for t sufficiently large and that $\lim_{t\to\infty} D_t = D$.

In order to prove some of these statements, we need again the notion of super- and subsolutions. We say $\bar{u}(x,t)$ is a supersolution ot (4.1) on an interval $(0,T)$ where T may be infinite, if

$$\begin{cases} L\bar{u} \geq -\lambda^2 \bar{u}^p, & x \in \Omega, \ 0 < t < T; \\ \bar{u}(x,0) \geq u_0(x); \ \bar{u}(\partial\Omega,t) \geq 1, & 0 < t < T. \end{cases} \tag{4.2}$$

A subsolution $\underline{u}(x,t)$ satisfies (4.2) with all 3 inequalities reversed.

From the maximum principle we know that every solution of (4.1) satisfies $0 \leq u(x,t) \leq 1$. Now let $\underline{u}, \bar{u}$ be a subsolution and supersolution, respectively, and let $\underline{u}(x,t) \leq \bar{u}(x,t)$. Then it can be shown that there exists one and only one solution $u(x,t)$ of (4.1) with $\underline{u} \leq u \leq \bar{u}$. Since $v \equiv 0$ is a subsolution and $w \equiv 1$ is a supersolution, (4.1) has one and only one solution. Moreover if $\bar{u}$ is any supersolution with $\bar{u} \leq 1$ and $\underline{u}$ is any subsolution with $\underline{u} \geq 0$, then automatically $\underline{u} \leq \bar{u}$ and the solution $u(x,t)$ of (4.1) satisfies $\underline{u} \leq u \leq \bar{u}$.

In what follows, we shall assume

$$u_0(x) \geq U(x) \quad \text{and} \quad -\Delta u_0 \geq -\lambda^2 u_0^p, \quad \text{in } \Omega. \tag{H.1}$$

We then have the results (see [17]):

a) $U(x) \leq u(x,t) \leq u_0(x)$. Indeed $u_0(x)$ is a supersolution and $U(x)$ is a subsolution of (4.2).

b) $u(x,t)$ is monotonically decreasing in time. Let $w(x,t) = u(x,t+\Delta t)$, then w satisfies the differential equation and boundary condition in (4.2). Moreover $w(x,0) = u(x,\Delta t) \leq u_0(x)$. Thus w is a subsolution of (4.2) and therefore $w(x,t) \leq u(x,t)$ so that it is clear that $u(x,t)$ is monotonically decreasing in time.

c) $u(x,t) \to U(x)$ as $t \to \infty$.

d) If $u(x,\tau) > U(x)$ for some $\tau > 0$, then $u(x,t)$ is strictly decreasing for $t < \tau$.

2 The formation of the dead core

When the steady problem corresponding to (4.1) does *not* have a dead core, which will be the case when $p \geq 1$ or when $0 < p < 1$ and λ is not too large, then the solution of (4.1) always lies above the steady state; that is $u(x,t) > U(x)$, $x \in \Omega, t > 0$.

Suppose now that $0 < p < 1$ and that λ in (4.1) is large enough so that the corresponding steady-state solution $U(x)$ has a dead core D. We always have $u(x,t) \geq U(x)$, but is $u(x,t) > U(x)$ for all t or does $u(x,t)$ itself develop a dead core D_t? We can get some hint of the answer by studying the limiting case $p = 0$, when the absorption is equal to $\lambda^2 H(u)$ with H the Heaviside function. This is clearly a case of stong absorption! We shall show that u cannot remain positive for all t if λ is sufficiently large. Our boundary value problem is

$$u_t - \Delta u = -\lambda^2 H(u), \quad x \in \Omega, t > 0; \quad u(x,0) = 1; \quad u(\partial\Omega, t) = 1.$$

If $u(x,t) > 0$ for $t < \tau$, then $H(u) = 1$ for $t < \tau$. Let us multiply the differential equation by the eigenfunction $\phi_1(x)$ (see (1.18)), where ϕ_1 has been normalized so that $\int_\Omega \phi_1(x) dx = 1$. Setting

$$E(t) = \int_\Omega u(x,t) \phi_1(x) dx,$$

we obtain

$$E'(t) + \lambda_1 E(t) = \lambda_1 - \lambda^2, \quad 0 < t < \tau; \quad E(0) = 1.$$

We therefore find that

$$E(t) = 1 - \frac{\lambda^2}{\lambda_1}(1 - e^{-\lambda t}), 0 < t < \tau.$$

It is clear that if $\lambda^2 > \lambda_1$, then $E(t)$ will ultimately become negative. In fact, $E(t)$ remains positive only until

$$\tau = \frac{1}{\lambda_1} |\log(1 - \frac{\lambda_1}{\lambda^2})|.$$

Thus, our solution must develop a dead core if $\lambda^2 > \lambda_1$ and our formula for τ gives an upper bound for the time of onset of the dead core.

Next we shall show that, if $0 < p < 1$, (4.1) develops a time-dependent dead core whenever the corresponding steady-state has a dead core, see [3], [18]. We shall deal first with the case $u_0(x) = 1$. If $x_0 \in \Omega$, it must be in the steady dead core $D(\lambda)$ for λ sufficiently large. We then have

Theorem 1 *Let $0 < p < 1, x_0 \in \Omega$ and $\lambda > \lambda_0$, where*

$$\lambda_0 := \inf(\lambda : x_0 \in D(\lambda)).$$

Then $x_0 \in D_t(\lambda)$ for $t \geq \frac{1}{(1-p)(\lambda - \lambda_0)}$.

Proof We construct a supersolution $\bar{u}(x,t)$ on $(0,T)$ with $\bar{u}(x_0, T) = 0$. It then follows that $u(x_0, T) = 0$ and $u(x_0, t) \equiv 0$ for $t \geq T$ (therefore $x_0 \in D_t(\lambda)$ for $t \geq T$). We set

$$\bar{u}(x,t) = U(x, \lambda_0) + B_p(\alpha t)$$

where $U(x, \lambda_0)$ is the steady-state solution for $\lambda = \lambda_0$ and $B_p(t)$ is given by (2.3) and α is to be selected. Now $B_p(\alpha t) \equiv 0$ for $t \geq \frac{1}{\alpha(1-p)}$ so that $\bar{u}(x_0, \frac{1}{\alpha(1-p)}) = 0$, $\bar{u}(x,t) > 0$ for $t < \frac{1}{\alpha(1-p)}$, $\bar{u}(x,0) \geq 1$ and $\bar{u}(\partial\Omega, t) \geq 1$. Therefore $\bar{u}$ will be the desired supersolution if it satisfies the remaining differential inequality in (4.2) on $0 < t < \frac{1}{\alpha(1-p)}$.

In this time interval, we find

$$L\bar{u} = -\alpha B_p^p(\alpha t) - \lambda_0^2 U^p \geq -(\alpha + \lambda_0^2)(B_p + U)^p = -(\alpha + \lambda_0^2)\bar{u}^p.$$

By taking $\alpha = \lambda^2 - \lambda_0^2$, the desired inequality and the theorem follow.

Remarks 1. Although λ_0 is not known explicitly in general (except in one-dimensional problems), we have $\lambda_0 d_0 \leq P_{p,n}$. Thus if $\lambda > P_{p,n}/d_0$, where d_0 is the distance from x_0 to $\partial\Omega$, we can be sure that x_0 is in the interior of the steady dead core corresponding to λ. The theorem then gives

$$x_0 \in D_t(\lambda) \quad \text{for } t \geq \frac{1}{(1-p)(\lambda - P_{p,n}/d_0)}.$$

2. If our interest is in the first time when $D_t(\lambda)$ is nonempty, we can take x_0 to be the center of the largest inscribed ball in Ω. If this radius is r_i, then $D_t(\lambda)$ is nonempty for $t \geq \frac{1}{(1-p)(\lambda - P_{p,n}/r_i)}$.

3. If $x_0 \in \partial D(\lambda)$, then $x_0 \notin D_t(\lambda)$ for any finite time (see [3]).

4. If $u_0(x) \leq 1$, then the solution $u(x,t)$ of the parabolic problem is smaller than the corresponding solution with initial value $\equiv 1$. Therefore the dead core $D_t(\lambda)$ is larger and our previous estimates hold.

For a more detailed study of how the time-dependent dead core D_t approaches the steady core D, consult the article by R. Ricci [21] and the corresponding references.

Chapter 5 – Gas-solid reactions

1 Formulation

Consider a gas diffusing through a porous solid and reacting with it isothermally and irreversibly. We assume the solid phase is immobile and is not structurally altered by the reaction. Let Ω be the domain occupied by the solid and let σ, γ be respective concentrations for the solid and the gas. We shall assume that initially there is no gas in Ω, but that thereafter the gas concentration on the boundary is maintained at a constant positive value. Mass balances for the solid-gas then give

$$\sigma_\tau = -k\sigma^m\gamma^p \tag{5.1a}$$

$$x \in \Omega, \tau > 0$$

$$\tilde{\epsilon}\gamma_\tau - D\Delta\gamma = k a \sigma^m \gamma^p \tag{5.1b}$$

$$\sigma(x,0) = \sigma_0, \quad \gamma(x,0) = 0, \quad \gamma(\partial\Omega, \tau) = \gamma_0. \tag{5.1c}$$

Here $\tilde{\epsilon}$ is the porosity, D a diffusion constant, k and ka are reaction rates and τ is the time.

We introduce nondimensional concentrations S, C by $\sigma = S\sigma_0, \gamma = C\gamma_0$ and a nondimensional time t by $t = k\sigma_0^{m-1}\gamma_0^p\tau$. Then (5.1a) becomes $S_t = -S^mC^p$, while after dividing by D, (5.1b) can be written as $\epsilon C_t - \Delta C = -\phi^2 S^m C^p$, where ϵ is a new porosity and ϕ^2 is the Thiele modulus. Equations (5.1a, b, c) then become

$$\begin{cases} S_t = -S^m C^p, & x \in \Omega, t > 0; \\ \epsilon C_t - \Delta C = -\phi^2 S^m C^p = \phi^2 S_t, & x \in \Omega, t > 0; \\ S(x,0) = 1, \ C(x,0) = 0, \ C(\partial\Omega, t) = 1. \end{cases} \tag{5.2}$$

Problem (5.2) and variants have been studied in [18]. The most complete treatment will appear in a forthcoming paper [6].

2 Preliminary results

It is clear from (5.2) that $S(\cdot, t)$ is *decreasing* and that $0 \leq C \leq 1, 0 \leq S \leq 1$. Thinking of $S^m(x, t)$ as being specified, the equation for C is

$$\epsilon C_t - \Delta C = -\phi^2 S^m(x,t)C^p; \ C(x,0) = 0, \ C(\partial\Omega, t) = 1. \tag{5.3}$$

Setting $E(x, t) = C(x, t + \Delta t)$, we see that E satisfies

$$\epsilon E_t - \Delta E = -\phi^2 S^m(x, t + \Delta t)E^p \geq -\phi^2 S^m(x,t)E^p \tag{5.4}$$

with $E(x, 0) = C(x, \Delta t) \geq 0, E(\partial\Omega, t) = 0$. Thus, E is a supersolution to (5.3) so that $E(x, t) \geq C(x, t)$ and therefore $C(\cdot, t)$ is *increasing*.

On the boundary (where $C = 1$) the equation for S gives $S(\partial\Omega, t) = B_m(t)$, with $B_m(t)$ given by (2.3). Note that if $m < 1, S(\partial\Omega, t) = 0$ for $t \geq \frac{1}{1-m}$, whereas from $m \geq 1, S(\partial\Omega, t) > 0$ for all t. At interior points, we can express S in terms of C by

$$S(x,t) = B_m\left(\int_0^t C^p(x,\tau)d\tau\right). \tag{5.5}$$

Integrating the equation for C from 0 to t and setting

$$\eta(x,t) = \int_0^t (1 - C)d\tau; \ \eta_t = 1 - C,$$

we obtain

$$\epsilon\eta_t - \Delta\eta = \epsilon + \phi^2[1 - S]; \ \eta(x,0) = 0, \ \eta(\partial\Omega, t) = 0. \tag{5.6}$$

Since S decreases in time, we note that $\eta(\cdot, t)$ increases and

$$0 \leq \eta(x,t) \leq (\epsilon + \phi^2)w(x), \tag{5.7}$$

where $w(x)$ is the *torsion function* satisfying

$$- \Delta w = 1, \quad x \in \Omega; \quad w(\partial \Omega) = 0. \tag{5.8}$$

In view of the fact that $C(\cdot, t)$ is increasing and that $\int_0^t (1 - C) d\tau$ is increasing in t and is bounded above by $(\epsilon + \phi^2) w(x)$, it follows that $C(x,t)$ must tend to 1 as $t \to \infty$, uniformly in x. Equation (5.5) shows that S tends to zero, uniformly in x. Thus, as expected, the solution of (5.2) approaches, as $t \to \infty$, the steady state $S \equiv 0, C \equiv 1$.

We can now describe the consumption of the solid in general terms:

a) if $m \geq 1, S(x,t) > 0$ for all x and t but $\lim_{t \to \infty} S(x,t) = 0$.

b) if $m < 1, S(x,t) > 0$ for $t < \frac{1}{1-m}$. At $t = \frac{1}{1-m}, S(\partial \Omega, t) = 0$ and $S(x,t) > 0$ in Ω. There is a finite time T at which the solid is completely consumed. This time is defined implicitly by

$$\min_{x \in \bar{\Omega}} \int_0^T C^p(x, \tau) d\tau = \frac{1}{1 - m}.$$

If $\frac{1}{1-m} < t < T$, the solid is fully converted near the boundary and is only partially converted in a region $D(t)$ which shrinks as t increases. Its boundary $\partial D(t)$ is a free boundary.

3 The pseudo-steady-state approximation

In many engineering problems the porosity ϵ is small (see, for instance, Froment-Bischoff [11]). This leads us to consider the following problem related to (5.2):

$$\begin{cases} \hat{S}_t = -\hat{S}^m \hat{C}^p, & x \in \Omega, t > 0; \\ -\Delta \hat{C} = -\phi^2 \hat{S}^m \hat{C}^p = \phi^2 \hat{S}_t, & x \in \Omega, t > 0; \\ \hat{S}(x,0) = 1, \quad C(\partial \Omega, t) = 1. \end{cases} \tag{5.9}$$

Note that $\hat{C}(x,0)$ cannot be specified as it is already determined as the solution of the elliptic problem

$$- \Delta \hat{C}(x,0) = -\phi^2 \hat{C}^p(x,0), \quad x \in \Omega; \quad \hat{C}(\partial \Omega, 0) = 1. \tag{5.10}$$

Clearly $0 \leq \hat{C}(x,0) < 1$ in Ω.

From the first equation in (5.9) we can express $\hat{S}$ in terms of $\hat{C}$ as before:

$$\hat{S}(x,t) = B_m \left(\int_0^t \hat{C}^p(x, \tau) d\tau \right). \tag{5.11}$$

We again introduce an integrated concentration

$$\hat{\eta}(x,t) = \int_0^t (1 - \hat{C}) d\tau,$$

which satisfies

$$- \Delta \hat{\eta} = \phi^2 (1 - \hat{S}), \quad x \in \Omega; \quad \hat{\eta}(\partial \Omega, t) = 0. \tag{5.12}$$

In [18] it is shown that $|\eta - \hat{\eta}| \leq \epsilon w(x)$. Here we content ourselves with proving the result when $p = 1$.

4 The validity of the pseudo-steady-state approximation

When $p = 1$, both (5.6) and (5.12) are scalar problems. Indeed, we have

$$S(x, t) = B_m(t - \eta), \quad \hat{S}(x, t) = B_m(t - \hat{\eta}), \tag{5.13}$$

so that

$$\epsilon \eta_t - \Delta \eta = \epsilon + \phi^2 [1 - B_m(t - \eta)], \quad \eta(x, 0) = 0, \quad \eta(\partial \Omega, t) = 0 \tag{5.14}$$

and

$$- \Delta \hat{\eta} = \phi^2 [1 - B_m(t - \hat{\eta})], \quad \hat{\eta}(\partial \Omega, t) = 0. \tag{5.15}$$

Equation (5.15) is, for each t, an elliptic equation for $\hat{\eta}$. It is easy to verify that $\eta(x, t)$ is a supersolution to (5.15) while $\eta - \epsilon w(x)$ is a subsolution. We therefore conclude that

$$\hat{\eta} \leq \eta \leq \hat{\eta} + \epsilon w(x), \quad \text{for all } x, t, \tag{5.16}$$

and that

$$\max_{x \in \bar{\Omega}} \left| \int_0^t (\hat{C} - C) d\tau \right| \leq \epsilon \|w\|, \tag{5.17}$$

where $\|w\|$ stands for the sup norm of $w(x)$. Both (5.16) and (5.17) show that the pseudo-steady-state solution provides an excellent approximation to (5.2) when ϵ is small. Observe that although C and $\hat{C}$ are appreciably different for small times, that difference quickly decreases as t increases.

5 Conversion estimates

Perhaps the most interesting practical piece of information in connection with (5.2) is the time required to achive a specified fractional consumption of the solid. We shall only consider the case $p = 1$ with $m < 1$ and ask for an estimate of the time T to full conversion for (5.2). From (5.13) we see that $S \equiv 0$ when $t - \eta \geq \frac{1}{1-m}$. Thus, the time T to full conversion is characterized by

$$T = \frac{1}{1 - m} + \max_{x \in \bar{\Omega}} \eta(x, T).$$

Using (5.7), we find

$$T \leq \frac{1}{1 - m} + (\epsilon + \phi^2) \|w\|.$$

For the pseudo-steady-state problem, the time $\hat{T}$ to full conversion can be calculated in closed form (in terms of $\|w\|$). Again $\hat{T}$ is characterized by

$$\hat{T} = \frac{1}{1-m} + \max_{z \in \bar{\Omega}} \hat{\eta}(x,T),$$

but now $\hat{\eta}(x,T)$ satisfies (see (5.15))

$$-\Delta\hat{\eta}(x,T) = \phi^2, \quad \hat{\eta}(\partial\Omega,T) = 0,$$

so that

$$\hat{\eta}(x,T) = \phi^2 w(x) \quad \text{and} \quad \hat{T} = \frac{1}{1-m} + \phi^2\|w\|. \tag{5.18}$$

Since $\eta \geq \hat{\eta}$, we also have $\hat{T} \leq T$ and, hence,

$$\hat{T} \leq T \leq \hat{T} + \epsilon\|w\|. \tag{5.19}$$

For the one-dimensional case where $\Omega : 0 < x < a$, we find $w(x) = \frac{x(a-x)}{2}, \|w\| = a^2/8$. From (5.18) we have

$$\hat{T} = \frac{1}{1-m} + \phi^2 a^2/8,$$

and (5.19) gives

$$\frac{1}{1-m} + \frac{a^2\phi^2}{8} \leq T \leq \frac{1}{1-m} + \frac{a^2\phi^2}{8} + \frac{\epsilon a^2}{8}.$$

References

[1] H. Amann, Fixed point equations and nonlinear eigenvalue problems in ordered Banach space, *SIAM Review* **18** (1976), 620–709.

[2] C. Bandle, R.P. Sperb and I. Stakgold, Diffusion-reaction with monotone kinetics, *Nonlinear Analysis* **8** (1984), 321–333.

[3] C. Bandle and I. Stakgold, The formation of the dead core in parabolic reaction-diffusion equations, *Trans. Amer. Math. Soc.* **286** (1984), 275–293.

[4] J.R. Cannon, *The One-Dimensional Heat Equation*, Addison-Wesley, 1984.

[5] J.I. Diaz, Nonlinear partial differential equations and free boundaries, *Research Notes in Mathematics* **106**, Pitman, 1985.

[6] J.I. Diaz and I. Stakgold, Mathematical analysis of the conversion of a porous solid by a distributed gas reaction, Proc. XI CEDYA/I Congr. Matematica Aplicada, Malaga 1989.

[7] L.C. Evans and B.F. Knerr, Instantaneous shrinking of the support of non-negative solutions to certain nonlinear parabolic equations and variational inequalities, *Illinois J. Math.* **23** (1979), 153–166.

[8] A. Friedman, *Partial Differential Equations of Parabolic Type*, Prentice-Hall, 1964.

[9] A. Friedman and M.A. Herrero, Extinction properties of semilinear heat equations with absorption, *J. Math. Anal. Appl.* **124** (1987), 530–546.

[10] A. Friedman and D. Phillips, The free boundary of a semilinear elliptic equation, *Trans. Amer. Math. Soc.* **282** (1984), 153–182.

[11] G.F. Froment and K.B. Bischoff, *Chemical Reactor Analysis and Design*, Wiley, 1979.

[12] R.E. Grundy and L.A. Peletier, The initial interface development for a reaction diffusion equation with power law initial data, Report 89-05, Mathematical Institute, University of Leiden, 1989.

[13] M.A. Herrero and J.L. Vazquez, Thermal waves in absorbing media, *J. Differential Equations* **74** (1988), 218–233.

[14] M.A. Herrero and J.J.L. Velázquez, On the dynamics of a semilinear heat equation with strong absorption, *Comm. Partial Differential Equations* **14** (1989), 1653–1715.

[15] S. Kamin and P. Rosenau, Propagation of thermal waves in an inhomogeneous medium, *Comm. Pure Appl. Math.* **34** (1981), 831–852.

[16] M.H. Protter and H. Weinberger, *Maximum Principles in Differential Equations*, Prentice-Hall, 1967.

[17] D.H. Sattinger, Topics in stability and bifurcation theory, *Lecture Notes in Mathematics* **309**, Springer (1973).

[18] I. Stakgold, Partial extinction in reaction-diffusion, *Conferenze del seminario di matematica*, Università di Bari, **224** (1987).

[19] I. Stakgold and L.E. Payne, Nonlinear problems in reactor analysis, in *Nonlinear Problems in the Physical Sciences and Biology* (I. Stakgold et al., eds.), Lecture Notes in Mathematics **322**, Springer (1973).

[20] D.V. Widder, *The Heat Equation*, Academic Press, 1975.

[21] R. Ricci, Large time behavior of the solution of the heat equation with nonlinear strong absorption *J. Differential Equations* **79** (1989), 1–13.

An Introduction to the Mathematical Theory of the Porous Medium Equation[1]

Juan Luis VAZQUEZ

Departamento de Matemáticas
Universidad Autonóma de Madrid
E-28049 Madrid
Spain

Abstract

These notes contain an introduction to the mathematical treatment of the porous medium equation $u_t = \Delta(u^m)$, one of the simplest examples of nonlinear evolution equation of parabolic type, which appears in the description of different natural phenomena related to diffusion, filtration or heat propagation. The notes begin with a discussion of the relevance of the equation and some of its applications. The main body of the notes is devoted to the study of the existence, uniqueness and regularity of a (generalized) solution for the two main problems, i.e. the initial-value problem and the Dirichlet boundary-value problem. Special attention is paid to the appearance of a free boundary, a consequence of the finite propagation property.

Introduction

The aim of these lectures is to provide an introduction to the mathematical theory of the so-called *Porous Medium Equation* (in short PME), i.e. the equation

$$u_t = \Delta(u^m), \qquad (0.1)$$

where $u = u(x, t)$ is a scalar function and m is a constant larger than 1. The space variable x takes values in $\mathbf{R}^d, d \geq 1$, while $t \in \mathbf{R}$. Physical considerations lead to the restriction $u \geq 0$, which is mathematically convenient and currently followed, but not essential.

The PME is an example of a nonlinear evolution equation, formally of parabolic type. In a sense it is the simplest possible nonlinear version of the heat equation. Written in divergence form

$$u_t = \operatorname{div}\left(D(u)\nabla u\right), \qquad (0.2)$$

we see that the diffusion coefficient $D(u)$ equals mu^{m-1}. It is clear that the equation is parabolic only at those points where $u > 0$, while the vanishing of D implies that it degenerates wherever $u = 0$. We say that the PME is a degenerate parabolic equation.

---

[1]Partially supported by EEC Project SC1–0019–C(TT).

M. C. Delfour and G. Sabidussi (eds.), *Shape Optimization and Free Boundaries*, 347–389.
© 1992 *Kluwer Academic Publishers*.

There are a number of physical applications where this very simple model appears in a natural way to describe processes involving diffusion or heat transfer. Maybe the best known of them is the description of the flow of an isentropic gas through a porous medium [M]. Another important application refers to heat radiation in plasmas, [ZR]. Other applications have been proposed in mathematical biology, in water infiltration, lubrication, boundary layer theory, and other fields.

In spite of the simplicity of the equation and of its applications, and due perhaps to its nonlinear and degenerate character, a mathematical theory for the PME has been developed only very recently. Though the techniques depart strongly from the linear methods used in treating the heat equation, it is interesting to remark that some of the basic techniques are not very difficult nor need a heavy machinery. What is even more interesting, they can be applied in, or adapted to, the study of many other nonlinear PDE's of parabolic type. The study of the PME can provide the reader with an introduction to some interesting concepts and methods of nonlinear science, like the existence of free boundaries and the occurrence of regularity thresholds.

To begin with the mathematical treatment of the PME, a first and fundamental example of solution was obtained around 1950 in Moscow by Zel'dovich and Kompaneets [ZK] and Barenblatt [Ba], who found and analyzed a solution representing heat release from a point source, i.e., a *source type solution*. In fact, the solution has the explicit formula

$$U(x,t) = t^{-\lambda} \left[C - k \frac{|x|^2}{t^{2\mu}} \right]_+^{\frac{1}{m-1}}, \tag{0.3a}$$

where $[s]_+ = \max\{s, 0\}$,

$$\lambda = \frac{d}{d(m-1)+2}, \quad \mu = \frac{\lambda}{d}, \quad k = \frac{\lambda(m-1)}{2md} \tag{0.3b}$$

and $C > 0$ is an arbitrary constant. This solution was subsequently found by Pattle [Pa] in 1959, and is often referred to in the literature as the *Barenblatt* and also as the *fundamental* or *source-type solution*, because it takes as initial data a Dirac mass: as $t \to 0$ we have $U(x,t) \to M\delta(x)$, where M is a function of the free constant C (and m and d).

An analysis of this example shows many of the important features which were encountered later in the general theory. Maybe the most important is the observation that the source-type solution has compact support in space for every fixed time, in physical terms that the disturbance propagates with finite speed. This is in strong contrast with one of the most contested properties of the classical heat equation, the infinite speed of propagation (a nonnegative solution of the heat equation is automatically positive everyhwere in its domain of definition). In a sense the property of finite propagation supports the physical soundness of the equation to model diffusion or heat propagation. The occurrence of this phenomenon is a consequence of the degeneracy of the equation, i.e., the fact that the coefficient D vanishes at the level $u = 0$.

The phenomenon of finite propagation gives rise to the appearance of a *free boundary* separating the regions where the solution is positive (i.e., where "there is gas", according to the standard interpretation of u as a gas density, see below), from the "empty region" where $u = 0$. Precisely, we define the free boundary as

$$\Gamma = \partial P_u \cap Q, \tag{0.4}$$

where Q is the domain of definition of the solution in space-time, $P_u = \{(x,t) \in Q : u(x,t) > 0\}$ is the positivity set, and ∂ denotes boundary. This free boundary or propagation front is an important and difficult subject of the mathematical investigation.

A second (and related) observation is that, though the source-type solution is continuous in its domain of definition $Q = \mathbf{R}^d \times \mathbf{R}_+$, it is not smooth at the free boundary, again a consequence of the loss of the parabolic character of the equation when u vanishes. In fact, the function u^{m-1} is Lipschitz continuous in Q with jump discontinuities on Γ (i.e., there exists a regularity threshold). On the contrary, the solution is C^∞-smooth in P_u. And we are interested in noting that though u is not smooth on Γ, nevertheless the free boundary is a smooth surface given by the equation

$$t = c|x|^{d(m-1)+2}, \tag{0.5}$$

where $c = c(C, m, d)$.

The systematic theory of the PME was begun by Oleĭnik and her collaborators around 1958 [OKC], who introduced a suitable concept of generalized solution and analyzed both the Cauchy and the standard boundary value problems in one space dimension. The work was continued by Sabinina, who extended the results to several space dimensions, and Kamin, who began the analysis of the asymptotic behaviour. Since the 70's the interest for the equation has touched many other scholars from different countries, notably Angenent, Aronson, Bénilan, Brézis, Caffarelli, Crandall, Dahlberg, di Benedetto, Friedman, Gilding, Kenig, Peletier and Pierre, to quote just a few names who made important contributions to the basic theory.

There exists today a relatively complete theory covering the subjects of existence and uniqueness of suitably defined generalized solutions, regularity, properties of the free boundary and asymptotic behaviour, for different initial and boundary-value problems. Many of these results have been extended to the natural generalizations of the equation, the simplest of them being the so-called *Fast Diffusion Equation*, which is the same equation (0.1) with exponent $m < 1$. More generally, we can consider the general *Filtration Equation*, namely

$$u_t = \Delta\Phi(u), \tag{0.6}$$

where Φ is an increasing function: $\mathbf{R}_+ \to \mathbf{R}_+$. Another extension avenue consists in considering the PME with data of changing sign. Then we have to write the power in a convenient way to account for negative values. The usual choice is

$$u_t = \Delta(|u|^{m-1}u). \tag{0.7}$$

Finally, we should mention that a parallel, sometimes divergent, sometimes convergent story applies to the other popular nonlinear degenerate parabolic equation

$$u_t = \text{div} \left(|\nabla u|^{p-2} \nabla u \right), \tag{0.8}$$

which has also attracted lots of attention from researchers.

These notes are an enlarged version of the course taught at the Université de Montréal in June–July of 1990, aimed at introducing the subject and its techniques to young researchers. The material has also been used for a graduate course at the Universidad Autónoma de Madrid. It is clearly impossible to cover the numerous developments that have occurred in this field in a short set of lectures. Therefore, a selection of topics was necessary. The leading idea in writing these notes has been that of providing an elementary introduction to the questions of existence, uniqueness and the main properties of the solutions, whereby everything is derived from basic estimates using standard Functional Analysis and well-known PDE results. The exposé begins with a reminder of the main applications. Chapters II and III deal respectively with the Dirichlet and Cauchy Problems. For reasons of space, the treatment is restricted to integrable data, a sound assumption on physical grounds.

We hope that the material will make it easier for the interested reader to delve into deeper or more specific literature. To guide this further study, there exist some expository works. The reader will find an account of many of the main results on the equation in the excellent survey paper by Aronson [Ar], written in 1986. A previous survey was published by Peletier [Pe] in 1981. In his book on Variational Principles and Free-Boundary Problems [F], 1982, Friedman devotes a chapter to the PME. These works contain details of the proofs and techniques. Another contribution, more in the form of a summary but including a discussion of related equations and a very extensive reference list is due to Kalashnikov [Ka] in 1987.

There is now a feeling that maybe the time is ripe for a complete version of the mathematics of the PME and related nonlinear parabolic equations and free-boundary problems. Actually, the outline here described originated as part of a bigger project with Don Aronson to supply a comprehensive and elementary introduction to the PME, in a way a natural continuation of his survey [Ar]. I am very happy to acknowledge my indebtedness to him for sharing with me his expertise in the field and for so many other reasons. I am also grateful to Ph. Bénilan, H. Brézis, L. Caffarelli and S. Kamin for their advice and suggestions.

Remark A comment about formula numbers: inside the same chapter formulas are described by their section and number; thus, (3.12) refers to the formula 12 in section 3; when referring to a different chapter the chapter number is added, thus (II.3.12). A similar notation applies to the numbering of Theorems, Propositions, Lemmas and Corollaries.

Chapter I Applications

1 Flow of a polytropic gas through a porous medium

According to Muskat [M] the flow of an ideal gas through a porous medium can be described in terms of the variables *density*, which we represent by u, *pressure*, represented by v, and *velocity*, represented by V, which are functions of space x and time t. These quantities are related by the following laws:

(i) *Mass balance*

$$\rho u_t + \nabla \cdot (uV) = 0. \tag{1.1}$$

(ii) *Darcy's Law*, [Dr], an empirical law which describes the dynamics of flows through porous media

$$\mu V = -k\nabla v. \tag{1.2}$$

(iii) *Equation of State*, which for perfect gases states that

$$v = v_0 u^\gamma, \tag{1.3}$$

where the exponent γ is 1 for isothermal processes and larger that 1 for adiabatic ones. The parameters ρ (the porosity), μ (the viscosity), k (the permeability) and v_0 (the reference pressure) are assumed to be positive and constant, which constitutes an admissible simplification in many practical instances. An easy calculation allows to reduce (1.1)–(1.3) to the form

$$u_t = c\Delta(u^m),$$

with $m = 1 + \gamma$ and $c > 0$ a constant, which can be easily scaled out, thus leaving us with the PME. Observe that in the above applications the exponent m is always equal or larger than 2. Mathematically the constants play no role, so it is now the custom to define the mathematical pressure by the expression

$$v = \frac{m}{m-1} u^{m-1}, \tag{1.4}$$

and write Darcy's law in the form

$$V = -\nabla v = -mu^{m-2}\nabla u. \tag{1.5}$$

Then the mass balance is just $u_t + \nabla \cdot (uV) = 0$. In all the formulas the operators $\nabla \cdot = \text{div}$, $\nabla = \text{grad}$ and Δ, the Laplacian, are supposed to act on the space variables $x = (x_1, \cdots, x_d)$.

Let us remark that the consideration of flows where ρ, μ and k are not constants, but functions of space and time, provides us with a natural generalization of the PME.

2 Heat transfer with temperature-dependent thermal conductivity

A second important application happens in the theory of heat propagation. The general equation describing such a process (in the absence of heat sources or sinks) takes the form

$$c\rho \frac{\partial T}{\partial t} = \text{div}\,(\kappa \nabla T), \tag{2.1}$$

where T is the temperature, c the specific heat (at constant pressure), ρ the density of the medium (which can be a solid, fluid or plasma) and κ the thermal conductivity. In principle all these quantities are functions of $x \in \mathbf{R}^3$ and $t \in \mathbf{R}$. In the case where the variations of c, ρ and κ are negligible, we obtain the classical heat equation. However, when the range of variation of the temperatures is large, say hundreds or thousands of degrees, such an assumption is not very reasonable. The simplest case of variable coefficients corresponds to constant c and ρ and variable κ, a function of temperature $\kappa = \phi(T)$. We then write (2.1) in the form

$$T_t = \Delta \Phi(T). \tag{2.2}$$

This is the generalized PME, called Filtration Equation in the Russian literature. The constitutive function Φ is given by

$$\Phi(T) = \frac{1}{c\rho} \int_0^T \phi(s)ds. \tag{2.3}$$

If the dependence is given by a power function

$$\kappa(T) = aT^n, \tag{2.4}$$

with a and $n > 0$ constants, then we get

$$T_t = b\Delta(T^m) \quad \text{with} \quad m = n + 1, \tag{2.5}$$

and $b = a/(c\rho m)$, thus the PME but for the constant b which is easily scaled out. In case we also assume that $c\rho$ is variable, $c\rho = \psi(T)$, we still obtain a generalized PME though we have to work a bit more. Thus, we introduce a new variable T' by the formula

$$T' = \Psi(T) \equiv \int_0^T \psi(s)ds. \tag{2.6}$$

We then obtain (2.2) for the variable T' but now

$$\Phi(T') = \int_0^T \phi(s)ds, \tag{2.7}$$

where T is expressed in terms of T' by inverting (2.6), $T = \Psi^{-1}(T')$. Again, if the dependences are given by power functions we obtain the PME with an appropriate exponent.

Zel'dovich and Raizer [ZR, Chapter X] propose the above model to describe heat propagation by radiation occurring in ionized gases at very high temperatures. According to them, a good approximtion of the process is obtained with the PME for an exponent m close to 6.

3 Other applications

The previous application shows how naturally the PME appears to replace the classical heat equation in processes of heat transfer (or diffusion of a substance) whenever the assumption of constancy of the thermal conductivity (resp. diffusivity) cannot be sustained, and instead it is reasonable to assume that it depends in a power-like fashion on the temperature (resp. density or concentration). Once the theory for the PME began to be known, a number of applications have been proposed.

A very interesting example concerns the spread of biological populations. The simplest law regarding a population consisting of a single species is

$$u_t = \text{div}\,(\kappa \nabla u) + f(u), \tag{3.1}$$

where the reaction term $f(u)$ accounts for the interaction with the medium, which is supposed to be homogeneous. According to Gurtin and McCamy [GMC], when populations behave so as to avoid crowding, it is reasonable to assume that the diffusivity κ is an increasing function of the population density, hence

$$\kappa = \phi(u), \quad \phi \text{ increasing.} \tag{3.2}$$

A realistic assumption in some particular cases is $\phi(u) = a\,u$. Disregarding the reaction term we obtain the PME with $m = 2$.

Of course, a complete study must take into account at least the reaction terms, and very often, the presence of several species. This leads to the consideration of nonlinear reaction-diffusion systems of equations of parabolic type containing lower order terms, whose diffusive terms are of PME type. Such equations and systems constitute therefore an interesting possibility of generalization of the theory of the PME. Among the many works on the subject let us mention the early papers of Aronson and Weinberger [AW] and Aronson, Crandall and Peletier [ACP].

Chapter II   The homogeneous Dirichlet problem

In this chapter we consider the first boundary-value problem to the PME in a spatial domain $\Omega \subset \mathbf{R}^d, d \geq 1$, which is bounded and has a smooth boundary. We also consider homogeneous Dirichlet boundary conditions in order to obtain a simple problem for which a fairly complete theory can be easily developed.

A consequence of the degeneracy of the equation is that we do not expect to have classical solutions of the problem when the initial data take on the value $u = 0$, say, in an open subset of Ω.

Therefore we need to introduce an appropriate concept of *generalized* solution of the equation. At the same time we have to define in what sense the initial and boundary conditions are taken. In many cases this latter information can be built into the definition of generalized solution.

There are different ways of defining generalized solutions, the most usual idea being that of multiplying the equation by suitable test functions, integrating by parts some or all of the terms and asking from the solution a regularity that allows this expression to make sense. In this case we say that the solution is a *weak* solution.

In any case, the concept of generalized solution changes the meaning of the term *solution*, so we have be careful to ensure that the new definition makes good sense. First of all, the new solutions must be so defined that they include all classical solutions whenever they exist. Moreover, a concept of generalized solution will be useful if the problem becomes well-posed for a reasonably wide class of data, i.e., if a unique such solution exists for each set of data in a given class and if it depends continuously on the data in the appropriate topologies. As we will see, it can happen that several concepts of generalized solution arise naturally. It is then important to check that they agree in their common domain of definition (i.e., for data which are compatible with two or more of them). Selecting one of them as the preferred definition depends of several factors, the most important being in principle that of having the largest domain. However, one could consider a more restrictive definition which still covers the applications in mind if it involves simpler statements or more natural concepts, or when it leads to simpler proofs of its basic properties.

In this chapter we introduce a suitable concept of weak solution and prove the existence and uniqueness of a weak solution for all initial data in $L^{m+1}(\Omega)$. We then extend our definition to encompass data in $L^1(\Omega)$. By means of appropriate estimates we also establish the main properties of these solutions. In particular, we show that the solution satisfies the equation in a strong sense.

Though a strong solution of the Dirichlet Problem will, in general, not be a classical solution, it is a continuous function in $Q = \Omega \times (0, \infty)$, with a uniform Hölder modulus of continuity away from $t = 0$.

Finally, we establish the existence of a special solution of the form $\tilde{U}(x, t) = f(x)t^{-\alpha}$ with decay rate $\alpha = 1/(m-1)$. This solution is unique and acts as an absolute upper bound for all solutions of the Dirichlet Problem. The existence of such a solution is a typical *nonlinear effect*, which is not possible in the linear theory.

Notations They are rather standard. As usual, $\mathbf{R}_+ = (0, \infty)$. For a subset E of a metric space, $\overline{E}$ denotes its closure. For vectors $\boldsymbol{u}$ and $\boldsymbol{v} \in \mathbf{R}^d$ the scalar product is denoted by $\boldsymbol{u} \cdot \boldsymbol{v}$. If $\Omega \subset \mathbf{R}^d$ is the domain where the spatial variable lives, then $\partial\Omega$ denotes its boundary, while Q is the cylinder $\Omega \times \mathbf{R}_+$, and for $0 < T < \infty$ we write $Q_T = \Omega \times (0, T)$ and $Q^T = \Omega \times (T, \infty)$. The lateral boundary of Q is denoted by $\Sigma = \partial\Omega \times [0, \infty)$, while $\Sigma_T = \partial\Omega \times [0, T]$. Integrals without limits are understood to extend to the whole domain under consideration, Ω, Q or Q_T.

Concerning functional spaces, $C(\Omega), C^k(\Omega)$ and $C^\infty(\Omega)$ denote the spaces of continuous, k-times differentiable and infinitely differentiable functions in Ω, $C_c^\infty(\Omega)$ denotes the C^∞-smooth functions with compact support in Ω, and $\mathcal{D}'(\Omega)$ the space of distributions. For $1 \le p \le \infty$ we denote the usual Lebesgue spaces by $L^p(\Omega)$ with norm $||\cdot||_p$, while $H^1(\Omega)$ and $H_0^1(\Omega)$ are the usual Sobolev spaces, and the subscript *loc* refers to local spaces. The same applies to functions defined in Q or Q_T or their closures. $C^{2,1}(Q)$ denotes those functions which are twice differentiable in the space variables and once in time. For a function $u(x,t)$ we use the notation $u(t)$ to denote the function-valued map $t \mapsto u(\cdot,t)$.

Finally, the symbol $[\cdot]_+$ means positive part, i.e., $\max\{\cdot,0\}$, and the function sign_0^+ is defined as

$$\text{sign}_0^+(s) = 1 \text{ for } s > 0, \quad \text{sign}_0^+(s) = 0 \text{ for } s \le 0.$$

1 Weak solutions

We look for solutions $u = u(x,t)$ of the problem

$$u_t = \Delta(u^m) \quad \text{in} \quad Q_T, \tag{1.1}$$

$$u(x,0) = u_0(x) \quad \text{in} \quad \Omega, \tag{1.2}$$

$$u(x,t) = 0 \quad \text{in} \quad \Sigma_T, \tag{1.3}$$

where $m > 1$ and u_0 is a nonnegative, locally integrable function defined in Ω, a bounded domain in $\mathbf{R}^d, d \ge 1$, with boundary $\Gamma = \partial\Omega \in C^{2+\alpha}, \alpha \in (0,1)$. The time T can be finite or infinite. Though we will obtain solutions for all $T > 0$, i.e. with $T = \infty$, it is interesting for technical reasons to allow $T < \infty$.

First of all we introduce a suitable concept of weak solution for problem (1.1)–(1.3). Following [OKC] and [Sa] we propose

Definition 1 A nonnegative function u defined in Q_T is said to be a *weak solution* of problem (1.1)–(1.3) if

i) $u^m \in L^2(0, T : H_0^1(\Omega))$;

ii) u satisfies the identity

$$\iint_{Q_T} \{\nabla(u^m) \cdot \nabla\varphi - u\varphi_t\}dx\,dt = \int_\Omega u_0(x)\varphi(x,0)dx \tag{1.4}$$

for any function $\varphi \in C^1(\overline{Q}_T)$ which vanishes on Σ and for $t = T$.

In the above definition u_0 should belong at least to $L^1(\Omega)$ for (1.4) to be well defined.

Observe that the equation is satisfied only in a weak sense since we do not assume that the derivatives appearing in equation (1.1) are actual functions, but merely exist in the sense of distributions. In fact, by specializing φ to the test function space $C_c^\infty(\Omega)$, we observe that $u_t = \Delta(u^m)$ in $\mathcal{D}'(Q_T)$. Moreover, the boundary condition (1.3) is hidden in the functional space $H_0^1(\Omega)$. Finally, the initial condition (1.2) is built into the integral formulation (1.4), and is actually satisfied in a very weak sense. Let us show as an example of the scope of the above definition how another natural way of defining a weak solution is included in Definition 1.

Proposition 1 *Let* u *be a nonnegative function defined in* Q_T *and such that*

i) $u^m \in L^2(0, T : H_0^1(\Omega))$;

ii) u *satisfies the identity*

$$\iint \{\nabla(u^m) \cdot \nabla\varphi - u\varphi_t\} dx dt = 0 \tag{1.5}$$

for any function $\varphi \in C_c^\infty(Q_T)$;

iii) *for every* $t > 0$ *we have* $u(t) \in L^1(\Omega)$ *and* $u(t) \to u_0$ *as* $t \to 0$ *in* $L^1(\Omega)$.

Then u *is a weak solution to* (1.1)–(1.3) *according to Definition 1.*

Proof Suppose that u is as in the statement. We have to prove that (1.4) holds. It is very easy to see that (1.5) continues to hold when $\varphi \in C^1(Q_T)$ with $\varphi = 0$ on the boundary of Q_T (Hint: approximate φ with $\varphi_\varepsilon \in C_c^\infty$ and pass to the limit).

Now if φ is as in (1.4) we take a cut-off function $\varsigma \in C^\infty(\mathbf{R}), 0 \leq \varsigma \leq 1$, such that $\varsigma(s) = 0$ for $s < 0, \varsigma(s) = 1$ for $s \geq 1$ and $\varsigma' \geq 0$, and let $\varsigma_n(t) = \varsigma(nt)$. Applying (1.5) with test function $\varphi(x, t)\varsigma_n(t)$ gives

$$\iint_Q \{\nabla(u^m) \cdot \nabla\varphi - u\varphi_t\}\varsigma_n dx dt = \iint_Q u\varphi\varsigma_{n,t} = \iint_{Q_{1/n}} u\varphi\varsigma_{n,t}$$

$$= \iint_{Q_{1/n}} (u - u_0)\varphi\varsigma_{n,t} + \iint_{Q_{1/n}} u_0(x)\varphi(x,t)\varsigma_{n,t}(t).$$

Fix $\varepsilon > 0$ and let n be so large that $||u - u_0||_1 \leq \varepsilon$ for $0 \leq t \leq 1/n$. Then the first integral in the last member can be estimated as $\varepsilon||\varphi||_\infty \int \varsigma_{n,t} dt = \varepsilon||\varphi||_\infty$ which vanishes as $n \to \infty$, $\varepsilon \to 0$. As for the last term, we get

$$\iint_{Q_{1/n}} u_0(x)\varphi(x,t)\varsigma_{n,t}(t) dx dt = \int_\Omega u_0(x)\varphi\left(x, \frac{1}{n}\right) dx$$

$$- \iint_{Q_{1/n}} u_0\varphi_t\varsigma_n dx dt \to \int_\Omega u_0(x)\varphi(x,0) dx$$

as $n \to \infty$, which proves (1.4). □

Of course, a classical solution of problem (1.1)–(1.3) is automatically a weak solution of the problem. Moreover, though the explicit source-type solution $U(x,t) = U(x,t;C)$ of the Introduction is not a weak solution because of its singular initial data and because the boundary data are not necessarily 0, we can obtain from it weak solutions by the following method. Take $x_0 \in \Omega$, let $\tau > 0$, and let the constant C in U be small enough. Then the function

$$w(x,t) = U(x - x_0, t + \tau; C) \qquad (1.6)$$

is a weak solution of the Dirichlet Problem (1.1)–(1.3) in any time interval $(0,T)$ in which the free boundary lies inside of Ω, i.e., if

$$\text{dist}(x_0, \partial\Omega)^{m+1} \geq c(T + \tau),$$

cf. (0.5). Observe that w is not a classical solution.

2 Uniqueness of weak solutions

The uniqueness of weak solutions as defined above is very easily settled by means of an interesting trick, consisting in using a specific test function.

Theorem 2 *Problem* (1.1)–(1.3) *has at most one weak solution.*

Proof Suppose that we have two such solutions u_1 and u_2. By (1.4) we have

$$\iint\limits_{Q_T} (\nabla(u_1^m - u_2^m) \cdot \nabla\varphi - (u_1 - u_2)\varphi_t)dxdt = 0 \qquad (2.1)$$

for all test functions φ. We want to use as a test function the one introduced by Oleĭnik

$$\eta(x,t) = \begin{cases} \int_t^T (u_1^m(x,s) - u_2^m(x,s))ds, & \text{if } 0 < t < T \\ 0, & \text{if } t \geq T, \end{cases} \qquad (2.2)$$

where $T > 0$. Even if η does not have the required smoothness we may approximate it (by mollification) with smooth functions η_ε for which (1.4) will hold. Since

$$\eta_t = -(u_1^m - u_2^m) \in L^2(Q_T), \qquad (2.3)$$

$$\nabla\eta = \int_t^T (\nabla u_1^m - \nabla u_2^m)ds \in L^2(Q_T), \qquad (2.4)$$

and moreover $\eta(t) \in H_0^1(\Omega)$ and $\eta(T) = 0$, we may pass to the limit $\varepsilon \to 0$, and (1.4) will still hold for η. Hence

$$\iint (u_1^m - u_2^m)(u_1 - u_2) - \iint (\nabla(u_1^m - u_2^m)) \cdot (\int_t^T (\nabla u_1^m - \nabla u_2^m)ds) = 0.$$

Integration of the last term gives

$$\iint\limits_{Q_T} (u_1^m - u_2^m)(u_1 - u_2)dxdt + \frac{1}{2}\int_\Omega (\int_0^T (\nabla u_1^m - \nabla u_2^m)ds)^2 dx = 0. \qquad (2.5)$$

Since both terms are nonnegative, we conclude that $u_1 = u_2$ a.e. in Q. $\qquad\square$

As a consequence of the uniqueness of weak solutions we have

Corollary 3 *There exist initial data for which the problem does not admit a classical solution.*

Proof This is a rather standard argument. Firstly, we note that a classical solution of problem (1.1)–(1.3) is necessarily a weak solution in our sense. Secondly, we remark that $w(x,t)$ defined in (1.6) is a weak and nonclassical solution. By the uniqueness result there cannot be any other weak solution of (1.1)–(1.3) with the same data, hence no classical solution exists. $\qquad\square$

3 Existence of a weak solution. Energy estimate

In a first approach we establish existence under the assumption that the data belong to the space $L^{m+1}(\Omega)$.

Theorem 4 *Suppose that $u_0 \in L^{m+1}(\Omega), u_0 \geq 0$. Then problem (1.1)–(1.3) has a weak solution with infinite time interval, $T = \infty$.*

The proof will be divided into several steps. First, we will consider the case of a smooth function u_0 vanishing on the border and prove the result obtaining at the same time an important estimate. This estimate will allow us to solve the general problem by an approximation technique.

First step *We assume that u_0 is a nonnegative and C^∞-smooth function with compact support in Ω.*

We begin by constructing a sequence of approximate initial data u_{0n} which does not take the value $u = 0$, so as to avoid the degeneracy of the equation. In our case we may simply put

$$u_{0n}(x) = u_0(x) + \frac{1}{n}. \qquad (3.1)$$

We now solve the problem

$$(u_n)_t = \Delta(u_n^m) \quad \text{in } Q, \qquad (3.2)$$
$$u_n(x,0) = u_{0n}(x) \quad \text{in } \overline{\Omega}, \qquad (3.3)$$
$$u_n(x,t) = \frac{1}{n} \quad \text{on } \Sigma. \qquad (3.4)$$

In view of the data we expect the solution to be bounded by

$$\frac{1}{n} \leq u_n(x,t) \leq M + \frac{1}{n} \quad \text{in } \overline{Q} \tag{3.5}$$

by the Maximum Principle, where $M = \sup(u_0)$. Therefore, we are dealing in practice with a uniformly parabolic problem. Actually, problem (3.2)–(3.4) has a unique solution $u_n \in C^{2,1}(\overline{Q})$. The rigorous justification uses a trick consisting in replacing equation (3.2) by

$$(u_n)_t = \text{div}(a_n(u)\nabla u), \tag{3.2'}$$

where $a_n(u)$ is a positive and smooth function, $a_n(u) > c > 0$, and $a_n(u) = m u^{m-1}$ in the interval $[1/n, M + 1/n]$. This equation is not degenerate and a unique solution u_n of (3.2'), (3.3), (3.4) exists in the space $C^{2,1}(\overline{Q})$ by the standard quasilinear theory, cf. [LSU] (Chapter 6) and it satisfies (3.5). Moreover, by repeated differentiation and interior regularity we are able to conclude that $u_n \in C^\infty(Q)$. Now, due to the definition of a_n, equations (3.2) and (3.2') coincide on the range of u_n. In this way problem (3.2)–(3.4) is solved in a classical sense, and the degeneracy of the equation has been avoided.

Moreover, again by the Maximum Principle,

$$u_{n+1}(x,t) \leq u_n(x,t) \quad \text{in } \overline{Q} \tag{3.6}$$

for all $n \geq 1$. Hence we may define the function

$$u(x,t) = \lim_{n\to\infty} u_n(x,t), \quad (x,t) \in \overline{Q}. \tag{3.7}$$

Then u_n converges to u in $L^p(\Omega)$ for every $1 \leq p < \infty$. In order to show that this u is the weak solution of problem (1.1)–(1.3) we will need some estimates. First of all, from (3.5) we get

$$0 \leq u \leq M \quad \text{in } \overline{Q}. \tag{3.8}$$

We now control the spatial gradient $\nabla(u^m)$. Multiply equation (3.2) by $\varphi_n = u_n^m - \left(\frac{1}{n}\right)^m$ and integrate by parts in Q_T to obtain

$$
\begin{aligned}
\iint_{Q_T} |\nabla u_n^m|^2 dx dt &= \int_\Omega \left(\frac{1}{m+1} u_{0n}^m(x) - \frac{1}{n^m}\right) u_{0n}(x) dx \\
&\quad - \int_\Omega \left(\frac{1}{m+1} u_n^m(x,T) - \frac{1}{n^m}\right) u_n(x,T) dx \\
&\leq \frac{1}{m+1} \int \left(u_0(x) + \frac{1}{n}\right)^{m+1} dx + \frac{1}{n^m}\left(M + \frac{1}{n}\right) \int_\Omega dx.
\end{aligned}
\tag{3.9}
$$

Since T is arbitrary, it follows that $\{\nabla u_n^m\}$ is uniformly bounded in $L^2(Q)$, and therefore a subsequence of it should converge to some limit ψ weakly in $L^2(Q)$. Since also $u_n^m \to u^m$ it follows that $\psi = \nabla u^m$ in the sense of distributions. The limit is uniquely defined so that the whole sequence must converge to it. Passing to the limit in (3.9), we get the following important estimate

$$(m+1) \iint_{Q_T} |\nabla u^m|^2 dx dt + \int u^{m+1}(x,T) dx \leq \int u_0^{m+1}(x) dx, \tag{3.10}$$

called the *Energy Estimate*. On the other hand, since $u_n \in C(\overline{Q})$, $u_n(x,t) = \frac{1}{n}$ on Σ and $0 \leq u \leq u_n$, we have

$$\lim_{(x,t)\to\Sigma} u(x,t) = 0$$

with uniform convergence. Hence $u^m(\cdot,t) \in H_0^1(\Omega)$ for a.e. $t > 0$.

Finally, since u_n is a classical solution of (1.1), it clearly satisfies (1.4) with u_0 replaced by u_{0n}. Letting $n \to \infty$ we obtain (1.4) for u. Therefore, u is a weak solution of (1.1)–(1.3).

Let us remark, to end this step, that if we have two initial data $u_0, \hat{u}_0$ such that $u_0 \leq \hat{u}_0$, then the above approximation process can be performed so as to produce ordered approximating sequences, $u_{0n} \leq \hat{u}_{0n}$. By the classical Maximum Principle, [LSU], we have $u_n \leq \hat{u}_n$ for every $n \geq 1$. Hence in the limit $u \leq \hat{u}$.

Second step *We assume that u_0 is bounded and vanishes near the boundary.*

The method of the previous step can still be applied. According to the quasilinear theory, cf. [LSU], now the approximate solutions $u_n \in C^\infty(Q) \cap C^{2,1}(Q \cup \Sigma)$ and are not continuous down to $t = 0$ unless the data are, but instead take the initial data in $L^p(\Omega)$ for every $p < \infty$. Passage to the limit offers no novelty. Comparison still applies.

Third step *General case.*

In the general case $u_0 \in L^{m+1}(\Omega)$ we take an increasing sequence of cutoff functions ς_k which vanish near Γ, and consider the sequence of approximations of the initial data

$$u_{0k}(x) = \min\{u_0(x)\varsigma_k(x), k\}. \tag{3.11}$$

Using Step 2 we solve problem (1.1)–(1.3) with initial data u_{0k} and obtain a unique weak solution u_k. By the comparison remark $u_{k+1} \geq u_k$ in Q. On the other hand, by estimate (3.10), $\{u_k\}$ is bounded uniformly in $L^\infty(0,\infty : L^{m+1}(\Omega))$, and ∇u_k^m is in $L^2(Q)$. Hence, $\{u_k\}$ converges a.e. to a function $u \in L^\infty(0,\infty : L^{m+1}(\Omega))$, ∇u_k^m converges weakly in $L^2(Q)$ to ∇u^m, and (3.10) holds for u. It follows from the Sobolev embedding that $u^m \in L^2(0,\infty : H_0^1(\Omega))$. Finally, equation (1.4) is satisfied. $\square$

An alternative proof, where Steps 2 and 3 are replaced by a single approximation step using a stability result, will be given at the end of §6. On the other hand, we see from the proof that the choice of $L^{m+1}(\Omega)$ as the space for the initial data depends essentially on estimate (3.10). A priori estimates are one of the most powerful and widely used tools in the study of P.D.E. This approach will be stressed in our treatment of the existence, uniqueness and qualitative properties of solutions to the different problems.

4 Absolute bound in sup norm

Before we proceed, we will derive another important estimate, the boundedness of the solutions for positive times. This bound will give us a needed control on u^m.

Proposition 5 *Every weak solution u of (1.1)–(1.3) is bounded in Q^τ for every $\tau > 0$. Moreover, we have an absolute decay estimate of the form*

$$u(x,t) \le C(m,d)R^{\frac{2}{m-1}}t^{-\frac{1}{m-1}}, \tag{4.1}$$

where $C(m,d) > 0$ and R is the radius of a ball containing Ω.

By *absolute* we mean that the bound does not depend on the data we are considering.

Proof Let us first consider the case where u_0 is continuous and vanishes on $\partial\Omega$. We will construct an explicit super-solution $z(x,t)$ with which to compare the approximate solutions u_n to (3.2)–(3.4).

In fact, we take a ball B of radius R strictly containing Ω, i.e., with $\Gamma \subset B$, and consider

$$z(x,t) = A(t+\tau)^{-\alpha}(1-bx^2)^\beta \tag{4.2}$$

for suitable constants $A, \tau, \beta, b > 0$. Setting $b = R^{-2}$ will make the function positive in $B \times (0,\infty)$, hence for all large n,

$$u_n(x,t) = \frac{1}{n} < z(x,t) \quad \text{in } \Sigma \tag{4.3}$$

if A, τ, α, β are kept fixed. Moreover, we will have

$$u_{0n}(x) \le z(x,0) \tag{4.4}$$

whenever τ is small enough. Finally, setting $\alpha = 1/(m-1)$ and $\beta = 1/m$, we obtain

$$z_t - \Delta(z^m) \ge 0 = (u_n)_t - \Delta(u_n^m) \tag{4.5}$$

whenever

$$A \ge R^{\frac{2}{m-1}}(2d(m-1))^{-\frac{1}{m-1}}. \tag{4.6}$$

With those choices, the classical Maximum Principle implies that $u_n(x,t) \le z(x,t)$ in Q, hence in the limit $u(x,t) \le z(x,t)$. Since τ could be arbitrarily small, we will let $\tau \to 0$ and finally get

$$u(x,t) \le At^{-\frac{1}{m-1}}\left(1-\left(\frac{x}{R}\right)^2\right)^{1/m} \tag{4.7}$$

By approximation, (4.7) holds for every weak solution. $\qquad\square$

5 Existence of classical solutions

Once a solution of the equation is constructed in some generalized sense, it is an important point to decide if it is indeed a classical solution. Though we know that in general this will not be the case, it can happen under additional requirements on the data. We prove in this section that when the initial data are smooth and positive inside Ω, so that the euqation is parabolic nondegenerate, we obtain a classical solution by essentially using the standard quasilinear theory.

Proposition 6 *Let $u_0 \in C(\overline{\Omega})$ be positive in Ω and vanish on Γ, and let u let be the corresponding weak solution. Then $u \in C^{\infty}(Q) \cap C(\overline{Q})$, u is positive in Q and vanishes on Σ.*

Proof The first step will be to prove that for every point $x_0 \in \Omega$ where $u_0(x_0) > 0$, we will have $u(x_0, t) > 0$ for every $t > 0$. This is done by the method of barriers, comparing u with a suitable source-type solution. Actually, if $B = B_r(x_0)$ is a ball of radius r where u_0 is positive, say $u_0(x) \geq c > 0$ for $x \in B$, we consider the Barenblatt function

$$\bar{u} = U(x - x_0, t + 1; C).$$

We may choose C small enough so that $u_0(x) \geq \bar{u}(x, 0)$ in B, and also that the support of $\bar{u}$ is contained in Q_T for a given $T > 0$. This support is of the form $S = \{(x, t) : c|x - x_0|^{m+1} < (t + 1)\}$ (cf. (0.5)), $\bar{u} \in C^{\infty}(S)$ and $\bar{u}$ vanishes on the lateral boundary of S.

Hence, by the classical Maximum Principle applied in $S \cap Q_T$ to $\bar{u}$ and a smooth approximation to u, we conclude that $u \geq \bar{u}$ in S, hence $u(x, t)$ is bounded uniformly away from 0 in a neighbourhood of the form $N = B_1 \times (0, T), B_1 = B_r(x_0)$.

Therefore, when taking the limit $u_n \to u$ in the approximation process of Theorem 4, we can apply in N the regularity theory of quasilinear nondegenerate parabolic equations, and conclude that $u \in C^{\infty}(N)$ and the initial data are taken continuously in B_1.

The fact that u vanishes continuously on Σ is a simple consequence of the approximation process (3.1)–(3.4). In fact $u \leq u_n, u_n \in C^{\infty}(\overline{Q})$ and $u_n(x, t) = \frac{1}{n}$ on Σ. □

Of course, if moreover u_0 is smooth, e.g. if $u_0 \in C^k(\Omega)$ for some $k > 0$, this regularity is reflected in the regularity of u near $t = 0$.

6 The basic L^1-estimate

This section contains another very important estimate which allows to develop an existence, uniqueness and stability theory in the space $L^1(\Omega)$.

Proposition 7 (L^1-Contraction Principle) *Let $u_0, \hat{u}_0$ be two initial data in $L^{m+1}(\Omega)$ and let $u, \hat{u}$ be their respective weak solutions. Then for every $t > \tau \geq 0$*

$$\int [u(x,t) - \hat{u}(x,t)]^+ dx \leq \int [u(x,\tau) - \hat{u}(x,\tau)]^+ dx \leq \int [u_0(x) - \hat{u}(x)]^+ dx, \qquad (6.1)$$

so in particular,

$$||u(t) - \hat{u}(t)||_1 \leq ||u(\tau) - \hat{u}(\tau)||_1 \leq ||u_0 - \hat{u}_0||_1. \qquad (6.2)$$

Proof Let $p \in C^1(\mathbf{R})$ such that $0 \leq p \leq 1, p(s) = 0$ for $s < 0, p'(s) > 0$ for $s > 0$, and consider the approximate solutions $u_n, \hat{u}_n$ to problem (3.2)–(3.4) with the same n. We have in Ω for $t > 0$

$$\int (u_n - \hat{u}_n)_t \, p(u^m - u^m) dx = \int \Delta(u_n^m - \hat{u}_n^m) p(u^m - u^m) dx$$

$$= -\int |\nabla(u_n^m - \hat{u}_n^m)|^2 p'(u^m - u^m) dx \leq 0.$$

(Observe that $p(u_n^m - \hat{u}_n^m) = 0$ on Σ.) Therefore, letting p converge to the sign function sign_0^+, and observing that $\dfrac{d}{dt}[u_n - \hat{u}_n]^+ = (u_n - \hat{u}_n)_t \, \mathrm{sign}_0^+(u_n - \hat{u}_n)$, cf. [GT], we get

$$\frac{d}{dt} \int [u_n - \hat{u}_n]^+ dx \leq 0,$$

which implies (6.1) for $u_n, \hat{u}_n$. Passing to the limit we obtain (6.1). To obtain (6.2), combine (6.1) applied first to u and $\hat{u}$, and then to $\hat{u}$ and u. $\qquad \square$

The above proof establishes the uniqueness of solutions of problem (1.1)–(1.3) by a technique (the L^1 technique) which is completely different from that of Theorem 4. It is interesting to remark that estimate (6.1) not only implies L^1-dependence of solutions on data, but also a *comparison* theorem.

Corollary 8 *If $u_0 \leq \hat{u}_0$ a.e. in Ω, then $u \leq \hat{u}$ a.e. in Q.*

Another consequence of the estimate is

Corollary 9 (Continuity in L^1) *The weak solution of (1.1)–(1.3) can be viewed as a continuous curve in $L^1(\Omega)$, i.e., $u \in C([0,\infty) : L^1(\Omega))$.*

Proof When $u_0 \in C(\Omega)$ is positive in Ω and vanishes on $\partial\Omega$, we have shown in §5 that $u \in C(\overline{Q})$, hence $u \in C([0,T] : L^1(\Omega))$. For general u_0, we approximate with functions $\hat{u}_0$ as above and write, using Proposition 7,

$$||u(\tau) - u_0||_1 \leq ||u(\tau) - \hat{u}(\tau)||_1 + ||u_0 - \hat{u}_0||_1 + ||\hat{u}(\tau) - \hat{u}_0||_1$$

$$\leq 2||u_0 - \hat{u}_0|| + ||\hat{u}(\tau) - \hat{u}_0||_1.$$

Therefore as $\hat{u}_0 \to u_0$ and $\tau \downarrow 0$ we get $u(\tau) \to u_0$. This settles the continuity at $t = 0$. To settle it at any other time $t > 0$ we may displace the origin of time and argue as before at the times t and $t + \tau$. $\qquad \square$

Remark *(The proof of Theorem 4 revisited)* To end this section, we give an alternative to steps 2 and 3 of the existence proof of Theorem 4, §3. Now that we know that solutions depend continuously on the data in $L^1(\Omega)$, we may approximate a general data $u_0 \in L^{m+1}(\Omega)$ with $u_{0n} \in C_c^\infty(\Omega)$, apply step 1 and pass to the limit using Proposition 7 and the Energy Estimate (3.10).

7 Solutions with data in $L^1(\Omega)$

The continuous dependence with respect to the L^1 norm allows us to extend our existence result and consider as data any nonnegative function $u_0 \in L^1(\Omega)$. The idea is to approximate the initial data with a sequence $u_{0n} \in L^\infty(\Omega)$ such that $u_{0n} \to u_0$ in $L^1(\Omega)$, obtain solutions u_n with these data, and use Proposition 7 to pass to the limit in u_n as $n \to \infty$. We thus obtain a function $u \in C([0,\infty) : L^1(\Omega))$ such that $u(0) = u_0$. The question is, is u a weak solution of (1.1)–(1.3) according to Definition 1?

To begin with, it turns out that in general u does not satisfy the condition $u^m \in L^2(0,\infty : H_0^1(\Omega))$, which is important in giving a sense to identity (1.4); therefore we must change our definition of weak solution. It happens that, thanks to the absolute bound, in passing to the limit $n \to \infty$ in the sequence u_n considered above we encounter difficulties in checking that u is a weak solution only near $t = 0$. A convenient definition to cover solutions with L^1 data is

Definition 2 A nonnegative function $u \in C([0,\infty) : L^1(\Omega))$ is said to be a *weak solution* of problem (1.1)–(1.3) if

i) $u^m \in L_{\text{loc}}^2(0,\infty : H_0^1(\Omega))$;

ii) u satisfies the identity

$$\iint_Q \{\nabla u^m \cdot \nabla \varphi - u\varphi_t\} dx dt = 0 \qquad (7.1)$$

for any function $\varphi \in C_c^1(Q)$;

iii) $u(0) = u_0$.

We immediately see that a weak solution in the sense of Definition 1 is a also a weak solution in the present sense if we can ensure that it belongs to the class $C([0,\infty) : L^1(\Omega))$. We will come back to the relation between both definitions later.

Theorem 10 *There exists a unique weak solution of problem* (1.1)–(1.3) *with given initial data* $u_0 \in L^1(\Omega), u_0 \geq 0$, *where weak solution is understood in the sense of Definition 2. The Contraction Principle holds for these solutions.*

Proof (i) *Existence.* We construct approximations u_n as before and pass to the limit using the L^1 and L^∞ estimates derived in Propositions 5 and 7 plus the energy estimate (3.10).

(ii) *Uniqueness.* Let u_1, u_2 be weak solutions of the problem with the same initial data u_0. Given $\varepsilon > 0$ there exists $\tau > 0$ such that $||u_1(t) - u_0||_1, ||u_2(t) - u_0||_1 < \varepsilon$ for $0 \le t \le \tau$. Consider now the functions $v_1(x,t) = u_1(x, t + \tau), v_2(x,t) = u_2(x, t + \tau)$. Both v_1 and v_2 satisfy the assumptions of Proposition 1, hence they are weak solutions of the same problem with initial data $u_1(x, \tau), u_2(x, \tau)$, resp. In particular, these initial data are bounded functions. Hence, by Proposition 7 we get for $t > \tau$,

$$||u_1(t) - u_2(t)||_1 = ||v_1(t - \tau) - v_2(t - \tau)||_1$$
$$\le ||v_1(0) - v_2(0)||_1 = ||u_1(\tau) - u_2(\tau)||_1 < \varepsilon.$$

We may now let $\varepsilon \to 0$ to get $u_1(t) = u_2(t)$ a.e. for every $t > 0$.

(iii) The validity of the Contraction Principle (Proposition 7) is just a consequence of the limit process.   □

We can now establish the relationship between both definitions.

Proposition 11 *A weak solution in the sense of Definition 1 such that $u \in C([0, T) : L^1(\Omega))$ is also a weak solution in the sense of Definition 2. Conversely, a weak solution in the new sense is also a weak solution according to Definition 1 if and only if $u_0 \in L^{m+1}(\Omega)$.*

Proof The first statement is obvious. This applies in particular whenever $u_0 \in L^{m+1}(\Omega)$. Thus, both definitions coincide in this case.

Now let u be a solution in the sense 2 with $u_0 \in L^1(\Omega)$. Then Definition 1 is satisfied if $\nabla u^m \in L^2(0, T : H_0^1(\Omega))$. It can be proved that the Energy Estimate holds with equality sign. But then the bound on ∇u^m is equivalent to the condition $u_0 \in L^{m+1}(\Omega)$.   □

Definition 2 has the advantage over Definition 1 that both the comparison and stability results proved in the last section are immediately seen to hold. It is also obvious according to this definition that, if $u(x,t)$ with data $u_0(x)$ is a solution and $\tau > 0$, then $v(x,t) = u(x, t - \tau)$ is the solution corresponding to data $v_0(x) = u(x, \tau)$.

8 Further regularity

Though solutions with data which are not strictly positive need not be classical solutions, they enjoy some interesting regularity properties that we derive as a consequence of new estimates. We recall that, as a consequence of estimate (3.10) the L^{m+1}-norm of a solution u is nonincreasing in time. By the same method we can obtain monotonicity in all the L^p norms, $1 < p < \infty$, by using other powers of u as test functions in the calculation. Thus we obtain the following monotonicity statement for the L^p-norms of the solutions.

Proposition 12 *For every $u_0 \in L^p(\Omega), p \geq 1$, we have $u(\cdot, T) \in L^p(\Omega)$ for any $T > 0$ and*

$$\|u(\cdot, t)\|_p \leq \|u_0\|_p. \tag{8.1}$$

Proof It is based on the estimate

$$\frac{4qm}{(q+m)^2} \iint_{Q_T} |\nabla (u^{\frac{q+m}{2}})|^2 + \frac{1}{q+1} \int u^{q+1}(x, T) dx \leq \frac{1}{q+1} \int u_0^{q+1}(x) dx, \tag{8.2}$$

valid for $q > 0$. To get the case $p = 1$ we pass to the limit as $q \to 0$.

The preceding estimate is not very important in our situation since we have a very precise decay result thanks to Proposition 5. More important is the inverstigation on regularity of the derivative u_t appearing as one of the members of the equation. Our results will lead in the next section to the concept of strong solution. We begin with an "energy" estimate for the time derivative.

Proposition 13 $(u^q)_t \in L^2(Q^\tau)$ *for every $q \geq (m+1)/2$ and every $\tau > 0$. The basic estimate is*

$$\frac{8m}{(m+1)^2} \iint_{Q_T} t |\frac{d}{dt}(u^{(m+1)/2})|^2 + T \int_\Omega |\nabla u^m(x, T)|^2 dx \leq \iint_{Q_T} |\nabla u^m|^2 dx dt, \tag{8.3}$$

valid for all $T > 0$. From this we obtain the decay rate

$$\iint_{Q^\tau} |(u^q)_t|^2 dx dt = O(t^{-\frac{2(q-1)}{m-1}}) \iint_{Q^\tau} |\nabla u^m|^2 dx dt. \tag{8.4}$$

Proof Let $w = u^{\frac{m+1}{2}}$ with $u = u_n$ solution of (3.2)–(3.4). We have

$$\left|\frac{dw}{dt}\right|^2 = \left(\frac{m+1}{2}\right)^2 u^{m-1}|u_t|^2 = \frac{(m+1)^2}{4m}(u^m)_t \Delta u^m,$$

and

$$\frac{1}{2}\frac{d}{dt}\int |\nabla u^m|^2 dx = \int \nabla u^m \cdot (\nabla u^m)_t dx = -\int \Delta u^m (u^m)_t dx,$$

since $(u^m)_t = 0$ on Σ. Therefore, integration in Q_T gives

$$\iint t \left|\frac{dw}{dt}\right|^2 = -\frac{(m+1)^2}{8m} \int_0^T t \left(\frac{d}{dt}\int |\nabla u^m|^2 dx\right) dt =$$

$$= -\frac{(m+1)^2}{8m} T \int |\nabla u^m(x, T)|^2 dx + \frac{(m+1)^2}{8m} \iint |\nabla u^m|^2.$$

This is estimate (8.3) for $u = u_n$. When u is any weak solution we proceed by approximation. We notice that the same argument applies in any time interval (τ, T) with $\tau > 0$. To estimate $(u^q)_t$ with $q > (m+1)/2$, we observe that

$$(u^q)_t = (2q/(m+1))u^{q-(m+1)/2}(u^{(m+1)/2})_t,$$

and recall that u is bounded in Q^τ by Proposition 5. Combining inequality (8.3) in (τ, T), with $T \to \infty$, with the L^∞-estimate (4.1), we get (8.4). $\qquad\qquad\square$

Remark (8.3) combined with (3.10) gives estimates of the first member in terms of $\int u_0^{m+1} dx$. Thus,

$$\iint_Q t \left| \frac{d}{dt}(u^{(m+1)/2}) \right|^2 dx dt \leq \frac{(m+1)^2}{8m} \int_\Omega u_0^{m+1}(x) dx.$$

Unfortunately, the preceding estimates do not allow for a direct control of the derivative u_t appearing in the equation. We obtain next an estimate for u_t.

Proposition 14 *We have in* $\mathcal{D}'(Q)$

$$u_t \geq -\frac{u}{(m-1)t}. \tag{8.5}$$

Proof (i) *First proof.* Let $u = u_n$ be one of the approximate solutions to problem (1.1)–(1.3). Consider the function

$$z = (m-1)tu_t + u.$$

It it a simple computation to show that z is a solution in Q of the equation

$$z_t = \Delta(mu^{m-1}z), \tag{8.6}$$

that $z(x,t) = u(x,t) \geq 0$ on Σ, and $z(x,0) \geq 0$ for all $z \in \Omega$. Hence, by the standard Maximum Principle $z(x,t) \geq 0$, which is equivalent to (8.5). In this case we obtain a pointwise inequality.

We now pass to the limit in (8.5) to obtain the estimate for any weak solution of (1.1)–(1.3). This can only be done on the weak or distributional form of the inequality, which is obtained by multiplying by a test function $\varphi \in C_c^\infty(Q), \varphi \geq 0$, and integrating by parts, i.e.,

$$\iint \left(\frac{1}{(m-1)t} u\varphi - u\varphi_t \right) dx dt \geq 0.$$

Therefore, the fact that (8.1) holds in the sense of distributions does not mean that u_t is a function. At least, since u is the limit of a sequence $\{u_n\}$ for which $(u_n)_t$ is locally bounded below uniformly in n, u_t is in principle a Radon measure.

(ii) *Second proof.* The reader may wonder how did we find the precise combination

$$z = u + (m-1)tu_t$$

to which the Maximum Principle can be applied. There is a beautiful and simple argument based on scaling which produces such a magic function. It runs as follows: given a smooth solution u and a constant $\lambda > 1$, we consider the function

$$\tilde{u}(x,t) = \lambda u(x, \lambda^{m-1}t). \tag{8.7}$$

This is again a solution of the PME. Moreover, for $\lambda > 1$ we have $\tilde{u}(x,0) = \lambda u(x,0) \geq u(x,0)$, hence by the Maximum Principle $\tilde{u} \geq u$ in Q. Now differentiate (8.7) with respect to λ at $\lambda = 1$. We get

$$0 \leq \frac{d}{d\lambda}\tilde{u}(x,t)|_{\lambda=1} = u(x,t) + (m-1)tu_t(x,t),$$

namely (8.5).   □

Remark We can extend (8.1) to an L^∞-estimate down $t = 0$ if Δu_0^m is bounded below. In fact, if

$$(m-1)\Delta u_0^m \geq -au_0 - b \tag{8.8}$$

for some constants $a, b > 0$, we may compare the functions $z_1 = (m-1)(at+1)u_t + au$ and $z_2 = -b$: both are solutions of (8.6) in Q and $z_1 \geq z_2$ on the parabolic boundary of Q. Hence, by the Maximum Principle which is again justified by approximation, we obtain $z_1 \geq z_2$, i.e.,

$$u_t \geq -\frac{au+b}{(m-1)(at+1)}. \tag{8.9}$$

In order to prove that u_t is actually an integrable function we have to translate the estimate for $(u^{(m+1)/2})_t$ into an estimate for u_t. This is rather technical. We use the following result.

Lemma 15 *Let K be a subset of $\mathbf{R}^d$ with finite measure, let $I = [t_0, t_1]$, and assume that v is a function defined in $K \times I$ that satisfies*

i) $v \in L^\infty(I : L^1(K)), v \geq 0, v_t \geq 0;$

ii) v^λ and $\dfrac{d}{dt}(v^\lambda) \in L^r(K \times I)$ for some $\lambda, r > 1.$

Then $\dfrac{d}{dt}v \in L^p(K \times I)$ for every $p \in [1, p_1)$, where

$$p_1 = \frac{r\lambda}{r(\lambda-1)+1} \in (1, r).$$

Proof Without loss of generality we may assume that $v \geq \varepsilon > 0$ in $K \times I$ by replacing v by $v + \varepsilon$ since our estimates will not depend on ε. Now, for any $p \in (1,r)$ and $\nu \in (0,p)$ we have

$$\left|\frac{dv}{dt}\right|^p = \left[\frac{1}{\lambda}\frac{dv^\lambda}{dt}\right]^\nu \left(v^{\sigma-1}\frac{dv}{dt}\right)^{p-\nu}$$

where $1 - \sigma = \nu(\lambda - 1)/(p - \nu)$. We choose ν such that $p - \nu + (\nu/r) = 1$, that is

$$\nu = \frac{(p-1)r}{r-1}.$$

Clearly, $0 < \nu < p$. Moreover, we obtain for σ the value

$$\sigma = 1 - \frac{r(p-1)(\lambda-1)}{r-p}$$

so that $\sigma > 0$ if $p < p_1$. With the assumption we have in $K \times I$

$$\iint \left|\frac{dv}{dt}\right|^p \le \frac{1}{\lambda^\nu} \left(\iint \left|\frac{dv^\lambda}{dt}\right|^r\right)^{\nu/r} \left(\iint v^{\sigma-1} \frac{dv}{dt}\right)^{p-\nu}.$$

Finally, the last integral is estimated as

$$\frac{1}{\sigma}\int v^\sigma \, dx \le \frac{1}{\sigma}(\text{meas } K)^{1-\sigma}\left(\int v \, dx\right)^\sigma.$$

$\square$

Corollary 16 *For any weak solution of problem* (1.1)–(1.3), $u_t \in L^p_{\text{loc}}(Q)$ *for any* $p \in [1, (m+1)/m)$ *and* $T > 0$.

Proof Again we may restrict ourselves to classical solutions by approximation. If u is the solution, then

$$v(x, t) = tu(x, t^{m-1})$$

satisfies the conditions of Lemma 15. Observe in particular that $v_t > 0$ is a consequence of (8.5). By Proposition 13 we may take $\lambda = \frac{m+1}{2}, r = 2$, hence $p_1 = (m+1)/m$. $\square$

As a consequence of Propositions 12, 14 and Corollary 16 we have

Corollary 17 *For any weak solution* $tu_t \in L^\infty(0, \infty : L^1(\Omega))$ *and*

$$\int u_t \, dx \le 0, \quad t\|u_t(t)\|_1 \le \frac{2}{m-1}\|u_0\|_1. \tag{8.10}$$

Proof In case u is smooth the first inequality follows from (8.1) for $p = 1$. Since $u_t = (u_t)^+ - (u_t)^-$, and $|u_t| = (u_t)^+ + (u_t)^-$, we have

$$\int (u_t)^+ \le \int (u_t)^- \quad \text{and} \quad \int |u_t| dx = \int (|u_t^+| + |u_t^-|)dx \le 2\int |u_t^-| dx.$$

We now use (8.5) to obtain (8.10). $\square$

Remark Again if Δu_0^m is bounded below as in (8.8) the bound (8.10)-right for u_t can be improved and $u_t \in L^\infty(0, \infty : L^1(\Omega))$.

9 Strong solutions

A nonnegative, locally integrable function u for which all the derivatives which appear in an equation are functions rather than distributions and such that the equation is satisfied a.e. in its domain is called a *strong* solution of that equation. For equation (1.1) these requirements amount to the following:

i) $u, u^m, u_t, \Delta(u^m) \in L^1_{loc}(Q)$,

ii) $u_t = \Delta(u^m)$ as locally integrable functions in Q.

A precise definition of strong solution for a problem like (1.1)–(1.3), asks for functional spaces which allow one to define in what sense the initial and boundary data are taken. Again a convenient choice of spaces should allow both for existence for a suitable class of data and, on the other side, for uniqueness.

In our case the estimates obtained in the previous section imply the following result.

Theorem 18 *For every $u_0 \in L^1(\Omega)$ the weak solution to problem (1.1)–(1.3) is a strong solution in the following sense:*

i) $u^m \in L^2(\tau, \infty : H^1_0(\Omega))$ *for every $\tau > 0$;*

ii) u_t *and* $\Delta u^m \in L^1_{loc}(0, \infty : L^1(\Omega))$ *and* $u_t = \Delta(u^m)$ *a.e. in Q;*

iii) $u \in C([0, T) : L^1(\Omega))$ *and* $u(0) = u_0$.

Conditions (i) and (iii) have been already established. As for (ii), we have proved that $u_t \in L^\infty_{loc}(0, \infty : L^1(\Omega)) \cap L^p_{loc}(Q)$ for $0 < p < p_1$. Returning to equation (7.1) with $\varphi \in C^\infty_c(Q)$, we conclude that $\nabla(u^m)$ has u_t as its divergence, hence $\Delta(u^m) \in L^p_{loc}(Q)$. By standard theory, all the second spatial derivatives of u^m belong to $L^p_{loc}(Q)$. Moreover, (1.1) is satisfied in Q. $\square$

We give next a summary of the additional properties of the solution:

Theorem 19 *The strong solution of problem (1.1)–(1.3) also satisfies*

(a) $u \in L^\infty(Q^\tau)$ *and (4.1) holds.*

(b) $\nabla(u^\gamma) \in L^2(Q^\tau)$ *for every $\gamma > m/2$, and (3.10), (8.1) and (8.2) hold.*

(c) $t u_t \in L^\infty(0, \infty : L^1(\Omega)) \cap L^p_{loc}(\Omega)$ *for $1 \le p < p_1$, and (8.3), (8.4), (8.5) and (8.10) hold.*

(d) *For any two solutions $u, \hat{u}$ we have* (6.1), (6.2). *In particular, $u_0 \leq \hat{u}_0$ implies $u \leq \hat{u}$ in Q.*

(e) *For every $t \geq \tau \geq 0$ and every $1 \leq p \leq \infty$ we have $\|u(t)\|_p \leq \|u(\tau)\|_p$.*

(f) *If $u_0 \in C(\Omega), u_0(x) > 0$ for $x \in Q$ and $u_0(x) = 0$ for $x \in \Gamma$, then u is a classical solution, positive in Q.*

Remark　The condition $u \in C([0, \infty) : L^1(\Omega))$ does not look essential in the definition, in the sense that we could replace it by $u \in L^1_{\text{loc}}(Q)$ (and then write $u(t) \to u_0$ in $L^1(\Omega)$ instead of $u(0) = u_0$ in (iii)) and get the same uniqueness result. Nevertheless, it is natural since we want to view our solution as a *continuous* curve in some functional space, in this case $t \in [0, \infty) \to u(t) \in L^1(\Omega)$. This view leads to the concept of *semigroup of transformations* with interesting consequences. Anyway, in our case it does not mean any extra condition since $u^m \in L^2_{\text{loc}}(0, \infty : H^1_0(\Omega))$ clearly implies $u \in L^2_{\text{loc}}(0, \infty : L^1_0(\Omega))$ which together with $u_t \in L^1_{\text{loc}}(0, \infty : L^1(\Omega))$ gives $u \in C((0, \infty) : L^1(\Omega))$. We make the assumption of continuity at $t = 0$ in order to satisfy the initial condition $u(0) = u_0$.

10 A comment on continuity

A typical result of the quasilinear elliptic and parabolic theories says that solutions of such equations which are bounded in some Lebesgue space, say L^2, and satisfy in a weak sense an equation with certain structural assumptions, are in fact Hölder continuous with Hölder exponents and constants depending only on the L^2 norm of the solution and the bounds in the structure assumptions. This is also true for the PME, notwithstanding the degeneracy of the equation: using essentially variants of the Moser iteration technique and some of the estimates established above, one can establish Hölder continuity in x and t for the solutions of the Dirichlet problem for the PME. The classical theory of quasilinear equations allows moreover to prove that the solution is C^∞-smooth in the open set where it is positive. Finally, if the initial data have certain smoothness properties, these are inherited by the solution at $t = 0$. We do not have time in these lectures to go into this interesting chapter of the theory, which can be found in [GP]. Continuity was first established for the Cauchy Problem in [CF].

11 The special solution in separated variables

In Proposition 5 we have obtained an absolute bound for the solutions of the Cauchy-Dirichlet Problem (1.1)–(1.3). This bound will be improved in this section. Indeed, we show that there exists a function $\tilde{U}$ which is the largest element in the class of functions which are weak solutions of the Dirichlet Problem in Q. We will call this solution the *maximal solution* of the Cauchy-Dirichlet Problem. Moreover, $\tilde{U}$ is a solution in separated-variables form. We have

Theorem 20 *There exists a unique function $\tilde{U}(x,t)$ which is a solution of the Dirichlet problem in Q_τ for every $\tau > 0$ and takes initial values $\tilde{U}(x,0) = \infty$. This function has the form*

$$\tilde{U}(x,t) = t^{-\frac{1}{m-1}} f(x). \tag{11.1}$$

Then $\tilde{U}$ is the maximal solution of the PME in Q with zero Dirichlet conditions, and $g = f^m$ is the unique positive solution of the nonlinear eigenvalue problem

$$\Delta g + \frac{1}{m-1} g^{\frac{1}{m}} = 0, \quad g \in H_0^1(\Omega). \tag{11.2}$$

Proof (i) For every integer $n \geq 1$ we solve the problem

$$(P_n) \begin{cases} (u_n)_t & = \Delta(u_n^m) \quad \text{in} \quad Q \\ u_n(x,0) & = n \qquad\quad \text{in} \quad \overline{\Omega}, \\ u_n(x,t) & = 0 \qquad\quad \text{on} \quad \Sigma. \end{cases}$$

Let u_n be the solution to this problem. Clearly, the sequence $\{u_n\}$ is monotone: $u_{n+1} \geq u_n$. We also know from Proposition 5 that for every n,

$$u_n(x,t) \leq Ct^{-1/(m-1)} \quad \text{in} \quad Q. \tag{11.3}$$

Therefore, we may pass to the limit and find a function

$$\tilde{U}(x,t) = \lim_{n \to \infty} u_n(x,t)$$

also satisfying (11.3). Now fix $\tau > 0$ and observe that, by (11.3), there exists $n_1 = n_1(\tau)$ such that for every $n \geq 1$

$$u_n(x,\tau) \leq n_1 = u_{n_1}(x,0).$$

By the Maximum Principle we conclude that

$$u_n(x,t+\tau) \leq u_{n_1}(x,t) \quad \text{in} \quad Q,$$

so that

$$\tilde{U}(x,t+\tau) \leq u_{n_1}(x,t) \quad \text{in} \quad Q. \tag{11.4}$$

As a monotone limit of bounded solutions u_n in Q_τ such that the u_n^m are bounded above by a function in $L^2(\tau,\infty : H_0^1(\Omega))$, it is straightforward to conclude that $\tilde{U}$ is a strong solution of the Cauchy-Dirichlet problem for the PME in any time interval (τ,∞). It is also clear that it takes on the value $\tilde{U}(x,0) = \infty$.

(ii) Let us now prove that $\tilde{U}$ has the form (11.1). To do that we introduce the transformation

$$(\mathcal{T}u)(x,t) = \lambda u(x,\lambda^{m-1}t), \quad \lambda > 0. \tag{11.5}$$

This transformation leaves the equation invariant. It is also clear that when applied to our latter sequence $\{u_n\}$ we get

$$(\mathcal{T}u_n)(x,t) = u_{\lambda n}(x,t) \quad \text{in} \quad Q \tag{11.6}$$

(check the initial and boundary values). Passing to the limit $n \to \infty$ in (11.6) we get

$$(T\tilde{U})(x,t) = \tilde{U}(x,t), \tag{11.7}$$

which holds for every $(x,t) \in Q$ and every $\lambda > 0$. Fixing (x,t) and setting $\lambda = t^{-1/(m-1)}$ we get (11.1) with $f(x) = \tilde{U}(x,1)$. The fact that $g = f^m$ satisfies (11.2)–(11.3) is also obvious.

(iii) Let us prove now that $\tilde{U}$ is larger than any solution of the Cauchy-Dirichlet Problem in Q. By Proposition 5 we know that all such solutions satisfy

$$u(x,\tau) \leq C(\tau) < \infty.$$

It follows from the Maximum Principle that

$$u(x,t+\tau) \leq \tilde{U}(x,t) \quad \text{in} \quad Q.$$

Letting now $\tau \to 0$ we get $u(x,t) \leq \tilde{U}(x,t)$ in Q as desired.

(iv) Finally, we prove the uniqueness of the solution with $u(x,0) = \infty$. Assume that v is another such solution. Since we assume that $v(x,t+\tau)$ is a weak solution of problem (1.1)–(1.3), $v(x,\tau)$ must be an element in $H_0^1(\Omega)$, hence $v(x,2\tau)$ is bounded (by Proposition 5), and by comparison with the sequence u_n we conclude that $v(x,t+2\tau) \leq u_n(x,t)$ in Q for some n large enough. Letting $\tau \to 0$ we get

$$v(x,t) \leq \tilde{U}(x,t). \tag{11.8}$$

On the other hand, a function v which has infinite initial values is of course larger than the solutions u_n, hence $v \geq \tilde{U}$. Therefore, $v = \tilde{U}$. $\qquad\qquad\square$

Comments

As we have explained above, solutions for the Cauchy, Dirichlet and Neumann Problems were first announced by Oleĭnik [O] and explained in detail in [OKC], published in 1958. The case of one space dimension was considered and a class of so-called generalized solutions was introduced. The uniqueness result, Theorem 2, follows the proof in [OKC]. A study of the properties of weak solutions to the Dirichlet Problem was done by Aronson and Peletier in [AP], who use a definition similar to our Definition 1. Hölder continuity of the solutions for this problem is proved by Gilding and Peletier in [GP]. Some of the estimates are more or less classical in nonlinear parabolic equations. The first proof of the control of u_t from below follows the proof of Caffarelli and Friedman in [CF], while an argument close to the second proof was used in [CVW] in the study of the regularity of the Cauchy Problem. Proposition 13 and Lemma 15 are due to Bénilan [Be]. These estimates are crucial in establishing that the solution is strong. The existence of the special solution (11.1) is established in [AP] by a different method, consisting in studying the elliptic equation (11.3). A more general result can be found in Dahlberg and Kenig [DK].

Chapter III   The Cauchy Problem. L^1 Theory

In this chapter we study the initial-value problem for the PME in d-dimensional space, $d \geq 1$, with integrable and nonnegative initial data, $u_0 \in L^1(\mathbf{R}^d)$, $u_0 \geq 0$. We establish existence and uniqueness of a strong solution for the Cauchy Problem as well as its main properties, among them the conservation of mass, the boundedness of the solutions for $t \geq \tau > 0$ and a version of the finite propagation property.

In this chapter we will use the symbols $Q = \mathbf{R}^d \times \mathbf{R}_+$ and $Q_T = \mathbf{R}^d \times (0, T)$.

1 Definition of strong solutions. Uniqueness

We consider the problem:

$$u_t = \Delta(u^m) \quad \text{in } Q \tag{1.1}$$

$$u(x, 0) = u_0(x) \quad \text{for } x \in \mathbf{R}^d, \tag{1.2}$$

where $m > 1$ and $u_0 \in L^1(\Omega)$, $u_0 \geq 0$. No difficulties arise in restricting time to the interval $0 \leq t \leq T$ and replacing Q by Q_T. Following the motivation of the previous chapter we will first give a suitable definition of strong solution for our initial-value problem and then prove existence, uniqueness and a series of basic properties of such solutions. Here we restrict ourselves to solutions which are integrable with respect to the space variables, or solutions with *finite mass*, and develop the corresponding L^1-theory.

Definition 1   We say that a nonnegative function $u \in C([0, \infty) : L^1(\mathbf{R}^d))$ is a *strong L^1-solution* of problem (1.1), (1.2) if

i) $u^m, u_t, \Delta(u^m) \in L^1_{\text{loc}}(0, \infty : L^1(\mathbf{R}^d))$;

ii) $u_t = \Delta(u^m)$ a.e. in Q;

iii) $u(0) = u_0$.

In the rest of the chapter, strong solution will always mean strong L^1-solution. Our first step in the study of strong solutions will be to establish the crucial L^1-order-contraction property, similar to Proposition II.7.

Lemma 1   *Let u_1, u_2 be two strong solutions of (2.1), (2.2) in Q_T. Then for every $0 < t_1 < t_2$ we have*

$$\int [u_1(x, t_2) - u_2(x, t_2)]_+ dx \leq \int [u_1(x, t_1) - u_2(x, t_1)]_+ dx. \tag{1.3}$$

Proof Let $p \in C^1(\mathbf{R}) \cap L^\infty(\mathbf{R})$ be such that $p(s) = 0$ for $s \leq 0$, $p'(s) > 0$ for $s > 0$ and $0 \leq p \leq 1$, and let $j(r) = \int_0^r p(s) ds$ be a primitive of p. Since p will be an approximation to the sign function

$$\operatorname{sign}_0^+(r) = 1 \text{ if } r > 0, \quad \operatorname{sign}_0^+(r) = 0 \text{ if } r \leq 0, \tag{1.4}$$

j will approximate the function $s \mapsto [s]_+$. Moreover, consider a cutoff function $\varsigma_0 \in C_c^\infty(\mathbf{R}^d)$ such that $0 \leq \varsigma_0 \leq 1$, $\varsigma_0(x) = 1$ if $|x| \leq 1$, $\varsigma_0(x) = 0$ if $|x| \geq 2$, and let $\varsigma = \varsigma_n(x) = \varsigma_0(x/n)$. As $n \to \infty$, $\varsigma_n \uparrow 1$.

We subtract the equations satisfied by u_1 and u_2, multiply by $p(u_1^m - u_2^m)\varsigma$ and integrate on $S = \mathbf{R}^d \times [t_1, t_2]$ to obtain, with $w = u_1^m - u_2^m$,

$$\iint (u_1 - u_2)_t p(w) \varsigma = \iint \Delta w p(x) \varsigma. \tag{1.5}$$

Now, approximate w by means of a smooth kernel sequence ρ_n. If $w_n = w * \rho_n$ ($*$ denotes convolution), we have $w_n \to w$, $\nabla w_n \to \nabla w$ and $\Delta w_n \to \Delta w$ in $L^1_{\text{loc}}(Q)$ and almost everywhere for a subsequence, so that $p(w_n) \to p(w)$ a.e. Moreover,

$$\iint p(w_n) \Delta w_n \varsigma + \iint p'(w_n) |\nabla w_n|^2 \varsigma + \iint p(w_n) \nabla w_n \cdot \nabla \varsigma = 0.$$

Letting $n \to \infty$ we observe that the second integral is uniformly bounded above. Moreover, by Fatou's lemma,

$$\iint p'(w) |\nabla w|^2 \varsigma \leq -\iint p(w) \Delta w \varsigma + \iint p(w) \nabla w \cdot \nabla \varsigma. \tag{1.6}$$

Hence, returning to (1.5) we get

$$\begin{aligned}
\iint (u_1 - u_2)_t p(w) \varsigma &\leq -\iint p'(w) |\nabla w|^2 \varsigma - \iint p(w) \nabla w \cdot \nabla \varsigma \\
&\leq -\iint p(w) \nabla w \cdot \nabla \varsigma = -\iint \nabla j(w) \cdot \nabla \varsigma \\
&= \iint j(w) \Delta \varsigma \leq \iint |w| \, |\Delta \varsigma|,
\end{aligned} \tag{1.7}$$

where integration is understood on S. Letting now p tend to sign_0^+ and observing that $\frac{d}{dt}[u_1 - u_2]_+ = \operatorname{sign}_0^+(u_1 - u_2)\frac{d}{dt}(u_1 - u_2)$, we get

$$\begin{aligned}
\int [u_1(x, t_2) - u_2(x_1, t_2)]_+ \varsigma \, dx &\leq \int [u_1(x, t_1) - u_2(x, t_1)]_+ \varsigma \, dx \\
&\quad + \|\Delta \varsigma\|_\infty \iint_{S \cap \{|x| > n\}} |w(x, t)| \, dx \, dt.
\end{aligned} \tag{1.8}$$

We let now $n \to \infty$ to obtain (1.3), since $w \in L^1(t_1, t_2 : L^1(\mathbf{R}^d))$ and $\|\Delta \varsigma_n\|_\infty = \|\Delta \varsigma_0\|_\infty / n^2$. $\qquad \square$

Again, as in Chapter II, we obtain *uniqueness* and *comparison* as simple consequences of this result.

Theorem 2 *Problem* (1.1), (1.2) *has at most one strong solution. If* u_1, u_2 *are strong solutions with initial data* u_{01}, u_{02} *resp., and* $u_{01} \leq u_{02}$ *are in* $\mathbf{R}^d$, *then* $u_1 \leq u_2$ *a.e. in* Q. *In particular, if* $u_{01} = u_{02}$ *a.e., then* $u_1 = u_2$ *a.e.*

Remark The proof of Lemma 1 actually uses the following requirements on $u : u, u^m$, $u_t, \Delta u^m \in L^1_{\text{loc}}(Q)$ and

$$\iint\limits_{S_n} u_1^m(x,t) dx dt = o(n^2) \text{ as } n \to \infty, \tag{1.9}$$

where $S_n = \{(x,t) : n \leq |x| \leq 2n, t_1 \leq t \leq t_2\}$ with $0 < t_1 < t_2$, which are weaker than our Definition 1. Therefore, Theorem 2 also holds under the above hypotheses, if (1.9) holds uniformly for $0 \leq t \leq t_2$ and if the initial data are taken continuously in $L^1_{\text{loc}}(\mathbf{R}^d)$. We shall use this remark later on.

Example Though the source-type solution $U(x,t)$ fails to be a strong solution of the Cauchy Problem because of the singularity of its initial data, any time-delayed version $u(x,t) = U(x,t+\tau)$ with $\tau > 0$ is indeed a strong solution.

2 Existence of solutions. Conservation of mass

We begin this section by constructing solutions for bounded initial data by using an approximation process and the results of the previous chapter. The existence result for general initial data in $L^1(\mathbf{R}^d)$ will follow once we show that every solution is bounded for $t \geq \tau > 0$, which will be done in §4.

Theorem 3 *For every nonnegative function* $u_0 \in L^1(\mathbf{R}^d) \cap L^\infty(\mathbf{R}^d)$ *there exists a strong solution of problem* (1.1), (1.2). *Moreover* $u_t \in L^p_{\text{loc}}(Q)$ *for* $1 \leq p < (m+1)/m$ *and*

$$u_t \geq -\frac{u}{(m-1)t} \text{ in } \mathcal{D}'(Q), \tag{2.1}$$

$$\|u_t(\cdot,t)\|_1 \leq \frac{2\|u_0\|_1}{(m-1)t}. \tag{2.2}$$

If $u_0 \in L^p(\mathbf{R}^d)$ *for* $1 \leq p \leq \infty$, *then* $u(t) \in L^p(\mathbf{R}^d)$ *and*

$$\|u(t)\|_p \leq \|u_0\|_p. \tag{2.3}$$

Moreover, the map $u_0 \mapsto u(t)$ *is an ordered contraction in* $L^1(\mathbf{R}^d)$.

Proof (i) We begin by assuming that u_0 is not only bounded and integrable over $\mathbf{R}^d$, but also that it is strictly positive, C^∞-smooth and all its derivatives are bounded in $\mathbf{R}^d$.

Under these conditions we construct a strong and classical solution. For that we consider the Dirichlet problem

$$(P_n) \begin{cases} u_t = \Delta(u^m) & \text{in} \quad Q_n = B_n(0) \times (0, \infty), \\ u(x, 0) = u_{0n}(x) & \text{for} \quad |x| \leq n, \\ u(x, t) = 0 & \text{for} \quad |x| = n, \ t \geq 0, \end{cases}$$

where $u_{0n} = u_0 \varsigma_n$, $\{\varsigma_n\}$ being a cutoff sequence with the following properties: $\varsigma_n \in C^\infty(\mathbf{R}^d)$, $\varsigma_n(x) = 1$ for $|x| \leq n - 1$, $\varsigma_n(x) = 0$ for $|x| \geq n$, $0 < \varsigma_n(x) < 1$ for $n - 1 < |x| < n$, the derivatives of the ς_n up to second order are bounded uniformly in $x \in \mathbf{R}^d$, and $n \geq 2$.

By the results of Chapter II (Theorem 4 and Proposition 6), (P_n) admits a unique classical solution $u_n \in C^\infty(Q_n) \cap C(\overline{Q}_n)$ and $u_n > 0$ in Q_n. In particular, u_{n+1} will be a classical solution of the PME in Q_n with positive boundary data and initial data larger than u_{0n}. We conclude from the classical Maximum Principle that $u_{n+1} \geq u_n$ in Q_n, i.e., the sequence $\{u_n\}$ is monotone. Moreover, we get from Chapter II uniform estimates for

(a) u_n in $L^\infty(0, \infty : L^p(B_n(0)))$, $1 \leq p \leq \infty$,

(b) $(u_n)_t$ in $L^\infty(0, \infty : L^1(B_n(0))) \cap L^p_{\text{loc}}(Q_n)$ for $1 \leq p < p_1$,

(c) u_n^m in $L^2(0, \infty : H_0^1(B_n(0)))$.

Since all of these estimates involve bounds which are independent of n, we may pass to the limit $n \to \infty$ and obtain a positive function $u \in L^\infty(0, \infty : L^p(\mathbf{R}^d))$ for every $p \in [1, \infty)$, such that $u_t, u^m, \Delta u^m$ belong to the same spaces to which $(u_n)_t, u_n^m, \Delta(u_n^m)$ belonged, and equation (1.1) holds in Q.

To check the smoothness of u, we first observe that in a neighbourhood $N \subset \overline{Q}$ of any point $(x_0, t) \in \overline{Q}$, $u_n(x, t)$ is defined and positive, say $u_n(x, t) \geq c > 0$ for every $(x, t) \in N$ if $n > n_0$. Since the sequence $\{u_n\}$ is monotone nondecreasing and bounded, the interior regularity theory for uniformly parabolic quasilinear equations gives uniform bounds for all the derivatives of $u_n, n \geq n_0$, in a smaller neighbourhood of (x_0, t). In the limit we conclude that $u \in C^\infty(\overline{Q})$. Moreover, for $t = 0$ we get $u(x, t) = u_0(x), x \in \mathbf{R}^d$.

We have proved that u is a classical solution of problem (1.1), (1.2). To comply with our definition of strong solution we still have to check the continuity of $u = u(t)$: $[0, \infty) \to L^1(\mathbf{R}^d)$. It is a consequence of the fact that $u \in L^\infty(0, \infty : L^1(\mathbf{R}^d))$ (by (II.8.1)) and $u_t \in L^\infty(0, \infty : L^1(\mathbf{R}^d))$ (cf. Remark to Corollary II.17), so that u is absolutely continuous from $[0, \infty)$ into $L^1(\mathbf{R}^d)$.

Estimates (2.1), (2.2), (2.3) are a consequence of similar estimates for the Dirichlet problem. In particular we have $0 \leq u(x, t) \leq ||u_0||_\infty$.

(ii) If $u_0 \in L^1(\mathbf{R}^d) \cap L^\infty(\mathbf{R}^d)$ does not fulfill the above requirements we approximate it by a sequence $\{u_{0n}\}$ of such functions. We may always do this in such a way that $||u_{0n}||_1 \leq ||u_0||_1$, $||u_{0n}||_\infty \leq ||u_0||_\infty$, $u_{0n} \to u_0$ in $L^1(\mathbf{R}^d)$. Let u_n be the solution with data u_{0n}. It follows from Lemma 1 that u_n converges in $C([0,\infty) : L^1(\mathbf{R}^d))$ to a function u and $u(0) = u_0$.

Again estimates (a), (b), (c) of the previous step will hold uniformly in n so that passing to the limit $n \to \infty$ produces a strong solution of (1.1), (1.2), which satisfies the estimates (2,1), (2.2), (2.3).   □

The solutions of the Cauchy problem (1.1), (1.2) have an important conservation property, not enjoyed by the solutions of the Dirichlet problem.

Proposition 4 (Conservation of total mass) *For every* $t > 0$

$$\int u(x,t)dx = \int u_0(x)dx. \tag{2.4}$$

Proof We take a cutoff function ς_n as in Theorem 3 and integrate by parts as follows:

$$\int u(x,t)\varsigma_n(x)dx - \int u_0(x)\varsigma_n(x)dx = \iint u_t \varsigma_n dxdt$$

$$= \iint \Delta u^m \varsigma_n dxdt = \iint u^m \Delta \varsigma_n dxdt \to 0 \quad \text{as } n \to \infty.$$

The calculation is justified if u is smooth and boundded. For general u it follows by approximation, using Lemma 1.   □

3 The fundamental estimate for the Cauchy problem

Perhaps the most significant novelty of the Cauchy problem is the existence of a lower bound for the Laplacian of the pressure. Indeed, we have

Proposition 5 *Let* $v = mu^{m-1}/(m-1)$. *Then*

$$\Delta v \geq -\frac{\lambda}{t} \quad \text{with} \quad \lambda = \frac{d}{d(m-1)+2}. \tag{3.1}$$

The inequality is understood in the sense of distributions in Q. This bound will be used so often that we consider it the *fundamental* estimate for the Cauchy problem. Estimates of the form $\Delta u \geq -C$ play a role in the theory of Hamilton-Jacobi equations, cf. e.g. [Li]. Such functions are called semi-superharmonic functions. Let us also remark that (3.1) is optimal in the sense that equality is actually attained by the source-type or Barenblatt solutions, which are a kind of worst case with respect to this bound, a fact which has interesting consequences.

As a consequence of (3.1) we have the following improvement of (2.1), (2.2).

Corollary 6 $u_t \in L^\infty_{loc}(0, \infty : L^1(\mathbf{R}^d))$ *and*

$$u_t \geq -\frac{\lambda u}{t} \quad in \quad \mathcal{D}'(R), \tag{3.2}$$

$$t\|u_t\|_1 \leq 2\lambda\|u_0\|_1. \tag{3.3}$$

Proof The first inequality is a consequence of

$$v_t = (m-1)v\Delta v + |\nabla v|^2 \geq (m-1)v\Delta v,$$

together with $v_t/v = (m-1)u_t/u$ and (3.1). For the second, one argues as in Corollary II.17. Again the calculations are justified for smooth solutions and hold in the limit for every solution. □

Proof of Proposition 5 (i) The formal derivation of the estimate is very simple. We first write the PDE satisfied by the pressure, v, i.e.,

$$v_t = (m-1)v\Delta v + |\nabla v|^2. \tag{3.4}$$

Then we write the equation satisfied by $p = \Delta v$ by differentiating (3.4) twice. We have

$$p_t = (m-1)v\Delta p + 2m\nabla v \cdot \nabla p + (m-1)p^2 + 2\sum_{i,j}\left(\frac{\partial^2 v}{\partial x_i \partial x_j}\right)^2.$$

Since

$$\sum_{i,j}(a_{ij})^2 \geq \sum_i(a_{ii})^2 \geq \frac{1}{d}\left(\sum_i a_{ii}\right)^2,$$

we get

$$\mathcal{L}(p) \equiv p_t - (m-1)v\Delta p - 2m\nabla v \cdot \nabla p - \left(m - 1 + \frac{2}{d}\right)p^2 \geq 0.$$

Here $\mathcal{L}$ is a quasilinear parabolic operator with smooth, variable coefficients, since we consider v as a fixed function of x and t. We now apply $\mathcal{L}$ to the trial function

$$P(x,t) = -\frac{C}{t+\tau} \tag{3.5}$$

and observe that $\mathcal{L}(P) \leq 0$ if and only if $C \geq \lambda = 1/(m-1+\frac{2}{d})$. We fix $C = \lambda$. By choosing τ small enough we may also obtain

$$p(x,0) \equiv \Delta v(x,0) \geq P(x,0) \equiv -\frac{C}{\tau}, \tag{3.6}$$

from which the classical Maximum Principle should allow us to conclude that $p \geq P$ in Q. Letting $\tau \to 0$ we would then obtain a pointwise inequality $\Delta v \geq -\lambda/t$.

(ii) The application of the Maximum Principle is justified when considering classical solutions of (3.4) such that $v, \nabla v$ and $p = \Delta v$ are bounded and v is bounded below away from 0, so that the equation is uniformly parabolic. Therefore, we need to construct *new approximate solutions*. This we do as follows. We may always restrict ourselves to initial data u_0 which are bounded, smooth and positive, thanks to Lemma 1. Consider now initial data

$$u_{0\varepsilon}(x) = u_0(x) + \varepsilon, \quad \varepsilon > 0. \tag{3.7}$$

According to [LSU] there exists exactly one function $u_\varepsilon \in C^\infty(\overline{Q})$ that solves (1.1) with initial data $u_{0\varepsilon}$, and $\varepsilon \le u_\varepsilon \le M + \varepsilon$, where $M = \|u_0\|_\infty$. Moreover, by interior regularity results all the derivatives of u_ε are bounded in Q. In particular, equation (1.1) is uniformly parabolic on u_ε. It follows that the fundamental estimate (3.1) holds for v_ε, the pressure of u_ε.

Now, if we prove that $v_\varepsilon \to v$ as $\varepsilon \to 0$ in $L^1_{\text{loc}}(Q)$, then (3.1) will still hold in the limit for v, though only in distribution sense, i.e.,

$$\iint \left(v\Delta\varphi - \frac{\lambda}{t}\varphi \right) dx dt \ge 0 \tag{3.8}$$

for every $\varphi \in C_c^\infty(Q), \varphi \ge 0$. Therefore, the proof is complete with the following convergence result.

Lemma 7 *As $\varepsilon \to 0$, $u_\varepsilon \to u$ locally uniformly in Q.*

Proof We first observe that by the Maximum Principle the family $\{u_\varepsilon\}$ is nonincreasing as $\varepsilon \downarrow 0$. It is also easy to establish that every u_ε is above the solution u with initial data u_0 (Hint: compare u_ε with the approximations u_n to u constructed in step 1 of Theorem II.4 in the domain Q_n and let $n \to \infty$). Since u is strictly positive in $\overline{Q}$ and $u_\varepsilon \ge u$, and thanks again to the interior regularity results, not only $\{u_\varepsilon\}$ converges to a function $\hat{u}$, but also the derivatives converge, so that $\hat{u}$ is a C^∞ solution of (1.1) in Q, $\hat{u}(\cdot,0) = u_0$ and $\hat{u} \ge u$.

To conclude that $\hat{u} = u$ we still need some control of u^m as $|x| \to \infty$, as in (1.9), to be able to apply Theorem 2. We use the following result.

Lemma 8 *For every ε and $t > 0$ we have*

$$\int (u_\varepsilon(x,t) - \varepsilon)dx \le \int u_0(x)dx. \tag{3.9}$$

Proof Formally, we have $\int u_{\varepsilon,t}dx = \int \Delta u_\varepsilon^m dx = 0$, hence

$$\int (u_\varepsilon(x,t) - \varepsilon)dx = \int (u_{0\varepsilon}(x) - \varepsilon)dx = \int u_0(x)dx.$$

More rigorously, we approximate u_ε with the solution $u_{\varepsilon n}$ of the following Dirichlet problem

$$\begin{cases} u_t = \Delta(u^m) & \text{in } Q_n \\ u(x,0) = u_{0n}(x) + \varepsilon & \text{for } |x| \le n \\ u(x,t) = \varepsilon & \text{for } |x| = n \text{ and } t \ge 0, \end{cases}$$

for which we argue as in Section II.3 and get a contraction formula as (II.6.1), which we apply to $u_{\epsilon n}$ and $\hat{u}_n = \epsilon$ to get (3.9) for $u_{\epsilon n}$. Letting $n \to \infty$ we obtain that $u_{\epsilon n}$ converges (the sequence is compact by the interior regularity theory) to a solution of (1.1) which is u_ϵ by uniqueness. In the limit (3.9) holds.

Going back now to the main argument, we let $\epsilon \to 0$ to obtain

$$\int \hat{u}(x,t)dx \leq \int u_0(x)dx.$$

It follows that $\hat{u}(t) \in L^\infty(0,\infty : L^1(\mathbf{R}^d)) \cap L^\infty(Q)$, hence by the Remark to Theorem 2 we conclude that $\hat{u} = u$ in Q. This ends the proof of the fundamental estimate. □

4 Boundedness of the solutions. Existence with general data

We are now in a position to prove that all solutions are bounded for $t \geq \tau > 0$, the so-called L^1-L^∞ smoothing effect.

Theorem 9 *For every $t > 0$ we have*

$$u(x,t) \leq c\|u_0\|_1^\alpha t^{-\lambda}, \tag{4.1}$$

where $\alpha = 2/(d(m-1)+2), \lambda = d/(d(m-1)+2)$ and $c > 0$ depends only on m and d.

The theorem can be derived as a consequence of the fundamental estimate (3.1), thanks to the following result.

Lemma 10 *Let g be any nonnegative, smooth, bounded and integrable function in $\mathbf{R}^d$ such that*

$$\Delta(g^r) \geq -K \tag{4.2}$$

for some r and $K > 0$. Then $g \in L^\infty(\mathbf{R}^d)$ and $\|g\|_\infty$ depends only on r, K, d and $\|g\|_1$ in the form

$$\|g\|_\infty \leq C(r,d)\|g\|_1^\rho K^\sigma, \tag{4.3}$$

with $\rho = 2/(2r+d)$ and $\sigma = d/(2r+d)$.

Given Lemma 10, it suffices to fix $t > 0$, and put

$$r = m-1, \quad g(x) = u(x,t) \quad \text{and} \quad K = \frac{\lambda(m-1)}{mt},$$

to obtain Theorem 9 in the case where the solution u is positive everywhere, hence smooth. The general case is done by approximation.

Proof of Lemma 10 Let $f(x) = g^r$. Then $\Delta f \geq -K$. Therefore, the function

$$F(x) = f(x) + \frac{K}{2d}|x|^2$$

in subharmonic in $\mathbf{R}^d$. Therefore, for every $x_0 \in \mathbf{R}^d$ and $R > 0$ we have

$$F(x_0) \le \fint_B F(x)dx, \tag{4.4}$$

where $B = B_R(x_0)$ and $\fint_B$ denotes average on B. The argument will continue in a different way for $r > 1$ and for $0 < r \le 1$.

(i) In the latter case, $r < 1$, we can use (4.4) to estimate f at an arbitrary point x_0 as follows:

$$f(x_0) \le \fint_B f(x)dx + \frac{K}{2d}\fint_B |x|^2 dx \le \left(\fint_B f^{1/r}dx\right)^r + \frac{KR^2}{2(d+2)}$$
$$\le \|g\|_1^r \left(\frac{1}{\omega_d R^d}\right)^r + \frac{KR^2}{2(d+2)}. \tag{4.5}$$

(ω_d denotes the volume of the unit ball). Minimization of the last expression with respect to $R > 0$ gives

$$f(x_0) \le C\|g\|_1^{\frac{2r}{rd+2}} K^{\frac{rd}{rd+2}},$$

which is equivalent to (4.3).

(ii) For $r > 1$ we modify the calculation as follows: we pick a point x_0 of maximum for g and estimate $g(x_0)$ as follows:

$$f(x_0) \le \fint_B g^r(x)dx + \frac{K}{2d}\fint_B |x|^2 dx \le g^{r-1}(x_0)\fint_B g dx + \frac{KR^2}{2(d+2)}$$
$$\le g^{r-1}(x_0)\|g\|_1 \frac{1}{\omega_d R^d} + \frac{KR^2}{2(d+2)}. \tag{4.6}$$

Putting $y = g(x_0)$ we can write (4.6) in the form

$$y^r \le Ay^{r-1} + B \quad \text{with} \quad A = c_1\|g\|_1 R^{-d}, B = c_2 KR^2,$$

which after an elementary calculation gives

$$y \le A + B^{1/r} = c_1\|g\|_1 R^{-d} + (c_2 KR^2)^{1/r}. \tag{4.7}$$

Minimization of this expression in R gives (4.3).   □

Formula (4.1) not only asserts that solutions with L^1 data are bounded for positive times, but also gives a very precise quantitative estimate of the bound. In fact, the exponents appearing in the formula can be derived from the general boundedness statement thanks to a scaling argument. Since this kind of argument has wider applicability we give here a proof of this implication, as a small diversion.

Lemma 11   *Suppose that for all solutions of* (1.1), (1.2) *with* $\|u_0\|_1 \le 1$ *we have* $\|u(\cdot,t)\|_\infty \le C = C(m,d) > 0$. *Then* (4.1) *necessarily holds.*

Proof Let u be any solution of (1.1), (1.2) with $||u_0||_1 = M > 0$. Now, if we consider the rescaled function

$$\hat{u}(x,t) = Ku(Lx, Tt),$$

with constants $K, L, T > 0$, $\hat{u}$ is again a solution of (1.1) if

$$K^{m-1}L^2 = T.$$

On the other hand $||\hat{u}_0||_1 = 1$ if

$$KM = L^\alpha.$$

Both equalities are satisfied for T arbitrary, $K = M^{-\alpha}T^\lambda$, $L = M^\beta T^\mu$ with $\beta = (m-1)\lambda/d$, $\mu = \lambda/d$. Under these conditions our assumptions say that $\hat{u}(x,1) \leq C$. Then

$$u(x,T) = K^{-1}\hat{u}(Lx, 1) \leq C/K = CM^\alpha T^{-\lambda}.$$

$\square$

It is interesting to remark that if we calculate the decay rate of the Barenblatt solution in *sup* norm we find that formula (4.1) holds with a certain precise constant. It can be shown that the constant corresponding to the Barenblatt solution is the *optimal* constant in inequality (4.1). This means that the Barenblatt solutions solve an extremal problem, that of maximizing $u(x,t)$ for given $t > 0$ and given $||u_0||_1 = M$. See in this respect [Va1].

The same techniques can be used to prove a more general version of the smoothing effect:

Proposition 12 *For every $t > 0$ and $1 \leq p < q \leq \infty$ we have*

$$||u(t)||_q \leq C||u_0||_p^\gamma t^{-\sigma} \tag{4.8}$$

whenever $u_0 \in L^p(\mathbf{R}^d)$. The constants C, γ and σ depend only m, p, q and d.

We leave it to the reader to fill in the details and also to calculate the explicit values of γ and σ, which are given again by a scaling argument.

We may now give our complete existence result.

Theorem 13 *There exists a strong solution u of problem (1.1), (1.2) for every $u_0 \in L^1(\mathbf{R}^d)$, $u_0 \geq 0$, $u \in C([0,\infty) : L^1(\mathbf{R}^d)) \cap L^\infty(\mathbf{R}^d \times (\tau, \infty))$ for every $\tau > 0$, and satisfies the estimates (2.1)–(2.4) and (3.1)–(3.3). If u_0 is strictly positive and continuous, then u is a classical solution of (1.1)*

Proof We only need to approximate u_0 with a sequence of functions $u_{0n} \in L^1(\mathbf{R}^d) \cap L^\infty(\mathbf{R}^d)$ converging to u_0, say $u_{0n}(x) = \max(u_0(x), n)$, apply the previous results to the solutions u_n and observe that since $||u_{0n}||_1 \leq ||u_0||_1$, the sequence $\{u_n(\cdot, t)\}$ is bounded in $L^\infty(\mathbf{R}^d)$ uniformly in n and $t \geq \tau > 0$. Therefore uniform estimates hold for u^m, u_t and Δu^m similar to the ones in Theorem 3 for $t \geq 2\tau > 0$, and we may pass to the limit $n \to \infty$

and obtain a strong solution u, which satisfies the above estimates. If u_0 is continuous and positive, u is a classical solution by local regularity theory as in Proposition II.5. $\square$

Proposition 14 *Let $u_0 \in C(\mathbf{R}^d) \cap L^1(\mathbf{R}^d)$ be strictly positive, and let u be the strong solution of* (1.1), (1.2). *Then $u \in C^\infty(Q) \cap C(\overline{Q})$ and is strictly positive in Q.*

Moreover, if u_0 is smooth this is reflected in the smoothness of u down to $t = 0$.

5 Finite speed of propagation. The free boundary

We have already remarked that the diffusivity $D(u) = mu^{m-1}$ vanishes in the PME at the level $u = 0$. This degeneracy causes an important phenomenon to occur, i.e., finite speed of propagation of disturbances from 0, or more briefly, *Finite Propagation* (F.P.). We have observed this phenomenon on the source-type solutions in the form of compact support of the solution at any time $t > 0$. We are now in a position to establish the same result for a wide class of solutions of the PME. We have:

Theorem 15 *Let u be the strong solution to* (2.1), (2.2) *with initial data $u_0 \in L^1(\mathbf{R}^d) \cap L^\infty(\mathbf{R}^d), u_0 \geq 0$ and such that u_0 is supported in a bounded set of $\mathbf{R}^d$. Then for every $t > 0$ the support of $u(\cdot, t)$ is a bounded set.*

The proof consists merely of noting that we can find a Barenblatt solution $U(x - x_0, t + \tau; M)$ such that

$$u_0(x) \leq U(x - u_0, \tau; M)$$

by suitably choosing x_0, τ and M. By Theorem 2 we get $u(x,t) \leq U(x - x_0, t + \tau; M)$, hence

$$\text{supp}(u(t)) \subset x_0 + B_{r(t+\tau)}. \tag{5.1}$$

$\square$

Much more precise versions of the Finite Propagation Property can be established.

It can be proved that all the solutions are in fact continuous, cf. [CF]. Then the *positivity set*

$$P = P_u = \{(x,t) \in Q : u(x,t) > 0\} \tag{5.2}$$

is an open set in $\mathbf{R}^{d+1}$ and so are its sections

$$P(t) = \{x \in \mathbf{R}^d : u(x,t) > 0\} \tag{5.3}$$

in $\mathbf{R}^d$. Of course, the support of $u(t)$ is the closure of $P(t)$ in $\mathbf{R}^d$. We have:

Proposition 16 *The family $\{P(t)\}_{t>0}$ is expanding, i.e., $P(t_1) \subset P(t_2)$ for every $0 < t_1 < t_2$.*

Proof It follows from estimate (3.2), which just means that the function $z(t) = u(x,t)t^\lambda$ is nondecreasing for every fixed $x \in \mathbf{R}^d$. Hence if $z(t_1) > 0$ and $t_2 > t_1$ we have $z(t_2) > 0$, i.e., $x \in P(t_2)$. □

Remark This property of expanding supports is sometimes called *retention property*, since u retains its positivity at any given point when time increases.

We may obtain a lower estimate for $P(t)$ similar to (5.1) as follows: we fix $\tau > 0$. Since u is continuous there exist $x_1 \in \mathbf{R}^d$ and M_1 and $\tau_1 > 0$ such that

$$u(x,\tau) \geq U(x - x_1, \tau_1; M_1).$$

By the comparison theorem it follows that for every $t \geq \tau$ we have $u(x,t) \geq U(x - x_1, t + \tau_1 - \tau; M_1)$ hence

$$P(t) \supset x_1 + B_{r(t+\tau_1-\tau)}, \tag{5.4}$$

which gives the desired lower bound.

The boundary of the positivity set in Q, $\Gamma = \partial P \cap Q$, called the *free boundary* or *interface*, is a very important object since it represents the region separating the "occupied region", $[u > 0]$, from the "empty region", $[u = 0]$. As a first result on the behaviour of the interfaces we can combine estimates (5.1) and (5.4) to get the following asymptotic expression:

Proposition 17 *There exist constants $c_1, c_2 > 0$ such that*

$$c_1 t^{\lambda/d} \leq |x| \leq c_2 t^{\lambda/d} \tag{5.5}$$

holds for every $(x,t) \in \Gamma$ if t is large enough.

Proposition 17 implies in particular that every point of the space is eventually reached by the diffusing substance, a property that was not obvious a priori. It also gives an estimate of the speed of penetration of the substance into the empty region with exact exponent in the dependence of the radius on time. An exact asymptotic value of the constant can also be obtained, i.e., we have $|x| \sim ct^{\lambda/d}$ with c depending only on m, d and the mass $\|u_0\|_1$. Actually, it can be proved that as $t \to \infty$ not only the solution but also its interface converge to the unique source-type solution $U(x,t;C)$ which has the same mass as u, i.e., C is determined from $\int U(x,t;C)dx = \int u(x,t)dx$, see [FK], [Va2] and [KV]. In this way the asymptotic sizes can be determined in first approximation by just copying from an explicit formula.

6 Local comparison

Theorem 2 allows us to compare solutions of the Cauchy problem. However, in many cases we will be interested in corresponding functions which either are defined in a subdomain of Q or are not exact solutions of (1.1). We will present here a variant of Lemma 1 which covers such situations.

A function u defined in a subdomain S of Q is called a (strong) *supersolution* of (1.1) in S if u, u^m, u_t and $\Delta u^m \in L^1_{loc}(Q)$ and $u_t \geq \Delta u^m$ a.e. in S. A *subsolution* is defined in a similar way, only $u_t \leq \Delta u^m$. We have:

Lemma 18 *Let Ω be a bounded subset of $\mathbf{R}^d$ with C^1 boundary, let $S = \Omega \times I \subset Q$, with $I = (t_1, t_2)$, and let u_1 be a subsolution, u_2 a supersolution of (1.1) in S. Assume moreover that u_1 and u_2 are continuous in $\overline{S}$ and $u_1 \leq u_2$ on $\partial\Omega \times I$. Then for every $t \in [t_1, t_2]$*

$$\int [u_1(x,t) - u_2(x,t)]_+ dx \leq \int [u_1(x,t_1) - u_2(x,t_2)]_+ dx \qquad (6.1)$$

In particular, if $u_1(\cdot, t) \leq u_2(\cdot, t_1)$ in Ω, we have $u_1 \leq u_2$ in S.

Proof It follows the main lines of Lemma 1. We first select functions p and ζ as follows: $p \in C^1(\mathbf{R}) \cap L^\infty(\mathbf{R})$, $0 \leq p \leq 1$, $p(s) = 0$ for $s \leq \varepsilon$ and $p'(s) > 0$ for $s > \varepsilon$ with ε small and positive, $\zeta \in C_c^\infty(\Omega)$ is a cutoff function such that $0 \leq \zeta \leq 1$. Moreover we may choose ζ in such a way that whenever $x \in \Omega$, $\zeta(x) < 1$ and $t \in I$ we have $u_1(x,t) \leq u_2(x,t) + \varepsilon$.

We subtract the inequalities satisfied by u_1 and u_2, multiply by $p(u_1^m - u_2^m)\zeta$ and integrate over S. Arguing as in Lemma 1 and observing that $p(w)\nabla\zeta$ vanishes identically, we get

$$\iint (u_1 - u_2)_t p(u_1 - u_2)\zeta \leq 0$$

from which we easily obtain (6.1) in the limit $\zeta \to 1$, $p(s) \to \text{sign}(s)$. $\qquad\square$

Remark We may combine Lemmas 1 and 18 to provide comparison for a subsolution and a supersolution defined in unbounded domains, for instance when $\Omega = (-\infty, 0)$ in one space dimension, or $\Omega = \mathbf{R}^d - B$, where B is a ball. We need to impose conditions on the initial and lateral boundary as in Lemma 18 plus integrability on the supersolution as $|x| \to \infty, t > 0$, like (1.9).

Comments

As explained in the preceding chapter, pioneering work is due to Oleĭnik and collaborators. Sabinina [Sa] made the extension to several dimensions. The fundamental estimate is due to Aronson and Bénilan [AB]. The authors point out its optimality by checking it on the Barenblatt solutions and use the estimate in establishing existence of a strong solution of the Cauchy Problem with L^1-data. The boundedness of the solutions was first obtained by Véron [Ve] and Bénilan [Be]. The proof given here and based on the fundamental estimate is new (and considerably shorter). The control of the growth of the support as $t \to \infty$ (formula (5.5)) was first obtained in $d = 1$ by Knerr [Kn]. Sharp results are due to Vazquez [Va2] in $d = 1$ and [CVW] for $d > 1$, while the large-time behaviour of the solution was described by Friedman and Kamin [FK]; cf. [KV] for recent results.

A very complete reference list on the literature concerning different aspects of the PME and related equations can be found in Kalashnikov's survey paper [Ka].

References

[Ar] D.G. Aronson, The Porous Medium Equation, in *Nonlinear Diffusion Problems* (A. Fasano and M. Primicerio, eds.), Lecture Notes in Mathematics **1224**, Springer Verlag, New York, 1986, 1–46.

[AB] D.G. Aronson & P. Bénilan, Régularité des solutions de l'équation des milieux poreux dans $\mathbf{R}^N$, *C.R. Acad. Sci. Paris A* **288** (1979), 103–105.

[ACP] D.G. Aronson, M.G. Crandall & L.A. Peletier, Stabilization of solutions of a degenerate nonlinear diffusion problem, *Nonlinear Anal.* **6** (1982), 1001–1022.

[AP] D.G. Aronson & L.A. Peletier, Large-time behaviour of solutions of the porous medium equation in bounded domains, *J. Differential Equations* **39** (1981), 378–412.

[AW] D.G. Aronson & H.F. Weinberger, Nonlinear diffusion in population genetics, combustion and nerve impulse propagation, in *Partial Differential Equations and Related Topics*, Lecture Notes in Mathematics, Springer Verlag, New York, 1975, 5–49.

[Ba] G.I. Barenblatt, On some unsteady motions of a liquid or a gas in a porous medium, *Prikl. Mat. Mekh.* **16** (1952), 67–78 (in Russian).

[Be] P. Bénilan, Opérateurs accrétifs et semi-groupes dans les espaces $L^p (1 \leq p \leq \infty)$, *France-Japan Seminar*, Tokyo, 1976.

[Be2] P. Bénilan, A strong regularity L^p for the solution of the porous media equation, Research Notes in Mathematics **89** (C. Bardos et al., eds.) Pitman, Boston, 1985, 39–58.

[CF] L.A. Caffarelli & A. Friedman, Continuity of the density of a gas flow in a porous medium, *Trans. Amer. Math. Soc.* **252** (1979), 99–113.

[CVW] L.A. Caffarelli, J.L. Vazquez & I.N. Wolanski, Lipschitz continuity of solutions and interfaces of the N-dimensional porous medium equation, *Indiana Univ. Math.* **36** (1987), 373–401.

[DK] B.E. Dahlberg & C.E. Kenig, Non-negative solutions of the initial-Dirichlet problem for generalized porous medium equations in cylinders, *J. Amer. Math. Soc.* **1** (1988), 401–412.

[Dr]   H. Darcy, *Les fontaines publiques de la ville de Dijon*, V. Dalmont, Paris, 1856, 305–401.

[F]   A. Friedman, *Variational Principles and Free Boundaries*, Wiley and Sons, 1982.

[FK]   A. Friedman & S. Kamin, The asymptotic behaviour of a gas in an n-dimensional porous medium, *Trans. Amer. Math. Soc.* **262** (1980), 551–563.

[GT]   D. Gilbarg & N.S. Trudinger, *Elliptic Partial Differential Equations of Second Order*, Springer Verlag, New York, 1977.

[GP]   B. Gilding & L.A. Peletier, Continuity of solutions of the porous medium equation, *Ann. Scuola Norm. Sup. Pisa* **8** (1981), 657–675.

[GMC]   M.E. Gurtin, R.C. McCamy, On the diffusion of biological populations, *Math. Biosci.* **33** (1977), 35–49.

[Ka]   A. S. Kalashnikov, Some questions on the qualitative theory of nonlinear, second-order degenerate parabolic equations, *Uspekhi Mat. Nauk* **42**, 2(1987), 135–254.

[KV]   S. Kamin & J.L. Vazquez, Asymptotic behaviour of the solutions of the porous medium equation with changing sign, *SIAM J. Math. Anal.* **22** (1991), 34–45.

[Kn]   B.F. Knerr, The porous medium equation in one dimension, *Trans. Amer. Math. Soc.* **234** (1977), 381–415.

[LSU]   O.A. Ladyzhenskaya, V.A. Solonnikov & N.N. Ural'tseva, *Linear and Quasilinear Equations of Parabolic Type*, Transl. Math. Monographs **23**, Amer. Math. Soc., Providence, 1968.

[Li]   P.L. Lions, *Generalized Solutions of Hamilton-Jacobi Equations*, Research Notes in Mathematics **69**, Pitman, Boston, 1982.

[M]   M. Muskat, *The Flow of Homegeneous Fluids Through Porous Media*, McGraw-Hill, New York, 1937.

[O]   O. Oleĭnik, On some degenerate quasilinear parabolic equations, *Seminari dell'Istituto Nazionale di Alta Matematica 1962-63*, Oderisi, Gubbio (1964), 355–371.

[OKC]   O. Oleĭnik, S.A. Kalashnikov & Y.L. Czhou, The Cauchy problem and boundary-value problems for equations of the type of unsteady filtration, *Izv. Akad. Nauk SSSR, Ser. Mat.* **22** (1958), 667–704 (Russian).

[Pa]   R.E. Pattle, Diffusion from an instantaneous point source with concentration dependent coefficient, *Quart. J. Mech. Appl. Math.* **12** (1959), 407–409.

[Pe]   L.A. Peletier, The Porous Medium Equation, in *Applications of Nonlinear Analysis in the Physical Sciences* (H. Amann et al., eds.), Pitman, London, 1981, 229–241.

[Sa]   E.S.Sabinina, On the Cauchy problem for the equation of nonstationary gas filtration in several space variables, *Dokl. Akad. Nauk SSSR* **136** (1961), 1034–1037.

[Va1]   J.L. Vazquez, Symétrisation pour $u_t = \Delta\phi(u)$ et applications, *C. R. Acad. Sci. Paris Sér. I* **295** (1982), 71–74.

[Va2]   J.L. Vazquez, Asymptotic behaviour and propagation properties of the one-dimensional flow of gas in a porous medium, *Trans. Amer. Math. Soc.* **277** (1983), 507–527.

[Ve]   L. Véron, Coercivité et propriétés régularisantes des semi-groupes non linéaires dans les espaces de Banach, *Ann. Fac. Sci. Toulouse Math.* **1** (1979), 171–200.

[ZK]   Ya. B. Zel'dovich, A.S. Kompaneets, Theory of heat transfer with temperature dependent thermal conductivity, in *Collection in Honour of the 70th Birthday of Academician A.F. Ioffe*, Izdvo. Akad. Nauk SSSR, Moscow, 1959, 61–72.

[ZR]   Ya. B. Zel'dovich, Yu. P. Raizer, *Physics of Shock Waves and High-Temperature Hydrodynamic Phenomena* II, Academic Press, New York, 1966.

Asymptotic Behaviour Near Extinction Points for a Semilinear Equation with Strong Absorption[1]

Juan J.L. VELÁZQUEZ

Departamento de Matemática Aplicada
Facultad de Matemáticas
Universidad Complutense
E-28040 Madrid
Spain

Abstract

We discuss the asymptotic behaviour of solutions and their interfaces near extinction points for the problem: $u_t = u_{xx} - u^p$, $-\infty < x < +\infty$, $t > 0$; $u(x,0) = u_0(x)$, $-\infty < x < +\infty$, where $u_0(x)$ is continuous, non-negative and bounded.

In this note I want to report on recent joint work with M.A. Herrero on the Cauchy problem

$$u_t - u_{xx} + u^p = 0; \quad -\infty < x < +\infty, \quad t > 0, \tag{1.1}$$

$$u(x,0) = u_0(x); \quad -\infty < x < +\infty, \tag{1.2}$$

where

$$0 < p < 1, \tag{1.3a}$$

$u_0(x)$ is a continuous, nonnegative and bounded function which is not identically zero. $\qquad$ (1.3b)

By standard results, (1.1), (1.2) has then a unique, nonnegative classical solution $u(x,t)$. As a consequence of assumption (1.3a), this solution exhibits some features which are absent when $p \geq 1$. For instance, there exists $T > 0$ such that $u(x,t) \not\equiv 0$ if $t < T$, but $u(x,t) \equiv 0$ for any $t \geq T$ (cf. [K]). T is then called the extinction time of $u(x,t)$, the set

$$E = \{x \in \mathbb{R} : \text{ there exist sequences } \{x_n\} \to x, t_n \to T$$
$$\text{as } n \to \infty \text{ such that } u(x_n, t_n) > 0 \text{ for any } n\}$$

---

[1]Partially supported by CICYT Grant PB86-0112-C02022 and EEC Contract SC1-0019-C.

M. C. Delfour and G. Sabidussi (eds.), Shape Optimization and Free Boundaries, 391–395.
© 1992 *Kluwer Academic Publishers.*

is termed the extinction set, and points in E are called extinction points. On the other hand, solutions may develop dead cores, i.e., regions where $u(x,t) = 0$, even when $u_0(x)$ is everywhere positive; cf. for instance [BF], [EK], [BS], [RK], [FH], [HV1], [CMM], ...

In recent years, the question of describing the behaviour of solutions near the extinction time has received considerable interest. Let us describe but a few results in this direction. If $u_0(x)$ has a single maximum and $\lim_{|x| \to \infty} u_0(x) = 0$, it follows from work in [EK] and [CMM] that

(i)   The extinction set E consists of a single point, $E = \{x_0\}$ for some $x_0 \in \mathbb{R}$,

(ii)   For any $t \in (0, T]$, the positivity set $\Omega_+(t) = \{x : u(x,t) > 0\}$ has the form $\Omega_+(t) = (\varsigma_1(t), \varsigma_2(t))$, where for $i = 1, 2$ the functions $x = \varsigma_i(t)$ are continuous and such that $\lim_{t \uparrow T} \varsigma_i(t) = x_0$.

The curves $x = \varsigma_i(t)$ are usually called interfaces or free boundaries. On the other hand, it follows from the results in [FH] that, if $u_0(x)$ is compactly supported, with a single maximum at $x = 0$, symmetric (i.e., $u_0(x) = u_0(-x)$ for any x) and satisfying some geometrical assumptions (cf. (1.8) in [FH]), the following results hold true:

$$\lim_{t \uparrow T}(T - t)^{-\frac{1}{1-p}} u(x,t) = (1 - p)^{\frac{1}{1-p}}, \tag{1.4}$$

uniformly on sets

$$|x| \le C(T - t)^{1/2} \quad \text{with } C > 0;$$

there exist constants c_1 and c_2 such that, for any t close enough to T,

$$\{x : |x| \le c_1(T - t)^{1/2}\} \subset \Omega_+(t) \subset \{x : |x| \le c_2(T - t)^{1/4}\}. \tag{1.5}$$

Notice that, for any $T > 0$

$$u_T(x,t) = ((1 - p)(T - t))_+^{\frac{1}{1-p}}, \tag{1.6}$$

where $(s)_+ = \max\{s, 0\}$, is an explicit solution of (1.1) which vanishes as $t = T$.

To formulate our first result, we need to introduce some notation. First, for $m = 1, 2, \ldots$, we define polynomials $H_m(y)$ as follows

$$H_m(y) = c_m \tilde{H}_m(y), \quad \text{where } c_m = (2^{m/2}(4\pi)^{1/4}(m!)^{1/2})^{-1} \tag{1.7}$$

and $\tilde{H}_m(y)$ is the standard m^{th} Hermite polynomial.

Let x_0 be an arbitrary real number. Following [GP], [GK], we perform the change of variables

$$u(x,t) = (T - t)^{\frac{1}{1-p}} \phi(y, \tau), \quad y = (x - x_0)(T - t)^{-1/2} \quad , \tau = -\ln(T - t) \tag{1.8}$$

and linearize about $\overline{\phi} = (1-p)^{\frac{1}{1-p}}$ (which corresponds to (1.6) in the new variables), by setting

$$\phi(y,\tau) = (1-p)^{\frac{1}{1-p}} + \psi(y,\tau). \tag{1.9}$$

When then have

Theorem 1 *Let $u(x,t)$ be a solution of (1.1)–(1.3) which vanishes at $t = T$ and is different from $u_T(x,t)$ in (1.6). Let $x = x_0$ be an extinction point of u, and let $\psi(y,\tau)$ be the function defined by (1.8), (1.9). Then the following alternative holds. Either*

$$\psi(y,\tau) = -\frac{(4\pi)^{1/4}(1-p)^{\frac{1}{1-p}}}{\sqrt{2p}} \cdot \frac{H_2(y)}{\tau} + o\left(\frac{1}{\tau}\right) \quad \text{as } \tau \to \infty, \tag{1.10}$$

or there exist $C > 0$ and an even integer $m \geq 4$, such that

$$\psi(y,\tau) = -Ce^{(1-\frac{m}{2})\tau} H_m(y) + o(e^{(1-\frac{m}{2})\tau}) \quad \text{as } \tau \to \infty, \tag{1.11}$$

where in (1.10), (1.11) convergence holds in $C^{k,\alpha}_{loc}$ for any $\alpha \in (0,1)$ and $k \geq 1$.

We next focus on the case where (1.10) takes place, and prove

Theorem 2 *Let $u(x,t), T$ be as in the statement of Theorem 1, and let $x = x_0$ be an extinction point of u. Assume also that (1.10) holds. We then have*

$$\lim_{t \uparrow T}(T-t)^{-\frac{1}{1-p}}u(x_0 + \xi((T-t)|\log(T-t)|)^{1/2}, t) =$$

$$(1-p)^{\frac{1}{1-p}}\left[1 - \frac{(1-p)\xi^2}{4p}\right]_+^{\frac{1}{1-p}}, \tag{1.12}$$

uniformly on sets $|\xi| \leq R$ with $R > 0$. Moreover, for t close enough to T, there exist continuous functions $\varsigma_1(t), \varsigma_2(t)$, such that

$$u(x,t) > 0 \quad \text{in } (x_0 - \varsigma_1(t), x_0 + \varsigma_2(t)), \tag{1.13a}$$

and for some $\delta > 0$,

$$u(x,t) = 0$$

in $[x_0 - \varsigma_1(t) - \delta, x_0 - \varsigma_1(t)]$, or $[x_0 + \varsigma_2(t), x_0 + \varsigma_2(t) + \delta]$, and

$$\lim_{t \uparrow T} \frac{\varsigma_i(t)^2}{(T-t)|\log(T-t)|} = \left[\frac{4p}{1-p}\right] \quad \text{for } i = 1,2. \tag{1.13b}$$

The corresponding result when (1.11) is satisfied reads as follows

Theorem 3 *Let $u(x,t), T$, be as in the statement of Theorem 1, and let $x = x_0$ be an extinction point of u. Assume now that (1.11) holds. We then have*

$$\lim_{t \uparrow T}(T - t)^{-\frac{1}{1-p}} u(x_0 + \xi(T - t)^{1/m}, t) =$$

$$(1 - p)^{\frac{1}{1-p}} \left(1 - c_m(1 - p)^{-\frac{1}{1-p}} C \xi^m\right)_{+}^{\frac{1}{1-p}}, \tag{1.14}$$

where c_m and C are as in (1.7), (1.11), uniformly on sets $|\xi| \le R$ with $R > 0$. Moreover, for t close enough to T, there exist continuous functions $S_1(t), S_2(t)$, such that

$$u(x,t) > 0 \quad in \;\; (x_0 - S_1(t), x_0 + S_2(t)), \tag{1.15a}$$

and for some $\delta > 0$,

$$u(x,t) = 0$$

if $x \in [x_0 - S_1(t) - \delta, x_0 - S_1(t)]$, or $x \in [x_0 + S_2(t), x_0 + S_2(t) + \delta]$, and

$$\lim_{t \uparrow T} \frac{S_i(t)}{(T - t)^{1/m}} = \left[\frac{(1-p)^{\frac{1}{1-p}}}{C c_m}\right]^{1/m} \quad for \;\; i = 1, 2. \tag{1.15b}$$

Notice that, as a consequence of Theorems 1–3, if $u(x,t) \not\equiv u_T(x,t)$ for any $T > 0$, the extinction set of u consists of isolated points. This fact has been previously proved in [CMM] under the assumption that $u_0(x)$ has compact support.

We finally address the actual occurrence of the behaviours described previously, and prove

Theorem 4 (a) *Assume that $u_0(x)$ has a single maximum. Then there exist $x_0 \in \mathbb{R}$ and $T > 0$ such that (1.12) and (1.13) hold.*

(b) *There exists an initial value $u_0(x)$ and a constant $C > 0$ such that (1.14) and (1.15) hold with $m = 4$.*

To our knowledge, the asymptotics described in Theorems 1–3 were conjectured for the first time in [GHV], where the possible behaviours of solutions and interfaces near an extinction point were formally obtained by the method of matched asymptotic expansions. We conjecture that, for any even number m with $m \ge 6$, there exist initial values $u_0(x)$ and constants $C > 0$ such that (1.14) and (1.15) hold for such m, but we have been unable to prove this fact so far.

For the proofs, the reader is referred to [HV2].

References

[BF]   H. Brézis and A. Friedman, Estimates on the support of solutions of parabolic variational inequalities, *Illinois J. Math.* **20** (1976), 82–98.

[BS]   C. Bandle and I. Stakgold, The formation of the dead core in a parabolic reaction-diffusion equation, *Trans. Amer. Math. Soc.* **286** (1984), 275–293.

[CMM] X. Chen, H. Matano and M. Mimura, Finite-point extinction and continuity of interfaces in a nonlinear diffusion equation with strong absorption, *J. Reine Angew. Math.*, to appear.

[EK]   L. C. Evans and B.F. Knerr, Instantaneous shrinking of the support of nonnegative solutions to certain nonlinear parabolic equations and variational inequalities, *Illinois J. Math.* **23** (1979), 153–166.

[FH]   A. Friedman and M.A. Herrero, Extinction properties of semilinear heat equations with strong absorption, *J. Math. Anal. Appl.* **124** (1987), 530–546.

[GHV] V.A. Galaktionov, M.A. Herrero and J.J.L. Velázquez, The structure of solutions near an extinction point in a semilinear heat equation with strong absorption: a formal approach, in: *Nonlinear Diffusion Equations and Their Equilibrium States* (N.G. Lloyd et al., eds.), Birkhäuser, to appear.

[GK]   Y. Giga and R.V. Kohn, Asymptotically self-similar blow-up of semilinear heat equations, *Comm. Pure Appl. Math.* **38** (1985), 297–319.

[GP]   V.A. Galaktinonov and S.A. Posashkov, Application of new comparison theorems in the investigation of unbounded solutions of nonlinear parabolic equations, *Diff. Urav.* **22** (7) (1986), 1165–1173.

[HV1] M.A. Herrero and J.J.L. Velázquez, On the dynamics of a semilinear heat equation with strong absorption, *Comm. Partial Differential Equations* **14** (12) (1989), 1653-1715.

[HV2] M.A. Herrero and J.J.L. Velázquez, Approaching an extinction point in one-dimensional heat equations with strong absorption, *J. Math. Anal. Appl.*, to appear.

[K]   A.S. Kalashnikov, The propagation of disturbances in problems of nonlinear heat conduction with strong absorption, *USSR Comput. Math. and Math. Phys.* **14** (1974), 70-85.

[RK]   Ph. Rosenau and S. Kamin, Thermal waves in an absorbing and convecting medium, *Phys. D.* **8** (1983), 273–283.

Introduction to Shape Optimization Problems and Free Boundary Problems

J.P. ZOLÉSIO

Institut Non Linéaire de Nice
Université de Nice, Parc Valrose
F-06034 Nice Cédex
France

Abstract

We are concerned with existence results in shape optimization as well as with necessary conditions for optimality. In the first section we give existence results for a weak shape formulation of Bernoulli-like free boundary problems for stationary potential flows. In the second section it is shown how the Bounded Perimeter-constraint can apply to give an existence result for control in the Transient Wave Equation. The third section deals with the very definition of shape derivatives and with results on the structure of the derivatives. The fourth section deals with the shape variational free boundary problem associated with the Stokes stationary fluid. It underlines that the free boundary condition cannot be achieved in such a linearized modelling. Also, we give existence and continuity results obtained by a penalty approach (via transmission "two-fluid" problems) which apply also to unilateral problems. Finally, the last section extends an existence result for eigenvalues of the Laplace operator.

Introduction

The theory of Optimal Control deals in particular with the control of solutions to Boundary Value Problems (BVP), the control being in general a relevant parameter or function occurring in a boundary condition. We consider here the situation where a part of the boundary is itself a control variable. When the BVP under consideration is time dependent we shall denote by $Q = I \times \Omega$ the domain of evolution, where $I = [0, T]$ and Ω is a bounded smooth domain in $\mathbf{R}^n$. In these lectures we shall always assume the following geometrical constraint on Ω: there exists a bounded smooth domain D in $\mathbf{R}^n$ such that Ω is a subset of D.

In general the boundary $\partial\Omega$ of Ω will be decomposed in two pieces: $\partial\Omega = \Gamma_0 \cup \Gamma$ where Γ_0 is the fixed (given or known) part of the boundary $\partial\Omega$ and Γ is the variable or controlled part.

In addition, we denote by $y = y(\Omega)$ or $y(\Gamma)$ (depending on the context) the solution of a BVP on Ω. As in Control Theory terminology, we shall refer to this BVP as the *state*

397

M. C. Delfour and G. Sabidussi (eds.), Shape Optimization and Free Boundaries, 397–457.

equation and call y the *state of the system*. The associated *cost function* will be of the form

$$J(\Omega) = j_\Omega(y(\Omega)),$$

where j_Ω is a mapping defined on a space of functions $H(\Omega)$ on the domain Ω.

Given a family $\mathbf{O}_{\mathrm{ad}}$ of admissible subsets Ω of D, we shall define for each element Ω of $\mathbf{O}_{\mathrm{ad}}$ the tangent space $\mathbf{T} = \mathbf{T}_\Omega(\mathbf{O}_{\mathrm{ad}})$ as the set of non-autonomous admissible vector fields V whose flow mapping $T_t(V)$ maps the set Ω onto $\Omega_t = T_t(V)(\Omega)$ which belongs to the family $\mathbf{O}_{\mathrm{ad}}$ of admissible subsets of D. Denote by $\mathbf{H}$ the union of all the Banach spaces $H(\Omega)$ when Ω ranges over the family $\mathbf{O}_{\mathrm{ad}}$. Then the cost functional J can be written as

$$J = j \circ y,$$

where

j maps $\mathbf{O}_{\mathrm{ad}}$ into $C^0(\mathbf{H}, \mathbf{R})$, with j_Ω in $C^0(H(\Omega), \mathbf{R})$;

y maps $\mathbf{O}_{\mathrm{ad}}$ into $\mathbf{H}$, with $y(\Omega)$ in $H(\Omega)$.

The derivative $dJ(\Omega; V)$ of J at an element Ω of $\mathbf{O}_{\mathrm{ad}}$ in the direction V, V in $\mathbf{T}_\Omega(\mathbf{O}_{\mathrm{ad}})$, will be defined with the help of the *shape derivatives* $y'(\Omega; V)$ and $j'_{\Omega;V}$ which we shall introduce in section 2 of the notes. In that section we shall introduce the *Speed Method* [16], [18], [5], [4], [8] which unifies all kinds of transformations T_t of the closure $cl(D)$ of D into itself, in particular transformations depending on parameters and finite dimensional approximations, large deformations, ...

The first section deals with four basic examples of shape optimization problems:

1) the wave equation in Q with a given homogeneous (for simplicity) boundary condition of Dirichlet or Neumann type on Γ – the cost J being a quadratic observation of y on Γ_0;

2) the wave equation in $Q_D = I \times D$ with an interface Γ in D on which a transmission condition is specified;

3) a steady state water wave free boundary value problem in $\mathbf{R}^3$ involving the non-linear Bernoulli condition;

4) a free boundary problem with non-linear eigenvalue.

We shall briefly study the existence question for the minimum of J over the admissible family $\mathbf{O}_{\mathrm{ad}}$. For simplicity we shall consider the functional j_Ω defined on a fixed part of $\partial\Omega$ so that the existence question is mainly related to the choice of topologies on $\mathbf{O}_{\mathrm{ad}}$, and the choice of the family V itself, that is, the constraint placed on the domains. The second problem is the most accessible one for it can be "relaxed" to the closed convex subset

$$K = \{\lambda \in L^2(D) : 0 \le \lambda(x) \le 1, \text{ a.e. } x \text{ in } D\}$$

of $L^2(D)$. The family $\mathbf{O}_{ad}$ can be considered as being a subset of K as we identify Ω with its characteristic function 1_Ω; then in this problem we choose

$$\mathbf{O}_{ad} = \text{Char} = \{\lambda \in K : \lambda = \lambda^2 \text{ a.e. in } D\}.$$

The third problem will be relaxed to the set

$$\mathbf{O}_{ad} = \{\lambda \in \text{Char} : \lambda \text{ is in } BV(D)\}.$$

In other words, the domain Ω will not be an open subset of D but only a measurable set in D for the second problem, and a measurable subset of D with finite perimeter in D (Cacciopoli domains) in the third problem. We shall see that the perimeter $P_D(\Omega)$ provides a shape-variational formulation for the mean curvature. This term comes from the surface tension that should not be neglected for existence purposes. The cost function will take the form

$$J(\Omega) = E(\Omega) + \sigma P_D(\Omega).$$

For this Bernoulli free boundary problem we also give an exsitence result by using the saddle point of a Lagrangian L defined over $\mathbf{O}_{ad} \times H(D)$ for a perturbed rheology [29]; this approximation could by useful for the "magneto-forming" free boundary problem which is a very similar problem. In the fourth problem the surface tension is not necessary for we have a global non-convex, non-differentiable variational formulation on D which leads to existence. This formulation is a generalization of the Auchmuty principle which in some sense includes the free boundary in the non-differentiability.

The existence question in the first problem, which is the usual situation concerning identification problems, is more delicate. Depending on the kind of boundary condition imposed on G, the problem can be relaxed to different families $\mathbf{O}_{ad}$. The continuity of the mapping $\Omega \to y(\Omega)$ will be obtained using a penalized approximation $y_\epsilon(\Omega)$ of $y(\Omega)$ which is the solution of a transmission problem with the same behavior as problem 2. For each ϵ and Ω, the convergence $y_\epsilon(\Omega)|_\Omega \to y(\Omega)$ holds. For each $\epsilon > 0$, the mapping $\Omega \to y_\epsilon(\Omega)$ is continuous on K, but in the limit we recover the continuity of $\Omega \to y(\Omega)$ if the family $\mathbf{O}_{ad}$ does have an additional property: there exists $M > 0$ such that for any Ω in $\mathbf{O}_{ad}$ one can find an extension operator $P_\Omega : H(\Omega) \to H(D)$ such that the norm of the linear operator satisfies the condition $\|P_\Omega\| \leq M$. This condition is satisfied when the elements of the family $\mathbf{O}_{ad}$ are Lipschitzian domains satisfying a uniform cone condition or when they are smooth open domains in D with a uniformly bounded mean curvature H. These conditions which are pointwise conditions on the boundaries should be relaxed to global ones.

Prior to the study of the question of existence for the free boundary problems 3 and 4, the main difficulty is their modelling as *shape variational problems* of the form

$$(P) : \inf\{J(\Omega) : \Omega \text{ in } \mathbf{O}_{ad}\}.$$

As we have said, we shall give two different cost functionals J for the third problem. The situation is as follows: when solving a shape optimization problem such as the first two problems, the first order necessary condition

$$dJ(\Omega; V) = 0 \text{ for all } V$$

provides a set of equations which are characteristic of a free boundary problem. For example,

$$
\begin{aligned}
\Box\, y(\Omega) &= 0 \text{ in } Q \\
\Box\, p(\Omega) &= 0 \text{ in } Q \\
\partial_n y(\Omega) &= 0 \text{ on } \Gamma \\
\partial_n p(\Omega) &= f(y) \text{ on } \Gamma \\
g(\Gamma) &= \text{grad } y(\Omega) \cdot \text{ grad } p(\Omega) + \cdots = 0 \text{ on } \Gamma,
\end{aligned}
\tag{0.1}
$$

where p is the solution of the adjoint problem and $g(\Gamma)$ is the density of the shape gradient on Γ.

When dealing with a free boundary problem, say

$$
\begin{aligned}
\Delta y(\Omega) &= 0 \text{ in } \Omega \\
\partial y(\Omega) &= 0 \text{ on } \Gamma
\end{aligned}
\tag{0.2}
$$

$$\frac{1}{2}|\text{grad } y(\Omega)|^2 \; + \; G + \sigma H = \text{ const. on } \Gamma,$$

the problem is to find a cost functional J such that (0.2) appears as being the necessary condition associated with J. Of course one can think to bootstrap: find $\underline{J}, \underline{J}(\Omega) = \underline{h}_\Omega(y(\Omega), p(\Omega))$, such that (0.1) is solved by the necessary condition on $\underline{J}$.

The first section will deal with the third problem (0.2). We briefly recall that *the derivative of an infimum, with respect to a parameter, is the infimum of the derivative over the set of optimal solutions* [19] [2]. We shall apply this result to the functional

$$J(\Omega) = E(\Omega) + \sigma P_D(\Omega),$$

where E is the energy associated to the variational boundary value problem whose y is a solution of the minimization problem:

$$E(\Omega) = \min\{E(\Omega, z) : z \in H(\Omega)\}.$$

Introducing an ϵ-approximation (a two-phase or two-fluid problem in D, Γ being an interface, and introducing the non-local operator β defined by

$$\beta(y)(x_0) = \text{ meas } \{x \text{ in } D : y(x) \le y(x_0)\},$$

we shall give two existence results for the perturbed models [27].

The fourth section will extend the shape differential calculus introduced in the second section. We shall turn to the second order shape derivative and in particular to shape acceleration, which is the situation when the solution $y(\Omega)$ of the state equation is a vector field on Ω. We consider then for any admissible field V the shape derivative $y'(\Omega; V)$;

but as $y(\Omega)$ is itself an admissible field, i.e., an element in $\mathbf{T}_\Omega(\mathbf{O}_{\mathrm{ad}})$, we can consider $y'(\Omega; y(\Omega))$ which is by definition the *shape acceleration* $y''(\Omega)$. This term is characterized, assuming smoothness results of the solution $y(\Omega)$, when $y(\Omega)$ is the solution of a quasi-steady Norton-Hoff flow. As an application we obtain numerical schemes of higher order of accuracy for large deformations (see T. Tiihonen and J.P. Zolésio [13]). The second order shape derivative of a cost functional J, $J''(\Omega; V, W)$, for two admissible fields V and W is defined as the shape derivative at Ω in the direction W of the functional $\Omega \to dJ(\Omega; V)$. This second order shape derivative is expressed in terms of a kernel $\mathcal{H}$ which is a distribution on $\Gamma \times \Gamma$, the *shape Hessian*. It is connected with the shape derivative of the density gradient $g(\Gamma)$ (the gradient G is a distribution on D of finite order with support in Γ and zero transverse order). We shall consider the second order necessary condition for Ω corresponding to the minimization of the functional

$$J = E + \sigma P_D(\Omega).$$

In particular if we consider vector fields V and W arising from potentials A and B defined on D such that

$$A = B = \text{const on } \Gamma,$$

$$\partial_n A = v, \ \partial_n B = w \text{ on } G,$$

$$\langle D^2 A \cdot n, n \rangle = \langle \epsilon(V) \cdot n, n \rangle = 0, \quad \langle D^2 B \cdot n, n \rangle = 0 \text{ on } \Gamma,$$

then

$$\text{div}_\Gamma(V|_\Gamma) = (\text{div}(V))|_\Gamma,$$

and we obtain a second order derivative in the direction (V, W) which is symmetric in V, W. For a general study of the second order derivatives we refer to M.C. Delfour and J.P. Zolésio [4,5,6]. As an exercise we shall consider the functional $J = E + \sigma P_D$ for an incompressible steady Stokes flow $y = (y_1, y_2, y_3)$ with boundary condition $y(\Omega) \cdot n = 0$ on Γ (streamline condition). We shall give an explicit expression for the tangential necessary condition arising from the minimization of the functional J over the admissible family $\mathbf{O}_{\mathrm{ad}} = \{\lambda \text{ in Char} : \lambda \text{ is in } BV(D)\}$. This condition is a tangential differential equation on Γ which is not, as in the third example, the expected physical boundary condition.

The shape derivative of a solution $y(\Omega)$ to a non-linear BVP is in general obtained with the help of some extra information on the sign of $y(\Omega)$ on some part of the domain Ω or of the boundary Γ (in fact, in the place where the cost functional *observes the solution* $y(\Omega)$). This extra information ensures that the linear adjoint problem is well posed in some Hilbert space $\mathcal{H}(\Omega)$ which is now different from the Banach space $H(\Omega)$. For such an example we refer to the radiator problem considered in M.C. Delfour and J.P. Zolésio [3]. For the non-linear steady Norton-Hoff flow this extra information is obtained as a physical assumption which states that there exists a direction e_i in $\mathbf{R}^3$ in which $y(\Omega) \cdot e_1$ is positive, $y(\Omega) \cdot e_1 \geq c > 0$ on Ω, and $y(\Omega) \cdot e_1$ is larger than $y(\Omega) \cdot e_i$ for $i \neq 1$. For non-linear problems involving unilateral conditions such as variational inequalities, we refer to J. Sokolowski and J.P. Zolésio [8]. Finally we shall just mention that in these notes the shape calculus of derivatives for continuous problems also contains in some sense the derivatives of discretized problems. This is done by making an appropriate choice of the admissible velocity field V_{M_i,e_j} such that the flow mapping T_t associated with V_{M_i,e_j} moves exactly the nodes M_i in the direction e_j , keeping the other nodes fixed and respecting the deformation of the elements. Such velocity fields can be constructed when the solution of the BVP under consideration is approximated by conformal finite elements (cf. J.P. Zolésio [22] in the case of piecewise linear finite elments or P^1 elements), and for Lagrange elements (cf. M. Souli and J.P. Zolésio [10]). Then the derivative is of the form $dJ_h(\Omega_h, V_{M_i,e_j})$, where the index h indicates that the cost function J_h is associated with the finite element discretization of the problem and cost functional. This is the exact derivative of $J_h(\Omega_h)$ with respect to the j-th coordinate of the node M_i. When the finite elements are not conformal, the basis functions are not convected by the flow T_t associated to the vector field V_{M_i,e_j}. As a consequence an extra term occurs (see T. Tiihonen and J.P. Zolésio [13] for an example of non-conformal finite elements).

The fourth section will be devoted to the non-linear free boundary eigenvalue problem. This problem belongs to the restricted class of free boundary problems in which Γ is an interface in D, $y(\Omega)$ being the solution of a well posed BVP in D, with a boundary condition on ∂D, and possibly a transmission condition on the interface Γ (which says that some quantity $\langle K \cdot \text{grad } y, n \rangle$ is preserved through Γ). Here the boundary conditions occur in a different way than in problems (0.1) and (0.2), for they take the following form:

$$
\begin{aligned}
A \cdot y(\Omega) &= (1 - \lambda_\Omega)f + \lambda_\Omega R \quad \text{in } \mathcal{D}'(D) \\
\Gamma &= \mathcal{F}(y(\Omega))^{-1}(0),
\end{aligned}
\tag{0.3}
$$

where λ_Ω stands for the characteristic function of the measurable set Ω in D. This kind of free boundary condition is the easy one. The main fact is that, after some technical changes, the free boundary Γ appears as a level set of some function. It would be necessary to recall some elements of geometrical measure theory on a measurable set D (Federer's theorem, ...) and some results concerning the level curves $u^{-1}(t)$ of a smooth positive function u defined on D and equal to zero on the boundary ∂D. Mainly we use the fact: $u^{-1}(t) = T_t(V)(\partial D)$, where the field V is given by $V = |\text{grad } u|^{-2}\text{grad } u$ and also that we have

$$
\begin{aligned}
\int_D f(x)dx &= \int_0^\infty dt \int_{T_t(V)(\partial D)} f|\langle V(t), n_t \rangle|d\mu_t \\
&= \int_0^\infty dt \int_{u^{-1}(t)} f||\text{grad } u||^{-1}d\mu_t.
\end{aligned}
$$

In the situation where $R = 0$, there is a very easy existence result. Following J.P. Zolésio [16,17]. Consider the problem: to find $(\Omega, y\lambda)$ such that

$$
\begin{aligned}
\Delta y(\Omega) &= 0 \quad \text{in } \Omega^c = D\backslash(\Omega \cup \Gamma) \\
\Delta y(\Omega) &= \lambda\beta(y) \quad \text{in } \Omega \\
y &\text{ in } C^1(cl(D)) \\
y &> 0 \text{ in } \Omega, \ y < 0 \text{ in } \Omega^c, \ y = 0 \text{ on } \Gamma, \ y = -\lambda \text{ on } \partial D.
\end{aligned}
\tag{0.4}
$$

Such free boundary problems have been considered by R. Temam, J. Mossino, J. Mossino and J.P. Zolésio and we refer to [20] for some extended shape results concerning the *shape differential equation formulation* for the Grad-Mercier equations, where in (0.4), $\beta(y)$ is replaced by $\partial_{ss}y^* \circ \beta(y)$, where y^* is the monotone rearrangement of y, and with the following boundary conditions:

$$y = \partial_n y = 0 \quad \text{on } \partial D, \partial_s y^*(0) = \partial_s y^*(\text{meas}(D)).$$

1. Existence results

1.1 Examples

In this section we consider four basic examples.

Example 1 Let D be a bounded smooth domain in $\mathbf{R}^n$, Γ_0 a part of the boundary ∂D of D, and Ω a smooth open domain in D (say with differentiable boundary, simply connected, lying on one side of its boundary Γ). Its boundary $\partial\Omega$ is made of two parts, $\partial\Omega = \Gamma_0 \cup \Gamma$, where

Γ is the free part of the boundary,

H the d'Alembertian operator, $Hy = \partial_{tt}^2 y - \Delta y$,

y_0, y_1 are given in $H^1(D) \times L^2(D)$, and

y_m is the measured acoustic pressure on $I \times \Gamma_0$.

Consider the following two boundary value problems:

$H \cdot y(\Omega) = 0$ in Q $\qquad\qquad$ $H \cdot y(\Omega) = 0$ in Q

$y(0, \cdot) = y_0, \partial_t y(\Omega)(0, \cdot) = y_1$ in $\Omega(N)$ $\quad$ $y(0, \cdot) = y_0, \partial_t y(\Omega)(0, \cdot) = y_1$ in $\Omega(D)$

$\partial_n y(\Omega) = 0$ on Σ $\qquad\qquad$ $y(\Omega) = 0$ on Σ.

The associated cost functionals are

$$2J_n(\Omega) = \iint_{I \times \Gamma_0} [y_m - y]^2 dt\, d\mu \qquad\qquad 2J_d(\Omega) = \iint_{I \times \Gamma_0} [(\partial_t y) - z_m]^2 dt\, d\mu.$$

The set of admissible fields is

$$\mathbf{O}_{\text{ad}} = \{V \in C^1([0,T]; C^1(D, \mathbf{R}^n)) : V(t,0) = 0 \text{ on } \Gamma_0\},$$

and y_m and z_m are given elements which are usually associated with measurements or observed variables.

Example 2 It is similar to the first example but the wave problem is now in the evolution domain $I \times D$ so that Γ appears as an interface on which we consider classical *transmission* conditions:

$$\left.\begin{array}{rcl} H \cdot y(\Omega) & = & 0 \text{ in } I \times (D \backslash \Gamma) \\ y(0, \cdot) & = & y_0, \ \partial_t y(\Omega)(0, \cdot) = y_1 \text{ in } \Omega \\ \partial_n y(\Omega) & = & 0 \text{ on } \Sigma_0 = I \times \Gamma_0 \\ k\partial_n(y(\Omega)_{|\Omega}) & = & \partial_n(y(\Omega)|_{\Omega}c), \text{ on } I \times \Gamma \end{array}\right\} \textbf{(TN)}$$

$$\left.\begin{array}{rcl} H \cdot y(\Omega) & = & 0 \text{ in } I \times (D \backslash \Gamma) \\ y(0, \cdot) & = & y_0, \ \partial_t y(\Omega)(0, \cdot) = y_1 \text{ in } \Omega \\ y(\Omega) & = & 0 \text{ on } \Sigma \times \Gamma_0 \\ k\partial_n(y(\Omega)_{|\Omega}) & = & \partial_n(y(\Omega)|_{\Omega}c), \text{ on } I \times \Gamma \end{array}\right\} \textbf{(TD)}$$

Example 3 This is a steady water wave problem. The geometry is the same as in the previous examples but the time t and the interval I disappear. The speed vector field V of the fluid particle arises from a potential y which is harmonic in Ω (the volume occupied by the fluid). On the internal boundary Γ the streamline and Bernoulli conditions are imposed:

$$\begin{array}{rcl} \Delta y(\Omega) & = & 0 \text{ in } \Omega \\ \partial_n y(\Omega) & = & 0 \text{ in } \Gamma \\ \partial_n y(\Omega) & = & f \text{ on } \Gamma_0 \\ \frac{1}{2}|\text{grad } y(\Omega)|^2 + G & = & c + \sigma H \text{ on } \Gamma, \ c \text{ is a constant,} \end{array}$$

where G is a positive smooth function defined on D and f is defined smoothly on ∂D with its support, supp f, contained in the interior of Γ_0.

The cost and the admissible family will be

$$J(\Omega) = E(\Omega) + \sigma P_D(\Omega),$$

$$\mathbf{O}_{\text{ad}} = \{\lambda \text{ in Char } : \lambda \text{ in } BV(D), \text{ meas } (\Omega) = \int_D \lambda dx = a\}.$$

Example 4 This is a free boundary problem: to find $\mu > 0$, Ω in D and y in $C^1(cl(\Omega)) \cap H^2(D)$ such that

$$\begin{array}{rcl} -\Delta y & = & \mu \beta(y) \text{ a.e. in } \Omega \\ -\Delta y & = & 0 \text{ a.e. in } D \backslash \Omega \\ y & = & -\mu \text{ on } \Gamma = \partial \Omega \end{array}$$

$$\int_\Gamma \partial_n y \, ds = I, \quad I > 0 \text{ given.}$$

The cost will be of the following form $J(\Omega) = E_c(D \backslash \Omega) + E(\Omega)$.

1.2 Classification of problem

In Example 4 the state $y(\Omega)$ and the cost $J(\Omega)$ are defined when Ω is a measurable set in D. In Examples 2 and 3 the boundary Γ of the open set Ω must be smooth enough, say C^1 (which is not the minimal smoothness), to make sense of the strong formulation of the Bernoulli boundary condition which involves the normal field n on Γ. But in the variational formulation Ω must be a measurable set such that the characteristic function λ belongs to $BV(D)$: then as far as existence is concerned we shall consider the relaxed problem. In the first example we shall also use a relaxed formulation of the problem but we shall need additional constraints to get an existence result.

1.3 Topologies

The simplest tool is to consider the open subsets of D as $\Omega = D \backslash A$, where $A = cl(A)$, and define the convergence $\Omega_n \to \Omega$ if $A_n \to A$ in the Hausdorff metric for which the family of closed subsets in D is compact (see §3.16 in J. Dieudonné [30]). Two other topologies are obtained via the characteristic functions. We have the following properties.

Proposition 1.1 *Let λ_n belong to* Char *and f in $L^2(D)$ such that $0 \le f \le 1$ a.e. in D. Then the following statements are equivalent:*

   (i) $\lambda_n \to f$ *weakly in $L^2(D)$;*
   (ii) *there exists $p \ge 1$ such that $\lambda_n \to f$ weakly in $L^p(D)$;*
   (iii) $\lambda_n \to f$ *weakly-* in $L^\infty(D)$;*
   (iv) $\lambda_n \to f$ *in $D'(D)$.*

Definition 1.2 For characteristic functions we say that

λ_n converges weakly to f, $(\lambda_n \rightharpoonup f)$, if one of the properties (i) to (iv) is satisfied.

λ_n converges to f, $(\lambda_n \to f)$, if the convergence is in the $L^2(D)$ norm, or equivalently in the L^p norm, $1 \le p < \infty$ (since $\lambda^p = \lambda$).

Remark 1.3
$$||\lambda - \mu||_\infty = \begin{cases} 1 & \text{if } \lambda - \mu \ne 0, \\ 0 & \text{if } \lambda - \mu = 0. \end{cases}$$

Concerning the weak topology it is well known that it is metrizable as being the weak topology of the bounded convex set $K = \{f \text{ in } L^2(\Omega) : 0 \le f \le 1 \text{ a.e.}\}$ and K is compact.

Remark 1.4 If $\lambda_n \rightharpoonup f$ and f belongs to Char, then $\lambda_n \to f$ (for we have the convergence of the L^2 norms).

Proposition 1.2 *Let* $\{\Omega_n\}$ *(or more precisely the characteristic functions) converge weakly to* f, *with* Ω_n *an open subset of* D. *For any subsequence (still denoted by* $\{\Omega_n\}$) *which converges in Hausdorff metric to an open set* Ω *in* D *we have* $D\backslash\Omega = \mathrm{supp}\,(1-f)$. $(1-f)$ *being considered in* $\mathcal{D}(D)'$ *its support is the complement of the largest open set* O *on which* $1-f$ *is not a.e. equal to zero).*

Remark 1.5 In general f is not in Char (Char is not weakly closed). Thus $\Omega_n \to \Omega$ in the Hausdorff distance, $\Omega_n \rightharpoonup f$ but $f \neq \lambda_\Omega$. This is consistent with the fact that the mapping $f \to \lambda_{\mathrm{supp}(f)}$ is not continuous from $L^p(D)$ to $\mathcal{D}'(D)$.

Remark 1.6 The state $y(\Omega)$ in the four examples is not continuous for the Hausdorff topology, even the volume $|\Omega| = \mathrm{meas}\,(\Omega)$ is not H-continuous. To see this consider the following sequence of domains in D:

$$\Omega_n = \{x \in D =]0,1[:\ \sin(n\pi x) > 0\}.$$

Then

$$\Omega_n \to D \text{ in the Hausdorff metric topology,}$$

while

$$|\Omega_n| = 0.5 \text{ and } \chi_{\Omega_n} \rightharpoonup 0.5.$$

Remark 1.7 Char is closed, but not compact in $L^2(D)$.

In the second example the mapping $\Omega \to y(\Omega)$, from Char into $H^1(D)$, is continuous, and it can be continuously extended to K. It is well known that, even for elliptic problems, the solution y is not continuous with respect to the coefficients (in the symbol) equipped with the weak $L^2(D)$ topology. Then to bypass this difficulty we introduce the space

$$BPS = \{\text{measurable sets } \Omega \text{ in } D\ :\ \lambda_\Omega \text{ is in } BV(D)\},$$

$$\|\lambda_\Omega\|_{BV(D)}. = |\Omega| + \|\mathrm{grad}\,\lambda_\Omega\|_{M^0(D)}.$$

Proposition 1.3 *Let* $\{\lambda_n\}$ *be a sequence in BPS such that* $\|\lambda_n\|_{BV(D)} \leq M$ *for some fixed constant* $M > 0$. *Then there exists a subsequence* $\{\lambda_n\}$ *and* λ *in BSP, such that* $\lambda_n \to \lambda$ *in* $L^1(D)$, $\mathrm{grad}\,\lambda_n$ *converges weakly to* $\mathrm{grad}\,\lambda$, *and* $\|\lambda\|_{BV(D)} \leq M$.

Concerning the strong version of Example 1 it is necessary to consider the family of smooth domains, and the continuity of the mapping $\Omega \to y(\Omega)$ will be obtained on the subfamily verifying the additional condition: there exists $M > 0$ such that for all Ω in $\mathbf{O}_{\mathrm{ad}}$ there exists an extension operator $P(\Omega)$ belonging to $\mathcal{L}(H^1(\Omega), H^1(D))$, with $\|P\| \leq M$. It turns out that this condition is verified for Lipschitzian domains endowed with a *uniform cone condition* (that uniform cone condition leading to compactness in the $L^2(D)$-topology).

For the homogeneous Dirichlet condition we have several relaxations of the problem since the extension operator is obvious.

For many problems it is necessary to consider stronger topologies on smooth domains. For instance for shells let Γ be the mean surface of the shell, $D_\Gamma\, n$ the tangential differential of the normal field n,

$$D_\Gamma\, n = (^*\mathrm{grad}_\Gamma\, n_1, \ldots, ^*\mathrm{grad}_\Gamma\, n_N)$$

and $\epsilon_\Gamma(W)$ the symmetrized of $D_\Gamma W$. The displacement of the shell is denoted by (u, w) in $E(\Gamma) = \{u \text{ in } H^1(\Gamma)^2 \ : \ u = u_\Gamma\} \times H^2(\Gamma)$. It is the element of E_0 which minimizes over E_0 the energy functional

$$\mathcal{E} = h\mathcal{E}_m + \frac{h^3}{12}\mathcal{E}_f + \text{ some linear terms,}$$

where m refers to the membrane, f to the flexion, and

$$\mathcal{E}_m = \frac{1}{2}\int_\Gamma ||\epsilon_\Gamma(u) + w\epsilon_\Gamma(n)||^2 d\Gamma,$$

$$\mathcal{E}_f = \frac{1}{2}\int_\Gamma ||\epsilon_\Gamma(D_\Gamma n \cdot u - \,^*\mathrm{grad}\,_\Gamma w)||^2 d\Gamma.$$

The flexion energy $\mathcal{E}_f$ requires second order tangential derivatives of the normal field which should be smooth enough, say in $C^2(\Gamma)$. Using smooth boundaries and associated distances and compactness we can easily obtain existence results and shape derivatives $u'(\Gamma, V)$, $w'(\Gamma, V)$ for smooth admissible fields V. Now this kind of existence result and associated necessary condition are not very useful for they are assosiated with constraints on which it is impossible to project (which was already true for the uniform cone condition). The reader is refered to [28].

1.4 Existence results for weak shape formulations of free boundary problems

We consider partial differential equations which are well posed (the solution exists and is unique in some function space) in a domain Ω of $\mathbf{R}^n$ when suitable boundary conditions are imposed on the boundary $\partial\Omega$. In general for a second order operator (e.g. elliptic, parabolic or hyperbolic in Ω or in the evolution domain $Q = I \times \Omega$) the boundary value problem (BVP) is well posed with one boundary condition. For instance the homogeneous Dirichlet condition $y = 0$ on $\partial\Omega$, or the homogeneous Neumann condition $\partial_n y = 0$ on $\partial\Omega$.

A free boundary problem is a BVP in which the two conditions (if the operator under consideration is second order) are simultaneously imposed. When the domain Ω is given, this problem has no solution for if the solution y verifies $y = 0$ on $\partial\Omega$ there is no reason for which the normal derivative $\partial_n y$ should be zero too. The only hope to find a solution to this problem is then to consider the boundary $\partial\Omega$ itself as an unknown of the problem, or at least the part of $\partial\Omega$ on which the extra boundary condition is imposed. A free boundary problem is then a boundary value problem in which the unknown is $(y, \partial\Omega)$.

To begin with, we consider the family of free boundary problems in which the additional condition concerning the boundary $\partial\Omega$ takes the very special form $y > 0$ in Ω, $y < 0$ outside of Ω, where the BVP is now solved in a domain D of $\mathbf{R}^N$, the boundary $\partial\Omega$ is an interface in D and the normal derivative $\partial_n y$ is continuous across $\partial\Omega$. We shall say that these free boundary problems are of the *second kind* since the extra condition which determines the interface is a continuity condition (without prescribed value for $\partial_n y$ on $\partial\Omega$).

We call a *weak formulation* a condition solved by a non-smooth solution y which under smoothness assumptions on y gives the (strong) free boundary conditions.

1.4.1 Free interface

In this subsection we discuss a simplified version of a problem studied in Chapter 7 of [16]. D is a bounded smooth domain of $\mathbf{R}^N$ and the unknown domain is a measurable set Ω in D whose boundary $G = cl(\Omega) \cap cl(D\backslash\Omega)$ is considered to be an interface for a BVP defined in the domain D.

1.4.1.1 Free interface with continuity condition

In this section we are interested in an interface Γ which is a level curve of the solution u to the boundary value problem. Consider the following problem for D and a function $f \geq 0$ in $L^2(D)$: to find a measurable set Ω in D and a function u in $H_0^1(D) \cap H^1(D)$ such that

$$-\Delta u(x) = \begin{cases} f(x) \text{ a.e. } x \text{ in } \Omega \\ 0 \text{ a.e. } x \text{ in } D\backslash\Omega \end{cases} \tag{1.1}$$

$$\text{meas}\{x \; : \; u(x) = 1\} = 0 \tag{1.2}$$

$$u(x) > 1 \text{ in } \Omega, \quad u(x) < 1 \text{ in } D\backslash\overline{\Omega}. \tag{1.3}$$

In other words, if $\Omega = \{x \text{ in } D \; : \; u(x) > 1\}$ and X_Ω is the characteristic function of Ω we can write the problem (1.1)-(1.3) equivalently as follows:

$$-\Delta u = X_\Omega f, \quad \Omega = \{x \; : \; u(x) > 1\}, \quad \text{meas}\{x \; : \; u(x) = 1\} = 0. \tag{1.4}$$

We consider the energy functional $W : H_0^1(D) \to \mathbf{R}$ defined by

$$W(\phi) = \int_D \left[\frac{1}{2}|\text{grad } \phi|^2 - f(\phi - 1)^+\right] dx. \tag{1.5}$$

The Hadamard semiderivative $W'(\phi; \gamma)$ exists for each ϕ and γ in $H_0^1(D)$ and is given by

$$W'(\phi; \gamma) = \int_D \text{grad } \phi \cdot \text{grad } \gamma \, dx - \int_D f X_{\phi>1} \gamma \, dx - \int_D f X_{\phi=1} \gamma^+ \, dx.$$

Obviously W is weakly lower semi-continuous and coercive on $H_0^1(D)$ so that it reaches its minimum on $H_0^1(D)$. Let Φ be a local minimum of W. The first order optimality necessary condition can be written as follows:

$$\int_D \text{grad } \phi \cdot \text{grad } \gamma \, dx - \int_D f X_{\phi>1} y \, dx \geq \int_D f X_{\phi=1} \gamma^+ \, dx. \tag{1.6}$$

Choosing $\pm\gamma$ in (1.6) and adding the two equalities we obtain at any local minimum Φ of W

$$\int_D f X_{\phi=1} |\gamma| dx = 0 \quad \text{for any } \gamma \text{ in } H_0^1(D). \tag{1.7}$$

From (1.7) it follows easily that, assuming meas$\{x \, : \, \phi(x) = 1\} = 0$, (1.2) holds. Then from (1.6) we get that u is a solution of the problem (1.1)-(1.3). We get the following results.

Proposition 1.0 *Let f be given in $L^2(D)$ verifying $f > 0$ a.e. in D. Then W reaches its minimum on $H_0^1(D)$. Let u be a local minimum of W. Then meas$\{x \, : \, u(x) = 1\} = 0$ and u is a solution to the problem (1.1)-(1.3).*

Proposition 1.1 *Assume that f is given in $L^p(D), p > N/2$, and verifies $f > 0$ a.e. in D. Then W reaches its minimum on $H_0^1(D)$. Let u be a local minimum of W, then meas$\{x \, : \, u(x) = 1\} = 0$ and u is a solution to the problem (1.1)-(1.3). Moreover since u belongs to $W^{2,p}(D)$ the set $\Omega = \{x \in D \, : \, u(x) > 1\}$ is open in D.*

Remark 1.2 In the special situation where $f \geq 0$ a.e. over D, W can be written as follows:

$$W(\phi) = \min \left\{ \int_D \left(\frac{1}{2} |\text{grad } \phi|^2 - f\mu\phi \right) dx \, : \, \mu \text{ in } M \right\}, \tag{1.8}$$

where

$$M = \{\mu \, : \, 0 \leq \mu(x) \leq 1, \text{ a.e. } x \text{ in } D\}. \tag{1.9}$$

Thus the minimization of W over $H_0^1(D)$ is equivalent to the following problem:

$$\min \left\{ \int_D \left(\frac{1}{2} |\text{grad } \phi|^2 - f\mu\phi \right) dx \, : \, \phi \text{ in } H_0^1(D), \, \mu \text{ in } M \right\}. \tag{1.10}$$

Remark 1.3.1 Assume that f belongs to $W^{s,\infty}(D), 0 < s < \frac{1}{2}$. In the problem (1.10), as $(\phi - 1)$ is an element of $H^1(D)$, the multiplier μ can be chosen in $H^{-s}(D), 0 < s < \frac{1}{2}$, and, as the unit ball of that Hilbert space is strictly convex, it turns out that in fact for each f in $H_0^1(D)$ there exists a unique minimizer μ in the unit ball of $H^{-s}(D)$.

Remark 1.3.2 The minimization problem

$$(P) \quad \min\{W(\phi) : \phi \in H_0^1(D)\}$$

possesses a solution for any f in $H^{-1}(D)$. That problem can be written as a shape variational problem in the following way. For any measurable set Ω in D consider

$$J_1(\Omega) = \min\left\{ \int_{D\backslash\Omega} \frac{1}{2}|\nabla\phi|^2 dx : \phi \in H_0^1(D),\ \phi \geq 0 \text{ a.e. in } D \right\}$$

$$J_2(\Omega) = \min\left\{ \int_{\Omega} \left(\frac{1}{2}|\nabla\psi|^2 - f\psi \right) dx : \psi \in H_0^1(D),\ \psi \geq 0 \text{ a.e. in } D \right\}.$$

The shape formulation of (P) is the following one

$$(P_s) \quad \min\{J_1(\Omega) + J_2(\Omega) : \Omega \subset D,\ \Omega \text{ measurable}\}.$$

If ϕ and ψ are respectively solutions of J_1 and J_2, then $u = \psi - \phi$ is a solution of (P). Conversely, if u is a solution of (P) then $\phi = (u-1)^+$ and $\phi = (u-1)^-$ are solutions of J_1 and J_2.

A priori these two problems are then equivalent but now in the form of P_s we can impose some constraints on the set Ω. For example given a positive number $\alpha, 0 < \alpha <$ meas (D), consider

$$(P_s^\alpha) \quad \min\{J_1(\Omega) + J_2(\Omega) : \Omega \subset D,\ \text{meas } (\Omega) = \alpha\}.$$

That problem cannot be equivalently formulated as a variational problem of the form P (or as a variational inequality similar to (1.6)). After another introductory subsection 1.4.1.2 we shall consider the case (P^α) for a more interesting free boundary problem involving the Bernoulli free boundary condition.

1.4.1.2 Homogeneous Dirichlet condition: The Caffarelli existence of weak solutions

We now turn to the situation of the free boundary problem: to find Ω in D and a function y on Ω such that $y = 0$ on $\partial\Omega$ and $\partial_n y = Q^2$ (Neumann condition), where Q is given over D.

In 1980, H.W. Alt and L.A. Caffarelli [1] introduced the following functional:

$$J(\phi) = \int_D \left[\frac{1}{2}|\text{grad } \phi|^2 - f\phi \right] dx + \int_D Q^2 X_{\phi>0} dx \tag{1.11}$$

to be minimized over

$$K = \{u \in H_0^1(D) : u(x) \geq 0 \text{ for a.e. } x \text{ in } D\}. \tag{1.12}$$

The existence result is based on the following lemma.

Lemma 1.14 *Let $\{u_n\}$ and $\{X_n\}$ be two converging sequences. Assume that $u_n \to u$ in $L^2(D)$ and that X_n is a characteristic function, $X_n(1 - X_n) = 0$, weakly converging in $L^2(D)$ to an element λ. Then we have*

$$(1 - X_n)u_n = 0 \text{ for all } n \text{ implies } \lambda \geq X_{u \neq 0}. \tag{1.13}$$

Proof We have $(1 - X_n)u_n = 0$ and in the limit we get $(1 - \lambda)u = 0$. Then on the set $\{x : u(x) \neq 0\}$, we have $\lambda = 1$. On the other hand as a weak limit of characteristic functions λ lies between 0 and 1.

Proposition 1.15 *Let f and Q be two elements of $L^2(D)$. Then there exists u in K which minimizes the functional J over the positive cone K of $H_0^1(D)$. Moreover, the set $\Omega = \{x \in D : u(x) > 0\}$ and the element $u|_\Omega$ are a weak solution of the free boundary problem*

$$-\Delta u = f \text{ in } \Omega, \quad u = 0, \quad \partial_n u = Q^2 \text{ on } \partial\Omega. \tag{1.14}$$

Proof Let $u_n \in K$ be a minimizing sequence of the functional J over the convex set K. We denote by X_n the characteristic function of the set $\{x \in D : u(x) > 0\}$ which is in fact the same subset. Then $\{x \in D : u(x) \neq 0\}$. It is immediate to verify that the sequence $\{u_n\}$ remains bounded in $H_0^1(D)$. So we denote by $\{u_n\}$ a subsequence which weakly converges in $H_0^1(D)$ to an element u of K. This convergence holds in $L^2(D)$ so that Lemma 1.14 applies and we get $\lambda \geq X_{u \neq 0}$ for any weak limiting element of the sequence X_n (which is bounded in $L^2(D)$).

Let j denote the minimum of J over K. Then $J(u_n)$ converges to j but in the weak limit we get

$$\int_D \left[\frac{1}{2}|\text{grad } u|^2 - fu\right] dx \leq \liminf \int_D \left[\frac{1}{2}|\text{grad } u_n|^2 - fu_n\right] dx$$

and

$$\int_D X_{u \neq 0} Q^2 dx \leq \int_D \lambda Q^2 dx = \lim \int_D X_n Q^2 dx. \tag{1.15}$$

Finally be adding these two inequalities we get $J(u) \leq j$.

Remark 1.16 Obviously in the majorization (1.15) Q^2 should be a positive function.

Remark 1.17 This minimization problem can be written as a shape optimization problem as follows: for any measurable subset Ω of D define the Sobolev space $H_0^1(\Omega)$ in the following way:

$$H_0^1(\Omega) = \{u \in H_0^1(D) : u(x) = 0 \text{ a.e. } x \text{ in } D \backslash \Omega\},$$

and the positive cone

$$H_0^1(\Omega)_+ = \{u \in H_0^1(\Omega) \ : \ u(x) \geq 0 \text{ a.e. } x \text{ in } D\}.$$

Then we consider the shape optimization problem

$$\inf\{E(\Omega) \ : \ \Omega \text{ is a measurable subset in } D\}, \tag{1.16}$$

where the energy functional E is given by

$$E(\Omega) = \min\left\{ \int_\Omega \left(\frac{1}{2}|\text{grad } \phi|^2 - f\phi \right) dx + \int_\Omega Q^2 dx \ : \ \phi \in H_0^1(\Omega)_+ \right\}. \tag{1.17}$$

Endowed with the norm of $H_0^1(D)$, $H_0^1(\Omega)$ is a Hilbert space so that for any measurable subset Ω in D the problem (1.16)-(1.17) does have a unique solution y in the closed convex case $H_0^1(\Omega)_+$ and we have the following equivalence between problems (1.16)-(1.17) and the minimization of J over K.

Proposition 1.18 *Let u be a minimizing element of J over K. Then $\Omega = \{x \in D \ : \ u(x) > 0\}$ is a solution of problem (1.16)-(1.17) while the solution to (1.17) coincides with u on Ω. Conversely, if Ω and y are a solution of (1.16)-(1.17), Ω being a measurable set in D and y in $H_0^1(\Omega)_+$, the element u defined by*

$$u(x) = y(x), \quad x \in \Omega \quad \text{and} \quad u(x) = 0, \quad x \in D\backslash\Omega,$$

lies in K and minimizes J over K.

1.4.1.3 Shape existence of weak solutions, the general case

1.4.1.3.1 Problem without constraint

The problem (1.16)-(1.17) can be relaxed as follows: given any f in $L^2(D)$, G in $L^1(D)$

$$(P_o) \quad \inf\{E(\Omega) \ : \ \Omega \text{ measurable in } D\} \tag{1.18}$$

where the energy functional is defined by

$$E(\Omega) = \min\left\{ \int_\Omega \left[\frac{1}{2}|\text{grad } \phi|^2 - f\phi \right] dx + \int_\Omega G \ dx \ : \ \phi \in H_0^1(\Omega) \right\}. \tag{1.19}$$

The Hilbert space $H_0^1(\Omega)$ is defined in Remark 1.17 for any measurable subset Ω in D. From a classical result by G. Stampacchia [11] we know that for any element u in $H_0^1(\Omega)$ we have grad $u(x) = 0$ a.e. in $D\backslash\Omega$ so that the functional E can be rewritten as follows:

$$E(\Omega) = \min\left\{ E(\Omega, \phi) + \int_\Omega G \ dx \ : \ \phi \in H_0^1(\Omega) \right\}, \tag{1.20}$$

where for any ϕ in $H_0^1(\Omega)$

$$E(\Omega, \phi) = \int_D \left[\frac{1}{2}|\text{grad }\phi|^2 - f\phi\right] dx = \int_\Omega \left[\frac{1}{2}|\text{grad }\phi|^2 - f\phi\right] dx.$$

We have the following existence result for problem (P_o).

Theorem 1.19 *For any f in $L^2(D)$ and $G = Q^2$ in $L^1(D)$, there exists a solution (at least one) to problem (P_o).*

Proof Let Ω_n be a minimizing sequence for problem (P_o), and for each n let u_n be the (unique) solution to the probelm (1.20). If X_n is the characteristic function of the measurable set Ω_n we have u_n in $H_0^1(\Omega_n)$ that implies $(1 - X_n)u_n = 0$. On the other hand the sequence u_n remains bounded in $H_0^1(D)$ (taking $\phi = 0$ in (3.3) we get

$$\int_D \left(\frac{1}{2}|\text{grad }u_n|^2 - fu_n\right) dx \leq 0$$

and the conclusion follows from the equivalence of the $H^1(D)$ and $H_0^1(D)$ norms). We can assume that X_n weakly converges in $L^2(D)$ to an element λ and u_n weakly converges in $H_0^1(D)$ to an element u. From Lemma 1.14 we get $\lambda \geq X_{u \neq 0}$ a.e. in D.

Define $\Omega = \Omega(u) := \{x \in D : u(x) \neq 0\}$. Then u belongs to $H_0^1(\Omega(u))$ and we have $\text{meas}(\Omega) = \text{meas}\{x \in D : \lambda(x) = 1\} \leq \alpha$ (as we have $\alpha = \text{meas}\{x : \lambda(x) = 1\} + \text{meas}\{x : 0 \leq \lambda(x) < 1\}$). In the limit we get

$$\int_\Omega \left[\frac{1}{2}|\text{grad }u|^2 - fu\right] dx = \int_D \left[\frac{1}{2}|\text{grad }u|^2 - fu\right] dx$$
$$\leq \liminf \int_D \left[\frac{1}{2}|\text{grad }u_n|^2 - fu_n\right] dx \quad (1.21)$$

and

$$\int_\Omega G\, dx \leq \int_D \lambda G\, dx = \lim \int_{\Omega_n} G\, dx. \quad (1.22)$$

By adding (1.21) and (1.22) we get that Ω minimizes E and u minimizes $E(\Omega, \cdot)$.

1.4.1.3.2 Constraint on the measure of the domain Ω

The problem (1.16)-(1.17) can also be relaxed as follows: given any f in $L^2(D)$, G in $L^1(D)$ and a real number α, $0 < \alpha < \text{meas}\,(D)$,

$$(P_0^\alpha) \quad \inf\{E(\Omega) : \text{meas}\,(\Omega) = \alpha,\ \Omega \subset D\}. \quad (1.23)$$

We have the following existence result for the problem (P_0^α):

Theorem 1.20 *For any f in $L^2(D)$, $G = 0$ and any real number α, $0 < \alpha < \text{meas}(D)$, there exists a solution (at least one) to problem (P_0^α).*

Proof Let $\{\Omega_n\}$ be a minimizing sequence for the problem (P_o), and for each n let u_n be the (unique) solution to the problem (1.20). If X_n is the characteristic function of the measurable set Ω_n we have u_n in $H_0^1(\Omega_n)$ which implies $(1 - X_n)u_n = 0$. On the other hand the sequence $\{u_n\}$ remains bounded in $H_0^1(D)$ (taking $\phi = 0$ in (1.20) we get

$$\int_D \left(\frac{1}{2}|\text{grad } u_n|^2 - f u_n \right) \, dx \leq 0$$

and the conclusion follows from the equivalence of the $H^1(D)$ and $H_0^1(D)$ norms). We can assume that X_n weakly converges in $L^2(D)$ to an element λ and u_n weakly converges in $H_0^1(D)$ to an element u. In the limit we get $\int_D \lambda(x)dx = \alpha$. From Lemma 1.14 we get $\lambda \geq X_{u\neq 0}$ a.e. in D.

Define $\Omega = \Omega(u) := \{x \in D : u(x) \neq 0\}$. Then u belongs to $H_0^1(\Omega(u))$ and we have $\text{meas}(\Omega) = \text{meas}\{x \in D : \lambda(x) = 1\} \leq \alpha$ (as we have $\alpha = \text{meas}\{x : \lambda(x) = 1\} + \text{meas}\{x : 0 \leq \lambda(x) < 1\}$). In the limit we get

$$\int_\Omega \left(\frac{1}{2}|\text{grad } u|^2 - fu \right) dx \; = \int_D \left(\frac{1}{2}|\text{grad } u|^2 - fu \right] dx$$
$$\leq \liminf \int_D \left(\frac{1}{2}|\text{grad } u_n|^2 - f u_n \right) dx \tag{1.24}$$

and

$$\int_D \lambda G \, dx = \lim \int_{\Omega_n} G \, dx \tag{1.25}$$

so that

$$\int_\Omega \left(\frac{1}{2}|\text{grad } u|^2 - fu \right) dx + \int_D \lambda G \, dx \leq \inf\{E(\Omega) : \text{meas}(\Omega) = \alpha\}$$

as $G \geq 0$ we get $E(\Omega) \leq \inf\{E(\Omega) : \text{meas}(\Omega) = \alpha\}$; but Ω does not satisfy the constraint on the measure.

Note that for any measurable set Ω' such that Ω' is between Ω and D we have

$$\int_{\Omega'} \left(\frac{1}{2}|\text{grad } u|^2 - fu \right) \, dx = \int_\Omega \left(\frac{1}{2}|\text{grad } u|^2 - fu \right) \, dx$$

so that in (1.24) Ω can be enlarged to any such Ω'. The inclusion of Ω in Ω' implies the inclusion of $H_0^1(\Omega)$ in $H_0^1(\Omega')$ so that in (1.20) with G being non-positive over D ($G \leq 0$ a.e. x in D), it readily follows that $E(\Omega') \leq E(\Omega)$. To conclude the proof we just have to select Ω' with $\text{meas}(\Omega') = \alpha$. That measurable set Ω' is admissible and minimizes the cost function in (1.23) and we have $E(\Omega') = E(\Omega) = \inf\{E(\Omega) : \text{meas}(\Omega) = \alpha\}$.

Corollary 1.21 *Assume f in $L^2(D)$ and $f = Q^2$ in $L^1(D)$. Then the following problem does have an optimal solution*

$$(P_o^{\alpha-}) \quad \inf\{E(\Omega) \ : \ \text{meas} \ (\Omega) \le \alpha, \quad \Omega \text{ in } D\}. \tag{1.26}$$

Proof It is similar to the proof of Theorem 1.20. The minimizing sequence is chosen such that meas $(\Omega_n) \le \alpha$, so that in the weak limit we get meas $(\Omega) \le \displaystyle\int_D \lambda(x)dx \le \alpha$.

1.4.1.4 Shape weak existence with bounded perimeter sets. Dirichlet condition

Problems $(P_o), (P_o^\alpha)$ and $(P_o^{\alpha-})$ do have optimal solutions but as they are associated with the homogeneous Dirichlet condition, u in $H_0^1(\Omega)$, the optimal domain Ω is not allowed in general to contain holes; that is to say, roughly speaking, that the topology of Ω is a priori given. In many examples it turns out that the solution u is physically interpreted as a potential so that the homogeneous Dirichlet condition does not turn out to be adequate. The physical condition is that the potential u should be constant on each connected component of the boundary $\partial\Omega$ in D. When Ω is a simply connected domain then $\partial\Omega$ has a single connected component so the constant can be chosen as zero. But in general this constant can be fixed only on one connected component; on the others the constant should be an unknown of the problem. To illustrate the reason for which holes cannot occur in the previous problems consider a simple example. D is the square $]0, 1[^2$, $f = 1$ and $G = 0$. For any smooth domain Ω in D we have

$$E(\Omega) = -\frac{1}{2} \int_\Omega u(x)dx$$

and it can easily be verified that $\inf\{E(\Omega) \ : \ \Omega \text{ measurable in } D\}$ is achieved at $\Omega = D$. In particular if one modifies this optimal domain by substracting a closed subset E such that its Hausdorff measure $|E|_{\mathcal{H}^{n-1}} = 0$ but with a positive capacity cap$(E) > 0$, such as for example when E is a line, then it is possible to construct a sequence $\{\Omega_n\}$ which converges to D but such that the optimal solutions $u_n = u(\Omega_n)$ do not converge to $u(D)$ but to $u(D \backslash E)$, for the homogeneous Dirichlet condition in $H_0^1(D \backslash E)$ implies that u be zero on E. In other words the mapping $X_\Omega \to u(\Omega)$ is not continuous from $L^2(D)$ to $H_0^1(D)$. Nevertheless the infimum in the problem is achieved. However, at least when f is positive, no hole is allowed in the optimal solutions.

These considerations justify the introduction of the following Hilbert space: for any measurable set Ω in D define

$$H_0^1(\Omega) = \{\phi \in H_0^1(D) \ : \ (1 - \chi_\Omega) \, \text{grad} \, \phi(x) = 0 \text{ a.e. in } D\}. \tag{1.27}$$

When Ω is an open subset in D such that $D \backslash cl(D)$ is not simply connected, we know from classical results of theory of distributions that f is constant on each connected component of $D \backslash cl(\Omega)$. This constant is zero on the components whose boundary contains part of ∂D.

The minimization problems $(P_o), (P_o^{\alpha})$ and $(P_o^{\alpha -})$ associated with that Hilbert space fail (in the sense that the previous techniques for existence of optimal Ω fail). The main reason is that Lemma 1.14 is fasle when u_n is replaced by grad u_n weakly converging in $L^2(D)^n$. The trick is to recover the equivalent of Lemma 1.14 by imposing the strong $L^2(D)$ convergence of the sequence $\{u_n\}$. In practice $\{u_n\}$ stands for the sequence of characteristic functions X_{Ω_n} of a minimizing sequence. To obtain the strong $L^2(D)$ convergence of a subsequence we add a constraint on the perimeters. We consider the family of bounded perimeter sets in D

$$BPS(D) = \{\text{subsets } \Omega \text{ of } D \,:\, X_\Omega \text{ belongs to } BV(D)\}.$$

The norm of X_Ω is given by

$$\|X_\Omega\|_{BV(D)} = \text{meas } (\Omega) + \|\text{grad } X_\Omega\|_{M^0(D)}.$$

The norm of grad X_Ω in the Banach space $M^0(D)$ is given by

$$\|\text{grad } X_\Omega\|_{M^0(D)}$$
$$= \sup\{\langle \text{grad } X_\Omega, g \rangle \,:\, g \in C_{\text{comp}}(D; \mathbf{R}^N), \quad \|g(x)\| \le 1, \, x \text{ in } D\}$$

where

$$\langle \text{grad } X_\Omega, g \rangle = -\lim \int_\Omega \text{div } (g_n) dx,$$

g_n being a sequence in $C^\infty_{\text{comp}}(D; \mathbf{R}^N)$ which converges to g in $C_{\text{comp}}(D; \mathbf{R}^N)$. (It can easily be verified that the limit is independent of the choice of such a sequence $\{g_n\}$). Finally, as D is assumed to be a bounded domain in $\mathbf{R}^N$, we get

$$BPS(D) = \{\Omega \,:\, \Omega \subset D, \text{ and }$$

$$\sup\{\int_\Omega \text{div}(g)dx \,:\, g \in C^\infty_{\text{comp}}(D; \mathbf{R}^N), \|g(x)\| \le 1, \, x \text{ in } D\} < \infty\}$$

and the *perimeter* of Ω in D is given by

$$P_D(\Omega) = \sup \left\{ \int_\Omega \text{div}(g)dx \,:\, g \in C^\infty_{\text{comp}}(D; \mathbf{R}^N), \|g(x)\| \le 1, \, x \text{ in } D \right\}.$$

When Ω is a smooth subdomain of D the $n - 1$ dimensional measure of its boundary is given by

$$|\partial\Omega|_{\chi^{n-1}} = P_D(\Omega) + |\partial(\Omega \cap D)|_{\chi^{n-1}}.$$

Lemma 1.22 *Let $\{\Omega_n\}$ be a sequence in $BPS(D)$ such that $P_D(\Omega) \le M$. Then there exists Ω in $BPS(D)$ and a subsequence, still denoted by $\{\Omega_n\}$, such that the characteristic functions converge in $L^1(D) \,:\, X_n \to X$ in $L^1(D)$ as $n \to \infty$, and for any g in $C_{\text{comp}}(D; \mathbf{R}^N)$,*

$$\langle \text{grad } X_{\Omega_n}, g \rangle \to \langle \text{grad} X_\Omega, g \rangle$$

and

$$P_D(\Omega) \le \liminf P_D(\Omega_n).$$

Proof This is a simple adaptation of the classical "compact embedding" of $BV(D)$ in $L^1(D)$, see for example R. Temam [12]. Define the perimeter for all measurable subsets Ω in D follows:

$$P_D(\Omega) = \sup\left\{\int_\Omega \text{div}(g)dx \; : \; g \in C^\infty_{\text{comp}}(D; \mathbf{R}^n), \|g(x)\| \le 1, x \text{ in } D\right\}.$$

It belongs to $\mathbf{R} \cup \{+\infty\}$. We introduce the following problem, for any $\sigma > 0$

$$(P_\sigma^\alpha) \quad \inf\{E_\sigma(\Omega) \; : \; \text{meas}\,(\Omega) = \alpha, \Omega \subset D\} \tag{1.28}$$

where

$$E_\sigma(\Omega) = E(\Omega) + \sigma P_D(\Omega) \tag{1.29}$$

$$E(\Omega) = \min\left\{\int_\Omega \left(\frac{1}{2}|\text{grad }\phi|^2 - f\phi\right) dx + \int_\Omega G\,dx \; : \; \phi \text{ in } H_0^1(\Omega)\right\}. \tag{1.30}$$

Theorem 1.23 *Let f belong to $L^2(D)$, G to $L^1(D)$, $\sigma > 0$, $0 \le \alpha < \text{meas}\,(D)$. Then the problem (P_σ^α) possesses (at least) one optimal solution Ω in $BPS(D)$.*

Proof Let Ω_n be a minimizing sequence for (P_σ^α). Taking $\phi = 0$ in (1.30) we get $P_D(\Omega_n) \le \sigma^{-1}(E(D) + \int_\Omega |G|dx)$. The we consider the subsequence $X_n = X_{\Omega_n}$ described by Lemma 1.22. For each n, let u_n in $H_0^1(\Omega_n)$ be the unique minimizer of (1.30). That sequence remains bounded in $H_0^1(D)$ so that we assume that it is weakly converging to an element u in $H_0^1(D)$. From $(1 - X_n)\,\text{grad }u_n = 0$ a.e. in D we get in the limit $(1 - X_\Omega)\,\text{grad }u = 0$ a.e. in D, so that the limiting element u belongs to $H_0^1(\Omega)$. We have

$$\int_\Omega \left(\frac{1}{2}|\text{grad }u|^2 - fu\right) dx = \int_D \left(\frac{1}{2}|\text{grad }u|^2\right) dx - \int_\Omega fu\,dx$$
$$\le \liminf \int_D \left(\frac{1}{2}|\text{grad }u_n|^2\right) dx - \int_{\Omega_n} fu_n dx \tag{1.31}$$

and

$$\int_\Omega fu\,dx = \lim \int_{\Omega_n} fu_n\,dx, \quad \int_\Omega G\,dx = \lim \int_{\Omega_n} G\,dx \tag{1.32}$$

so that

$$\int_\Omega \left(\frac{1}{2}|\text{grad }u|^2 - fu\right) dx + \int_\Omega G\,dx + \sigma P_D(\Omega) \tag{1.33}$$
$$\le \inf\{E_\sigma(\Omega) \; : \; \text{meas}\,(\Omega) = \alpha\}.$$

An important case is when the "forcing term" f is not a distributed function but a boundary term. When working with smooth domains Ω, it is easy to included a Neumann condition $\partial_n u = g$ on a part Γ_0 of $\partial\Omega$ but now Ω is just a measurable subset in D. We shall relax that Neumann condition by using a weak formulation as follows: we set $f = 0$ and Σ_0 being a fixed given subset of ∂D we consider $\Gamma_0 = \Sigma_0 \cap cl(\Omega)$. Of course Γ_0 can be empty but the minimization with respect to the set Ω will force Γ_0 to have a strictly positive $N-1$ dimensional measure.

Define

$$(\mathcal{P}\Sigma_\sigma^\alpha) \quad \inf\{E\Sigma_\sigma(\Omega) \; : \; \text{meas }(\Omega) = \alpha, \Omega \text{ in } D\}, \tag{1.34}$$

where

$$E\Sigma_\sigma(\Omega) = E\Sigma(\Omega) + \sigma P_D(\Omega), \tag{1.35}$$

$$E\Sigma(\Omega) = \min \left\{ \int_\Omega \frac{1}{2}|\text{grad }\phi|^2 dx \right.$$
$$\left. + \int_{\Sigma_0 \cap cl(\Omega)} g\phi d\mu + \int_\Omega G \, dx \; : \; \phi \in H^1\Sigma_0(\Omega) \right\} \tag{1.36}$$

and

$$H^1\Sigma_0(\Omega) = \{\phi \in H^1(D) : \quad \phi|_{\partial D} = 0 \text{ a.e. in } \partial D \backslash (\Sigma_0 \cap cl(\Omega)),$$
$$(1 - X_\Omega) \text{ grad } \phi = 0 \text{ a.e. in } D\}. \tag{1.37}$$

The subset Σ_0 of ∂D is chosen in such a way that on $H^1\Sigma_0(\Omega)$ the norm of $H_0^1(D)$ is equivalent to the norm of $H^1(D)$.

Theorem 1.24 *Let G and g be given, respectively in $L^1(D)$ and $L^2(\Sigma_0)$. Then there exists a solution (at least one) Ω in $BPS(D)$ to the problem $(\mathcal{P}\Sigma_\sigma^\alpha)$.*

Proof Let $\{\Omega_n\}$ be a minimizing sequence. For each n let u_n be the unique solution to the problem (1.36). The function u_n belongs to $H^1\Sigma_0(\Omega)$ so that

$$\int_\Omega \left(\frac{1}{2}|\text{grad }u_n|^2\right) dx = \int_D \left(\frac{1}{2}|\text{grad }u_n|^2\right) dx$$

and

$$\int_{\Sigma_0 \cap cl(\Omega_n)} g u_n d\mu = \int_{\partial D} g u_n d\mu. \tag{1.38}$$

Taking $\phi = 0$ in (1.36) we get

$$\left(\int_D |\text{grad }u_n|^2 dx\right)^{\frac{1}{2}} \leq c\|g\|_{L^2(\partial D)}.$$

We can then extract a subsequence which converges weakly in $H^1(D)$ to an element u. As in the proof of Theorem 1.23 we can consider that Ω_n converges to Ω, Ω in BPS and u is in $H^1\Sigma_0(\Omega)$.

The main point is to note that

$$
j(\Omega_n) \; := 2\left(E\Sigma(\Omega_n) - \int_{\Omega_n} G\,dx\right) = \int_{\Sigma_0 \cap cl(\Omega_n)} gu_n d\mu
$$

$$
= -\int_D |\mathrm{grad}\,u_n|^2 dx.
$$

$$(1.39)$$

Obvioulsy if the set $w_n := \Sigma_0 \cap cl(\Omega_n)$ is empty then at the minimum $j = \lim j_n$ will be zero. After an additional extraction of a subsequence, we can assume that for all n,

$$
|\Sigma_0 \cap cl(\Omega_n)|_{\mathcal{H}^{n-1}} > 0.
$$

$$(1.40)$$

From the two equalities (1.38) we get in the limit:

$$
\int_\Omega \left(\frac{1}{2}|\mathrm{grad}\,u|^2\right) dx \; = \int_D \left(\frac{1}{2}|\mathrm{grad}\,u|^2\right) dx \le \liminf \int_D \left(\frac{1}{2}|\mathrm{grad}\,u_n|^2\right) dx
$$

$$
= \liminf \int_{\Omega_n} \left(\frac{1}{2}|\mathrm{grad}\,u_n|^2\right) dx
$$

$$(1.41)$$

$$
\int_{\Sigma_0 \cap cl(\Omega)} gu\,d\mu = \int_{\partial D} gu\,d\mu \le \liminf \int_{\partial D} gu_n d\mu = \liminf \int_{\Sigma_0 \cap cl(\Omega_n)} gu_n\,d\mu, \qquad (1.42)
$$

so that in the limit we get

$$
\int_\Omega \left(\frac{1}{2}|\mathrm{grad}\,u|^2\right) dx + \int_{\Sigma_0 \cap cl(\Omega)} gu d\mu + \sigma P_D(\Omega)
$$

$$
\le \inf\{E\Sigma_\sigma(\Omega) \; : \; \mathrm{meas}\,(\Omega) = \alpha,\ \Omega \text{ in } D\}
$$

and as u belongs to $H^1\Sigma_0(\Omega)$ we get that Ω is a solution to $(\mathcal{P}\Sigma_\sigma^\alpha)$ while u is the solution of (1.36).

2 Shape calculus and the wave equation

The objective of this section is to consider the necessary conditions related to the minimization of a distributed cost function for a system governed by the wave equation. In the first section we considered cost functions which are related to the energy of the system. This is true for the *compliance* in structural mechanics. In these situations the necessary condition for optimality does not require an adjoint state equation. Now we consider the general case of a cost functional for which the introduction of an adjoint equation will be necessary to obtain the shape gradient and the necessary condition for optimality. We briefly give here existence results concerning the wave equation, then we shall investigate in the next section the general concept of shape derivative and finally very briefly determine the shape derivative for the wave equation and the gradient of the cost functional under consideration. The original definition of shape derivatives was introduced in [16], [18].

2.1 Introduction

The identification of coefficients in partial differential equations is a well-known problem which has been intensively studied in the last ten years. A robust method is to introduce the minimization of a cost function $J(K)$, K being the parameter or a function to identify, $J(K) = |y(K) - y_m|$, where $y(K)$ is the solution of the PDE associated with K and y_m is the measurement of the state. It turns out that numerically it is much more efficient to solve the equation $\nabla J(K) = 0$ rather than to minimize $J(K)$. In general for the Laplace equation, div $(K\nabla u) = f$ and to obtain existence results one has to introduce some constraints on K. These constraints are of two kinds: smoothness, for example $K \in H^1(D)$, boundedness, for example $0 \leq \alpha \leq K \leq \beta$ for some fixed constants α and β. In this paper we shall consider the transient wave equation in a not necessarily bounded domain D.

Using the $BV(D)$ regularity and boundedness for the K's we shall prove in Proposition 2.10 that the functional J possesses a minimum. To get this result we shall need an estimate to obtain the boundness of $y(K)$ in $W^{1,\infty}(0, T; L^2(D)) \cup L^\infty(0, T; H^1(D)/\mathbf{R})$ and establish the continuity of $K \mapsto y(K)$ from $L^1(D)$ to weak-$*$ topologies.

In fact the identification of K is limited by the fact that only a part of K, $K|_{D_1}$, can be identified (D_1 is a bounded subset of D). For example if K_0 is given then K should be equal to K_0 in $D \backslash D_1$ and the cost function J will be regularized to $J(K) + \sigma |K_0 - K|_{BV(D_1)}$.

Let $D \subset \mathbf{R}^n$ be an unbounded domain with smooth boundary ∂D and let $I = [0, T]$ be the time interval. The space variable is $x \in D$ and $Q = I \times D$ is the evolution cylinder in $\mathbf{R}^{N+1}$. Let $K(x)$ be a given matrix function, $K_{ij}(x) \in L^\infty(D)$, $1 \leq i, j \leq N$. We consider for any $S \in L^1(I; L^2(D))$ the wave equation

$$\frac{\partial^2 y}{\partial t^2} - \text{div } (K\nabla y) = S \text{ in } Q \tag{2.1}$$

with Neumann boundary condition on the lateral boundary

$$K\nabla y \cdot n = g \text{ on } I \times \partial D = \Sigma. \tag{2.2}$$

The initial condition is of the form

$$y(0, x) = \frac{\partial y}{\partial t}(0, x) = 0 \text{ in } D. \tag{2.3}$$

We give an a priori estimate for the problem (2.1)-(2.3). We denote by $H = H(K)$ the wave operator

$$H \cdot \varphi = \frac{\partial^2 \varphi}{\partial t^2} - \text{div} (K \nabla \varphi)$$

for any matrix function $K \in L^2(D)^{N^2}$. Notice that in this example K is now a matrix of functions in D.

2.2 A priori estimate

Consider the vector space

$$K = \{\varphi \in H^1(Q) : H\varphi \in L^2(Q)\}.$$

It turns out that K is a closed subspace of $H^1(Q)$ depending on K. Using the Green formula we have the following expression.

Lemma 2.1 *For all* $(y, \varphi) \in K \times H^1(Q)^2$ *such that* $y(0) = \varphi(T) = 0$ *on* D, *we have*

$$\int_\Sigma (K\nabla y \cdot n)\varphi d\sigma = \int_Q \left[K\nabla y \cdot \nabla \varphi - \frac{\partial y}{\partial t}\frac{\partial \varphi}{\partial t} \right] dQ - \int_Q H y\varphi \, dQ. \qquad (2.4)$$

In view of (2.4) we have the following corollary.

Corollary 2.2 *For any* $y \in K, K\nabla y \cdot n$ *is defined as an element of* $H_{00}^{\frac{1}{2}}(\Sigma)'$.

In view of Corollary 2.2 it makes sense to look, with S and g given in $L^2(D)$ and $H_{00}^{\frac{1}{2}}(\Sigma)'$, for a solution y of the wave problem (2.1)-(2.3) in $K(Q)$.

In fact, for smooth data $S \in L^2(D)$ and $g = 0$, we can now give an priori estimate. For any t, $0 \le t \le T$, consider the energy term

$$E(t) = \frac{1}{2} \int_D K\nabla y \cdot \nabla y \, dx + \frac{1}{2} \int_D \left| \frac{\partial y}{\partial t} \right|^2 dx.$$

For a given y in $C^\infty(\overline{Q})$, its derivative is

$$E'(t) = \int_D Hy\frac{\partial y}{\partial t} dx + \int_{\partial D} K\nabla y \cdot n\frac{\partial y}{\partial t} ds.$$

By the Cauchy-Schwarz inequality

$$E'(t) \le \|Hy\|_{L^2(D)}\|\frac{\partial y}{\partial t}\|_{L^2(\partial D)} + \|\gamma_K \cdot y\|_{L^2(\partial D)}\|\frac{\partial y}{\partial t}\|_{L^2(\partial D)}$$

and when $\gamma_K y = K\nabla y \cdot n = 0$ on ∂D we obtain, in view of the fact that $E(0) = 0$, that for any t, $0 \le t \le T$,

$$E(t) \le \|Hy\|_{L^2(I;L^2(D))} \|\frac{\partial y}{\partial t}\|_{L^\infty(I;L^2(D))}$$

$$\le \|Hy\|_{L^1(I;L^2(D))} \max_{0 \le t \le T} E(t)^{\frac{1}{2}} \tag{2.5}$$

and then the following lemma.

Lemma 2.3 *For any function $y \in C^\infty(Q)$ such that $K\nabla y \cdot n = 0$ on ∂D we have, assuming that $K \gg \alpha I_d$,*

$$\alpha\|y\|_{L^\infty(I;H^1(D)/\mathbf{R})} \le \|Hy\|_{L^1(I;L^2(D))} \tag{2.6}$$

$$\alpha\|\frac{\partial y}{\partial t}\|_{L^\infty(I;L^2(D/\mathbf{R}))} \le \|Hy\|_{L^1(I;L^2(D))}. \tag{2.7}$$

Corollary 2.4 *The solution y of problem (2.1)-(2.3), with $S \in L^1(I;L^2(D))$ and $g = 0$, in the space $L^\infty(I;H^1(D)) \cap W^{1,\infty}(I;L^2(D))$ is unique.*

Remark 2.5 From (2.5) we could have, instead of (2.6), the following inequality:

$$\|K\nabla y\|_{L^\infty(I;L^2(D))} \le \|Hy\|_{L^1(I;L^2(D))}.$$

Now we turn to the study of the continuity property of the mapping $K \to y = y(K)$, where $y(K)$ is the solution of problem (2.1)-(2.3) for a given S in $L^1(I;L^2(D))$. Let $K_n \to K$ in $L^1(D)^N$ with the following conditions

$$\alpha I_d \ll K_n \ll M I_d \tag{2.8}$$

and denote by y_n the solution of problem (2.1)-(2.3) associated to K_n for S given in $L^1(I;L^2(D))$ and $g = 0$. In view of the estimates (2.5), (2.6) and (2.7) y_n and $\partial y_n/\partial t$ remain bounded. Then one can consider a subsequence (still denoted by $\{y_n\}$ for simplicity) such that

$$\forall \varphi \in H^1(D)/\mathbf{R}, \quad \int_D K_n\nabla y_n \cdot \nabla\varphi \, dx \to \int_D K\nabla y \cdot \nabla\varphi \, dx \tag{2.9}$$

weakly-* in $L^\infty(0,T)$,

$$\forall \phi \in L^2(D), \quad \int_D \frac{\partial y_n}{\partial t}\phi \, dx \to \int_D \frac{\partial y}{\partial t}\phi \, dx, \tag{2.10}$$

weakly-* in $L^\infty(0,T)$. Multiplying the problem (2.1)-(2.3) associated to K_n by a function $\varphi \in \mathcal{D}(Q)$ we obtain the weak form of the wave problem as

$$\forall \varphi \in \mathcal{D}(Q), \quad \int_Q K_n\nabla y_n \cdot \nabla\varphi \, dQ - \int_Q \frac{\partial y_n}{\partial t}\frac{\partial\varphi}{\partial t} dQ = \int_Q S\varphi \, dQ. \tag{2.11}$$

In view of (2.9), (2.10), (2.11) in the limit we get as $n \to \infty$:

$$\int_Q K\nabla y \cdot \nabla\varphi dQ - \int_Q \frac{\partial y}{\partial t}\frac{\partial \varphi}{\partial t} dQ = \int_Q S\varphi \, dQ. \tag{2.12}$$

Then $y = \lim_{n \to \infty} y_n$ is the solution of problem (2.1)-(2.3) associated with the matrix K and

$$y \in L^\infty(0, T; H^1(D)/\mathbf{R}) \cap W^{1,\infty}(0, T; L^2(\Omega))$$

with $H(K) \cdot y = S$ and $K\nabla y \cdot n = 0$ on Σ. (Here again the term $K\nabla y \cdot n$ is considered in the weak sense (see Corollary 2.2) to accomodate K's chosen in $L^\infty(D)^{N^2}$.)

Finally we obtain the following proposition.

Proposition 2.6 *Let K_n satisfy the condition (2.8) and $K_n \to K$ in $L^1(D)^{N^2}$. Then $y_n \to y(K)$ weakly-* in $L^\infty(0, T; H^1(D)/\mathbf{R})$ while $\dfrac{\partial y_n}{\partial t} \to \dfrac{\partial y}{\partial t}$ weakly-* in $L^\infty(0, T; L^2(D))$.*

We consider the identification problem associated with the wave equation in the spatial domain D. The general approach is to consider a cost functional $J = y \mapsto J(y)$ which is continuous on $L^2(Q)$. For example, given some measurements y_m of the solution, we have the quadratic cost

$$J(K) = \frac{1}{2}\int_0^T \int_D |y(t, x) - y_m(t, x)|^2 dt \, dx$$

and the minimization of J with respect to the matrix K is the identification problem of K.

In general the measurement y_m is not done for all x in D, but on a possibly smaller subdomain D_0 of D, and the cost function takes the form

$$J(K) = \frac{1}{2}\int_0^T \int_{D_0} |y - y_m|^2(t, x)|^2 dt \, dx. \tag{2.13}$$

2.3 Existence results

To obtain an existence for the minimization of the cost J over the set of matrices K we have to restrict the minimization to a compact family of matrices. The compactness being compatible with the continuity properties of the cost J and of the state $y(K)$ which is described in Proposition 1.6. Using the compact embedding of $H^1(0, T)$ in $L^2(0, T)$, Proposition 1.6 leads to the following proposition.

Proposition 2.7 *Let K_n satisfy the conditions (1.8) and $K_n \to K$ in $L^1(D)^{N^2}$. Then $y_n \to y$ strongly in $L^2(Q)$.*

Proof We have $\dfrac{\partial y_n}{\partial t} \rightharpoonup \dfrac{\partial y}{\partial t}$ weakly-* in $L^\infty(0,T;L^2(D))$. Then it also weakly converges in $L^2(0,T;L^2(D))$. But $y_n(0)$ is given by the initial data and then the result is standard.

Proposition 2.8 *Let J be any cost functional continuously defined on $L^2(Q)$. Let D_1 be a bounded subdivision in D with Lipschitzian boundary and*

$$BM(D_1) = \{matrices\ K \in L^\infty(D)^{N^2}\ : \alpha I \le K \le \beta I, K|_{D\setminus D_1} = K_0,$$
$$\|K\|_{BV(D_1;\mathbf{R}^{N^2})} \le M\}. \tag{2.14}$$

Then there exists an optimal matrix $K^* \in B(M)$ such that

$$J(K^*) \le J(K), \quad \forall K \in BM.$$

Proof The proof follows from the compact embedding of $BV(D_1)$ in $L^1(D)$, see R. Temam [12], and Proposition 2.7. More precisely let $\{K_n\}$ be a sequence lying in BM, then there exists a subsequence, still denoted by $\{K_n\}$, such that $K_n \to K$ in $L^1(D_1)$.

Remark 2.9 The fact that K is the identity in (2.13) could be changed for: $K|_{D\setminus D_1} = K_0$ where K_0 is any given element in $D\setminus D_1$ with $K_0 \in L^2(D\setminus D_1)$, $\alpha I \le K_0 \le \beta I$. This kind of hypothesis means that we are not identifying K in the large but only on a compact region D_1 (D_1 being itself arbitrary).

From the practical viewpoint the cost functional J can be regularized as follows:

$$\sigma > 0, \quad J_\sigma(K) = J(K) + \sigma\|K\|_{BV(D_1;\mathbf{R}^{N^2})}. \tag{2.15}$$

Proposition 2.10 *Let $\sigma > 0$ and J be continuously defined on $L^2(Q)$ and let*

$$K_{ad} = \{matrices\ K \in L^\infty(D;\mathbf{R}^{N^2})\ : \ \alpha I \le K \le \beta I, K = I_d\ in\ D\setminus D_1\}. \tag{2.16}$$

Then there exists an optimal element $K^* \in K_{ad}$, i.e.,

$$J_\sigma(K^*) \le J_\sigma(K), \forall\ K \in K_{ad}. \tag{2.17}$$

Remark 2.11 Note that, K was equal to a given K_0 in the large we should modify J_σ as follows:

$$J_\sigma(K) = J(K) + \sigma\|K - K_0\|_{BV(D_1;\mathbf{R}^{N^2})}. \tag{2.18}$$

In fact the norm in $BV(D_0, \mathbf{R}^{N^2})$ could be replaced in the definition of K_{ad} and J_σ by any norm such that the boundedness of this norm implies the convergence of a subsequence in $L^2(D_0)$. In particular one could choose the $H^\epsilon(D_0)$ norm, $0 < \epsilon < \frac{1}{2}$, which allows for the discontinuities of K as in the case of the BV norm.

In the special case of interfaces the matrix $K(x)$ will be a characteristic function

$$K(x) = \alpha\chi_\Omega(x) + \beta(1 - \chi_\Omega(x))$$

where Ω is a measurable subset of D_0. The BV norm on K is then reduced to the perimeter of Ω relative to D_0:

$$P_{D_0}(\Omega) = \sup\left\{ \int_\Omega \operatorname{div}(g)\, dx \; : \; g \in C_0^1(D_0; \mathbf{R}^N), \sup\|g(x)\| \le 1 \right\}.$$

If the optimum domain is smooth enough, the necessary condition will provide the free boundary conditions that we shall investigate in section 2.3.

2.4 The shape optimization problem

With the matrix K in the form (2.14) the problem (1.20) can be slightly modified to the following form:

$$\min\{J(\Omega) + \sigma P_{D_0}(\Omega) \; : \; \Omega \subset D_0, \text{ meas }(\Omega) = \alpha\}$$

with

$$J(\Omega) = J((\alpha - \beta)\chi_\Omega + \beta)$$

and Proposition 2.10 leads to the existence result.

Proposition 2.12 *Let $\sigma > 0$ and J be defined by (2.16). Then there exists an Ω with bounded perimeter in D_0 which is a solution to the problem (2.15).*

We shall obtain the necessary conditions satisfied by the domain Ω. To do that we have to rewrite the functional J and the problem in the following Hamiltonian form:

$$J(\Omega) = \min_\varphi \sup_\psi \frac{1}{2} \int_0^T \int_{D_0} (\varphi - y_m)^2 dt\, dx + \iint_Q [K\nabla\varphi\nabla\psi - \partial_t\varphi\partial_t\psi]dt\, dx$$
$$- \iint_Q S\varphi\psi\, dt\, dx + \sigma P_{D_0}(\Omega).$$

The arguments in the minimum and maximum are taken in a fixed Hilbert space (in the sense that the Hilbert space does not depend on K). Then the differentiability results of such a min max can be easily applied since the saddle point is a singleton. The derivative of J is then the derivative with respect to t, keeping the state and co-state functions fixed. It is then a simple exercise to obtain the shape gradient and the associated necessary condition for the optimality of the cost J.

3 Shape derivative

3.1 Notation

Let D be an open set in $\mathbf{R}^N$ and $\mathcal{O}$ a family of subsets of D, $\mathcal{O} \subset P(D)$. For example

$$\mathcal{O}_k = \{\Omega \;:\; \partial\Omega = \Gamma \text{ is a } C^k \text{ manifold}, \Omega \text{ located on one side of } \Gamma\}$$

$$\mathcal{O}_\infty = \{\Omega \;:\; \partial\Omega = \Gamma \text{ is a } C^\infty \text{ manifold}, \Omega \text{ located on one side of } \Gamma\}$$

$$\mathcal{O}_{\text{Lip}} = \{\Omega \;:\; \partial\Omega = \Gamma \text{ is Lipschitzian}, \Omega \text{ located on one side of } \Gamma\}$$

$$\mathcal{O}_0 = \{\Omega \;:\; \Omega \text{ is measurable}\}.$$

$$\mathcal{U}_{\text{ad}} = \{\text{admissible velocities}\} \subset F = \text{Fréchet space.}$$

Consider a domain functional and its shape derivative:

$$J : \mathcal{O} \mapsto \mathbf{R}, \quad dJ(\Omega; V), \quad V \in \mathcal{U}_{\text{ad}}.$$

In this section we shall use the following notation:

$$\mathcal{U}_{\text{ad}} = \{V \in F \;:\; V(t, x) \cdot n(x) = 0 \text{ on } \Gamma = \partial D\}$$

or occasionally

$$\mathcal{U}_{\text{ad}} = \{V \in F \;:\; \forall t \in [0, \tau], \; \text{div } V(t) = 0\},$$

where

$$F = C^0([0, \epsilon[; C^k(\overline{D}, \mathbf{R}^N)).$$

In the sequel we assume that k can take the value $+\infty$. Given $\Omega \in \mathcal{O}_k$, let y denote the solution of a BVP in Ω. This defines a map

$$y : \mathcal{O}_k \to H = \bigcup_{\Omega \in \mathcal{O}_k} H^1(\Omega).$$

Moreover,

$$y' : \mathcal{O}_k \times \mathcal{U}_{\text{ad}} \to \bigcup_{\Omega \in \mathcal{O}_k} L^2(\Omega),$$

will denote the map associated with the shape derivative $y'(\Omega; V)$, and

$$y'' \;:\; \mathcal{O}_k \times \mathcal{U}_{\text{ad}} \times \mathcal{U}_{\text{ad}} \to \bigcup_{\Omega \in \mathcal{O}_k} L^2(\Omega),$$

the map associated with the second order shape derivative $y''(\Omega; V, W)$. Finally

$$dJ(\Omega, V) = \langle G(\Omega), V(0) \rangle_{\xi_{\overline{\Omega}}^k(\mathbf{R}^N)' \times \xi^k(\mathbf{R}^n)}$$

will be the Eulerian shape derivative where G is the shape gradient which was studied in J.P. Zolésio [18], [19].

3.2 Introduction

D is an open set in $\mathbf{R}^N$. Let $\mathcal{O}$ be a family of measurable sets in D and for any Ω in $\mathcal{O}$ let $H(\Omega)$ be a Banach space of functions defined on Ω. We define

$$H = \bigcup_{\Omega \in \mathcal{O}} H(\Omega) \tag{3.1}$$

and consider a given mapping

$$\Omega \mapsto y(\Omega) \; : \; \mathcal{O} \to H \tag{3.2}$$

such that

$$y(\Omega) \in H(\Omega). \tag{3.3}$$

From now on we shall assume that $H(\Omega)$ is a closed subspace of $L^2(\Omega)$ so that H can be continuously embedded in $L^2(D)$.

The problem we address in this section is to make sense of the derivative of the mapping y. The difficulty arises from the fact that $\mathcal{O}$ and H are not linear spaces.

In the most regular situation that we shall consider, $\mathcal{O}$ will be a complete metric topological space while H will be a fiber bundle on $\mathcal{O}$. In this situation the condition (1.3) which says that $y(\Omega)$ lies in the fiber $H(\Omega)$ means that y is a section of the fiber bundle H.

As $\mathcal{O}$ and H are topological spaces the continuity of y is easily defined. To handle the first variation of y the topological set $\mathcal{O}$ should be endowed with a C^1 manifold structure. In fact for our purposes it suffices to provide a well-defined sense for the tangential linear space $T_\Omega \mathcal{O}$ and give a meaning to the derivative

$$y'(\Omega, \cdot) = \frac{\partial y}{\partial \Omega}(\Omega, \cdot)$$

as a mapping defined from $T_\Omega \mathcal{O}$ into $L^2(\Omega)$.

To begin with we shall consider the smooth case where $\mathcal{O}$ is the family of domains of class C^∞.

3.3 C^∞ domains and shape derivatives

We assume that D is a bounded domain in $\mathbf{R}^N$ with boundary of class C^∞ and set

$$\mathcal{O}_\infty = \{\Omega : \Omega \text{ open}, \ \Omega \subset D, \text{ with smooth boundary } \Gamma \text{ which is} \tag{3.4}$$
$$\text{a } C^\infty \text{ manifold, } \Omega \text{ is located on one side of } \Gamma\}$$

so that $\mathcal{O}_\infty$ contains D itself. For each Ω in $\mathcal{O}_\infty$,

$$H(\Omega) \text{ is a closed subspace of } H^1(\Omega). \tag{3.5}$$

For each admissible vector field $V \in \mathcal{U}_{ad}$,

$$\mathcal{U}_{ad} = \{V \in C^0([0, \epsilon(V)[; C^\infty(\overline{D}, \mathbf{R}^N)) : V(t, x) \cdot n(x) = 0 \text{ on } \partial D\}, \tag{3.6}$$

where n denotes the unitary normal field on ∂D, we consider the perturbation of each element Ω of $\mathcal{O}_\infty$

$$\Omega_t = T_t(V)(\Omega). \tag{3.7}$$

To bypass the fact that H is not a linear space we consider the largest one, $H(D)$, and to define the element $\frac{\partial}{\partial t} y(\Omega_t)_{t=0}$ which is the candidate for the shape derivative $y'(\Omega; V)$ at Ω in the direction of the admissible field V, we look for a smooth extension $Y(t, \cdot)$ defined for each $t \in [0, \epsilon(V)[$ on the largest space $H(D)$.

The idea is then to consider the derivative $\frac{\partial}{\partial t} Y(0, \cdot)$ as being the shape derivative. To make this notion more precise we need the following basic result.

Proposition 3.1 *Assume that $y(\Omega_t)$ is such that there exists an extension Y verifying the following two conditions:*

$$Y \in C^1([0, \epsilon(V)[; L^2(D)) \cap C^0([0, \epsilon(V)[; H(D)); \tag{3.8}$$

$$Y(t, \cdot)|_{\Omega_t} = y(\Omega_t). \tag{3.9}$$

Then $\frac{\partial}{\partial t} Y(0, \cdot)|_\Omega$, the restriction of $\frac{\partial}{\partial t} Y(0, \cdot)$ to the domain Ω, does not depend on the choice of the smooth extension Y verifying (3.8) and (3.9).

Proof Let Y_1 and Y_2 be two functions verifying (3.8) and (3.9), then let $Y = Y_2 - Y_1$. From (3.9) we have $Y(t, \cdot)|_{\Omega_t} = 0$. It follows that for any $\varphi \in \mathcal{D}(\overline{D})$ and $t \in [0, \epsilon(V)[$ we have

$$\int_{\Omega_t(V)} \varphi(x) Y(t, x) dx = 0, \quad \int_{\Gamma_t(V)} \varphi(x) Y(0, x) V(0, x) \cdot n(x) d\Gamma = 0$$

and taking the derivative with respect to t we get

$$\int_\Omega \varphi(x) \frac{\partial}{\partial t} Y(0, x) dx = 0. \tag{3.10}$$

As (3.10) is valid for each φ we conclude that $\frac{\partial}{\partial t} Y(0, \cdot)$ is almost everywhere equal to zero on Ω. $\qquad \square$

Definition 3.2 A mapping $y : \mathcal{O}_\infty \to H, \Omega \mapsto y(\Omega) \in H(\Omega)$, is *shape differentiable* at Ω in the direction of the admissible field $V, V \in \mathcal{U}_{ad}$, if there exists an extension Y defined on $[0, \epsilon(V)[\times H(D)$ and verifying (3.8) and (3.9). The shape derivative is defined as being the element of $L^2(\Omega)$ characterized by

$$y'(\Omega; V) = \frac{\partial}{\partial t} Y(0, \cdot)|_\Omega. \tag{3.11}$$

We now have several structural results concerning the shape derivative.

Proposition 3.3 *Assume that* $y : \mathcal{O}_\infty \to H$, $\Omega \mapsto y(\Omega) \in H(\Omega)$ *is shape differentiable at each* Ω *in* $\mathcal{O}_\infty$ *(and in each direction* $V \in \mathcal{U}_{\mathrm{ad}}$*). Assume that the mapping* $V \mapsto y'(\Omega; V)$ *is continuous from* $\mathcal{U}_{\mathrm{ad}}$ *into* $L^2(\Omega)$*. Then* y' *just depends on the autonomous field* $V(0)$*:*

$$y'(\Omega; V) = y'(\Omega; V(0)). \tag{3.12}$$

Proof Define the sequence $\{V_n\}$ by

$$V_n(t, x) = \begin{cases} V(t, x), & \text{if } 0 \le t \le \frac{1}{n} \\ V(\frac{1}{n}, x), & \text{if } t \ge \frac{1}{n}. \end{cases}$$

V_n belongs to $\mathcal{U}_{\mathrm{ad}}$ and $V_n \to V(0)$, as $n \to \infty$ in the Banach space $C^0([0, \epsilon(V)[; C^\infty(\overline{D}, \mathbf{R}^N))$. Following the continuity assumption we get $y'(\Omega; V_n) \to y'(\Omega, V(0))$ in $L^2(\Omega)$. But for $0 \le t \le \frac{1}{n}$ we have $\Omega_t(V_n) = \Omega_t(V)$ so that if we call Y_n an extension associated to $\Omega_t(V_n)$ we get $Y_n(t, \cdot) = Y(t, \cdot)$ when $0 \le t \le \frac{1}{n}$ and then $\frac{\partial}{\partial t} Y_n(0, \cdot) = \frac{\partial}{\partial t} Y(0, \cdot)$ so that we have

$$y'(\Omega; V_n) = y'(\Omega; V) \to y'(\Omega; V(0)).$$

$\square$

Then from Proposition 3.3 we know that if the shape derivative is continuous with respect to the direction V then it is sufficient to characterize it with autonomous vector fields $V \in \mathcal{U}_{\mathrm{ad}} \cap C^\infty(\overline{D}; \mathbf{R}^N)$, i.e., independent of t. In fact, as $\Omega \in \mathcal{O}_\infty$ the boundary is smooth and the outward unit normal field n to Ω on Γ is itself smooth,

$$n \in C^\infty(\Gamma; \mathbf{R}^N). \tag{3.13}$$

It follows from that consideration that actually the shape derivative does not depend on $V(0)$ but only on its normal component on Γ.

Proposition 3.4 *Let* $y : \mathcal{O}_\infty \to H$, $\Omega \mapsto y(\Omega) \in H(\Omega)$ *be a shape differentiable mapping such that* $V \mapsto y'(\Omega; V)$ *is linear and continuous,*

$$y'(\Omega, \cdot) \in \mathcal{L}(C^\infty(\overline{D}, \mathbf{R}^N) \cap \mathcal{U}_{\mathrm{ad}}, L^2(\Omega)). \tag{3.14}$$

Then there exists a linear continuous mapping

$$\tilde{y}'(\Omega; \cdot) \in \mathcal{L}(C_D^\infty(\Gamma); L^2(\Omega)), \tag{3.15}$$

where

$$C_D^\infty(\Gamma) = \{v \in C^\infty(\Gamma), \ v|_{\Gamma \cap \partial D} = 0\} \tag{3.16}$$

such that for all $V \in \mathcal{U}_{\mathrm{ad}}$,

$$y'(\Omega; V) = \tilde{y}'(\Omega; V(0) \cdot n), \tag{3.17}$$

where $v = V(0) \cdot n$ is the normal component of the autonomous vector field $V(0)$ on Γ; it is an element of $C_D^\infty(\Gamma)$.

Proof The proof proceeds in several steps.

Lemma 3.5 *Let $V \in C^\infty(\overline{D}, \mathbf{R}^N) \cap \mathcal{U}_{\mathrm{ad}}$ such that*

$$v = V \cdot n = 0 \quad \text{on } \Gamma. \tag{3.18}$$

Then $y'(\Omega; V) = 0$.

Proof The condition (3.18) implies that the transformation $T_t(V)$ is the flow of the field V. Let the boundary Γ be globally invariant: $T_t(\Gamma) = \Gamma$ for each t. For simplicity we give here a short self-contained proof in which V need not be autonomous. So consider a vector field V belonging to $C^0([0, \epsilon[; C^\infty(\overline{D}; \mathbf{R}^N)) \cap \mathcal{U}_{\mathrm{ad}}$ and verifying (3.18).

Let $X_0 \in \Gamma$ and $c : B \to c(B)$ a smooth one to one mapping such that $X \in c(B_0) \subset \Gamma$, where

$$
\begin{aligned}
B &= \{x \in \mathbf{R}^N : \|x\| \le 1\}, \\
B_0 &= \{x = (x', x_N) \in B : x_N = 0\};
\end{aligned}
$$

x is now the local coordinate of Γ in the neighborhood of the point X of Γ. We have $c(B_0) = c(B) \cap \Gamma$ and c, c^{-1} are of class C^∞ defined from B onto $c(B)$.

The normal field $n(X)$ at X to Γ can be written (in the neighborhood of X_0)

$$n(X) = \|{}^*(Dc)^{-1}(x)e_n\|^{-1} \, {}^*(Dc)^{-1}(x)e_n,$$

where $e_n = (0, \ldots, 0, 1) \in B$.

Condition (3.18) can now be written

$$[(Dc)^{-1}(x)V(t, c(x))] \cdot e_n = 0.$$

We set

$$W(t, x) = (Dc)^{-1}(x)V(t, x)$$

which is a field defined on $[0, \epsilon[\times B$. It is immediate to verify that $W(t, x)e_n = 0$ for all x in B_0,

$$W(t, x)e_n = V(t, X) \cdot [{}^*(Dc)^{-1}(x)e_n] = 0.$$

Consider for any $\xi \in B_0$ the flow x of the field W:

$$\frac{\partial z}{\partial t}(t, \xi) = W(t, z(t, \xi)), \ z(0, \xi) = \xi$$

and define y, $t \in [0, \epsilon[$, $X \in \Gamma$,

$$y(t, X) = c(z(t, c^{-1}(X))).$$

It is immediate to verify that y so defined is also a solution of the problem

$$\frac{\partial y}{\partial t}(t, X) = V(t, y(t, X)), \quad y(0, X) = X$$

but by construction (with the help of c) $y(t, X)$ lies on the surface Γ. We get

$$
\begin{aligned}
\frac{\partial y}{\partial t}(t, X) &= \frac{\partial}{\partial t}[c(z(t, c^{-1}(X)))] \\
&= (Dc)(z(t, c^{-1}(X)))\frac{\partial}{\partial t}z(t, c^{-1}(X)) \\
&= (Dc)(z(t, c^{-1}(X)))(Dc)^{-1}(z(t, c^{-1}(X))V(t, y(t, X))).
\end{aligned}
$$

Finally we conclude that under the condition (3.18) the boundary Γ is globally invariant under the transformation $T_t(V)$. Therefore $\Omega_t = \Omega$ and $y(\Omega_t) = y(\Omega)$. Finally, the extension Y can be chosen independent of t, so that

$$\frac{\partial Y}{\partial t} = 0$$

and then $y'(\Omega; V) = 0$. $\qquad\square$

The second step in the proof of Proposition 3.4 is the following.

Lemma 3.6 *Let*

$$N(\Gamma) = \{V \in C^{\infty}(\overline{D}; \mathbf{R}^N) \cap \mathcal{U}_{\mathrm{ad}} : V \cdot n = 0 \text{ on } \Gamma\}. \tag{3.19}$$

Then the mapping

$$\theta : V \mapsto v = V \cdot n$$

is an isomorphism from $[C^{\infty}(D, \mathbf{R}^N) \cap \mathcal{U}_{\mathrm{ad}}]/N(\Gamma)$ *onto* $C_D^{\infty}(\Gamma)$, *the quotient space being endowed with the quotient Banach norm, as* $\overline{D}$ *is compact.*

Proof The mapping is well defined from the quotient space to $C_D^{\infty}(\Gamma)$: if V_1 and V_2 are equal in the quotient, i.e. $V_2 - V_1 \in N(\Gamma)$ then $V_2 \cdot n = V_1 \cdot n = v$ is uniquely determined.

To complete the proof we construct the inverse mapping. For any $v \in C_D^{\infty}(\Gamma)$ we have

$$v = \sum_{i=1}^{m} v_i \text{ where } v_i = v r_i$$

with $r_1, \ldots, r_m$ being a partition of unity,

$$0 \leq r_i \leq 1, \quad \sum_{i=1}^{m} r_i = 1, \quad r_i \in C^{\infty}(\Gamma)$$

with support $r_i \subset c_i(B_0)$. Following the notations used in the proof of the Lemma 3.5, c_i is a local coordinate system for the manifold Γ. Denote

$$\omega_i(x',0) = v_i(c_i(x'));$$

this element ω_i can be extended to B in the following way:

$$\tilde{\omega}_i(x',x_N) = \omega_i(x',0), (x',x_N) \in B.$$

Define

$$Pv = \sum_{i=1}^{m} r_i \tilde{\omega}_i \circ c_i^{-1}.$$

$Pv \in C^\infty(U)$ is an extension to a neighborhood U of the element v defined on Γ.

Let N_0 be an extension to $\overline{D}$ of the field n (normal field on Γ) with $N_0 \in C^\infty(\overline{D}, \mathbf{R}^N)$, support of $N_0 \subset U$, then $v \mapsto (Pv)N_0$ is the required inverse mapping θ^{-1}.

Proof of Proposition 3.4 Let π denote the canonical linear surjective mapping defined from $E = C^\infty(D; \mathbf{R}^N) \cap \mathcal{U}_{ad}$ onto $E/N(\Gamma)$. From Lemma 3.5 there exists an element $\hat{y}' \in \mathcal{L}(E/N(\Gamma); L^2(\Omega))$ such that

$$y'(\Omega;V) = \hat{y}'(\Omega; \pi(V)).$$

From Lemma 3.6 we can set, for each $v \in C_D^\infty(\Gamma)$,

$$\tilde{y}'(\Omega; v) = \hat{y}'(\Omega; \theta^{-1}(v)).$$

And finally we get

$$y'(\Omega;V) = \tilde{y}'(\Omega; \theta\pi(V))$$

and for simplicity we write $V \cdot n = \theta\pi(V)$.

We can summarize the situation by the following commutative diagram.

$$
\begin{array}{ccc}
E & \xrightarrow{y'(\Omega;\cdot)} & L^2(\Omega) \\[4pt]
\pi \downarrow & & \| \\[4pt]
E/N(\Gamma) & \xrightarrow{y'(\Omega;\cdot)} & L^2(\Omega) \\[4pt]
\theta \downarrow & & \| \\[4pt]
C_D^\infty(\Gamma) & \xrightarrow{\tilde{y}'(\Omega;\cdot)} & L^2(\Omega).
\end{array}
$$

Finally we obtain the structural result for the shape derivative y'.

Theorem 3.7 *Let O_∞ and H be defined as in section 3.1, and let y : $O_\infty \mapsto H$, $y(\Omega) \in H(\Omega)$ be a shape differentiable mapping such that the mapping $V \mapsto y'(\Omega;V)$ is linear and continuous on $\mathcal{U}_{ad}$. Then there exists a kernel $y'(\Omega) \in \mathcal{D}'(\Gamma) \times L^2(\Omega)$ such that*

$$\forall V \in \mathcal{U}_{ad}, \ y'(\Omega;V)(X) = \langle y'(\Omega)(X,\cdot), \ V(0) \cdot n(\cdot) \rangle_{\mathcal{D}'(\Gamma) \times \mathcal{D}(\Gamma)}. \qquad (3.20)$$

Proof We consider the bilinear continuous mapping

$$(v,\varphi) \mapsto \int_\Omega \tilde{y}'(\Omega,v)(x)\varphi(x) \, dx$$

defined from $C_D^\infty(\Gamma) \times L^2(\Omega)$ to $\mathbf{R}$. By L. Schwartz's kernel theorem, there exists a kernel $y'(\Omega) \in (C_D^\infty(\Gamma) \times L^2(\Omega))'$ such that

$$\int_\Omega \tilde{y}'(\Omega;v)(x)\varphi(x)dx = \langle y'(\Omega), \varphi \otimes v \rangle_{\mathcal{D}'(\Gamma \times \Omega) \times \mathcal{D}(\Gamma \times \Omega)}. \qquad (3.21)$$

3.4 Second order shape derivative

We consider now a mapping y : $O_\infty \to H, y(\Omega) \in H(\Omega)$ which is shape differentiable and enjoys the following regularity property:

$$\forall V \in \mathcal{U}_{ad}, \ \forall \Omega \in O_\infty, \ y'(\Omega;V) \in H^1(\Omega).$$

Definition 3.8 Let V and W be two autonomous vector fields, $V, W \in C^\infty(\overline{D}, \mathbf{R}^N) \cap \mathcal{U}_{ad}$. The *second shape derivative*, if it exists, is given by

$$y''(\Omega;V,W) = z_V'(\Omega;W),$$

where $z_V : O_\infty \to H = \bigcup_{\Omega \in O_\infty} H^1(\Omega)$ is the mapping defined by $z_V(\Omega) = y'(\Omega;V)$.

When we assume the linearity with respect to W we get the same structural result for y'' that we got for y'; more precisely we have the following.

Proposition 3.9 *V and W being two autonomous vector fields, assume that $W \mapsto y''(\Omega;V,W)$ is linear and continuous in $C^\infty(\overline{D}, \mathbf{R}^N)$. Then there exists an element $\tilde{y}''(\Omega;V) \in \mathcal{L}(C^\infty(\Gamma); L^2(\Omega))$ such that $y''(\Omega;V,W) = \tilde{y}''(\Omega;V)W \cdot n$.*

Proof As $y''(\Omega;V,W) = z_V'(\Omega;W)$ we apply the Proposition 3.4. $\qquad \square$

But from (3.19) we get

$$\int_\Omega \tilde{y}''(\Omega;\omega)(x)\varphi(x)dx = \langle Z_V'(\Omega), \varphi \otimes w \rangle_{\mathcal{D};(\Gamma \times \Omega) \times \mathcal{D}(\Gamma \times \Omega)}$$

where $Z_V'(\Omega)$ is a kernel which depends on the autonomous vector field V.

3.5 The shape derivative of the wave equation

We consider the problem (2.22), (2.23), (2.24) and we recall here the variational technique developed in J.P. Zolésio [18] to obtain the shape derivative $y'(\Omega; V)$. In a first step we show the existence of the material derivative

$$\dot{y}(\Omega; V) = \left[\frac{d}{dt} y(\Omega_t) \circ T_t(v) \right]_{t=0}. \tag{3.22}$$

Then from the existence of $\dot{y}$ we obtain the existence of the shape derivative $y' = \dot{y} - \nabla y \cdot V$ which can be characterized by taking the derivative of the weak formulation of y.

In the next section we consider the situation where the field v is the solution of the stationary Stokes system (linearized Navier-Stokes). When the domain Ω is perturbed by the solution v (which is the physical speed vector field of the fluid), i.e. $\Omega_t = T_t(v)(\Omega)$, then we obtain the following shape derivative $v'(\Omega; v)$ of v at Ω in the direction v that we call shape acceleration that was characterized in Zolésio [24] and numerically in T. Tiihonen and J.P. Zolésio [14].

3.6 Structure of the shape Hessian

Let V and W be two autonomous fields, and

$$dJ(\Omega; V) = \int_\Gamma gV \cdot n \, d\Gamma$$

g being the density gradient, so that the gradient is

$$G(\Omega) = \gamma_\Gamma^* \cdot (gn).$$

In the smooth case, g is the trace of some function $Q(\Omega)$ defined over Ω. Then we get

$$dJ(\Omega; V) = \int_\Omega \text{div}(QV) \, dx$$

and

$$d^2 J(\Omega; V, W) = \int_\Omega \text{div}[Q'V] \, dx + \int_\Gamma \text{div}(QV) \, W \cdot n \, d\Gamma,$$

where $Q' = Q'(\Omega; W)$ just depends on $W \cdot n$, as we know from (3.17). We easily get

$$
\begin{aligned}
d^2 J(\Omega; V, W) &= \int_\Gamma \left[Q'V \cdot n + \frac{\partial Q}{\partial n} V \cdot n \, W \cdot n \right] d\Gamma \\
&+ \int_\Gamma [\nabla_\Gamma Q \cdot V_\Gamma \, W \cdot n + Q \, \text{div} \, V \, W \cdot n] d\Gamma,
\end{aligned}
$$

where the first term depends only on the normal components of the two fields.

Let us now consider a critical domain (resp. a critical domain with the constraint on the given measure). Then the necessary condition leads to

$$g = Q|_\Gamma = 0 \text{ on } \Gamma \text{ (resp. constant)}$$

so that in both cases we get

$$Q \operatorname{div} V = 0 \text{ on } \Gamma.$$

Also, $\nabla_\Gamma Q$ is zero on the boundary so that the second term vanishes on any critical domain, and then the second derivative depends on the normal components.

4 Stokes' equation and general existence results with stronger constraints

In section 4.1 we shall derive the (necessary) free boundary condition associated with the Stokes fluid, assuming that the boundary and the velocity of the fluid are very smooth. The objective is simply to show that equation (4.58) is not the expected one.

In section 4.2 we consider the more restrictive situation, concerning the existence of a minimum for a general distributed cost functional, when constraints are imposed on the domains. These results also apply to variational inequalities.

4.1 Stokes' equation

For a linear incompressible stationary fluid we consider the energy $E(\Omega)$, Ω being the volume occupied by the fluid. The objective of the first section is to consider the minimization of $E(\Omega)$ and to underline the difficulties encountered to obtain existence results.

The objective of the second section is to study the necessary conditions which are satisfied when the minimum energy $E(\Omega)$ of a fluid is minimized or maximized with respect to the boundary Γ.

Ω will denote the volume of $\mathbf{R}^N$ occupied by the fluid, for $N = 2$ or 3, Γ is a part of its boundary that we choose as a control variable. The total boundary will be in general denoted by $\partial\Omega = \overline{\Gamma} \cup \overline{\Sigma}$, where Σ is an open part of $\partial\Omega$ and can be itself decomposed in several parts corresponding to several kinds of boundary conditions that will be imposed on the fluid of $\overline{\Sigma}$. Without loss of generality we shall suppose that Ω is contained in a fixed smooth domain D. We shall slightly reduce the generality by assuming that D is bounded which is necessary in some situations to make use of some compact imbeddings of Banach spaces of functions defined over D. The velocity of the fluid particle at a point

x of Ω, at time t, will be denoted by $u(t)(x) = u(t,x)$. When the flow is stationary, Γ does not depend on t and Γ is a streamline of the field u.

4.1.1 Minimization of the energy

Assuming that Γ is smooth enough so that the normal field $n(x)$ (going out of Ω) is defined, say Γ of class C^1, we have for all t and $x \in \Gamma$

$$n(x) \cdot u(t,x) = 0. \tag{4.1}$$

Consider the simple rheology as being the linear imcompressible Stokes model

$$\text{div } u(t,x) = 0 \text{ in } \Omega \tag{4.2}$$

$$-\Delta u + \nabla p = f \text{ in } \Omega \tag{4.3}$$

$$(\epsilon(u) \cdot n)_\Gamma = g \text{ on } \Gamma \tag{4.4}$$

where f is given in $L^2(D)$, g in $H^1(D)$, and H is the *mean curvature* of the boundary Γ. Γ is assumed smooth enough in this strong formulation, $\epsilon(u)$ is the *displacement tensor*, $2\epsilon(u) = Du + {}^*Du$, σ is the *surface tension*, a non-negative given number

$$\sigma \geq 0,$$

and $(\epsilon(u) \cdot n)_\Gamma$ stands here for the tangential component of the vector $\epsilon(u) \cdot n$ on Γ, using tensor notation. The weak formulation of problem (4.1)-(4.4) is as follows: we introduce the Hilbert space

$$H^1(\Omega) = \{u \in L^2(\Omega)^N : \text{div } \varphi = 0, \ \epsilon(u) \in L^2(\Omega)^{N^2}\}. \tag{4.5}$$

Using the classical Green's theorem we have that for all $\varphi, \phi \in C^\infty(\overline{\Omega}; \mathbf{R}^N)$

$$\int_\Omega \epsilon(\varphi)..\epsilon(\phi)dx + \int_\Omega \langle \text{div } (\epsilon(\varphi)), \phi \rangle \ d\Gamma = \int_{\partial\Omega} \langle \epsilon(\varphi) \cdot n, \phi \rangle \ d\Gamma; \tag{4.6}$$

it is obvious that $\epsilon(\varphi) \cdot n$ is defined, as an element of $H^{-\frac{1}{2}}(\partial\Omega)^N = (H^{\frac{1}{2}}(\partial\Omega)^N)'$, as soon as $\varphi \in H^1(\Omega)$ and $\text{div}(\epsilon(\varphi)) \in L^2(\Omega)^N$. (This fact follows from the extension of (4.6) when φ and ϕ are in $H^1(\Omega)^N$, $\text{div}(\epsilon(\varphi)) \in L^2(\Omega)^N$. Let

$$H_0^1(\Omega) = \{\varphi \in H^1(\Omega) : \varphi \cdot n = 0 \text{ on } \Gamma, \varphi = 0 \text{ on } \Sigma\}.$$

The weak solution u of (4.1)-(4.4) is the unique minimizer in H_0^1 of

$$E(\Omega) = \min_{u \in H_0^1(\Omega)} \frac{1}{2} \int_\Omega \epsilon(u)..\epsilon(u) \ dx - \int_\Omega f \cdot u \ dx + \int_\Gamma g \cdot u \ d\sigma. \tag{4.7}$$

From the well-known Korn inequality on $H^1(\Omega)$,

$$a(u,u) = \left(\int_\Omega \epsilon(u)..\epsilon(u) dx \right)^{\frac{1}{2}}$$

is an equivalent norm to the $H_0^1(\Omega)$ norm.

The objective of this section is to minimize $E(\Omega)$ with respect to Ω. We shall first define $E(\Omega)$ when Ω is not a smooth domain in D but simply a measurable subset of D. The first objective is then to define $H_0^1(\Omega)$ when Ω is a measurable subset of D.

In weak form the condition (4.1) on Γ can be written (assuming Σ empty)

$$U \in H_0^1(D), \forall \, \phi \in C^1(\overline{D}), \quad \int_\Omega U \cdot \nabla\phi \, dx = 0. \tag{4.8}$$

More precisely we have the following lemma.

Lemma 4.1 *Let Ω be a domain in D with Lipschitzian boundary Γ and let U be an element in $H_0^1(D)$ and $u = u|_\Omega$ (the restriction to Ω). Then u belongs to $H_0^1(\Omega)$ if and only if U satisfies (4.8).*

Proof If u belongs to $H_0^1(\Omega)$, then $u \cdot \nabla\phi = \operatorname{div}(\phi u)$ and by Stokes' theorem

$$\int_\Omega \operatorname{div}(\phi U) dx = \int_\Gamma U \cdot n\phi \, dx = 0.$$

In view of the minimization of the energy we introduce the perimeter of Ω relative to D,

$$P_D(\Omega) = \|\nabla\chi_\Omega\|_{M^0(D)},$$

where $M^0(D)$ is the Banach vector space of bounded measures on D (cf. for instance R. Temam [12], J.P. Zolésio [23,21]). The problem

$$\inf_{E \cap \Omega \subset D} E(\Omega) + \sigma P_D(\Omega) \tag{4.9}$$

with $g = 0$ is then equivalent to the following one:

$$\inf_{E \cap \Omega \subset D} \inf_{u \in H_0^1(D)} \sup_{\phi \in H_0^1(D)} e(\Omega, u, \phi) + \sigma P_D(\Omega), \tag{4.10}$$

where

$$e(\Omega, u, \phi) = \int_\Omega \left(\frac{1}{2}|\epsilon(u)|^2 - f_E \cdot u + u \cdot \nabla\phi \right) dx. \tag{4.11}$$

Let

$$Z(\Omega, u) = \sup_{\phi \in H_0^1(D)} e(\Omega, u, \phi). \tag{4.12}$$

Lemma 4.2 *If*

$$\chi_{\Omega_n} \to \chi_\Omega \text{ in } L^1(D) \text{ and } u_n \to u \text{ in } H_0^1(D), \tag{4.13}$$

then $Z(\Omega, u) \leq \liminf_n Z(\Omega_n, u_n)$.

Proof For each ϕ in $H_0^1(D)$ we have $e(\Omega_n, u_n, \phi) \to e(\Omega, u, \phi)$ as $n \to \infty$; then taking the supremum over ϕ we get (4.13).

The main difficulty arises from the fact that minimizing sequences (Ω_n, u_n) for the problem (4.10) are such that $\{\chi_{\Omega_n}\}$ is bounded in $BV(D, \mathbf{R}^N)$ but only $\{\chi_{\Omega_n}\epsilon(u_n)\}$ is bounded in $L^2(D, \mathbf{R}^{N^2})$ (and $\epsilon(u_n)$ is not a priori bounded in $L^2(D, \mathbf{R}^{N^2})$). One simple possibility to overcome this difficulty is to consider for some given $\epsilon > 0$ the following perturbed problem

$$(P_\epsilon) \quad \inf_{\substack{E \subset \Omega \subset D \\ u \in H_0^1(D)}} \sup_{\phi \in H_0^1(D)} e_\epsilon(\Omega, u, \phi) + \sigma P_D(\Omega) \tag{4.14}$$

with

$$e_\epsilon(\Omega, u, \phi) = e(\Omega, u, \phi) + \epsilon e(D, u, \phi). \tag{4.15}$$

The boundary condition (4.4) will be changed to a transmission condition involving the parameter ϵ on Γ. The fluid is now in the whole domain D but Γ is an interface. For this situation we get the following existence result.

Proposition 4.3 *There exists* $(\Omega, u(\Omega)) \in \{\Omega \ : \ E \subset \Omega \subset D, P_D(\Omega) < \infty\} \times H_0^1(D)$ *such that* $u = U(\Omega)|_\Omega \in H_0^1(\Omega)$ *and* $\forall \phi \in H_0^1(D), \forall \Omega', E \subset \Omega' \subset D, \forall u' \in H_0^1(D)$

$$e_\epsilon(\Omega, u, \phi) \leq \sup\{e_\epsilon \Omega', u', \varphi) : \varphi \in H_0^1(D)\}.$$

4.1.2 Necessary condition satisfied by stationary domains

In this section we are dealing with the Eulerian derivative of the energy functional with respect to the domain Ω. The energy functional $E(\Omega)$ is associated with a stationary linear Stokes flow; Ω is the volume of $\mathbf{R}^N$, for $N = 2$ or 3, occupied by the flow and at each point x of $\Omega, u(x) = (u_1(x), \dots, u_n(x))$ is the velocity vector of the particle located at x. The fluid is assumed to be incompressible: that is, we assume that

$$\text{div}(u(x)) = 0, \ x \in \Omega. \tag{4.16}$$

The boundary $\partial\Omega$ of Ω is usually decomposed into several parts on which appropriate and possibly different types of boundary conditions are imposed. In general $\partial\Omega = \overline{\Sigma} \cup \overline{\Gamma}$, where u is specified on the fixed part Σ, while Γ is the *free part* of the boundary. Without any loss of generality for the results in this section, we set $\Sigma = \emptyset$ (empty set) and we introduce a forcing term $f \in L^2(D)^N$, two fixed smooth bounded domains E and D in $\mathbf{R}^N$, such that $E \subset D$ with $0 < \text{meas}\,(E) < \alpha < \text{meas}\,(D)$. Consider the *set of admissible domains* Ω in $\mathbf{R}^N$ such that

$$E \subset \Omega \subset D \tag{4.17}$$

$$\text{meas } (\Omega) = \alpha. \tag{4.18}$$

Then the strong formulation of the Stokes equation is

$$-\Delta u + \nabla p = \chi_E f \text{ in } \Omega, \tag{4.19}$$

where p is the pressure and χ_E the characteristic function of the set E.

$$u \cdot n = 0 \text{ on } \Gamma \tag{4.20}$$
$$(\epsilon(u) \cdot n)_\Gamma = 0 \text{ on } \Gamma \tag{4.21}$$

and the free boundary condition would be that $\epsilon(u) \cdot n \cdot n$ is specified explicitly or is the solution of some tangential problem on $\Gamma \backslash (\partial D \cup \partial E)$–see (4.54)-(4.56).

For each smooth domain Ω we introduce the following Hilbert space

$$H(\Omega) = \{u \in H^1(\Omega)^N : \text{div } u = 0, \ u \cdot n = 0 \text{ on } \partial\Omega\} \tag{4.22}$$

and the functional

$$J_\Omega(u) = \frac{1}{2} \int_\Omega \epsilon(u)..\epsilon(u) \, dx - \int_E f \cdot u \, dx. \tag{4.23}$$

From the Korn inequality we know that J_Ω is coercive on $H(\Omega)$ and that there exists a unique minimizing element u of J_Ω over $H(\Omega)$. It turns out from (4.6) that u is a weak solution to problem (4.1), (4.4), (4.5).

The energy functional is then

$$E(\Omega) = J_\Omega(u) = \min\{J_\Omega(v) : v \in H(\Omega)\}. \tag{4.24}$$

Lemma 4.4 *Let V be an admissible field, i.e., $V \in C^0([0,\epsilon[; C^1(D, D))$ with $V \cdot n = 0$ on ∂D. Let $\Omega_t = T_t(V)(\Omega)$ be the perturbed domain. Then the following transformation*

$$u \mapsto u_t = [(\det(DT_t))^{-1} DT_t \cdot u] \circ T_t^{-1} \tag{4.25}$$

is a linear isomorphism from $H(\Omega)$ onto $H(\Omega_t)$.

Proof It can easily be verified that

$$(\text{div } u) \circ T_t = (\det(DT_t))^{-1} \text{div}((\det DT_t)(DT_t)^{-1} u \circ T_t) \tag{4.26}$$

and also that on the boundary Γ_t of Ω_t we have

$$n_t(\|M(DT_t) \cdot n\|^{-1} M(DT_t) \cdot n) \circ T_t^{-1} \text{ on } \Gamma_t, \tag{4.27}$$

where n_t is the exterior unitary normal field to Γ_t and $M(A)$ is the cofactor matrix of A,

$$M(A) = \det A \, {}^*A^{-1}.$$

Form (4.25), (4.27) we get, with $J_t = \det(DT)_t$, $(u_t \cdot n_t) \circ T_t$ proportional to $\langle M(DT_t) \cdot n, DT_t \cdot u \rangle$, that is, proportional to $\langle \, {}^*DT_t^{-1} \cdot n, DT_t \cdot u \rangle = \langle n, DT_t^{-1} \cdot DT_t u \rangle$. Then $(u_t \cdot n_t) \circ T_t = a(t) u \cdot n$, and it can be verified that for t small enough, $0 \le t \le \epsilon$, the function $a(t)$ is strictly positive on $\overline{D}$.

From (4.25) and (4.26) we get

$$(\text{div } u_t) \circ T_t = J_t^{-1} \text{ div } [J_t (DT_t)^{-1} J_t^{-1} DT_t \cdot u],$$

that is,

$$(\text{div } u_t) \circ T_t = J_t^{-1} \text{ div } u. \tag{4.28}$$

Again, as J_t^{-1} is strictly positive (and continuous) on $\overline{D}$, this concludes the proof of the lemma.

Corollary 4.5 *Given u_t defined by (4.25) we have*

$$E(\Omega_t) = \min_{u \in H(\Omega)} J_{\Omega_t}(u_t). \tag{4.29}$$

In the sequel we shall write

$$E(\Omega_t) = \min_{u \in H(\Omega)} F(t, u) \tag{4.30}$$

with

$$
\begin{aligned}
F(t, u) &= J_{\Omega_t}(u_t) \\
&= \frac{1}{2} \int_{\Omega_t} |\epsilon[(J_t^{-1} DT_t \cdot u) \circ T_t^{-1}]|^2 \, dx - \int_E \langle f, (J_t^{-1} DT_t \cdot u) \circ T_t^{-1} \rangle \, dx.
\end{aligned} \tag{4.31}
$$

We now assume that Ω and f are smooth enough so that $u_t, t \ge 0$, is smooth, say $u_t \in H^2(\Omega_t)^N \cap H(\Omega_t)$. In that case the result on the differentiation with respect to the parameter t of a minimum applies (see J.P. Zolésio [19], M.C. Delfour and J.P. Zolésio [2], and we get from (4.30)

$$dE(\Omega; V) = \frac{d}{dt} E(\Omega_t)|_{t=0} = \frac{\partial}{\partial t} F(0, u). \tag{4.32}$$

Then we concentrate our study on the computation of the term $\frac{\partial}{\partial t} F(0, u)$, u being assumed smooth. From (4.31) we know that this derivative will involve two integrals, a boundary integral and a volume integral over Ω. Only the first one requires smoothness of u. In that form the necessary condition associated with any stationary domain will immediately yield a free boundary condition in the strong form

$$
\begin{aligned}
\frac{\partial}{\partial t} F(0, u) = & \int_\Omega \epsilon(u)..\epsilon(\partial u) \, dx - \int_\Omega f \partial u \, dx \\
& + \int_\Gamma \frac{1}{2} |\epsilon(u)|^2 V(0) \cdot n \, d\Gamma - \int_\Gamma f \cdot u \, V(0) \cdot n \, d\Gamma,
\end{aligned} \tag{4.33}
$$

where

$$\partial u = \frac{\partial}{\partial t} u_t \Big|_{t=0} \in H^1(\Omega)^N \tag{4.34}$$

is the element of $H^1(\Omega)^N$ such that $u_t = u + t\,\partial u + O(t)$, where $O(t) \in H^1(\Omega)^N$, $\|O(t)\|/t \to 0$, as $t \to 0$.

We have the following characterization:

Lemma 4.6

$$\partial u = \text{div}\, V u + DV \cdot u - Du \cdot V. \tag{4.35}$$

Proof Identity (4.34) follows from the classical results, see J.P. Zolésio [16, 18], J. Sokolowski and J.P. Zolésio [8], and assuming smoothness we have

$$\frac{d}{dt} f(t, T_t(x)) \Big|_{t=0} = \frac{\partial}{\partial t} f(0, x) + \nabla_x f(0, x) \cdot V(0, x)$$

and

$$\left(\frac{d}{dt} J_t\right)_{t=0} = \text{div}\, V(0); \quad \left(\frac{d}{dt} DT_t\right)_{t=0} = DV(0).$$

Remark 4.7 As the derivatives with respect to t and to the shape variable x commute we get

$$\frac{\partial}{\partial t} \epsilon(u_t) \Big|_{t=0} = \epsilon(\partial u). \tag{4.36}$$

Corollary 4.8 *Let N_0 be a smooth unitary extension of the normal field n; if $\Gamma = \partial\Omega$ is of class $C^k, k \geq 1$, there exists such an $N_0 \in C^{k-1}(\mathcal{U})$, for some smooth neighborhood $\mathcal{U}$ of Γ. Consider speed vector fields V in $\mathcal{U}$ of the form*

$$V(t, x) = v(t, x) N_0(x). \tag{4.37}$$

Then the normal component of ∂u on Γ is given by

$$\partial u \cdot n = \text{div}_\Gamma(vu), \tag{4.38}$$

where div_Γ is the tangential divergence (see J.P. Zolésio [18], J. Sokolowski and J.P. Zolésio [8]) defined for a vector field e on Γ as

$$\text{div}_\Gamma e = \text{div}\, E - \langle \epsilon(E) \cdot n, n \rangle \tag{4.39}$$

for some extension E of e. (It turns out that the right hand side of (4.39) is independent of the choice of the extension E defined on an arbitrary neighborhood $\mathcal{U}$ of Γ).

Proof Let V be any admissible vector field. From (4.35), (4.20) we obtain

$$\partial u \cdot n = \langle DV(0) \cdot u, n \rangle - \langle Du \cdot V, n \rangle.$$

But when V takes the form (3.21) we get, see J.P. Zolésio [18],

$$DV(0) = v(0)DN_0 + n \cdot {}^*\nabla v(0) \text{ on } \Gamma,$$

and then

$$\partial u \cdot n = \nabla_\Gamma v(0) \cdot u + v(0) \operatorname{div}_\Gamma u = \operatorname{div}_\Gamma(v(0)u).$$

Remark 4.9 As u is a smooth divergence-free field, from (4.39) and (4.16) we get (writing (4.16) on the boundary)

$$\operatorname{div}_\Gamma(u) = -\langle \epsilon(u) \cdot n, n \rangle. \tag{4.40}$$

We adopt the fluid mechanics notation $\epsilon(u) \cdot n \cdot n$ for this term.

Finally we obtain a first expression for the Eulerian semiderivative of the energy $E(\Omega)$.

Proposition 4.10 *Given the speed field V satisfying (4.37), then*

$$
\begin{aligned}
dE(\Omega; V) = &\int_\Omega (\langle -\Delta u - f, \partial u \rangle_{\mathbf{R}^N}) \, dx \\
&+ \int_\Gamma \left(\langle \epsilon(u) \cdot n, \partial u \rangle_{\mathbf{R}^N} + \frac{1}{2}\epsilon(u)..\epsilon(u)v(0) - f \cdot uv(0) \right) d\Gamma.
\end{aligned} \tag{4.41}
$$

Using now the problem (4.16), (4.19), (4.20), (4.21), where u is assumed to be the strong solution, we get the following expression from (4.41).

Corollary 4.11

$$
\begin{aligned}
dE(\Omega; V) = &-\int_\Omega \partial u \nabla p \, dx - \int_\Gamma \operatorname{div}_\Gamma(u) \operatorname{div}_\Gamma(v(0)u) \, d\Gamma \\
&+ \int_\Gamma \frac{1}{2}\epsilon(u)..\epsilon(u)v(0)d\Gamma - \int_\Gamma f \cdot uv(0) \, d\Gamma.
\end{aligned} \tag{4.42}
$$

Proof From (4.21) we get $\epsilon(u) \cdot n = \epsilon(u) \cdot n \cdot n$, n on Γ, then from (4.39), (4.40):

$$\epsilon(u) \cdot n = -\operatorname{div}_\Gamma(u)n \text{ on } \Gamma; \tag{4.43}$$

using (4.19), (4.43) and (4.38) in (4.39) we get (4.42).

Lemma 4.12 *Assuming that the pressure is sufficiently smooth, we have*

$$\int_\Omega \partial u \cdot \nabla p \, dx = \int_\Gamma p \operatorname{div}_\Gamma(v(0)u) \, d\Gamma. \tag{4.44}$$

Proof From Lemma 4.1 we know that the element u_t belongs to $H(\Omega_t)$, and that $\operatorname{div}(u_t) = 0$ in Ω_t. Taking the derivative with respect to t, which commutes with the divergence operator we get, at $t = 0$, $\text{div}(\partial u) = 0$ in Ω. Then $\partial u \nabla p = \text{div}(p \partial u)$ and from Stokes' formula we have

$$\int_\Omega \partial u \cdot \nabla p \, dx = \int_\Gamma p \, \partial u \cdot n \, d\Gamma. \tag{4.45}$$

Corollary 4.13 *Assuming the speed field V such that (4.36) is verified and assuming Γ, u and p smooth enough we get the Eulerian semiderivative as expected by the structure theorem (see J.P. Zolésio [16]), i.e. as a boundary expression*

$$dE(\Omega; V) = -\int_\Gamma (p + \text{div}_\Gamma(u)) \, \text{div}_\Gamma(v(0)u) d\Gamma + \frac{1}{2} \int_\Gamma (\epsilon(u)..\epsilon(u) - f \cdot u)v(0)d\Gamma.$$

Lemma 4.14 *Assuming Γ, p and u smooth enough*

$$\int_\Gamma u \, \text{div}_\Gamma(u) \nabla_\Gamma v(0) d\Gamma = -\int_\Gamma \text{div}_\Gamma[(\text{div}_\Gamma(u)) \cdot u] v(0) d\Gamma. \tag{4.46}$$

Proof Identity (4.46) follows directly from the integration by parts formula on Γ, see J.P. Zolésio [18], J. Sokolowski and J.P. Zolésio [8]. It must be noticed that as $u \cdot n = 0$ the mean curvature H does not occur in (4.36); also the speed v must be zero at the boundary of Γ. In short, we assume here that the boundary Γ has no boundary (it is a compact manifold).

Proposition 4.15 *Assuming Γ, u smooth enough, we have*

$$dE(\Omega; V) = \int_\Gamma \left(\nabla_\Gamma(\text{div}_\Gamma u) \cdot u + \frac{1}{2}\epsilon(u)..\epsilon(u) - f \cdot u \right) v(0) d\Gamma$$
$$+ \int_\Gamma \nabla_\Gamma p \cdot u \, v(0) d\Gamma. \tag{4.47}$$

Proof We have

$$\text{div}_\Gamma(au) = a \, \text{div}_\Gamma u + (\nabla_\Gamma a) \cdot u \tag{4.48}$$

then from (4.46) we get:

$$dE(\Omega; V) = \int_\Gamma \text{div}_\Gamma(u \, \text{div}_\Gamma u) v(0) d\Gamma - \int_\Gamma (\text{div}_\Gamma u)^2 v(0) d\Gamma$$
$$+ \int_\Gamma \left(\frac{1}{2}\epsilon(u)..\epsilon(u) - f \cdot u \right) v(0) d\Gamma$$

and with (4.48),

$$dE(\Omega, V) = \int_\Gamma ((\text{div}_\Gamma u)^2 + u \cdot \nabla_\Gamma(\text{div}_\Gamma u) - (\text{div}_\Gamma u^2)v(0)d\Gamma + \dots$$

and we obtain (4.47).

Remark 4.16 The expression $u \cdot \nabla_\Gamma(\text{div } u) = u \cdot \nabla(\text{div } u)$ that we got in (4.47) is in fact a material derivative, not with respect to the speed field V that we introduced to generate virtual deformations of the domain Ω, but with respect to the physical velocity vector $u(x)$ of the fluid. Let $X_0 \in \Gamma$ be given and set

$$\frac{d}{dt}x(t, X_0) = u(x(t, X_0)), \quad x(0, X_0) = X_0; \tag{4.49}$$

we introduce the flow transformation $T_t(u)$ and

$$\gamma(t) = (\text{div}_\Gamma\ u)(x(t, X_0)) = (\text{div}_\Gamma\ u) \circ T_t(u)(X_0), \tag{4.50}$$

where u is the solution of the fluid problem. Assuming that Γ and u are sufficiently smooth we obtain (with the notation $x = x(t, X_0)$ to simplify):

$$\frac{d}{dt}\gamma(t) = \nabla_\Gamma(\text{div}_\Gamma u)(x(t)) \cdot u(x(t)), \tag{4.51}$$

that is,

$$\frac{d}{dt}(\text{div}_\Gamma\ u) \circ T_t(u) = (\nabla_\Gamma \text{ div } u) \circ T_t(u) \cdot u \circ T_t(u). \tag{4.52}$$

Theorem 4.17 (Necessary optimality condition). *Let Ω be a smooth stationary domain for the energy functional, that is,*

$$\text{for any admissible field } V, \ dE(\Omega; V) = 0. \tag{4.53}$$

Then $\epsilon(u) \cdot n \cdot n = -\text{div}_\Gamma u$ on Γ solves (the tangential differential equation): $\forall x_0 \in \Gamma$,

$$[\epsilon(u) \cdot n \cdot n](x(t)) = (\epsilon(u) \cdot n \cdot n)(x_0) + ct$$

$$- \int_0^t \left[\frac{1}{2}|\epsilon(u)|^2(x(s)) - (f \cdot \nabla p)(x(s)) \cdot u(x(s))\right] ds, \tag{4.54}$$

where $|\epsilon(u)|^2 = \epsilon(u)..\epsilon(u) = \sum_{i,j}(\epsilon_{ij})^2$ and c is a constant.

Proof We assume that Ω is a stationary domain for the functional E in the set of all admissible domains with prescribed measure. Then the field $V(t, \cdot)$ has to be chosen with free divergence so that the measure $\text{meas}(\Omega_t) = \text{meas}(\Omega)$ is given, from Stokes' formula, that is to say that the normal component $v(0)$ of the field $V(0)$ on Γ satisfies

$$\int_\Gamma v(0, x)d\Gamma(x) = 0. \tag{4.55}$$

From (4.55), (4.47) and (4.52) we get

$$\nabla_\Gamma(\operatorname{div}_\Gamma u) \cdot u + \frac{1}{2}|\epsilon(u)|^2 - f \cdot u = C \text{ on } \Gamma$$

$$\frac{d}{dt}[(\operatorname{div}_\Gamma u) \circ T_t(u)] = C - \frac{1}{2}|\epsilon(u)|^2 \circ T_t(u) - (f \cdot u) \circ T_t(u),$$

(4.56)

where C is a constant coming from (4.55), as we have the orthogonality to a closed subspace.

Remark 4.18 If the force f is zero on Γ, then from (4.54) we get that $t \to (\epsilon(u)\cdot n\cdot n)(x(t))$ is monotonicaly decreasing when the volume of the fluid is not prescribed.

Remark 4.19 If the volume of the domain Ω is not prescribed then the constant $c = 0$. We consider now the situations involving the surface tension σ. We introduce the functional

$$E_\sigma(\Omega) = E(\Omega) + \sigma P_D(\Omega)$$

(4.57)

where $P_D(\Omega)$ is the perimeter of Ω relative to D.

Theorem 4.20 *Let Ω be a smooth stationary domain for the functional E_σ. Then, assuming that $u = u(\Omega)$ is smooth enough, the term $\epsilon(u) \cdot n \cdot n$ solves the following problem:* $n_0 \in \Gamma$, $x(t) = x_0 + \int_0^t u(x(s))ds$,

$$[\epsilon(u) \cdot u \cdot n](x(t)) = (\epsilon(u) \cdot n \cdot n)(x_0) + ct$$

$$- \int_0^t \left[\sigma H(x(s)) + \frac{1}{2}|\epsilon(u)|^2(x(s)) + (\nabla p - f)(x(s))\right] ds,$$

(4.58)

where $H(x)$ is the mean curvature of the surface Γ at the point x.

Proof When Ω is a smooth domain of class C^2 we have $P_D(\Omega) = \int_\Gamma d\Gamma$, where $\Gamma = \partial\Omega \backslash \partial D$, and then the Eulerian derivative of $P_D(\Omega)$, with an admissible field V such that $V = 0$ on $\partial D \cap \partial\Omega$, is given by

$$dP_D(\Omega; V) = \int_\Gamma HV(0) \cdot n \, d\Gamma,$$

(4.59)

where H is the mean curvature of Γ and n is the outward unitary normal field on Γ to Ω. Then (4.58) follows from (4.54) and (4.59).

4.2 Existence results obtained by penalization methods

We consider now the shape optimization problems in which the admissible family of domains in D will be subject to stronger constraints. Up to now we have considered only the constraints on the volume and the perimeter of the admissible field. When we impose stronger constraints, the situation is a much easier one.

The guideline for this section is as follows. Let y_ϵ be the solution of a transmission problem in D (which approaches the solution of $y(\Omega), \epsilon \downarrow 0$). Then the continuity of the mapping $\Omega \to y_\epsilon(\Omega)$ is obtained from $L^2(D)$ in $H^1(D)$ (as an example).

When $\mathcal{O}_{ad} = \{\Omega \text{ admissible}\}$ possesses the uniform extension property, then $\Omega \to y(\Omega)$ is continuous from $\mathcal{O}_{ad} \cap L^2(D)_{l\alpha}$, and with any compactness result of $\mathcal{O}_{ad}$ in $L^2(D)$ one would obtain an existence result associated to any smooth cost functional J. As we shall see, this technique covers the situation of classical variational inequalities.

4.2.1 Continuity of the solution of a transmission problem

4.2.1.1 A technical result

Lemma 4.21 *Let D be a measurable set in $\mathbf{R}^N$ with finite measure. Let $\{f_n\}$ be a sequence of functions, $f_n \in L^\infty(D)$, such that there exists a positive constant M verifying*

$$\forall n, \quad \|f_n\|_{L^\infty(D)} \leq M$$

and such that

$$f_n \to f \text{ in } L^2(D).$$

Let $\{g_n\}$ be a sequence in $L^2(D)$ such that

$$g_n \rightharpoonup g \text{ weakly in } L^2(D).$$

Then we have the following convergence

$$\forall h \in L^2(D), \quad \int_D f_n g_n h \, dx \to \int_D f g h \, dx.$$

Proof Note that the sequence $\{f_n h\}$ is in $L^2(D)$. Effectively we have

$$(f_n h)^2 \leq M h^2 \in L^1(D). \tag{4.60}$$

There exists a subsequence, still denoted by $\{f_n\}$, which converges almost everywhere to f in D. Then we get

$$(f_n h)(x) \to (f h)(x) \text{ a.e. } x \text{ in } D. \tag{4.61}$$

It follows from (4.60), (4.61) and the classical Lebesgue theorem that we have

$$f_n h \to f h \text{ in } L^2(D). \tag{4.62}$$

The sequence $f_n h$ remains in a bounded ball of $L^2(D)$. Then in fact all the sequence $\{f_n h\}$ converges to fh in $L^2(D)$. (If not, by weak compactness of that ball we could obtain subsequences of $\{f_n h\}$ weakly converging to a different limit, which would be contradictory to (4.62).)

4.2.1.2 Continuity of the solution of the transmission problem

Let Ω_n be a sequence of measurable sets which converges to Ω in the sense that the characteristic functions converge in $L^1(D)$:

$$\chi_{\Omega_n} \to \chi_\Omega \text{ in } L^1(D). \tag{4.63}$$

Let f in $L^2(D)$ be given and consider y_n in $H_0^1(D)$ the solution to the following variational problem (given k_1 and k_2, say $0 < k_1 < k_2$)

$$\forall \varphi \subset H_0^1(D), \int_D (k_1 + (k_2 - k_1)\chi_{\Omega_n})\nabla y_n \cdot \nabla \varphi \, dx = \int_D f\varphi \, dx. \tag{4.64}$$

When Ω is a smooth open set in D, y_n is a weak solution to the following boundary value problem:

$$-\Delta y_n = f \text{ in } \Omega_n \cup \Omega_n^c \tag{4.65}$$

$$y_n = 0 \text{ on } \partial D \tag{4.66}$$

$$k_1 \frac{\partial y_n}{\partial \nu} = k_2 \frac{\partial}{\partial \nu} y_n \text{ on } \partial \Omega_n \tag{4.67}$$

where $\Omega_n^c = D \setminus \overline{\Omega}_n$, and ν is the outward unitary normal field on $\Gamma_n = \partial \Omega_n$ to Ω_n.

We introduce $y = y(\Omega)$, the solution to the boundary value problem associated with Ω:

$$y \in H_0^1(D), \ \forall \varphi \in H_0^1(D), \int_D (k_1 + (k_2 - k_1)\chi_\Omega)\nabla y \nabla \varphi \, dx = \int_D f\varphi \, dx. \tag{4.68}$$

Proposition 4.22 *The following weak convergence holds in $H_0^1(D)$:*

$$y_n \rightharpoonup y \text{ weakly in } H_0^1(D). \tag{4.69}$$

Proof The first step is to verify that the sequence y_n remains bounded in $H_0^1(D)$. For that purpose let $\varphi = y_n$ in (4.63). As

$$k_1 + (k_2 - k_1)\chi_{\Omega_n} \geq k_1 > 0, \tag{4.70}$$

we get

$$\|y_n\|_{H_0^1(D)}^2 \leq k_1^{-1}\|f\|_{L^2(D)} \|y_n\|_{L^2(D)}.$$

Using the Poincaré inequality

$$\|y_n\|_{L^2(D)}^2 \leq \lambda_1^{-1}\|y_n\|_{H_0^1(D)}^2 \tag{4.71}$$

(where λ_1 is the first eigenvalue of the Laplace operator $-\Delta$ in $H_0^1(D)$) we obtain

$$\|y_n\|_{H_0^1(D)} \leq (k_1\sqrt{\lambda_1})^{-1}\|f\|_{L^2(D)}$$

so that the sequence y_n is bounded in $H_0^1(D)$.

The second step consists of considering a weakly converging subsequence which, for simplicity, we still denote by y_n:

$$y_n \rightharpoonup \text{ weakly in } H_0^1(\Omega).$$

For any element φ given in $H_0^1(D)$ we shall verify that the left hand side of (4.64) converges to the left hand side of (4.68). This follows from Lemma 4.21 applied as follows: $\chi_n = k_1 + (k_2 - k_1)\chi_{\Omega_n}$ converges in $L^2(D)$ to $\chi = k_1 + (k_2 - k_1)\chi_\Omega$, while it remains bounded in $L^\infty(D): ||\chi_n||_{L^\infty(D)} \leq k_2$, $h = \partial_{x_i}\varphi$ is an element of $L^2(D)$ while $\partial_{x_i}y_n$ weakly converges in $L^2(D)$ to $\partial_{x_i}y$. Then Lemma 4.21 applies and (χ_Ω, y) verifies the problem (4.68) so that the limiting element y is the solution of the boundary value problem associated with the limiting set Ω. □

We now consider the continuity of the mapping $\chi_\Omega \to y$ from $L^1(D)$ in $H_0^1(D)$ (i.e. for the norm of $H_0^1(D)$). For each n, we have

$$
\begin{aligned}
k_1||y_n - y||_{H_0^1(D)}^2 &\leq \int_D [k_1 + (k_2 - k_1)\chi_{\Omega_n}]|\nabla(y_n - y)|^2 dx \\
&= \int_d (k_2 - k_1)(\chi_{\Omega_n} - \chi_\Omega)\nabla y \cdot \nabla(y_n - y)dx \qquad (4.72) \\
&\leq ||y_n - y||_{H_0^1(D)} \int_D [(k_2 - k_1)(\chi_{\Omega_n} - \chi_\Omega)]^2 |\nabla y|^2 dx.
\end{aligned}
$$

In the last integral the integrand converges to zero almost everywhere and is dominated by $k_2^2|\nabla y|^2$ which belongs to $L^1(D)$. Then by the classical Lebesgue convergence theorem we get that $||y_n - y||$ goes to zero.

Proposition 4.23 *The following strong convergence holds in $H_0^1(D): y_n \to y$ strongly in $H^1(D)$.*

Remark 4.24 The continuity modulus of the mapping $\chi_\Omega \mapsto y$ defined from $L^1(D)$ to $H_0^1(D)$ is given through (4.72) by the following estimate

$$||y_n - y||_{H_0^1(D)} \leq k_1^{-1}k_2 \left(\int_D (\chi_{\Omega_n} - \chi_\Omega)|\nabla y|^2 dx\right)^{\frac{1}{2}}. \qquad (4.73)$$

In the situation where Ω and y are smooth enough so that y belongs to $W^{1,\infty}(D)$, then from (4.73) we would obtain

$$||y_n - y|| \leq c \text{ meas } (\Omega_n \Delta \Omega), \qquad (4.74)$$

where $\Omega_n \Delta \Omega = (\Omega_n \backslash \Omega) \cup (\Omega \backslash \Omega_n)$ is the symmetric difference, and $d(\Omega_n, \Omega) = $ meas $(\Omega_n \Delta \Omega) = ||\chi_{\Omega_n} - \chi_\Omega||_{L^1(D)}$ is a distance. In that situation the mapping $\Omega \mapsto y$ could be Lipschitzian. But in general we have no regularity results on Ω or y, so that such a property is not available. Of course that property is satisfied when we restrict our study to families of smooth domains.

4.2.2 Stokes flow

We consider now the following vectorial situation:

$$H = \{u \in H^1(D)^N \;:\; u \cdot n = 0 \text{ on } \partial D, \text{div } u = 0 \text{ a.e. in } D\} \qquad (4.75)$$

and the bilinear form $a(\Omega, \cdot, \cdot)$ on $H \times H$ defined as follows:

$$a(\Omega, u, v) = \int_D [k_1 + (k_2 - k_1)\chi_\Omega]\epsilon(u)..\epsilon(v)\, dx \qquad (4.76)$$

where $2\epsilon(u) = Du + {}^*Du$ is the symmetrized part of the Jacobian matrix Du, while the contraction is defined by $\epsilon(u)..\epsilon(v) = \epsilon_{ij}(u)\epsilon_{ij}(v)$. For given f in $L^2(D)^N$ and $\Omega \subset D, \Omega$ a measurable subset, we consider the element $u = u(\Omega)$ of H defined by the following variational problem:

$$\forall v \in H, \quad a(\Omega, u, v) = \int_D f \cdot v\, dx. \qquad (4.77)$$

Using Stokes' theorem we can verify that u is a weak solution to the following boundary value problem:

$$- \text{div}([k_1 + (k_2 - k_1)\chi_\Omega]\epsilon(u)) = f + \nabla p, \quad \text{in } \Omega \qquad (4.78)$$

with boundary conditions

$$u \cdot n = 0 \text{ on } \partial D \qquad (4.79)$$

$$\text{div } u = 0 \text{ in } D, \qquad (4.80)$$

where $p \in \mathcal{D}'(D)$ is the pressure distribution and u is the volocity of the fluid occupying the domain D. When the interface $\Gamma = \partial\Omega$ is smooth, the condition (4.78) can be equivalently written as follows:

$$k_2(\epsilon(u_{|\Omega}) \cdot n - \langle\epsilon(u_{|\Omega}) \cdot n, n\rangle n) = k_1(\epsilon(u_{|\Omega^c}) \cdot n - \langle\epsilon(u_{|\Omega^c}) \cdot n, n\rangle n) \text{ on } \Gamma \qquad (4.81)$$

$$\langle(k_2\epsilon(u_{|\Omega}) - k_1\epsilon(u_{|\Omega^c}))n, n\rangle + p_{|\Omega} - p_{|\Omega^c} = \text{const on } \Gamma, \qquad (4.82)$$

where n is the exterior unitary normal field on Γ.

In other words, the two conditions (4.81) and (4.82) characterize the jump across the interface of the tangential component of $k\epsilon(u) \cdot n$ and of $k\langle\epsilon(u) \cdot n, n\rangle + p$. The first jump is zero along Γ while the second one is constant along the interface.

In the two open domains Ω and $\Omega^c = D\backslash\overline{\Omega}$ we get u solution of the Stokes problem

$$k\Delta u = f + \nabla p \text{ in } \Omega \cup \Omega^c, \qquad (4.83)$$

where $k = k_1$ in Ω^c and $k = k^2$ in Ω. The continuity of the mapping $\chi_\Omega \mapsto u$ from $L^1(D)$ to H is obtained exactly by the same technique as the one used in the proof of Proposition 4.22.

Proposition 4.25 *H being considered as a closed subspace of $H^1(D)^N$, we know by the Korn inequality that $[\int_D \epsilon(u)..\epsilon(u)dx]^{\frac{1}{2}}$ is an equivalent norm on H. Then the mapping $\chi_\Omega \mapsto u(\Omega)$ is continuous from $L^1(D)$ to H.*

4.2.3 Continuity of unilateral problems

We consider the obstacle problem in $H_0^1(D)$. Let

$$K = \{\varphi \in H_0^1(D) \ : \ \varphi \geq g \text{ a.e. in } D\}, \tag{4.84}$$

where g is given in $L^2(D)$. Let f be another element given in $L^2(D)$. Then consider the element $y = y(\Omega)$ of K obtained as the projection of f onto K for the scalar product

$$a(\Omega, y, z) = \int_D [k_1 + (k_2 - k_1)\chi_\Omega]\nabla y \cdot \nabla z \, dx. \tag{4.85}$$

In other words, y is the unique element in K verifying the following variational inequality

$$\forall z \in K, \quad a(\Omega, y, z - y) \geq \int_D f(z - y)dx. \tag{4.86}$$

We now consider a sequence Ω_n of measurable sets in D and Ω such that the characteristic functions χ_{Ω_n} converge in $L^1(D)$ to χ_Ω. For each n, $y_n = y(\Omega_n)$ is the element of K such that

$$\forall z_n \in K, \quad a(\Omega_n, y_n, z_n - y_n) \geq \int_D f(z_n - y_n)dx. \tag{4.87}$$

With $z = y_n$ in (4.86) and $z_n = y$ in (4.87) and adding the two inequalities we get

$$\int_D [k_1 + (k_2 - k_1)\chi_{\Omega_n}]\nabla(y - y_n)\nabla(y_n - y)dx$$

$$+ \int_D (k_2 - k_1)(\chi_\Omega - \chi_{\Omega_n})\nabla y \nabla(y_n - y)dx \geq 0,$$

that is,

$$k_1\|y_n - y\|^2 \leq \int_D [k_1 + (k_2 - k_1)\chi_{\Omega_n}]|\nabla(y_n - y)|^2 dx$$

$$\leq -(k_2 - k_1)\int_D (\chi_\Omega - \chi_{\Omega_n})\nabla y \cdot \nabla(y_n - y)dx$$

$$\leq k_2\|y_n - y\| \left(\int_D |\chi - \chi_n|^2|\nabla y|^2 dx\right)^{1/2}$$

so that we arrive at the following result.

Theorem 4.26 *The sequence $\{y(\Omega_n)\}$ converges to $y(\Omega)$ in $H_0^1(D)$ as $n \to \infty$. More precisely we have*

$$\|y_n - y\|_{H_0^1(D)} \leq \frac{k_2}{k_1} \left(\int_{\Omega_n \Delta \Omega} |\nabla y|^2 dx \right)^{1/2}.$$

5 Optimization of eigenvalues

We briefly investigate the minimization (with respect to the domain) of the first eigenvalue of $-\Delta$ in $H_0^1(\Omega)$, that Sobolev space being defined, when Ω is just a measurable set, as in section 1.2. In section 5.2 we consider the free boundary value problem associated with that eigenvalue: given the function g to find y such that

$$-\Delta y = \begin{cases} \lambda(y - g) \text{ on } \Omega = \{y > g\} \\ 0 \text{ on } D \backslash \overline{\Omega} \end{cases} \tag{5.0}$$
$$y = 0 \text{ on } \partial D.$$

The free boundary $\Gamma = \{y = g\}$ has zero measure and the problem (5.0) has a variational formulation similar to the one of Auchmuty which is in fact a particular and easier situation for the free boundary eigenvalue problem related to the H. Grad equation [32] developed in J.P. Zolésio [17], [15], [18], [16].

5.1 Optimization of the first eigenvalue

Let D be a bounded domain with a Lipschitzian boundary. We consider the first eigenvalue $\lambda(\Omega) = \lambda$ of the Laplace operator $-\Delta$ in $H_0^1(\Omega)$ associated with an arbitrary measurable subset Ω of D given by

$$-\frac{1}{2\lambda} = \min \left\{ \frac{1}{2} \int_D |\nabla \varphi|^2 - \left(\int_D \varphi^2 dx \right)^{\frac{1}{2}} : \varphi \in H_0^1(\Omega) \right\}, \tag{5.1}$$

where $H_0^1(\Omega)$ is defined as in section 1.4.1:

$$H_0^1(\Omega) = \{\varphi \text{ in } H_0^1(D) : \varphi = 0 \text{ q.e. in } D \backslash \Omega\}. \tag{5.2}$$

We consider the following minimization problem: given α, $0 < \alpha < \text{meas}(D)$,

(P_α) to find $\Omega \subset D$, $\text{meas}(\Omega) \leq \alpha$ such that
$\lambda(\Omega) \leq \lambda(O)$, for any $O \subset D$, $\text{meas}(O) = \alpha$.

We obtain the following existence result.

Proposition 5.1 *There exists a domain Ω which is the solution of problem (P_α).*

Proof We consider a minimizing sequence $\{\Omega_n\}$ to problem P_α. For each n let y_n in $H_0^1(\Omega_n)$ be a solution to the problem (5.1). Such a minimizing y_n exists since the functional (5.1) is coercive and weakly lower semi-continuous on $H_0^1(\Omega)$ while being non-convex. As in section 1.4.1 we notice that the integrals on D are in fact integrals over the set Ω_n for, as y_n belongs to $H_0^1(\Omega_n)$, y_n and ∇y_n are zero almost everywhere in $D\backslash\Omega_n$. By a coercivity argument the sequence y_n remains bounded in $H_0^1(D)$. Then one can extract a subsequence, still denoted by y_n, and y in $H_0^1(D)$ such that the following weak convergence in $H_0^1(D)$ holds: $y_n \rightharpoonup y$. We can also assume that the sequence has been chosen such that the convergence holds quasi everywhere in D. The sequence of characteristic functions $\chi_n = \chi_{\Omega_n}$ weakly converges (for example in $L^2(D)$) to an element μ, with $0 \le \mu \le 1$.

Define the set Ω as

$$\Omega = \{x \in D \; : \; y(x) \ne 0\}. \tag{5.3}$$

We obtain $\mu = 1$ on Ω (for $(1 - \chi_n)y_n = 0$ a.e. implies in the limit $(1 - \mu)y = 0$ a.e.). Then $\text{meas}(\Omega) \le \int_D \mu \, dx \le \alpha$ and we get

$$\int_D |\nabla y|^2 dx = \int_\Omega |\nabla y|^2 dx \quad \text{and} \quad \int_D y^2 dx = \int_\Omega y^2 dx$$

so that

$$-\frac{1}{2\lambda} = \frac{1}{2}\int_\Omega |\nabla y|^2 - \left(\int_D y^2 dx\right)^{1/2} \le \liminf_{n\to\infty} -\frac{1}{2\lambda(\Omega_n)}.$$

We shall prove that $\lambda = \lambda(\Omega)$. First notice that

$$-\frac{1}{2\lambda(\Omega)} \le -\frac{1}{2\lambda} \quad \text{for } y \in H_0^1(\Omega)$$

then we get

$$-\frac{1}{2\lambda(\Omega)} \le -\frac{1}{2\lambda} \le \liminf_{n\to\infty} -\frac{1}{2\lambda(\Omega_n)}$$

so that Ω is a solution to the problem (P).

5.2 Free boundary problem associated with the first eigenvalue

We consider now at the following problem (P) a generalized Auchmuty principle which leads to a non-linear eigenvalue problem associated with the first eigenvalue in a subdomain Ω, whose boundary–which is free–is characterized by a level set of a function. This problem is similar to the one studied in section 1.4.1.1. That kind of free boundary problem is a simplified version of a more general situation related to the H. Grad plasma physics equation, which is mentioned in the introduction and for which we refer to J.P. Zolésio [16], [17].

The main property is that here, as well as in section 1.4.1.1 the solution is shown to be without level step, that is: $\text{meas}\{x \in D \; : \; y(x) = 0\} = 0$.

Consider the following problem

$$(P) \quad \min_{\varphi \in H_0^1(D)} \int_D \frac{1}{2} |\nabla \varphi|^2 dx - \left(\int_D [(\varphi - g)^+]^2 dx \right)^{\frac{1}{2}}, \tag{5.4}$$

where g belongs to $L^2(D)$ such that

$$\int_D g^+ dx > 0. \tag{5.5}$$

The result is that (P) has optimal solutions y which are solutions to the following non-linear eigenvalue problem:

$$-\Delta y = \lambda(y - g)^+ \quad \text{in } D \tag{5.6}$$
$$y = 0 \quad \text{on } D \tag{5.7}$$

with

$$\lambda = \left(\int_D ((y - g)^+)^2 dx \right)^{-\frac{1}{2}}. \tag{5.8}$$

This problem (5.6)-(5.8) is in fact a free boundary value problem.

Define

$$\Omega = \{x \; : \; y(v) > g\}$$

which is an open set when y belongs to $H^1(D)$. Then (y, Ω) is a solution of the following problem:

$$-\Delta y = \lambda(y - g) \text{ in } \Omega \tag{5.9}$$
$$-\Delta y = 0 \text{ in } D \backslash \bar{\Omega} = \Omega^c \tag{5.10}$$
$$y = g \text{ on } \Gamma \tag{5.11}$$

$$\frac{\partial y}{\partial n}\Big|\Omega = \frac{\partial y}{\partial n}\Big|\Omega^c \text{ on } \Gamma. \tag{5.12}$$

More precisely we have the following result.

Theorem 5.2 *There exist minimizers to the problem P. Let y be one of them, then under assumption (5.5) we have*

$$\frac{1}{\lambda} = \left(\int_D ((y - g)^+)^2 \right)^{\frac{1}{2}} > 0 \tag{5.13}$$

and y verifies (5.6), (5.7).

If g belongs to $H^1(D)$, then $y \in H^3(D) \subset C^1(\overline{D})$ (for $N = 2$ or 3), then Ω is an open subset of D and (5.11), (5.12) hold in the classical sense.

Proof The functional in (P) is well-known to be Gâteaux-differentiable so that the necessary condition for y can be written as

$$\forall \Psi \in H_0^1(D),$$

$$\int_D \nabla y \nabla \Psi dx = \left(\int_D [(y-g)^+]^2 dx \right)^{-1/2} \int_D (y-g)^+ \Psi dx$$

from which the problem (5.6), (5.7) follows. When g is in $H^1(D)$, then $(y-g)^+$ is also in $H^1(D)$ and from classical elliptic regularity results y is in $H^3(D) \cap H_0^1(D)$. To conclude let us remark that in (P) if $\varphi = 0$ the functional is a strictly negative number and (5.13) follows.

Remark 5.3 The eigenvalue problem (P) is closely related to the Auchmuty principle but now in (P) the value of the minimum cannot be related explicitly to the eigenvalue λ as it is done in the Auchmuty principle.

Let

$$J(y) = \int_D \frac{1}{2} |\nabla y|^2 dx - \left(\int_D [(y-g)^+]^2 dx \right)^{\frac{1}{2}}.$$

Then from the problem (5.6)-(5.8) whose y is a solution we get

$$
\begin{aligned}
J(y) &= \frac{1}{2} \int_D -\Delta y\, y\, dx - \lambda^{-1} \\
&= \frac{1}{2} \lambda \int_D (y-g)^+ y - \lambda^{-1} \\
&= \frac{1}{2} \lambda \int_D ((y-g)^+)^2 dx + \frac{1}{2} \lambda \int_D (y-g)^+ g\, dx - \lambda^{-1} \\
&= -\frac{1}{2} \lambda^{-1} + \frac{1}{2} \lambda \int_D (y-g)^+ g\, dx
\end{aligned}
$$

while in the usual Auchmuty principle we have

$$J(y) = -\frac{1}{2} \lambda^{-1}.$$

We would be back in that situation if we take $g = 0$ and $y = -a < 0$ on ∂D.

References

[1] H.W. Alt and L.A. Caffarelli, *Existence and Regularity for a Minimum Problem with a Free Boundary*, Courant Institute, 1980, preprint.

[2] M.C. Delfour and J.P. Zolésio, Shape sensitivity analysis via MinMax differentiability, *SIAM J. Control Optim.* **26** (1988), 834–862.

[3] M.C. Delfour and J.P. Zolésio, Shape sensitivity analysis via a penalization method, *Ann. Mat. Pura Appl. (4)* **151** (1988), 179–212.

[4] M.C. Delfour and J.P. Zolésio, Anatomy of the shape Hessian, *Ann. Mat. Pura Appl. (4)* **158** (1991), 315–339.

[5] M.C. Delfour and J.P. Zolésio, Velocity method and Lagrangian formulation for the computation of the shape Hessian, *SIAM J. Control Optim.* **29** (1991), 1414–1442.

[6] M.C. Delfour and J.P. Zolésio, Structure of shape derivatives for nonsmooth domains, *J. Funct. Anal.* **104** (1992), 1–33.

[7] J. Mossino and J.P. Zolésio, Formulation variationnelle de problèmes issus de la physique des plasmas, *C.R. Acad. Sci. Paris Série A* **285** (1977), 1003–1007.

[8] J. Sokolowski and J.P. Zolésio, *Introduction to Shape Optimization*, Computational Mathematics vol. 16, Springer Verlag, Berlin, Heidelberg, New York, 1992.

[9] M. Souli and J.P. Zolésio, Le contrôle de domaine dans le problème de résistance de vague non linéaire en hydrodynamique, *Ann. Sci. Math. Québec* **15** (1991), 203–214.

[10] M. Souli and J.P. Zolésio, Shape derivative of discretized problems, *Comput. Methods Appl. Mech. Engrg.* (to appear).

[11] G. Stampacchia, *Équations elliptiques du second ordre à coefficients discontinus*, Séminaire de Mathématiques Supérieures **16**, Les Presses de l'Université de Montréal, Montréal, 1966.

[12] R. Temam, *Problèmes mathématiques en plasticité*, Gauthier-Villars, Paris, 1983.

[13] T. Tiihonen and J.P. Zolésio, Gradient with respect to nodes for non-isoparametric finite elements, in: *Boundary Control and Boundary Variations* (J.P. Zolésio, ed.), Lecture Notes in Control and Inform. Sci. **100**, Springer Verlag, Berlin, Heidelberg, New York, 1988, 311–317.

[14] T. Tiihonen and J.P. Zolésio, Shape acceleration: a second order accurate incremental method for large deformations, in: *Fifth Symp. on Control of Distributed Parameter Systems* (A. El Jai and M. Amouroux, eds.), Pergamon Press, Oxford, New York, 1989, 91–94.

[15] J.P. Zolésio, Un résultat d'existence de vitesse convergente en optimisation de domaine, *C.R. Acad. Sci. Paris Série A* **283** (1976), 855–859.

[16] J.P. Zolésio, *Identification de domaines par déformation*, Thèse de doctorat d'état, Université de Nice, 1979.

[17] J.P. Zolésio, Solution variationnelle d'un problème de valeur propre non linéaire et frontière libre issu de la physique des plasmas, *C.R. Acad. Sci. Paris Série A* **288** (1979), 911–914.

[18] J.P. Zolésio, The Material Derivative (or Speed) Method for shape optimization, in: *Optimization of Distributed Parameter Structures*, vol. II (E.J. Haug and J. Céa, eds.), Sijthoff and Nordhoff, Alphen aan den Rijn, 1981, 1089–1151.

[19] J.P. Zolésio, Semiderivatives of repeated eigenvalues, in: *Optimization of Distributed Parameter Structures*, vol. II (E.J. Haug and J. Céa, eds.), Sijthoff and Nordhoff, Alphen aan den Rijn, 1981, 1457–1473.

[20] J.P. Zolésio, Domain variational formulation of free boundary problems, in: *Optimization of Distributed Parameter Structures*, vol. II (E.J. Haug and J. Céa, eds.), Sijthoff and Nordhoff, Alphen aan den Rijn, 1981, 1152–1194.

[21] J.P. Zolésio, Numerical algorithms and existence results for a Bernoulli-like steady state free boundary problem in: *Large Scale Systems, Theory and Applications*. North-Holland, Amsterdam, 1984.

[22] J.P. Zolésio, Les dérivées par rapport aux noeuds des triangulations et leurs applications en optimisation de domaines, *Ann. Sci. Math. Québec* **8** (1984), 97–120.

[23] J.P. Zolésio, Shape variational solution for a Bernoulli-like steady free boundary problem, in: *Distributed Parameter Systems* (F. Kappel et al., eds.), Lecture Notes in Control and Inform. Sci. **102**, Springer-Verlag, Berlin, Heidelberg, New York, 1987, 333–343.

[24] J.P. Zolésio, Shape derivatives and shape acceleration, in: *Control of Partial Differential Equations* (A. Bermúdez, ed.), Lecture Notes in Control and Inform. Sci. **114**, Springer-Verlag, Berlin, Heidelberg, New York, 1989, 309–318.

[25] J.P. Zolésio, Weak shape formulation of free boundary problem, in: *Stabilization of Flexible Structures* (J.P. Zolésio et al., eds.), Lecture Notes in Control and Inform. Sci. **147**, Springer-Verlag, Berlin, Heidelberg, New York, 1991.

[26] J.P. Zolésio, Shape formulation of free boundary problems with non-linearized Bernoull condition, in: *Boundary Control and Boundary Variations* (J.P. Zolésio, ed.), Lecture Notes in Control and Inform. Sci. **178**, Springer-Verlag, Berlin, Heidelberg, New York, 1992, 362–388.

[27] J.P. Zolésio, Weak shape formulation of non-linearized Bernoulli free boundary problem, *Ann. Scuola Norm. Sup. Pisa Cl. Sci. (4)*, to appear.

[28] J.P. Zolésio, Dynamical shells, in: *Boundary Control and Boundary Variations* (J.P. Zolésio, ed.) Proc. IFIP Conf., Sophia-Antipolis, June 1992, to appear.

[29] J.P. Zolésio, Existence result in shape optimization for perturbed models, to appear.

[30] G. Auchmuty, Dual variational principles for eigenvalue problems, in: *Nonlinear Functional Analysis and its Applications* (F. Browder, ed.), American Mathematical Society, Providence, 1986, 55–71.

[31] J. Dieudonné, *Élément d'analyse*, 1. *Fondements de l'analyse moderne*, Gauthier-Villars, Paris 1969, Bordas 1979 (Transl. from *Foundations of Modern Analysis*, Academic Press, New York and London, 1960), §3.16.

[32] H. Grad, P.N. Hu, D.C. Stevens, Adiabatic evolution of plasma equilibrium, *Bull. Amer. Phys. Soc.* **19** (1974), 865.

Index